Request Your Complimentary Copies of *Organic Chemistry, 3/e* and *Experimental Organic Chemistry* Today

D1795801

Use this card to request your complimentary copy of the new edition of Vollh
Organic Chemistry: Structure and Function.

Use this card to request your complimentary copy of:

Mohrig, et al., *Experimental Organic Chemistry, A Balanced Approach: Macroscale and Microscale Flexible Connector Microscale Glassware version*—available August 1998!
Ground Glassware Version—available now!

Sample experiments and techniques from the flexible connector microscale glassware version of *Experimental Organic Chemistry* follow.

MOHRIG, et al.

VOLLHARDT & SCHORE

MOHRIG column

For a complimentary review copy of the flexible connector version of ***Experimental Organic Chemistry*** please complete and return this postpaid card. You may contact your local representative or email your request to facultyservices@sasmp.com

☐ Yes! Please send me *Experimental Organic Chemistry* using flexible connector glassware, 0-7167-3330-7
☐ Yes! Please send me *Experimental Organic Chemistry* using ground glassware, 0-7167-2818-4

Apply peel-off label or fill in your name and address:

Name _____

School _____
<place address label here>

Address _____

City _____ State _____ Zip _____

E-mail _____

Current text in use _____
(author/title)

Annual enrollment _____ Decision date _____

Course length: ☐ One semester ☐ Two semesters ☐ Quarters

Course name and number _____

Department _____

Office Hours: _____ Best time to call _____

Phone _____

Likelihood of change: ☐ High ☐ Medium ☐ Low

What do you like/dislike about your current book? _____

ST	SCH	REP	JOB
			7034

VOLLHARDT & SCHORE column

For a complimentary review copy of **Vollhardt & Schore's *Organic Chemistry, 3/e*** please complete and return this postpaid card. You may contact your local representative or email your request to facultyservices@sasmp.com

☐ Yes! Please send me *Organic Chemistry, 3/e*, 0-7167-2721-8

Apply peel-off label or fill in your name and address:

Name _____

School _____
<place address label here>

Address _____

City _____ State _____ Zip _____

E-mail _____

Current text in use _____
(author/title)

Annual enrollment _____ Decision date _____

Course length: ☐ One semester ☐ Two semesters ☐ Quarters

Course name and number _____

Department _____

Office Hours: _____ Best time to call _____

Phone _____

Likelihood of change: ☐ High ☐ Medium ☐ Low

What do you like/dislike about your current book? _____

ST	SCH	REP	JOB
			7035

BUSINESS REPLY MAIL

FIRST CLASS MAIL   PERMIT NO. 7953   NEW YORK, NY

POSTAGE WILL BE PAID BY ADDRESSEE

W. H. FREEMAN AND COMPANY

C/O SCIENTIFIC AMERICAN/ST. MARTIN'S COLLEGE DESK
345 PARK AVENUE SOUTH
NEW YORK, NY 10160-1039

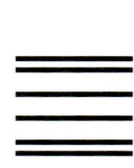

BUSINESS REPLY MAIL

FIRST CLASS MAIL   PERMIT NO. 7953   NEW YORK, NY

POSTAGE WILL BE PAID BY ADDRESSEE

W. H. FREEMAN AND COMPANY

C/O SCIENTIFIC AMERICAN/ST. MARTIN'S COLLEGE DESK
345 PARK AVENUE SOUTH
NEW YORK, NY 10160-1039

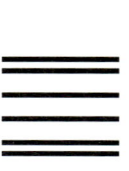

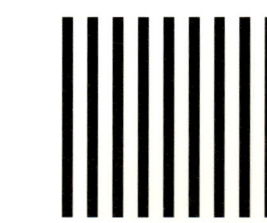

Dear Instructor,

This is an introduction to our innovative and remarkably versatile new organic chemistry laboratory manual. Here you will find one experiment (Experiment 9.1 "Free Radical versus Ionic Addition of Hydrobromic Acid to Alkenes"), one project (Project 3 "Synthesis and Dehydration of 3,3-Dimethyl-2-Butanol"), plus the techniques your students will need to complete them—all using flexible connector microscale glassware. A version of *EXPERIMENTAL ORGANIC CHEMISTRY* that calls for ground microscale glassware is currently available. A version using flexible connector microscale glassware will be available August 1998—in time for your fall classes.

Please accept our invitation to tear out these perforated pages and use this experiment and project in your laboratory classroom. We realize the importance of class-testing a laboratory manual before committing to it. All the experiments in *EXPERIMENTAL ORGANIC CHEMISTRY* have been class-tested by students at Carleton College, Vassar College, and Bowling Green State University. If you would like to class-test other experiments this Spring, please choose from the procedures listed in the Table of Contents and contact your sales representative or notify our faculty services department through electronic mail (facultyservices@sasmp.com).

In addition to being rigorously class-tested, *EXPERIMENTAL ORGANIC CHEMISTRY* comes with a CD-ROM showing selected techniques. Studies show that students meet greater success in the laboratory when they see techniques performed ahead of time. The *EXPERIMENTAL ORGANIC CHEMISTRY, A Balanced Approach: Macroscale and Microscale CD-ROM* is designed to increase your students' success and efficiency in the laboratory.

On the flip side of this introductory experiment and project, please find a preview of the new Third Edition of our bestselling organic chemistry text, *ORGANIC CHEMISTRY: Structure and Function* by K. Peter C. Vollhardt and Neil E. Schore. The preview book includes the Preface, the complete Table of Contents, and the first eight chapters of the new edition.

For complete copies of *ORGANIC CHEMISTRY,* and both versions of *EXPERIMENTAL ORGANIC CHEMISTRY,* please fill out and return the postpaid card on the preceding page, contact your sales representative, or notify our faculty services department.

To receive periodic updates about all our organic chemistry products, please e-mail us at organicchemistrylist@whfreeman.com and sign up.

Sincerely,

Kimberly Manzi,
Marketing Manager, Chemistry
W. H. Freeman & Company

Contents

Part 1 Experiments

M = macroscale procedure

m = microscale procedure

🔒 = discovery based

💿 = CD ROM

Contents

Contents

Contents

Part 2 Projects

Part 3 Organic Qualitative Analysis

Part 4   Spectrometric Methods

Part 5 Techniques of the Organic Laboratory

Contents   11

ADDITION REACTIONS OF ALKENES

Combine theory and experiment to compare ionic and radical addition products, isomers formed in a dimerization reaction, and the stereochemistry of ionic addition. Use gas chromatography, spectrometry, and melting-point determination to analyze the reaction products.

A common reaction of alkenes is that of 1,2-addition, where the $C=C$ is broken and an atom or group of atoms adds to each carbon. A wide range of reagents (AB) can add by different mechanisms to the double bond of alkenes:

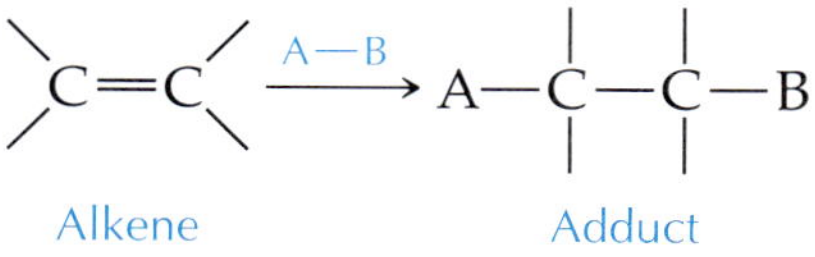

Alkene   Adduct

One of the more common addition processes is ionic. The majority of ionic addition reactions of simple alkenes involves initial attack of the double bond by the electrophilic portion (E^+) of the ionic reagent (ENu):

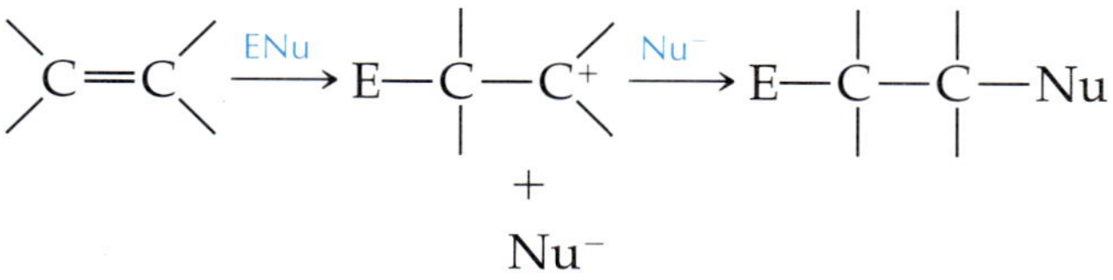

Ionic additions to double bonds are initiated by electrophilic attack, because the alkene π-bond is electron rich and thus is electrostatically vulnerable to attack by positive reagents, such as electrophiles. In general, the more stable the carbocation intermediate is, the faster it forms.

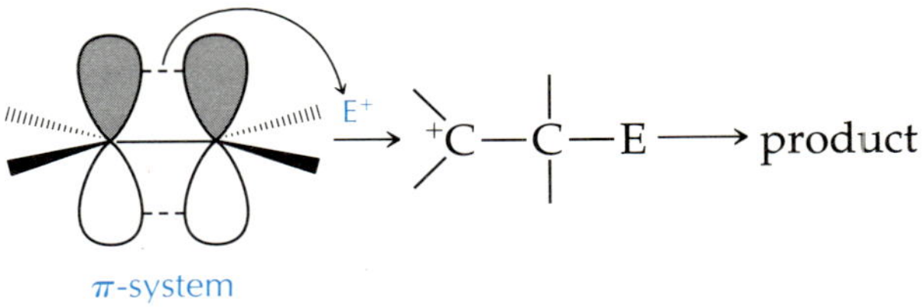

$$\xrightarrow{E^+} \quad \diagdown C - C - E \longrightarrow \text{product}$$

π-system

A wide range of electrophiles can initiate such addition reactions, including the proton (H$^+$) and halonium ions (X$^+$) contributed by Br$_2$, Cl$_2$, and I$_2$. We will be examining proton-initiated reactions in Experiments 9.1 and 9.2, and bromine addition in Experiment 9.3.

9.1
Free-Radical versus Ionic Addition of Hydrobromic Acid to Alkenes

Compare the controlling factors and contrast the results in free-radical and ionic additions to alkenes.

Understanding the addition of HBr to alkenes requires an appreciation and comparison of two types of mechanisms, ionic additions and radical additions:

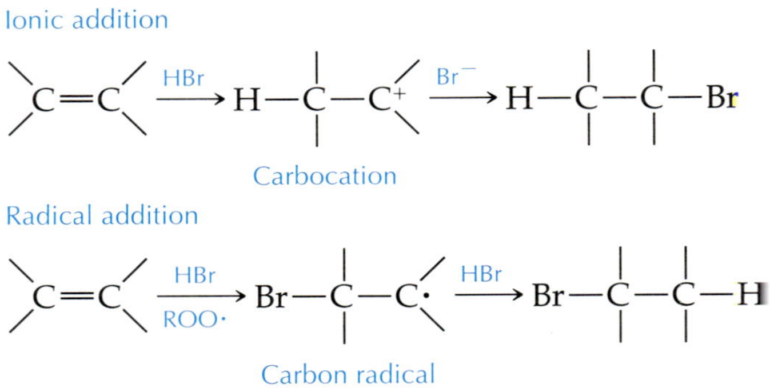

Ionic addition

$$\diagdown C = C \diagup \xrightarrow{\text{HBr}} H - C - C^+ \xrightarrow{Br^-} H - C - C - Br$$

Carbocation

Radical addition

$$\diagdown C = C \diagup \xrightarrow[\text{ROO·}]{\text{HBr}} Br - C - C· \xrightarrow{\text{HBr}} Br - C - C - H$$

Carbon radical

Ionic addition of hydrohalic acids (HX) to double bonds can be pictured as an initial proton attack on an alkene π-system to form a carbocation, followed by entrapment of that carbocation by halide ion to produce an organic halogen compound:

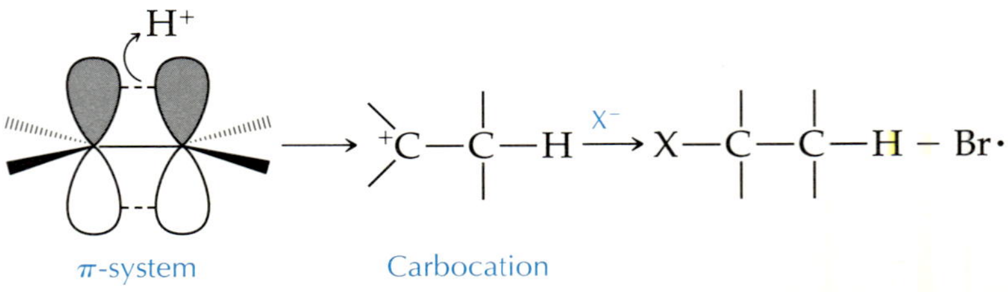

H$^+$

$$\longrightarrow \quad {}^+C - C - H \xrightarrow{X^-} X - C - C - H - Br·$$

π-system   Carbocation

Thus, the fact that HCl adds to propene to produce only 2-chloropropane is readily explained by the formation of the more stable secondary carbocation. The alternative primary carbocation, $CH_3CH_2CH_2^+$, is far too unstable to be formed. Therefore, when the chloride ion traps the positively charged intermediate, only the secondary, 2-chloropropane product is formed:

$$H_2C\!=\!CH \overset{HCl}{\longrightarrow} \begin{cases} H_3C\!-\!\overset{H}{\underset{CH_3}{C^+}} \overset{Cl^-}{\longrightarrow} H_3C\!-\!\overset{Cl}{\underset{CH_3}{CH}} \\ \text{Secondary cation} \quad\quad \text{2-Chloropropane} \\ \\ H_3C\!-\!\overset{CH_2}{\underset{CH_2^+}{CH_2}} \overset{Cl^-}{\longrightarrow} H_3C\!-\!\overset{CH_2}{\underset{CH_2Cl}{CH_2}} \\ \text{Primary cation} \quad\quad \text{1-Chloropropane} \end{cases}$$

When HBr is added to an alkene, a free-radical chain reaction can compete with the ionic reaction described earlier. This process is identifiable during the addition of HBr to propene by the fact that 1-bromopropane (rather than the 2-bromo isomer) is obtained:

$$H_2C\!=\!\underset{CH_3}{CH} + Br\cdot \longrightarrow H_2C\!-\!\underset{\underset{CH_3}{|}}{\overset{\overset{H}{|}}{C}}\!\cdot\;\underset{Br}{} \overset{HBr}{\longrightarrow} H_2C\!-\!\underset{\underset{CH_3}{|}}{\overset{\overset{H}{|}}{C}}\!-\!H + Br\cdot \;\underset{Br}{}$$

The bromine radical bonds to the terminal carbon of propene because this structure produces the more highly branched and more stable radical intermediate.

The addition of HX to an unsymmetrical alkene occurs in a Markovnikov fashion when the HX reagent adds a proton to the alkene carbon that already has more attached hydrogens. Thus, Markovnikov addition of HBr to propene gives 2-bromopropane. If, on the other hand, 1-bromopropane is formed, the reaction is said to have occurred in an anti-Markovnikov fashion.

Peroxides can be used to induce anti-Markovnikov addition. In fact, the oxygen in the air can, by its paramagnetic character, form alkyl peroxides or hydroperoxide from hydrocarbons or HBr:

Vladimir V. Markovnikov, 1838–1904, was a Russian chemist at Odessa and later at Moscow University who carried out early studies of such additions.

$$\cdot\overset{..}{\underset{..}{O}}\!-\!\overset{..}{\underset{..}{O}}\cdot$$

Paramagnetic oxygen molecule

$$O_2 + RH \longrightarrow R\cdot + H\!-\!O\!-\!O\cdot$$

$$O_2 + R\cdot \longrightarrow R\!-\!O\!-\!O\cdot$$

Prior to 1933, chemists did not understand why addition of HBr to alkenes often occurred with inconsistent results. Sometimes Markovnikov addition was the major process, and sometimes anti-Markovnikov addition predominated. The pioneering work of M. S. Kharasch and F. W. Mayo lead to the discovery that, unless the reagents for this reaction were scrupulously purified, HBr addition could indeed yield mixed results. Small amounts of dissolved oxygen can give rise to peroxides as just described, and these would lead to anti-Markovnikov addition. Highly purified reagents and the presence of free-radical inhibitors give rise to largely Markovnikov addition. Thus, if ROO· is our general radical-initiation species, the mechanism of anti-Markovnikov addition is the following chain reaction:

Initiation

$$R—O—O· + HBr \longrightarrow Br· + ROOH$$

Propagation

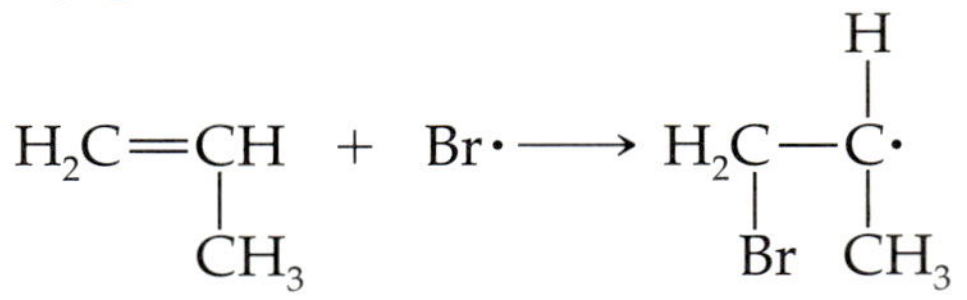

$$H_2C{=}CH(CH_3) + Br· \longrightarrow H_2C(Br)—\overset{H}{\underset{CH_3}{C}}·$$

Termination

$$H_2C(Br)—\overset{H}{\underset{CH_3}{C}}· + HBr \longrightarrow H_2C(Br)—\overset{H}{\underset{CH_3}{C}}—H + Br·$$

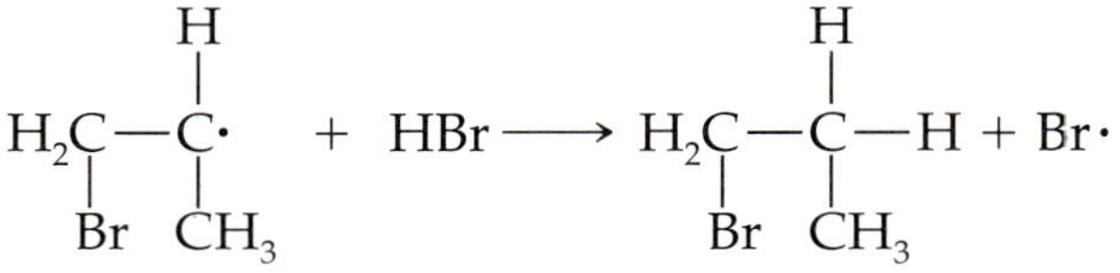

$$R· + R· \longrightarrow R—R$$

In Experiment 9.1, the addition of HBr to an alkene is carried out in three ways. The first and second reactions use a mixture of HBr in acetic acid to carry out the addition; the first reaction uses 1-hexene as the substrate, and the second uses 2-methyl-2-butene. In the first reaction, we can assume that the ratio of 2-bromohexane to 1-bromohexane corresponds to the ratio of the Markovnikov to anti-Markovnikov processes and thus, in turn, to the ratio of ionic to radical addition. In like fashion, the 2-methyl-2-butene reaction allows the use of relative amounts of Markovnikov versus *anti*-Markovnikov product as a measure of ionic versus free-radical processes. The third reaction involves very different conditions, but again we can use the relative amounts of Markovnikov versus *anti*-Markovnikov product as a measure of ionic versus free-radical processes.

More recently, newly developed reagents use heterogeneous conditions to ensure ionic addition to an alkene. You will use one of these methods in the third reaction. Specifically, when an alkene is dissolved in dichloromethane and treated with HX (or a

hydrohalic acid source) while in contact with alumina or silica gel, an efficient ionic addition takes place. In this experiment, you will treat 1-hexene in dichloromethane with oxalyl bromide in the presence of alumina. The protons of the HBr that eventually add to the alkene are obtained from the hydroxylated surface of the alumina, and the bromide ions are obtained from oxalyl bromide:

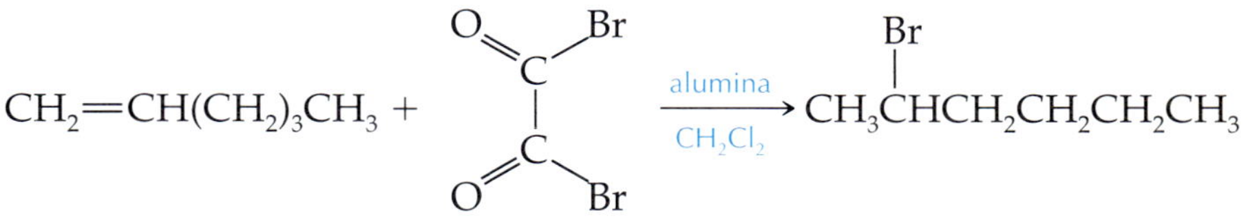
$CH_2{=}CH(CH_2)_3CH_3$ +
O
Br
C
C
O
Br
alumina
CH_2Cl_2
Br
$CH_3CHCH_2CH_2CH_2CH_3$

1-Hexene Oxalyl bromide 2-Bromohexane

An important aspect of the alumina is its pretreatment at 120°C. This process removes all the water loosely associated with the alumina surface but allows the water that is covalently bonded to the alumina to remain as a monolayer on the surface:

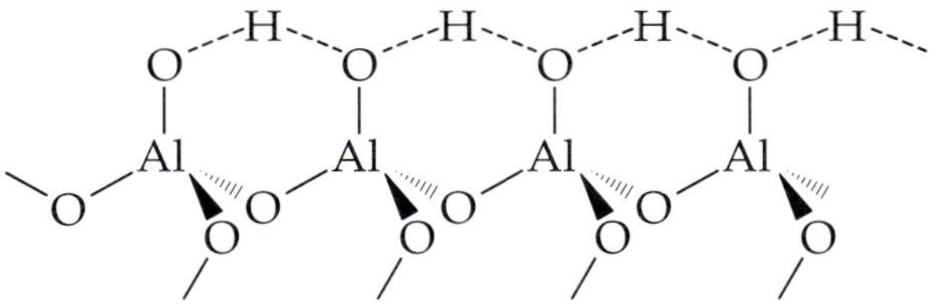
H
O
Al

It is this surface that promotes ionization of the HX that adds to the alkene. In this way the alumina promotes ionic addition to the double bond.

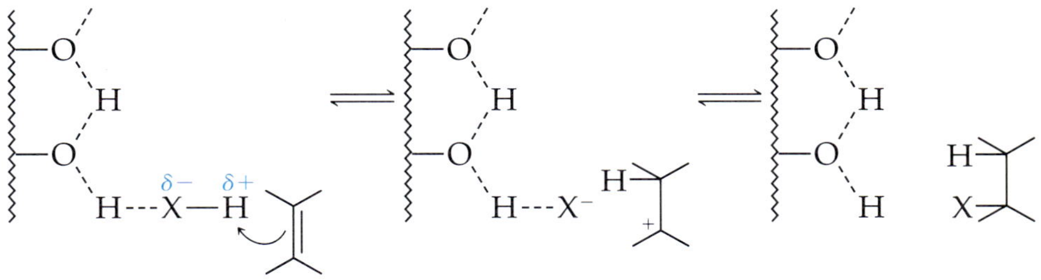
O
H
δ−
δ+
H---X—H
H---X⁻
+
X

microscale Procedure

Techniques Microscale Extractions: Techniques 4.5d, 4.5a, and 4.5b, Method B

Pasteur Filter Pipet: Technique 2.4

Gas Chromatography: Technique 11

Prelaboratory Assignment:
Name the possible products of
HBr addition to 1-hexene and
2-methyl-2-butene. Locate the
boiling points of these
products in a handbook such
as the CRC Handbook *or*
Lange's Handbook of
Chemistry *(see*
Introduction I.6).

SAFETY INFORMATION

Wear gloves while conducting all of these experiments.

1-Hexene and **2-methyl-2-butene** are very flammable and are skin irritants. Avoid contact with skin, eyes, and clothing. Use them in a well-ventilated area.

30% Hydrogen bromide in acetic acid and **oxalyl bromide** are toxic and corrosive. Avoid breathing the vapors. Avoid contact with skin, eyes, and clothing. Use these reagents in a hood and wear gloves.

Diethyl ether is very flammable. Use it in a hood and keep it away from flames or electrical heating devices.

Alumina is an eye and respiratory irritant. Avoid breathing any fine particles while weighing it.

Experiment 1

Use a 15-mL centrifuge tube with a tight-fitting cap as the reaction vessel. Add 0.35 mL of 1-hexene to the centrifuge tube. Working in a hood, add 1.0 mL of 30% HBr in acetic acid (5M) to the tube. Cap the tube and shake it frequently for 10 minutes. Occasionally loosen the cap to release any buildup of pressure.

After the reaction period, allow the phases to separate and remove the lower acetic acid layer with a Pasteur filter pipet [Pasteur filter pipet: Technique 2.4, microscale extraction: Technique 4.5d]. Place the acid layer in a 150 mL beaker containing 50 mL of water. Add 2.0 mL of ether and 2.0 mL of water to the organic phase remaining in the centrifuge tube. Cap the tube and shake it to mix the phases. Remove the lower aqueous layer, adding it to the beaker containing the previously removed acid layer. Wash the ether layer with 2.0 mL of 5% sodium bicarbonate solution. Remove the lower aqueous phase. Add anhydrous calcium chloride pellets to the remaining ether solution, cap the centifuge tube, and allow the solution to dry for 10 - 15 min.

Set the centrifuge tube in a
small beaker so it does not
tip over.

SAFETY PRECAUTION

There may be pressure inside the centrifuge tube from carbon dioxide while washing with $NaHCO_3$, open the cap slowly to relieve the pressure.

Prepare another Pasteur filter pipet. Transfer the ether/product solution to a clean test tube. If you are analyzing the product mixture on a capillary column gas chromatograph simply inject 1-μL of the solution into the chromatograph. If your laboratory is equipped with packed column chromatographs, you will probably need to evaporate some of the ether with a steam of nitrogen or on a steam bath before injecting 1-μL into the chromatograph. Consult your instructor about specific sample preparation techniques and sample size for your instruments. Use a nonpolar column such as a OV-1 or SE-30, with a column temperature of 60 - 70° C.

Experiment 2

Repeat the procedure for Experiment 1 using a dry 15-mL centrifuge tube and substituting 0.35mL of 2-methyl-2-butene for 1-hexene.

The first step of the extraction procedure differs from that of Experiment 1 because the organic (product) phase moves from the top layer to the bottom layer as the reaction proceeds. Remove the lower (organic) layer with a Pasteur filter pipet and transfer it to a clean 15-mL centrifuge tube [see Technique 4.5a]. Add 2.0 mL of ether and 2.0 mL of water to the organic phase. Continue as directed in the procedure for Experiment 1. Carefully pour the acetic acid/HBr solution remaining in the first centrifuge tube into the beaker containing the acetic acid/HBr from Experiment 1.

Experiment 3

> ── **SAFETY PRECAUTION** ──────
>
> Conduct this experiment in a hood until you have transferred the water/dichloromethane mixture to a centrifuge tube.

The alumina must be heated at 120°C for 48 h before using it in this experiment.

Weigh 110–120 mg of 1-hexene by adding it dropwise from a Pasteur pipet to a dry 25-mL Erlenmeyer flask. Cork the flask while you obtain the other reagents. Add 5 mL of dichloromethane, 2.5 g of alumina (Al_2O_3), and a magnetic stirring bar to the flask. Prepare an ice-water bath in a small beaker or crystallizing dish; set the ice bath on a magnetic stirrer and clamp the flask in the bath. Stir the mixture for 5 min to cool its contents. Wearing gloves, take a corked 10 × 75 mm test tube to the reagent-dispensing hood and place 1.0 mL of 2 M oxalyl bromide in CH_2Cl_2 in the

test tube; immediately recork the test tube. Using a Pasteur pipet, add the oxalyl bromide solution dropwise over a period of 1 min to the stirred reaction mixture. Continue stirring the mixture for an additional 10 min.

Gravity filter the reaction mixture, using a small funnel and fluted filter paper; collect the filtrate in a 25-mL Erlenmeyer flask [see Technique 4.7, Figure 4.13]. Retrieve the magnetic stirring bar and wash it. Add 5 mL of ice-cold water to the filtrate, place the stirring bar in the flask containing the filtrate, and stir the mixture until bubbling ceases (about 5 min). Pour the mixture into a 15-mL centrifuge tube with a tight cap.

Transfer the lower organic phase to another centrifuge tube, using a Pasteur filter pipet [see Technique 4.5b, Method 3]. Add 2.0 mL of 5% sodium hydroxide solution to the centrifuge tube containing the dichloromethane solution (organic phase). Cap the tube and shake it to mix the layers. If the phases do not separate cleanly, spin the tube for 1 min in a centrifuge. Transfer the lower organic phase to a clean 10-mL Erlenmeyer flask and dry the solution with anhydrous calcium chloride pellets for 10–15 min.

Separate the drying agent and analyze the product solution as directed in Experiment 1. Utilize the same gas chromatograph and instrument parameters you used for Experiment 1.

Identification of the Products

Because you are using a nonpolar column on the gas chromatograph, the products usually elute in order of increasing boiling point. Having known samples of the products available will allow you to verify the identity of the products by using the peak enhancement method [see Technique 11.6]. After you have taken a gas chromatogram of the product of Experiment 1, add 1 drop of 1-bromohexane to the ether (product) solution. Inject a 1-μL sample of this "enhanced" solution into the same chromatograph (at the same parameters) already used for Experiment 1. The peak whose height has increased relative to the others corresponds to the known compound that you added.

In the same way, add 1 drop of 2-bromo-2-methylbutane to the ether/product solution from Experiment 2 after you have analyzed it by GC. Carry out the chromatographic determination as described in the previous paragraph.

Compare the retention times of the peaks found in Experiment 3 with those for Experiment 1. What product(s) forms in Experiment 3?

Cleanup: Combine all the aqueous solutions from the extractions in the beaker containing the acetic acid / HBr extracts. Neutralize the acid by adding solid sodium carbonate in small portions. **(Caution: Foaming.)** Wash the neutralized solution down the sink or pour it into the container for aqueous inorganic waste. Pour the product solutions into the container for halogenated organic waste. Place the calcium chloride drying agent and alumina in the container for hazardous solid waste or in the container for inorganic waste.

Treatment of Data

1. Compare the results of Experiments 1 and 3. Explain any differences.

2. Compare the results of Experiments 1 and 2. Does the difference in structure of the substrates affect the results? Explain.

References

1. Brown, T. M.; Dronsfield, A. T.; Ellis, R. *J. Chem. Educ.* **1990,** *67,* 518.
2. Kropp, P. J.; Daus, K. A.; Crawford, R.; Tubergen, M. W.; Kepler, K. D.; Craig, S. L.; Wilson, V. P. *J. Am. Chem. Soc.* **1990,** *112,* 7433–7434.

Questions

1. In Experiment 3, what would you expect the other organic product to be?
2. Hydrogen chloride adds ionically to alkenes and, in fact, the competing free-radical reactions described for HBr additions do not occur for HCl additions. In each case, predict the course of the reaction of HCl with the following hydrocarbons by writing the structure of the product: (a) styrene (phenylethene); (b) isobutylene (2-methylpropene); (c) ethylene (ethene); (d) α-methylstyrene (1-methyl-1-phenylethene).
3. Predict the course of the reaction of HBr with the substrates listed in Question 2 by providing the structure of the organic product. Assume that a free-radical process applies in all cases.

4. Silica gel may be used to promote the ionic addition of hydrogen halides in much the same fashion as alumina does. Structures A and B are different representations of a silica gel surface. Surface A contains two hydroxyl groups on each silicon atom (geminal —OH groups) that lead to the hydrogen bonding shown; surface B does not yield such hydrogen bonding. (a) Give a simple explanation of why structure A allows hydrogen bonding and B does not. (b) The hydrogen-bonded hydroxyl groups of A (pK_a 5–7) are much more acidic than the hydroxyl groups of B (pK_a 9.5). The acidic groups are thus sufficiently reactive to promote the addition of hydrogen halides to alkenes (in other words, a structure with halide bonded directly to Si, analogous to that described earlier for alumina incorporating Al-halide bonds, is not necessary). Outline a mechanism for the addition of a general hydrogen halide (HX) to alkene, using silica gel.

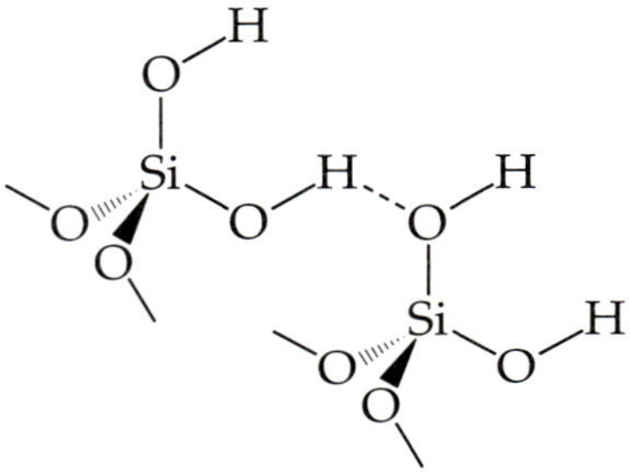

A 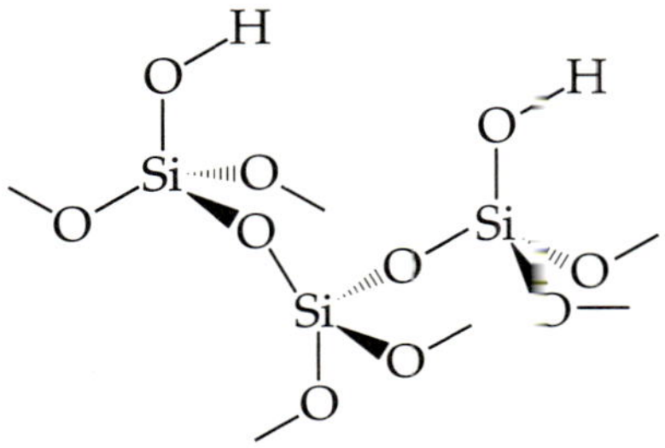   B

SYNTHESIS AND DEHYDRATION OF 3,3-DIMETHYL-2-BUTANOL

Project 3 involves a two-step reaction sequence, the first step of which is the reduction of the ketone, 3,3-dimethyl-2-butanone (also called pinacolone), to produce a secondary alcohol, 3,3-di-methyl-2-butanol. This reduction reaction is followed by acid-catalyzed dehydration of the secondary alcohol to yield a mixture of three alkenes, two of which form by carbon-skeleton rearrangement. The composition of the alkene mixture will be determined by gas chromatography.

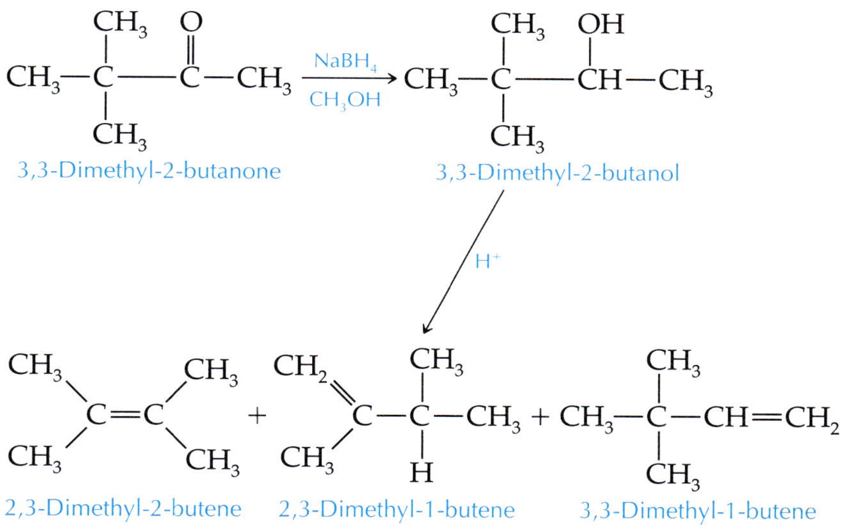

3,3-Dimethyl-2-butanone   3,3-Dimethyl-2-butanol

2,3-Dimethyl-2-butene   2,3-Dimethyl-1-butene   3,3-Dimethyl-1-butene

These reactions are two separate, sequential procedures. You work up and characterize the intermediate, 3,3-dimethyl-2-butanol, before carrying out the dehydration reaction.

Reduction reactions using sodium borohydride will be discussed briefly in the next section. You can also find information about sodium borohydride reductions in Experiment 19.2 and Project 4. Dehydration of alcohols with acid catalysts is described in Experiment 11. In this project we will focus on the carbocation aspects of the dehydration reaction as they relate to rearrangement.

3.1 Synthesis of 3,3-Dimethyl-2-Butanol by Reduction with Sodium Borohydride

$$4H_3C-\underset{\underset{CH_3}{|}}{\overset{\overset{CH_3}{|}}{C}}-\overset{O}{\overset{\|}{C}}\diagup^{CH_3} \;+\; NaBH_4 \;\;\xrightarrow[\text{methanol}]{NaOCH_3}\;\; 4H_3C-\underset{\underset{CH_3}{|}}{\overset{\overset{CH_3}{|}}{C}}-\underset{\underset{H}{|}}{\overset{\overset{OH}{|}}{C}}-CH_3 \;+\; NaBO_2$$

3,3-Dimethyl-2-butanone
(pinacolone)
bp 106°C
MW 100.2
density 0.801 g · mL⁻¹

Sodium borohydride
MW 37.83

3,3-Dimethyl-2-butanol
(pinacolyl alcohol)
bp 120.1°C
MW 102.2
density 0.812 g · mL⁻¹

Sodium borohydride ($NaBH_4$) has become a popular reagent for reducing the carbonyl groups of aldehydes and ketones. One of the reasons for this popularity is the fact that borohydride reductions occur readily within a reasonable length of time, yet $NaBH_4$ is far less dangerous to use than lithium aluminum hydride ($LiAlH_4$). Sodium borohydride reduces carbonyl groups significantly faster than it reacts with hydroxylic solvents, such as alcohols; whereas $LiAlH_4$ can react explosively with alcohols. BH_4^- is a good source of hydride ($:H^-$), which reduces the carbonyl group to an oxyanion intermediate by a nucleophilic addition pathway:

$$R-\underset{\underset{R'}{|}}{\overset{O}{\overset{\|}{C}}} + NaBH_4 \longrightarrow R-\underset{\underset{R'}{|}}{\overset{H}{\overset{|}{C}}}\diagup^{\bar{O}BH_3\;Na^+} \;\;\xrightarrow[^-OCH_3]{CH_3OH}\;\; R-\overset{H}{\underset{R'}{C}}\diagup^{OH}$$

The borohydride species that forms can further reduce another molecule of the ketone. Overall, one mole of sodium borohydride can reduce four moles of the ketone.

You can characterize your intermediate reduction product, 3,3-dimethyl-2-butanol, by its boiling point and its IR spectrum, and assess its purity by gas chromatographic analysis.

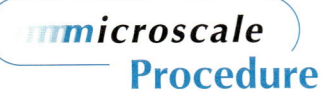
microscale Procedure

Techniques   Extraction: Technique 4.2
Microscale Distillation: Technique 7.3b
Gas Chromatography: Technique 11
IR Spectrometry: Spectrometric Method 1

3,3-Dimethyl-2-butanone (pinacolone) is flammable.

Sodium methoxide in methanol solution is flammable, corrosive, and moisture sensitive. Avoid contact with skin, eyes, and clothing.

Sodium borohydride is harmful if swallowed, inhaled, or absorbed through the skin. Avoid breathing the dust; and avoid contact with skin, eyes, and clothing. It decomposes to flammable, explosive hydrogen gas.

Methanol is volatile, flammable, and toxic.

Diethyl ether is very flammable. Use it only in a hood and keep it away from flames or electrical heating devices.

Hydrochloric acid solution (6 M) is a skin irritant.

Keep the container of sodium methoxide solution tightly closed and store the reagent in a desiccator.

$NaBH_4$ is moisture sensitive; keep the bottle tightly closed. Store the reagent in a desiccator.

Pour 4.0 mL of 3,3-dimethyl-2-butanone (pinacolone) and 5 mL of methanol into a 50 mL Erlenmeyer flask. Place the flask in an ice-water bath.

Combine 7.0 mL of methanol and 0.40 mL of 25 wt. % sodium methoxide in methanol in another Erlenmeyer flask. Add 0.75 g of sodium borohydride and stir until the large particles of sodium borohydride are completely dispersed (the mixture will be cloudy). Add the borohydride solution in small portions with a Pasteur pipet over a period of 10 minutes, swirling the reaction flask after each addition. Allow the reaction mixture to stand in the ice bath for an additional 20 minutes with occasional swirling of the flask.

Gaseous hydrogen is evolved during the addition of HCl, and the mixture will foam out of the flask if the acid is added too quickly.

With the flask still in the ice bath, add 10 mL of water and allow the mixture to stand for another 5 minutes. Add 2.25 mL of 6 M hydrochloric acid *dropwise* with swirling after each addition. If hydrogen is still being formed with the last portion of HCl, add an additional 0.25 - 0.50 mL of HCl.

Separate any precipitated borate salt from the product mixture by vacuum filtration. Add 12 mL of ether and 0.50 g of sodium chloride to the filtrate (or reaction mixture, if there was no precipi-

The extractions in this synthesis require the use of a 50- or 125-mL separatory funnel because the volumes exceed the capacity of a 15-mL centrifuge tube.

tate). Stir to dissolve the NaCl. Transfer the mixture to a separatory funnel. Shake the separatory funnel to extract the product into the ether [see Technique 4.2]. Drain the lower aqueous phase into a labeled Erlenmeyer flask and pour the upper organic phase into a labeled 25-mL Erenmeyer flask. Return the aqueous phase to the separatory funnel and extract it a second time with 8 mL of ether. Drain off the aqeous phase. Return the first ether solution to the separatory funnel. Wash the combined ether phase with 5 mL of water, then with 5 mL of saturated sodium chloride solution removing the aqueous phase each time. Pour the ether solution into a dry 50-mL Erenmeyer flask. Add anhydrous magnesium sulfate to the ether solution and allow the mixture to stand for at least 10 minutes [see Technique 4.6].

Filter the ether solution through a Pasteur pipet that has cotton packed in the upper portion of the tip [see Technique 4.7, Figure 4.15]. Clamp the filter pipet in an upright position and place a 5-mL round-bottomed flask under it. Using another Pasteur pipet transfer approximately 3-4 mL of the dried ether solution to the filter pipet. Working in a hood, blow off the ether with a stream of nitrogen until the volume is reduced to approximately one-third. Repeat the filtration and blowing off steps with portions of the remaining ether solution until all the product had been transferred to the 5-mL round-bottomed flask. Alternatively, the ether can be removed on a rotary evaporator [see Technique 4.8] and the remaning product transferred with a Pasteur pipet to a 5-mL round-bottomed flask for the final distillation.

When most of the ether is blown off, add a boiling stone to the flask containing the product. Assemble the apparatus for a simple distillation [see Technique 7.3b]. You will need two receiving vials, one for the fraction boiling below 100° C and the other, a tared vial, for collecting the product fraction. Place the receiving vial in a beaker of ice and water. Heat an aluminum block or a sand bath to about 100-110° C and distill the residual ether into the first receiving vial. When the rate of distillation slows or stops, increase the rate of heating until the aluminum block or sand bath is 140-150° C. Change to the tared receiving vial when the vapor temperature reaches 100° C. Collect the product fraction from 100-122° C.

SAFETY PRECAUTION

Be sure to stop the distillation before the boiling flask reaches dryness. Diethyl ether may form peroxides. If the flask goes to dryness, the temperature inside it rises rapidly and the peroxides could explode.

Tightly cap the vial containing your product and store it for use in Project 3.2.

Weigh your alcohol product and calculate the percent yield. Use gas chromatographic analysis to assess the percent purity of your product. Any starting material present in the product can be identified by the peak enhancement method [see Technique 11.6]. Obtain an IR spectrum of your product, if directed to do so by your instructor [see Spectrometric Method 1].

Cleanup: Pour the aqueous phase from the extractions into the container for aqueous inorganic waste. Place the solid borate residue and the spent drying agent in the container for inorganic waste. Pour the recovered ether and the residue left in the boiling flask into the container for flammable (organic) waste.

3.2 Acid-Catalyzed Dehydration of 3,3-Dimethyl-2-Butanol with Rearrangement

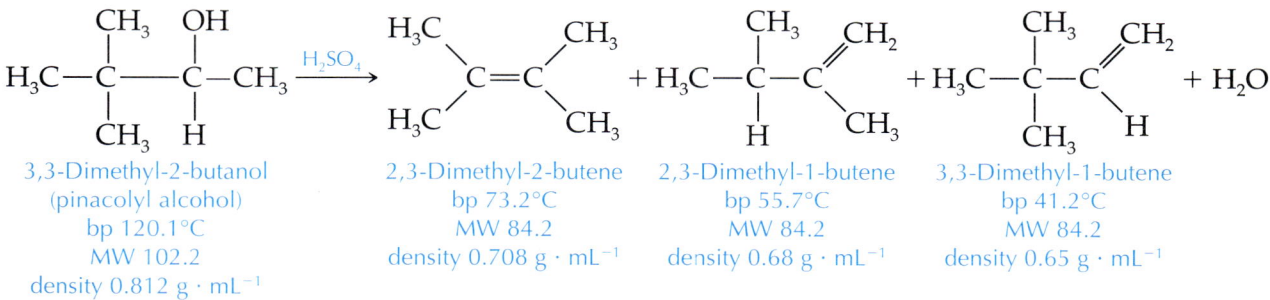

3,3-Dimethyl-2-butanol (pinacolyl alcohol) bp 120.1°C MW 102.2 density 0.812 g · mL^{-1}

2,3-Dimethyl-2-butene bp 73.2°C MW 84.2 density 0.708 g · mL^{-1}

2,3-Dimethyl-1-butene bp 55.7°C MW 84.2 density 0.68 g · mL^{-1}

3,3-Dimethyl-1-butene bp 41.2°C MW 84.2 density 0.65 g · mL^{-1}

As described in Experiment 11, dehydration promoted by sulfuric acid often involves carbocation intermediates. These intermediates are prone to rearrangement whenever a pathway is available. For example, a secondary carbocation can undergo a hydride or methyl migration and thereby rearrange to a more stable tertiary carbocation, as occurs in this experiment.

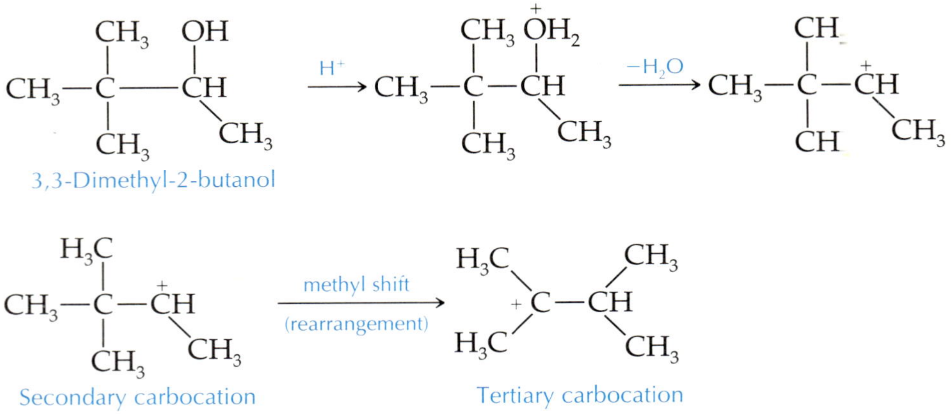

3,3-Dimethyl-2-butanol

Secondary carbocation   Tertiary carbocation

The tertiary carbocation then readily deprotonates to yield an alkene:

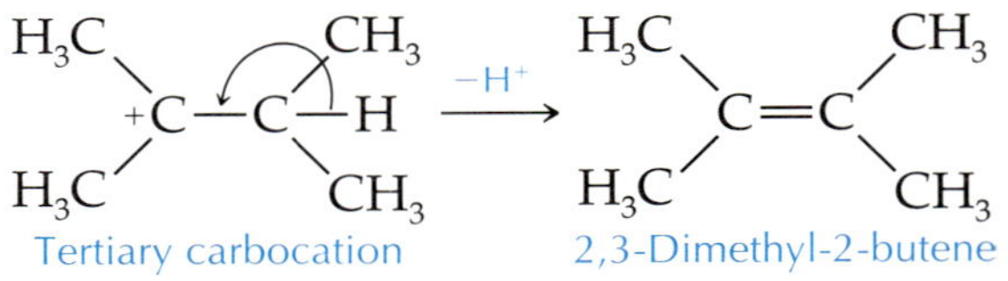

Tertiary carbocation   2,3-Dimethyl-2-butene

The same tertiary carbocation can deprotonate in a different fashion to form 2,3-dimethyl-1-butene:

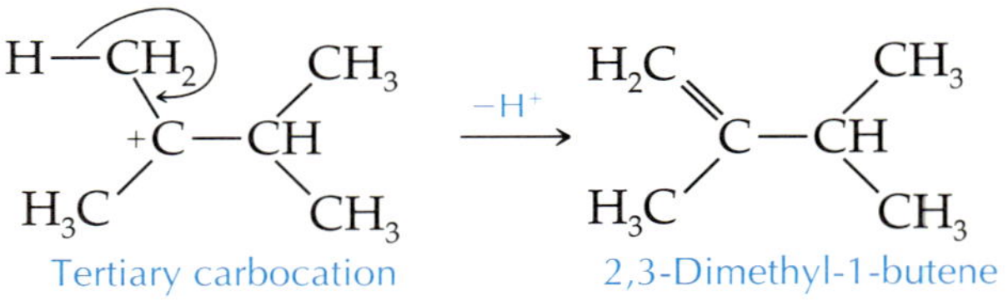

Tertiary carbocation   2,3-Dimethyl-1-butene

We might expect more 2,3-dimethyl-2-butene to form because it is more stable, having more alkyl groups directly attached to the sp^2 carbon atoms.

The secondary carbocation formed earlier in the mechanistic sequence can also deprotonate to form a third alkene, 3,3-dimethyl-1-butene:

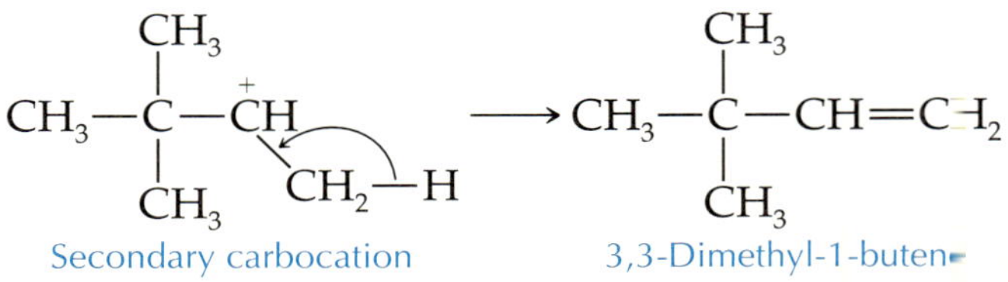

Secondary carbocation   3,3-Dimethyl-1-butene

The reaction actually produces a mixture of all three butenes. You can determine the relative amounts of each alkene by analyzing your product with gas chromatography. This information will allow you to investigate the relative likelihood of the competing reaction pathways in this acid-catalyzed dehydration reaction.

ꜱmicroscale Procedure

Techniques   Microscale Distillation: Technique 7.3b
Pasteur Filter Pipets: Technique 2.4
Gas Chromatography: Technique 11

SAFETY INFORMATION

3,3-Dimethyl-2-butanol and **heptane** are flammable.

Concentrated sulfuric acid is extremely corrosive and causes severe burns. Avoid contact with skin, eyes, and clothing.

The alkenes produced in this reaction have a strong, musty odor. Conduct the experiment in a hood, if possible.

Glass beads serve the same purpose as boiling stones, which tend to disintegrate in hot concentrated sulfuric acid.

Transfer all of the 3, 3-dimethyl-2-butanol that you synthesized in Project 3.1 to a 5-mL round bottomed flask with a Pasteur pipet. Add 3 drops of concentrated sulfuric acid for each gram of 3,3-dimethyl-2-butanol (proportionally adjust the amount of acid for your amount of 3,3-dimethyl-2-butanol). Put three glass beads in the flask. Assemble the apparatus for simple distillation using a vial placed in an ice-water bath as the receiver [see Technique 7.3b]. Heat the reaction flask gently with an aluminum heating block or a sand bath heated to 150° C. Distill the product mixture at a moderate rate until the vapor temperature reaches 100° C or until distillation ceases. A brown residue will remain in the distilling flask.

The alkene products evaporate quite easily at room temperature. Keep the vial they are in tightly capped.

Dry the product mixture with 150 mg of anhydrous potassium carbonate. Cap the vial so that the volatile alkenes do not evaporate during the drying process. Transfer the dried product to a tared (weighed) vial, using a Pasteur filter pipet [see Technique 2.4]. Weigh the product mixture and determine the percent yield.

Analyze the product mixture by gas chromatography [see Technique 11]. For a packed column instrument, use a 20% Carbowax 20M on Chromosorb P column at 75°C and a gas flow rate of 50 mL·min.$^{-1}$ (Ref. 1). Consult your instructor about sample preparation for a packed column gas chromatograph. The alkenes will elute in order of increasing boiling point.

For a capillary column gas chromatograph with a nonpolar column, such as polydimethylsiloxane, use a column temperature no higher than 50°C. Prepare the sample by dissolving 2 drops of the product mixture in 0.5 mL of heptane; inject 1 μL of this solution into the GC. The alkenes elute from a nonpolar column in order of increasing boiling point, with the solvent peak (heptane) occurring *after* the product peaks rather than before, as is usually the case.

Identification of the alkenes can be done by the peak enhancement method, because all three products are commercially available [see Technique 11.6]. Submit the remaining product to your instructor.

Determine the relative amounts of each alkene from the peak areas or the peak heights on your chromatogram. Explain the formation of all observed products and discuss the observed product distribution.

Cleanup: Carefully dilute the residue in the distilling flask with about 4 mL of water. (**Caution: The residue contains sulfuric acid.**) Add solid sodium carbonate in small portions until the acid is neutralized. (**Caution: Foaming.**) Pour the neutralized solution on a Buchner funnel to recover the glass beads. The filtrate may be washed down the sink or poured into the container for aqueous inorganic waste. Put the glass beads in the appropriate container. Pour the heptane solution remaining from the GC analysis into the container for flammable waste.

Reference

1. Sayed, Y.; Ahlmark, C. A.; Martin, N. H. *J. Chem. Educ.* **1989,** *66,* 174−175.

Questions

1. Why does the solvent peak (heptane) show a longer GC retention time than the alkene products? Why is heptane a better choice for the solvent than diethyl ether in this GC analysis?

2. Acid-catalyzed dehydration of an alcohol occurs under equilibrating conditions that favor the thermodynamically more stable product(s). Identify the thermodynamic product(s) and the kinetic product(s) of your dehydration reaction.

3. Why did the observed boiling point in the distillation in Project 3.2 rise above the boiling points of the alkene products?

4. Because acid-catalyzed dehydration is an equilibrium situation, how was the reaction forced to completion in Project 3.2?

5. Predict the products of the reaction:

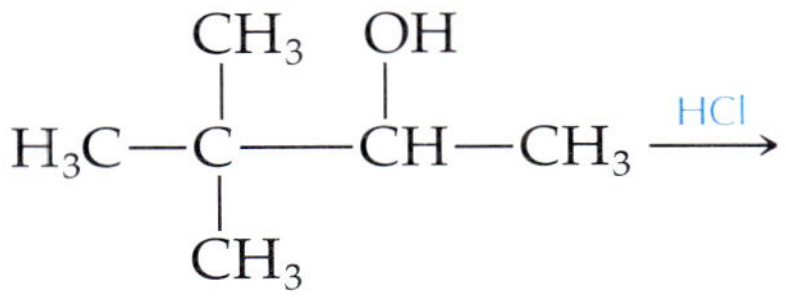

$$
H_3C-\underset{\underset{CH_3}{|}}{\overset{\overset{CH_3}{|}}{C}}-\underset{\overset{|}{OH}}{CH}-CH_3 \xrightarrow{\text{HCl}}
$$

Use a mechanism to explain your prediction.

6. Why might a poor yield of alkenes be realized if the dehydration were carried out with hydrochloric acid rather than H_2SO_4?

7. With the knowledge that $NaBH_4$ is a base, suggest a reaction that would make clear why the presence of any Brønsted acid of significant strength would be deleterious to a sodium borohydride reduction.

8. Methanol is known to undergo autoprotolysis to form $CH_3OH_2^+$ and CH_3O^-. Suggest how the Le Chatelier principle can be used to explain why added methoxide ion would adjust the autoprotolysis equilibrium reaction for methanol in a way favorable to carbonyl reduction.

9. Explain why the necessary reaction conditions would be successively milder (lower temperatures, lower acid concentrations) for the dehydration of the following alcohols: 1-heptanol, 2-heptanol, 1-methylcyclohexanol.

10. Write a mechanism for the following dehydration and rearrangement reaction:

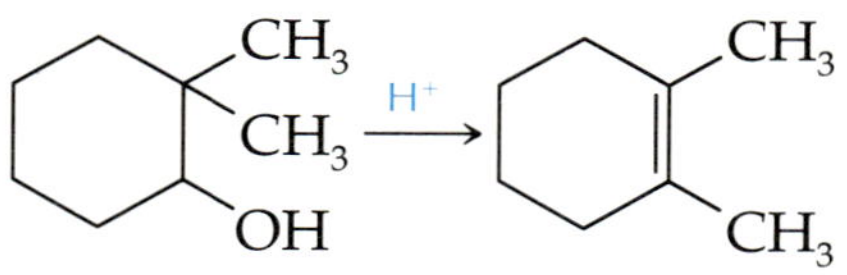

11. Write a mechanism for the following reaction:
12. Potassium carbonate is used as the drying agent for the alkene
product to prevent acid-catalyzed rearrangements. This is important because you need to measure the alkene product ratio arising
from the dehydration process, rather than from an isomerization
that occurs after dehydration. Sketch mechanisms that rationalize
the acid-catalyzed interconversion of 2,3-dimethyl-2-butene, 2,3-
dimethyl-1-butene, and 3,3-dimethyl-1-butene.

Techniques for Experiment 9 and Project 3, excerpted from *Experimental Organic Chemistry*—Flexible Connector Microscale Glassware Version

2.4
Pasteur Filter Pipets

Microscale laboratory work often requires the transfer of small volumes (1–5 mL) of liquids from one container to another. The Pasteur pipet serves as the tool of choice for these transfers. Volatile liquids, however, will frequently drip out of a Pasteur pipet during such transfers because the solvent vapors build up in the rubber bulb and create increased pressure that forces the liquid out of the pipet. Losses of even a few drops need to be avoided in microscale work. Placing a small plug of cotton in the tip of the Pasteur pipet provides enough resistance to this increased pressure to prevent dripping during the transfer of a volatile liquid. The cotton plug also allows the solution to be drawn out or expelled from the pipet at a slower rate so that better control can be maintained, especially during microscale extractions (discussed in Technique 4).

Airborne particles, such as dust and lint, also present problems when working with microscale volumes of liquids. The cotton plug serves to filter the solution each time it is transferred with a filter tip pipet. The filter pipet also removes a solution very efficiently from a drying agent (discussed in Technique 4.7).

Filter pipets are made by using a piece of wire that has a diameter slightly less than the inside of the capillary portion of the pipet to push a tiny piece of cotton into the tip of a Pasteur pipet (Figure 2.7). A piece of cotton of the appropriate size should offer only slight resistance to being pushed by the wire. If there is so much resistance that the cotton cannot be pushed to the tip of the pipet, then the piece is too large. Remove the wire and insert it through the tip to push the cotton back into the upper part of the pipet, and take a bit off the piece of cotton before putting it back into the pipet. The finished cotton plug should be 2–3 mm long

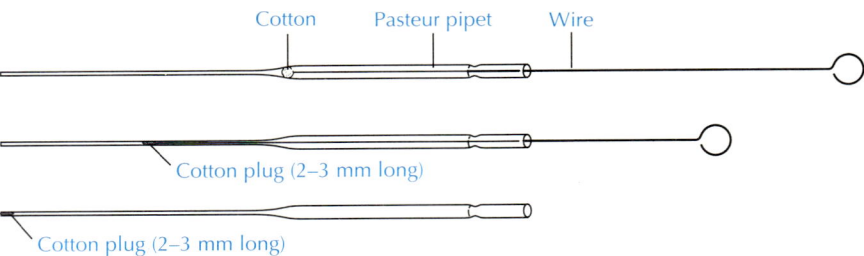

FIGURE 2.7 Preparing Pasteur filter pipet.

and should fit snugly but not too tightly. If the cotton is packed too tightly in the tip, liquid will not flow through it; if it fits too loosely, it may be expelled with the liquid. With a little practice, you should be able to prepare a filter pipet easily.

When using a Pasteur filter pipet (or any other pipet with a bulb attached), you should set the pipet in a test tube or Erlenmeyer flask to keep it upright. Laying the pipet on the bench top or other horizontal surface allows the rubber bulb to be contaminated with solvent or compounds dissolved in the solvent. (Figure 2.8).

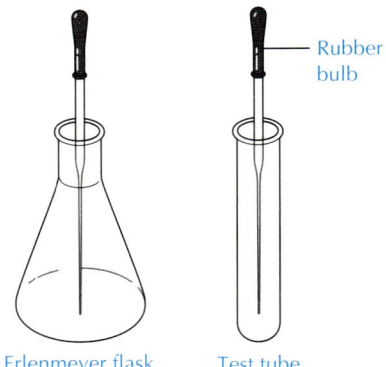

FIGURE 2.8 Correct ways to temporarily store Pasteur pipet when you are not using it to transfer liquid.

SAFETY PRECAUTION

Glass Pasteur pipets are puncture hazards. They should be handled and stored carefully. Dispose of Pasteur pipets in a manner that does not present a hazard to lab personnel or housekeeping staff. Check with your instructor about the proper disposal method in your laboratory.

4.5a Extracting an Aqueous Solution with an Organic Solvent More Dense Than Water

The microscale extraction of an aqueous solution with a solvent, such as dichloromethane (CH_2Cl_2), that is more dense than water is an example of this type of extraction. The product is transferred from the aqueous phase to the organic phase and the lower organic phase (dichloromethane) needs to be removed from the vial in order to separate the layers.

Set centrifuge tubes in a test tube rack to prevent tipping.

Place the aqueous solution and the specified amount of solvent in a centrifuge tube labeled centrifuge tube 1 (Figure 4.7). Tightly cap the tube and shake the mixture thoroughly. Loosen the cap slightly to release the pressure and allow the layers to separate completely.

Prepare a Pasteur filter pipet [see Technique 2.4]. Press the air from the rubber bulb and put the pipet into the centrifuge tube with the tip touching the bottom of the cone. Partially release the pressure on the bulb and draw the lower layer into the pipet until the inter-

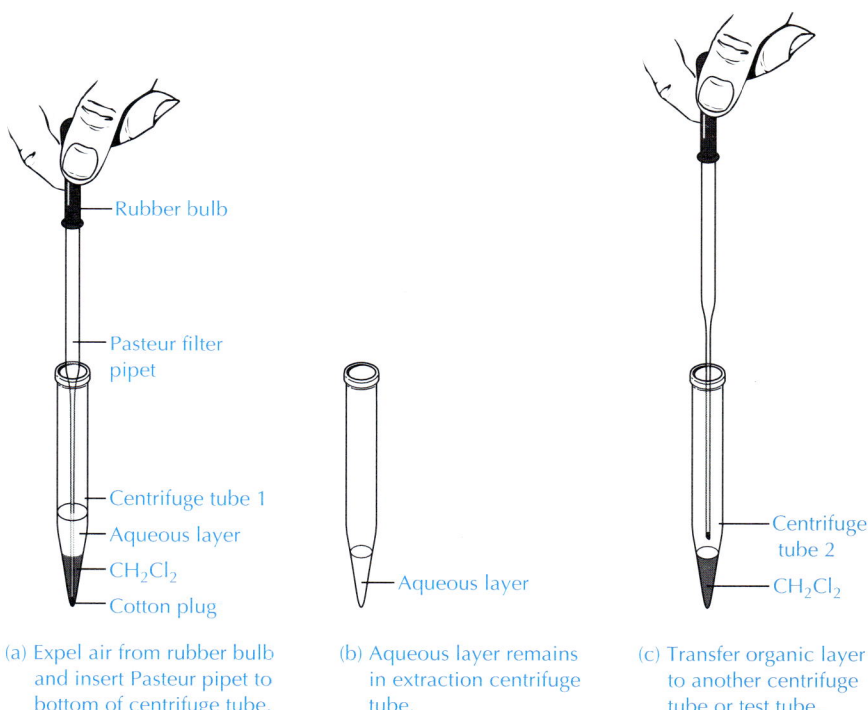

Rubber bulb

Pasteur filter pipet

Centrifuge tube 1
Aqueous layer
CH_2Cl_2
Cotton plug

Aqueous layer

Centrifuge tube 2
CH_2Cl_2

(a) Expel air from rubber bulb and insert Pasteur pipet to bottom of centrifuge tube.

(b) Aqueous layer remains in extraction centrifuge tube.

(c) Transfer organic layer to another centrifuge tube or test tube.

FIGURE 4.7 Extracting an aqueous solution with a solvent more dense than water.

Hold tube 2 close to tube 1 so that the transfer can be accomplished without loss of liquid.

face between the two layers is exactly at the bottom of the tube. Maintain a steady pressure on the rubber bulb while transferring the pipet to another centrifuge tube (centrifuge tube 2 in Figure 4.7c) or test tube. The aqueous layer remains in the extraction tube and can be extracted again with a second portion of CH_2Cl_2. The second portion of CH_2Cl_2 is added to tube 2 after the separation.

4.5b Washing an Organic Liquid That Is More Dense Than Water

When an organic phase (either a liquid product or a solution of organic solvent and product) more dense than the aqueous phase is being washed, the aqueous phase needs to be removed. There are two methods for doing this operation. Method A is preferable when a liquid organic product is not dissolved in a solvent, because losses on the glassware are minimized. Method B is easier to perform because the lower phase is always removed, but it requires several more transfers than Method A, plus two or three additional centrifuge tubes. . . .

Method B. Two centrifuge tubes are needed for this procedure; a third is useful (Figure 4.9). The lower organic phase is transferred to centifuge tube 2 in the first separation. Pour the aqueous phase remaining in tube 1 into a labeled test tube. Clean and dry centifuge tube 1 (not necessary if a third centifuge tube is available). Add the second portion of wash liquid (e.g., water, NaOH solution). Proceed with the second washing of the organic phase in tube 2 by capping the tube and shaking the two phases together. Allow the phases to separate and again remove the lower organic phase with a Pasteur filter pipet, placing the organic phase in the clean centrifuge tube 1 (or centrifuge tube 3). Repeat the procedure if additional washings are required. After the last washing, the organic phase will be in a clean centrifuge, ready for the next step of the experimental procedure. Treat the aqueous washes as directed in the cleanup procedure.

Tipping the centifuge tube helps you to see the interface.

Remember, as in any extraction, no material should be discarded until you are certain that you have the desired product in hand.

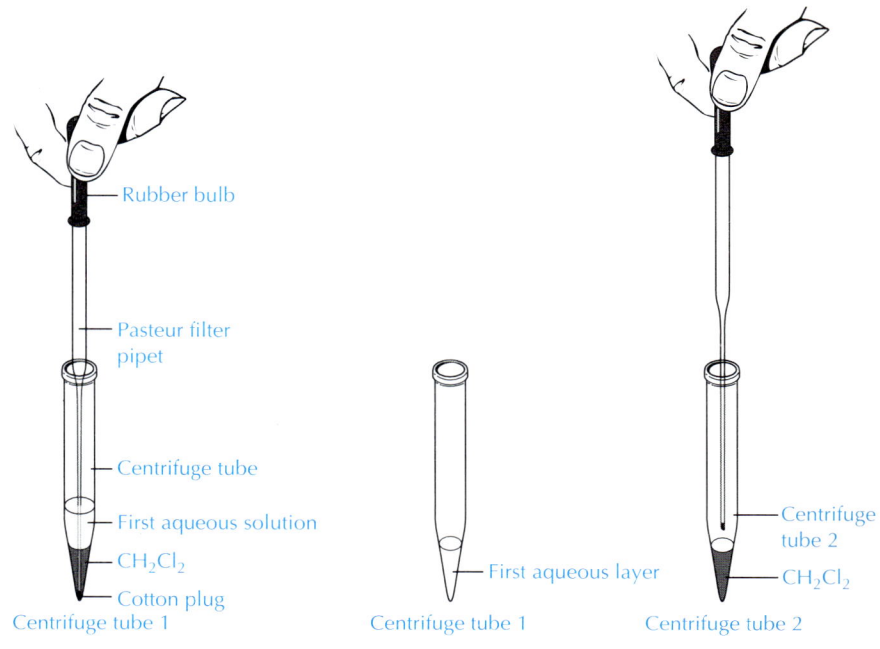

Rubber bulb

Pasteur filter pipet

Centrifuge tube

First aqueous solution

CH$_2$Cl$_2$

Cotton plug

Centrifuge tube 1

First aqueous layer

Centrifuge tube 1

Centrifuge tube 2

CH$_2$Cl$_2$

Centrifuge tube 2

(a) Expel air from rubber bulb and insert Pasteur pipet to bottom of tube. Draw lower layer into pipet.

(b) Transfer lower layer to centrifuge tube 2.

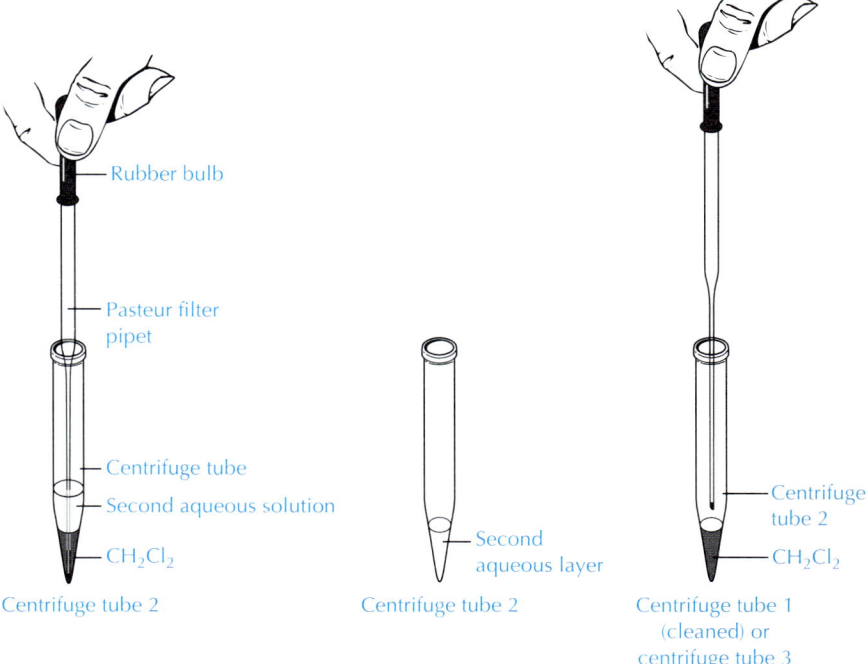

Rubber bulb

Pasteur filter pipet

Centrifuge tube

Second aqueous solution

CH$_2$Cl$_2$

Centrifuge tube 2

Second aqueous layer

Centrifuge tube 2

Centrifuge tube 2

CH$_2$Cl$_2$

Centrifuge tube 1 (cleaned) or centrifuge tube 3

FIGURE 4.9 Method B for washing an organic liquid that is more dense than water.

(c) After washing with second aqueous solution, draw lower organic layer into Pasteur pipet.

(d) Transfer organic layer to a clean dry centrifuge tube.

4.5d Washing an Organic Liquid That Is Less Dense Than Water

When washing an organic solution or product less dense than water, the lower or aqueous layer needs to be removed from the centrifuge tube. This operation is similar to the procedure in Section 4.5c for extracting an aqueous solution with a solvent less dense than water. The only difference is that the organic phase remains in the same centrifuge tube for the entire procedure and only the aqueous solutions are removed (Figure 4.11). Add the specific amount of water or aqueous solution, cap the centrifuge tube, and shake it to mix the phases. Open the cap to vent the tube and allow the layers to separate. Expel the air from the rubber bulb of a Pasteur filter pipet and insert the pipet into the bottom of the centrifuge tube. Draw the aqueous layer into the pipet and transfer it to a test tube. The upper organic phase remains in the centrifuge tube, ready for the next step, which may be another washing or drying with an anhydrous salt.

Remember, as in any extraction, no material should be discarded until you are certain that you have the desired product in hand.

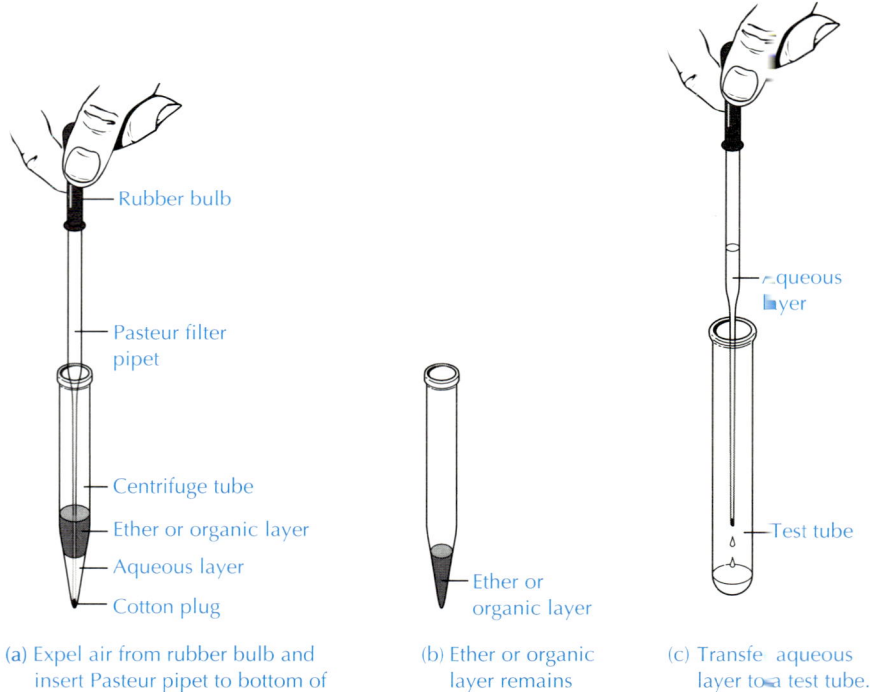

(a) Expel air from rubber bulb and insert Pasteur pipet to bottom of

(b) Ether or organic layer remains

(c) Transfer aqueous layer to a test tube.

Rubber bulb

Pasteur filter pipet

Centrifuge tube

Ether or organic layer

Aqueous layer

Cotton plug

Ether or organic layer

Aqueous layer

Test tube

FIGURE 4.11 Washing an organic liquid that is less dense than water.

⑦.3b Microscale Distillations

When the volume of a liquid to be distilled is only a few milliliters, a significant part of the sample would be used just to fill the short-part system described in Technique 7.3a with vapor (called the *holdup volume*). For small volumes, use a microscale distillation apparatus that is essentially a miniature version of the standard-taper short-path distillation apparatus. The apparatus consists of a 5-mL or 10-mL round-bottomed flask and a distillation head connected by a flexible connector with a support rod. The thermometer is held in place by the flexible thermometer adapter, as shown in Fig. 7.9. The distillate is collected in a small vial that is at least three-fourths submerged in a beaker of ice and water.

Place the round-bottomed flask in a 50-mL beaker while you are transferring liquid to it.

To carry out a microscale distillation, select a 5-mL or 10-mL round-bottomed flask appropriate for the volume or liquid to be distilled; the flask should be no more that two-thirds full. Using a Pasteur pipet, transfer the liquid to be distilled to the round-bottomed flask and add a magnetic stirring bar or a boiling stone. Attach the flexible connector with support rod to the flask and clamp the support rod to a vertical rod or ring stand. Fit the flexible thermometer adapter to the tope of the distilling head and carefully push a thermometer through the adapter.

--- SAFETY PRECAUTION ---

Holding the thermometer by the upper part of the stem while inserting it through the adapter could break the thermometer and force a piece of broken glass into your hand.

Grasp the thermometer close to the bulb and push it gently into the adapter a centimeter or two. Move your hand several centimeters up the thermometer stem and repeat the pushing motion. Continue this process until the thermometer is properly positioned.

The top of the thermometer bulb should be placed just below the sidearm as shown by the dotted line drawn across the distillation head in Figure 7.9. Carefully attach the distilling head to the flexible connector about the round-bottom flask. Place the receiving vial in a 30- or 50-mL beaker of ice and water, and position the vial under the outlet of the distillation head as far up as it will go. Put a sand bath or an aluminum heating block with a flask depression under the round-bottom flask.

Head the sand bath or aluminum block to a temperature of 20–50° above the boiling point of the liquid being distilled. After the liquid in the flask boils, you will notice a ring of condensate slowly moving up the flask and into the distillation head. The temperature observed on the thermometer will rise as the vapor reaches the thermometer bulb. For the distillation of liquids boiling above 120° C, it may be necessary to wrap the neck of the flask and the portion of the distillation head near the thermometer loosely in glasswool to slow the rate of heat loss. The distillation should be done at a slow enough rate for the vapor to condense in the chilled vial and not evaporate out of the system. Stop the distillation by raising the apparatus above the heat source before the flask goes to dryness.

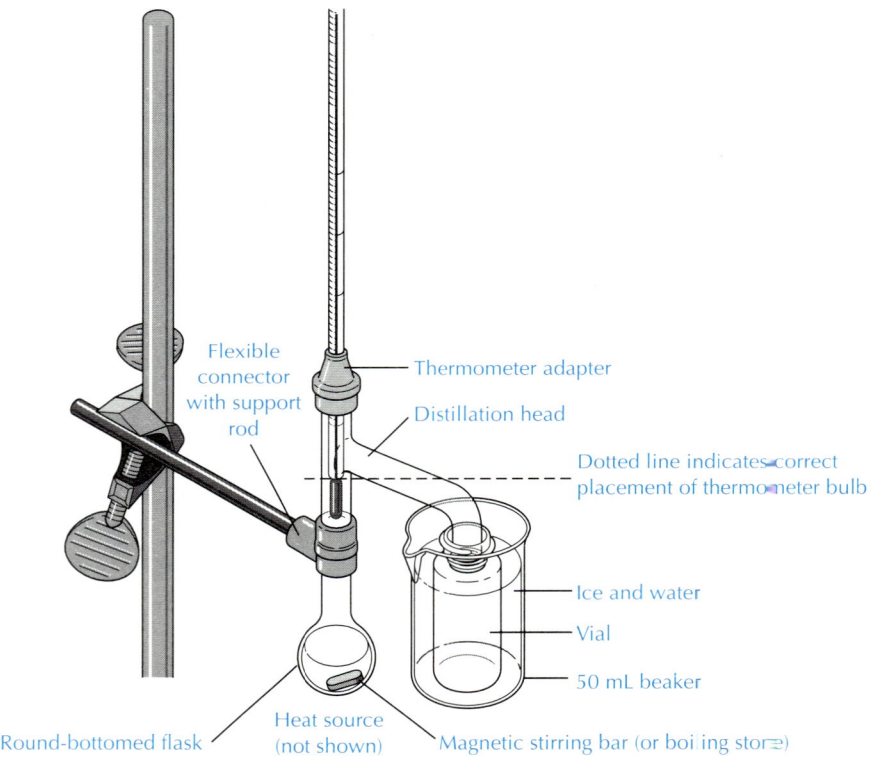

FIGURE 7.9 Microscale apparatus for a simple distillation.

ORGANIC CHEMISTRY

Structure and Function

ORGANIC CHEMISTRY: Structure and Function, 3/e will be available for your Fall 1998 courses

Dear Instructor:

Welcome to the new edition of *ORGANIC CHEMISTRY: Structure and Function*. We understand the importance of experiencing a text before using it in your course. In sending you this preview book containing the Preface, the complete Table of Contents, and the first eight chapters of the new edition, we are inviting you to do just that.

As you teach your course, compare these chapters to the coverage in your current text. You will see how the authors have extensively revised the material to make it **more accessible and student friendly**, as well as updating it for currency. Show *ORGANIC CHEMISTRY: Structure and Function* to your students to see what they think.

The approach is what makes *ORGANIC CHEMISTRY: Structure and Function* unique. Within each chapter and throughout the text, K. Peter C. Vollhardt and Neil E. Schore describe the structure of each new group of compounds before explaining how it functions. This approach helps students organize and interpret new concepts and relate them to what they've already learned.

On the flip side of this book is an introduction to our new organic chemistry laboratory manual—*EXPERIMENTAL ORGANIC CHEMISTRY, A Balanced Approach: Macroscale and Microscale*. We have included the Preface and the complete Table of Contents, as well as an experiment and project using flexible connector microscale glassware. A version of the text using ground glassware is already available. Here again, we want you to see for yourself how we've tailored our text to fit your teaching methods and the needs of your students.

For sample copies of these titles, please fill out and return the postpaid card in the back of this preview book. You may also contact your sales representative or email our faculty services department (facultyservices@sasmp.com). To receive periodic updates on *ORGANIC CHEMISTRY: Structure and Function* and information about all our organic chemistry products, please e-mail us at organicchemistrylist@whfreeman.com and sign up. The updates will start appearing in mid-December. In January, our new Web site (http://www.whfreeman.com/ochem) will begin broadcasting information on the new edition of *ORGANIC CHEMISTRY: Structure and Function*, as well as informing you of new educational resources germane to your field.

Third Edition

ORGANIC CHEMISTRY

Structure and Function

K. Peter C. Vollhardt
University of California at Berkeley

Neil E. Schore
University of California at Davis

W. H. Freeman and Company
NEW YORK

About the Cover: *Brevetoxin B (shown as a ball-and-stick model) is a potent marine neuro-toxin responsible for massive fish kills, mollusk poisoning, and human food poisoning along the coast of Florida, the Gulf of Mexico, and many other parts of the world. It is associated with the explosive growth, or bloom, of the algae dinoflagellate* Ptychodiscus brevis *under certain favorable conditions of temperature, salinity, and sunlight, causing a phenomenon called "red tide." The front cover depicts a red tide on May 6, 1976, approaching the coast of Oshima Island, Japan. The back cover shows space-filling and bond-line formulas of the unique structure of brevetoxin B consisting of a single carbon chain arranged in a rigid ladder-like framework and composed of 11 contiguous trans fused ether rings. Brevetoxin B was made in the laboratory by total synthesis in 1994, requiring 83 steps from 2-deoxyribose and 12 years of effort, spearheaded by Professor Kyriacos Costa Nicolaou (who also supplied the original photograph) at the Scripps Research Institute and the University of California at San Diego. For a discussion of such syntheses, see Chapter 8.*

Acquisitions Editor: Michelle Russel Julet

Development Editor: Randi Rossignol

Project Editor: Mary Louise Byrd

Marketing Manager: Kimberly Manzi

Cover Designer: Cambraia Magalhães

Text Design: Circa 86, Inc.

Illustration Coordinator: Bill Page

Illustrator: Network Graphics

Production Coordinator: Ellen Cash

Composition: York Graphic Services

Manufacturing: R. R. Donnelly & Sons

Vollhardt, K. Peter C.
    Organic chemistry: structure and function/K. Peter C.
Vollhardt, Neil E. Schore. — 3rd ed.
      p. cm.
    Includes index.
    ISBN 0-7167-2721-8
     1. Chemistry, Organic.   I. Schore, Neil Eric, 1948–   .
II. Title.
QD251.2.V65   1998   97–22455
547 — dc21   CIP

Printed in the United States of America

First printing, 1998

ORGANIC CHEMISTRY
Structure and Function

CONTENTS OVERVIEW

ORGANIC CHEMISTRY
Structure and Function

CONTENTS

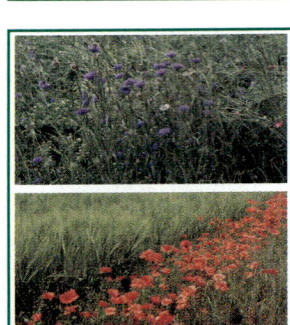
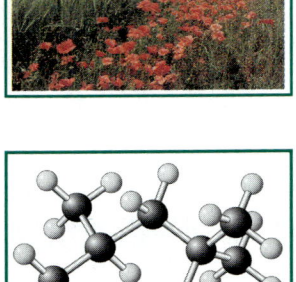

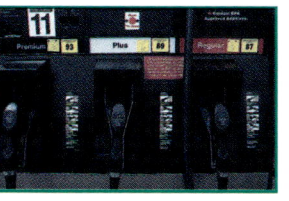

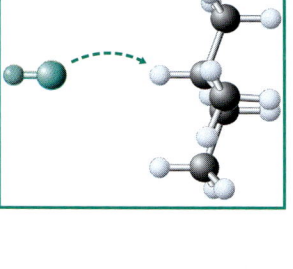

3   Reactions of Alkanes: Bond-Dissociation Energies, Radical Halogenation, and Relative Reactivity   93

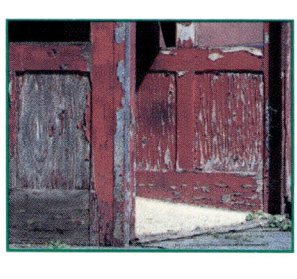

4   Cyclic Alkanes   129

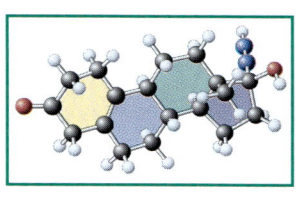

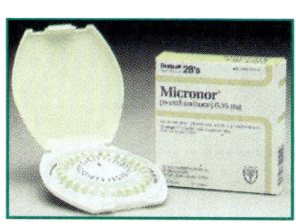

211

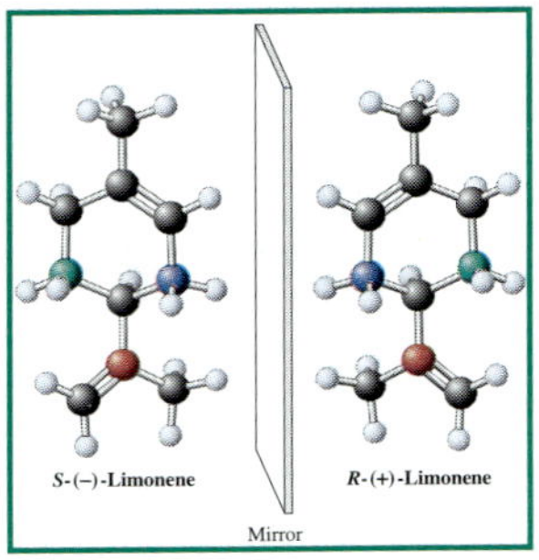

S-(−)-Limonene   *R*-(+)-Limonene

Mirror

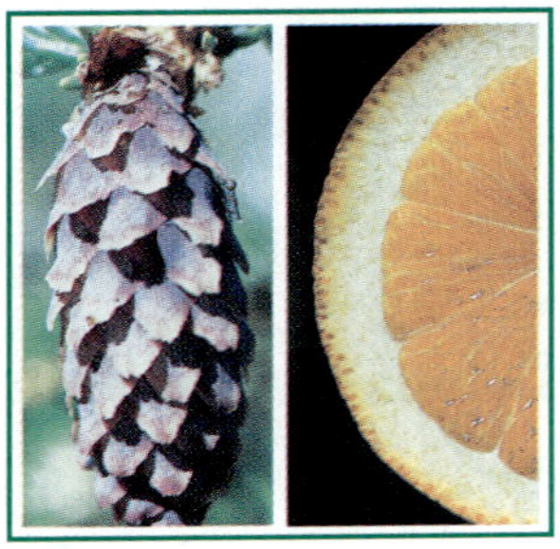

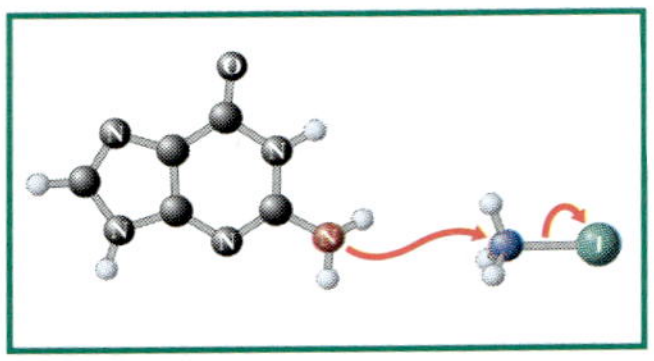

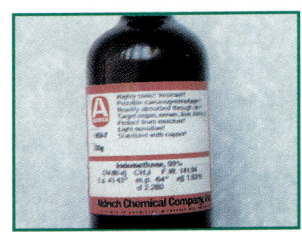

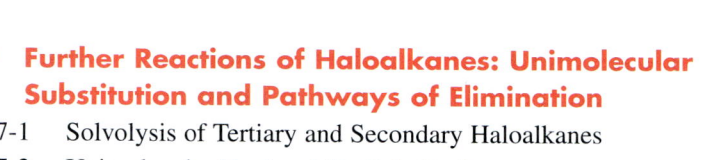
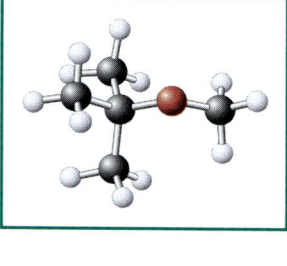

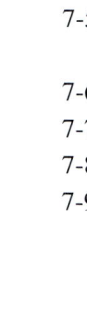

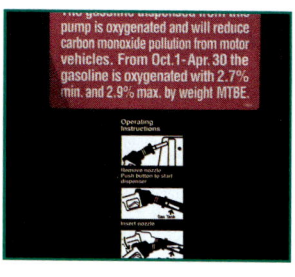

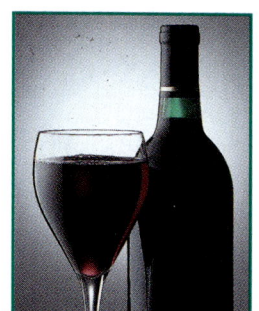

PREFACE

Structure and Function Motif

Too many students find that organic chemistry is an overwhelming parade of facts. Our goals are to dispel this notion and, more importantly, to help students learn and understand organic chemistry. The best way to do this is to provide a framework, or scaffolding, around which students can organize their thoughts. The framework that we provide is the accessible notion that understanding structure will lead to an understanding of function.

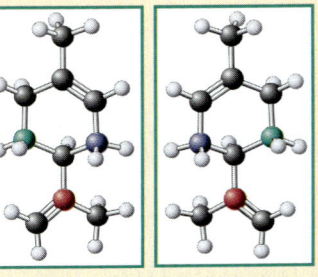

S-(−)-Limonene   R-(+)-Limonene

A Uniform Organization Emphasizes the Relation Between Structure and Function

Much like a language, in which grammar would be dangling without the "meat" of vocabulary, the text develops material as a juxtaposition of structure and function. Thus, Chapter 1 provides the fundamentals of structure and bonding, specifically as they will become useful in understanding organic chemistry. Chapter 2 then follows with an introduction to the structural features of the alkanes and how they "function" in the simplest sense, namely, conformational mobility. Chapter 3 relates bond-dissociation energies to the *lead reaction:* radical halogenation (functionalization) of alkanes. Chapter 4 repeats the motif of Chapter 2 but has cycloalkanes as the center of focus.

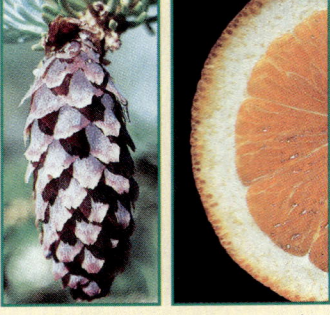

Silver fir tree cone   Orange peel

The structure of the haloalkanes and how it determines their fate in nucleophilic substitution and elimination reactions are the topics of Chapters 6 and 7. Each subsequent functional group is covered according to the same rhythm: naming, structure, spectroscopy, preparations, reactions, and biological and other applications.

Students Are Given the Structural "Tools" They Need to Understand Function

The interplay between structure and function that gives a hierarchy to individual chapters also confers a hierarchy on the text as a whole. This is why we introduce stereochemistry in Chapter 5. Students learn about stereochemical principles so that they are prepared to understand the substitution and elimination reactions of haloalkanes (Chapters 6 and 7) and the addition reactions of alkenes (Chapter 12). Moreover, this hierarchy allows the mechanistic discussion of all new important reactions to take place concurrently rather than being scattered in different places throughout the text. Such a unified presentation of mechanisms benefits the student enormously.

Alcohols (Chapters 8 and 9) with the simplest oxygen-containing function are treated early because their chemistry sets the stage for understanding their central role in synthesis. Similarly, carbocations (and their rearrangements; see Section 9-3) appear before a discussion of the Markovnikov rule, alkenes (Chapter 12) before conjugated polyenes (Chapter 14), and conjugated polyenes before aromatic systems.

Early Presentation of Spectroscopy

Our first edition broke ground by introducing spectroscopy right after alcohol chemistry. Early coverage, beginning with NMR in Chapter 10, offers opportunities to practice the application of spectroscopic methods to many kinds of compounds. After NMR, we cover IR- and UV-visible spectroscopy in Chapters 11 and 14 in the context of functional groups. Courses can include each of the principal types of spectroscopy in the first half of the text.

The Lead Reaction: Radical Halogenation of Methane

The first, more detailed discussion of a reaction and its mechanism, the "lead reaction," is presented very early in Chapter 3. For several reasons, the best (most logical) choice is the chlorination of methane. First, all chemical reactions require bond-making and bond-breaking. The radical halogenation of methane allows the introduction of the concepts of bond-dissociation energies and the stability and structure of the ensuing radicals. This leads to an understanding of the thermal stability of the simplest organic bonds, C–H and C–C, and, hence, of why organic materials are capable of existence. The students learn that to "activate" a C–H bond, a reactive agent is required. Second, the lead reaction, because it purposely does not include ionic species, can be analyzed thermodynamically by calculating enthalpies of the overall process, as well as individual steps. This exercise is fundamental and gives the student the basic tool of "eyeballing" the relative feasibility of all future transformations. It also serves as a first application of potential energy diagrams to a chemical process. Third, the generalization of the chlorination of methane to the halogenation of other alkanes permits the simple introduction of the concepts of reactivity and selectivity, a feeling for the statistics needed to deal with molecules endowed with several equally reactive sites, and practical applications of these principles.

Emphasis on Synthetic Strategy

Retrosynthetic analysis simplifies synthesis problems

Many compounds that are commercially available and inexpensive are also small, containing six or fewer carbon atoms. Therefore, the most frequent task facing the synthetic planner is that of building up a larger, complicated molecule from smaller, simple fragments. The best approach to the preparation of the target is to work its synthesis *backward* on paper, an approach called **retrosynthetic analysis*** (*retro*, Latin, backward). In this analysis, strategic

The importance of synthesis is stressed starting on page 3, and the considerations entering into the development of a good synthetic strategy and the avoidance of pitfalls are developed throughout the text. Since the innovative introduction of Section 8-9 on *retrosynthetic analysis* in the first edition, this aspect of the text has received much positive feedback from teachers and students. The present edition has added a slightly more explicit treatment of *linear* versus *convergent* synthesis to this section. *Multistep* partial and total syntheses are pointed out where appropriate in the various functional-group treatments. Particular emphasis is placed on *stereo-* and *regioselectivity* (Chapter 12 and Section 16-5), *biological* and *medicinal* relevance (e.g., Sections 9-11, 12-16, 18-12, and 19-4 and Chapters 24 through 26), and the importance of *materials* synthesis (e.g., Sections 12-13 through 12-15, 13-11, 14-10, and 21-12). These discussions are then extensively reinforced in the In-Chapter Exercises and End-of-Chapter Problems and highlighted in numerous Chemical Highlights.

Modern Coverage

Modern developments at the forefront of current research have been included in the main body of the text or in **Chemical Highlights**, in particular enantioselectivity, advances in molecular biology (e.g., structure recognition), medicine (drugs), and progress in materials science.

Rigorous (IUPAC and *Chemical Abstracts*) nomenclature is used whenever possible, and common names are added in parentheses when warranted. The reactions presented in the text are (with minor exceptions) drawn from the *original literature* with actual conditions and yields.

We recorded authentic *spectra* on modern computerized equipment. We have incorporated 300 MHz and higher frequency ^1H NMR spectra to demonstrate the higher resolution at such field strengths and, when necessary, to simplify the spectral patterns of compounds that produce overlapping signals at lower fields. We have retained 90 MHz as the resonance frequency for molecules that exhibit first-order behavior under these conditions, because the reproduction of their spectra is clearer, is easier to read, and does not require the use of expanded inserts.

CHEMICAL HIGHLIGHT 5-5

Why Is Nature "Handed"?

In this chapter, we have seen that many of the organic molecules in nature are chiral. More importantly, most natural compounds in living organisms not only are chiral, but also are present in only one enantiomeric form. An example of an entire class of such compounds consists of the *amino acids,* which are the component units of *polypeptides.* The large polypeptides in nature are called *proteins* or, when they catalyze biotransformations, *enzymes.*

Absolute Configuration of Natural Amino Acids and Polypeptides

Amino acid (R variable)

Polypeptide
Amino acid 1 · Amino acid 2 · Amino acid 3

Being made up of smaller chiral pieces, enzymes arrange themselves into bigger conglomerates that also are chiral and show handedness. Thus, much as a right hand will readily distinguish another right hand from a left hand, enzymes (and other biomolecules) have "pockets" that, by virtue of their stereochemically defined features, are capable of recognizing and processing only one of the enantiomers in a racemate. The differences in physiological activity of the two enantiomers of a

---

Emphasis on Problem Solving and Tools for Studying

In-Chapter **Exercises** provide immediate reinforcement of concepts as they are presented. All Exercises are answered at the back of the book. More than 750 End-of-Chapter Problems give students additional practice in problem-solving skills.

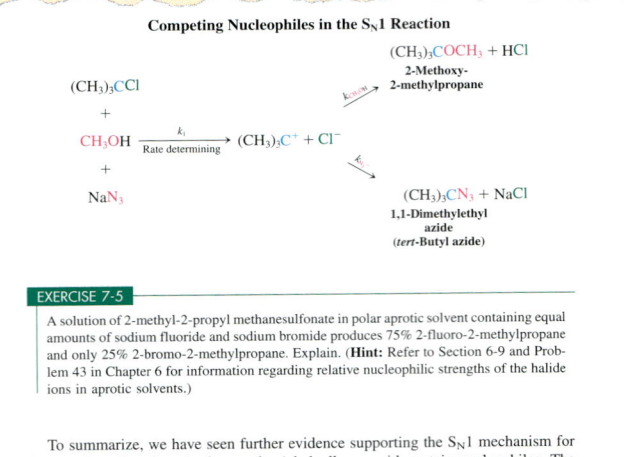

Competing Nucleophiles in the S_N1 Reaction

$(CH_3)_3CCl$

+

CH_3OH $\xrightarrow[\text{Rate determining}]{k_1}$ $(CH_3)_3C^+ + Cl^-$

+

NaN_3

$(CH_3)_3COCH_3 + HCl$
2-Methoxy-2-methylpropane

$(CH_3)_3CN_3 + NaCl$
1,1-Dimethylethyl azide
(*tert*-Butyl azide)

EXERCISE 7-5

A solution of 2-methyl-2-propyl methanesulfonate in polar aprotic solvent containing equal amounts of sodium fluoride and sodium bromide produces 75% 2-fluoro-2-methylpropane and only 25% 2-bromo-2-methylpropane. Explain. (**Hint:** Refer to Section 6-9 and Problem 43 in Chapter 6 for information regarding relative nucleophilic strengths of the halide ions in aprotic solvents.)

To summarize, we have seen further evidence supporting the S_N1 mechanism for the reaction of tertiary (and secondary) haloalkanes with certain nucleophiles. The

NEW REACTIONS

1. Bimolecular Substitution—S$_N$2 (Sections 6-3 through 6-10, 7-5)

Primary and secondary substrates only

$$H_3C \overset{H}{\underset{CH_2CH_3}{\overset{|}{\underset{|}{C}}}} - I \xrightarrow{\text{:Nu}^-} Nu - \overset{CH_3}{\underset{CH_2CH_3}{\overset{|}{\underset{|}{C}}}} H + I^-$$

Direct backside displacement with 100% inversion of configuration

2. Unimolecular Substitution—S$_N$1 (Sections 7-1 through 7-5)

Secondary and tertiary substrates only

$$CH_3 - \overset{CH_3}{\underset{CH_3}{\overset{|}{\underset{|}{C}}}} Br \xrightarrow{-Br^-} CH_3 - \overset{CH_3}{\underset{CH_3}{\overset{+}{C}}} \xrightarrow{\text{:Nu}^-} CH_3 - \overset{CH_3}{\underset{CH_3}{\overset{|}{\underset{|}{C}}}} Nu$$

Through carbocation: Chiral systems are racemized

IMPORTANT CONCEPTS

1. Secondary haloalkanes undergo slow and tertiary haloalkanes fast **unimolecular substitution** in polar media. When the solvent serves as the nucleophile, the process is called **solvolysis.**

2. The slowest, or rate-determining, step in unimolecular substitution is dissociation of the C–X bond to form

5. **Unimolecular elimination** to form an alkene accompanies substitution in secondary and tertiary systems.

6. High concentrations of strong base may bring about **bimolecular elimination.** Expulsion of the leaving group accompanies removal of a hydrogen from the neighboring carbon by the base. The stereochemistry

Chapters conclude with summaries of **New Reactions** and **Important Concepts** to aid students in learning.

NEW FEATURES OF THIRD EDITION

All chapters have been revised and updated. Text has been clarified, simplified, and corrected in accord with new literature and the comments of reviewers and students. Structural drawings have been carefully scrutinized for consistent and accurate representation, and many condensed formulas and Fischer projections have been replaced by line structures. We have also reviewed the use of color in reaction schemes.

New Chapter Introductions

When you hear or read the word *steroids,* two things probably come to mind immediately: athletes who illegally "take steroids" to develop their muscles and "the pill" used for birth control. But what do you know about steroids aside from this general association? What is their structure? How does one steroid differ from another? Where are they found in nature?

An example of a naturally occurring steroid is diosgenin, obtained from the Mexican yam and used as a starting material for the synthesis of several commercial steroids. Most striking is the number of *rings* in the compound.

The **introduction** to each chapter has been "spiced up" with thought-provoking questions relating the relevance of the chapter's material to everyday experience. These general questions find an answer on further reading, thus prompting the student's interest and participation.

Improved Presentation of Reaction Mechanisms

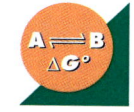 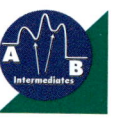

The presentation of reaction mechanisms has been improved by the increased use of arrows to better show electron flow. The introduction of **icons** for a "reaction" and its "mechanism" serves to emphasize the "vocabulary-grammar" duality

of the two types of schemes. In addition, we have added a new section on *electron-pushing arrows* (Section 6-4) to familiarize the student explicitly with this technique.

Curved-Arrow Representations of Several Common Types of Mechanisms

$$H-\ddot{O}:^- + -\overset{|}{\underset{|}{C}}-Cl \xrightarrow{\text{Nucleophilic substitution}} -\overset{|}{\underset{|}{C}}-OH + Cl^-$$

Compare with Brønsted acid-base reaction

$$-\overset{|}{\underset{|}{C}}-Cl \xrightarrow{\text{Dissociation}} -\overset{|}{\underset{|}{C}}{}^+ + Cl^-$$

Reverse of Lewis acid–Lewis base reaction

$$H-\ddot{O}:^- + \overset{|}{C}=O \xrightarrow{\text{Nucleophilic addition}} \overset{HO}{-\overset{|}{\underset{|}{C}}-O^-}$$

Only one of the two bonds between C and O is cleaved

Early Coverage of Acids and Bases

We have moved the discussion of acids and bases to Section 2-9 and expanded it to provide the student with an early review of this aspect of general chemistry as it applies to organic systems. This treatment now includes explicitly Lewis acids and bases and sets the student up for a general understanding of the similarity between such diverse processes as nucleophilic trapping of carbocations, solvation (e.g., of Grignard reagents or by crown ethers or ionophores), and for the role of metal halides in Friedel-Crafts alkanoylation.

Lewis Acid-Base Reactions

$$H^+ + :\ddot{O}-H \longrightarrow H-O-H$$

$$\underset{Cl}{\overset{Cl}{Cl-Al}} + :\underset{CH_3}{\overset{CH_3}{N-CH_3}} \longrightarrow \underset{Cl}{\overset{Cl}{Cl-Al}}-\underset{CH_3}{\overset{CH_3}{N^\pm-CH_3}}$$

$$\underset{F}{\overset{F}{F-B}} + :\underset{CH_2CH_3}{\overset{}{\ddot{O}-CH_2CH_3}} \longrightarrow \underset{F}{\overset{F}{F-B}}-\underset{CH_2CH_3}{\overset{}{\ddot{O}^\pm-CH_2CH_3}}$$

Unified View of Spectroscopy

We have added problems that unify the application of spectroscopic techniques in structure determination.

SOLUTION

First, we write the structure of the starting material (alcohol nomenclature, Section 8-1) and what we know about the reaction:

$$\xrightarrow[\text{2-Methyl-2-pentanol}]{} \quad \xrightarrow{\text{Dilute } H_2SO_4, \ 50°C}$$

This is an acid-catalyzed reaction of a tertiary alcohol (Section 11-9). Even though we may know enough to be able to make a sensible prediction, let us proceed by interpretation of the spectra first and see if the answer that we get is consistent with our expectations.

For the major product, a peak at 1660 cm^{-1} in the IR spectrum is in the range for the alkene C=C bond-stretching frequency (1620–1680 cm^{-1}, Table 11-3). Armed with this information, we turn to the NMR spectrum and immediately look for signals in the region

New Approaches to Problem Solving

SOLUTION

Before we start a random trial and error approach to solving this problem, it is better to take an inventory of what is given. First, we are given cyclohexane, and we note that this unit shows up as a substituent in tertiary alcohol A. Second, a total of seven additional carbons appears in the product, so our synthesis will require some additional stitching together of smaller fragments because we cannot use compounds containing more than four carbons. Third, target A is a tertiary alcohol, which should be amenable to the retrosynthetic analysis introduced in Section 8-9 (M = metal):

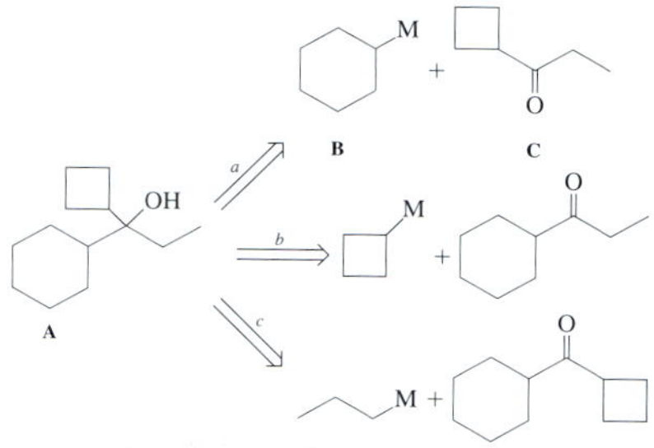

New **Chapter Integration Problems** are solved problems that emphasize concept integration both within and between chapters. The solution is worked out in a step-by-step manner, teaching the art of problem solving in general and specifically demonstrating how one set of learned skills builds on and interacts with preceding ones. Particular emphasis is placed on problem analysis, deductive reasoning, and logical conclusions.

Team Problem

47. Consider the general substitution-elimination reactions of the bromoalkanes.

$$R—Br \xrightarrow{Nu/Base} R—Nu + alkene$$

How do the reaction mechanisms and product formation differ when the structure of the substrate and reaction conditions change? To begin to unravel the nuances of bimolecular and unimolecular substitution and elimination reactions, focus on the treatment of bromoalkanes A through D under conditions (a) through (e). Divide the problem evenly among yourselves so that each of you tackles the questions of reaction mechanism(s) and qualitative distribution of product(s), if any. Reconvene to discuss your conclusions and come to a consensus. When you are explaining a reaction mechanism to the rest of the team, use curved arrows to show the flow of electrons. Label the stereochemistry of starting materials and products as *R* or *S*, as appropriate.

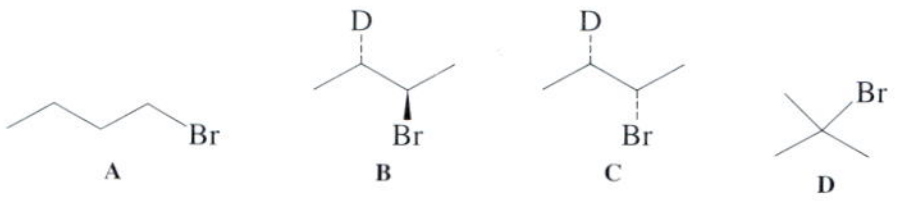

(a) NaN₃, DMF **(b)** LDA, DMF **(c)** NaOH, DMF **(d)** CH₃CO⁻Na⁺, CH₃COH
(e) CH₃OH

The **Team Problem** is also new to each chapter. Team Problems encourage discussion and collaborative learning among students. Although these problems could be assigned in a classroom setting, they are written so as to be perfectly workable in an unstructured, casual setting, such as a library, coffee shop, study hall, or home. The idea is to stimulate "cross-talk," an exchange of information and ideas, and support among students.

Preprofessional Problems

55. The enantiomer of

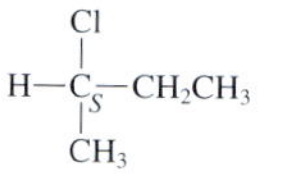

$$H—\underset{\underset{CH_3}{|}}{\overset{\overset{Cl}{|}}{C}}—CH_2CH_3$$

(a) is CH₃CH₂—C(R)(Cl)(CH₃)—H

(b) can exist only at low temperatures

(c) is nonisomeric

(d) is incapable of existence

Students who are planning careers in medicine or related field will appreciate the new **Preprofessional** multiple-choice problems that are typical of those that appear on the MCAT, GRE, and DAT.

Reaction Summary Road Maps

From Chapter 8 onward, the chemistry of each functional group is shown in condensed form through two types of **reaction summary road maps**, providing "the functional group at a glance." The first type depicts the function as the origin of multiple reaction arrows, each labeled with a particular reagent, ending in a specific product. Section numbers indicate where the transformation is discussed in the text, and color

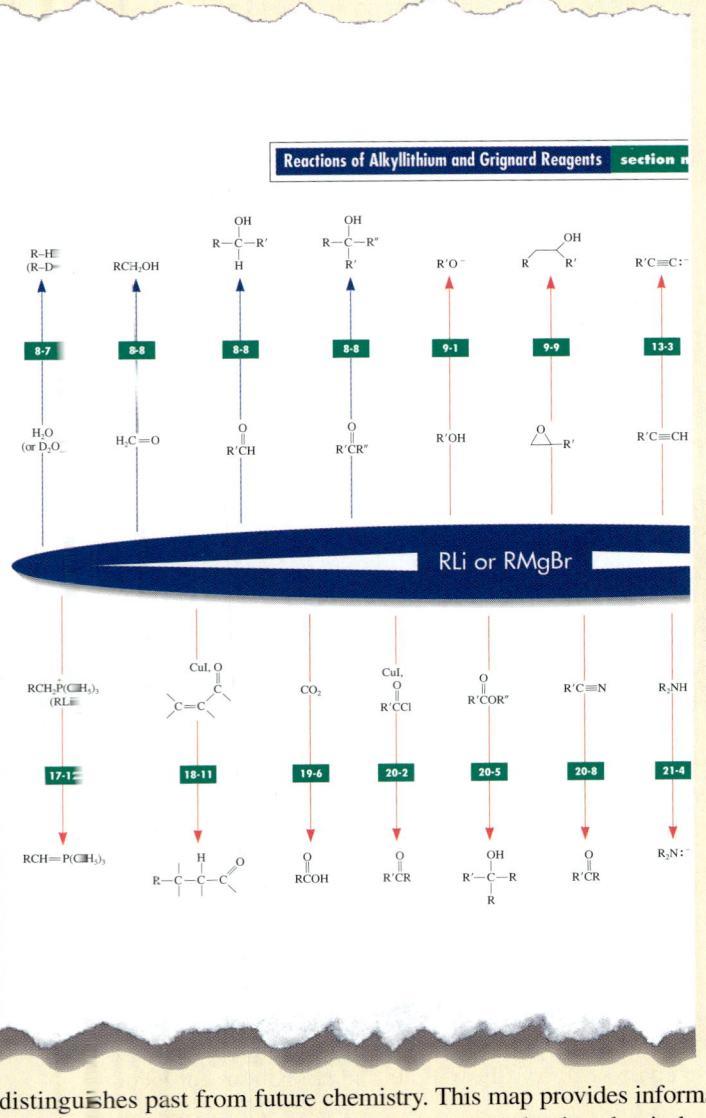

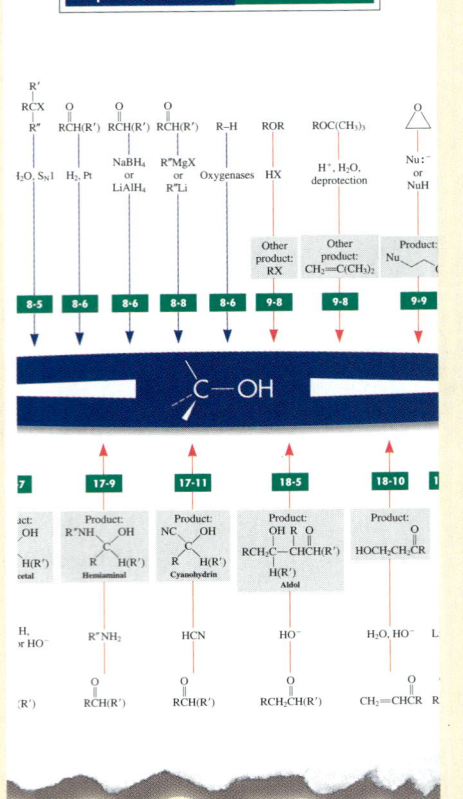

distinguishes past from future chemistry. This map provides information about the reactions of the functional group—that is, what it does. The second type of map is very similar, but the reaction arrows are reversed—that is, pointing toward the functionality. This map provides information about the function's possible origins—that is, what are its precursor functional groups. Thus, a specific reaction A→B may appear in two separate schemes, one with A and the other with B as the center. These maps are an important aid in synthetic-retrosynthetic analysis and a check on the student's "vocabulary" of synthetic methodology.

Computer-Generated Ball-and-Stick and Space-Filling Models

The first and second editions emphasized the importance of building molecular models as an aid in visualizing three-dimensional structure and dynamics. We have highlighted this emphasis by a third **icon** at numerous locations. In addition, we have now included computer-generated pictures of ball-and-stick and of space-filling models. These pictures encourage students to build actual models. They also provide students with lowest-energy conformations, guiding them in the construction of realistic assemblies. Finally, space-filling renditions create a more accurate impression of size, shape, and the extent of orbitals. Ball-and-stick model kits are available for purchase through the publisher.

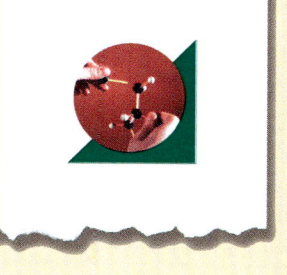

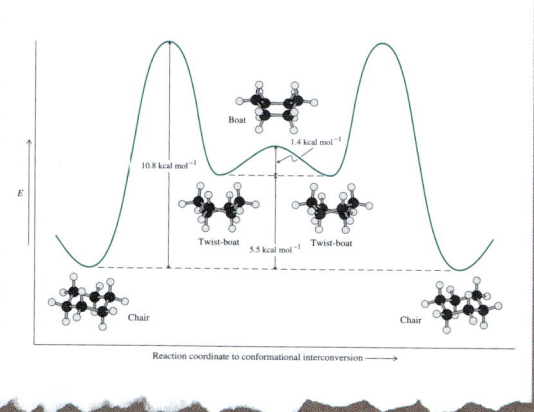

SUPPLEMENTS AND LABORATORY MANUALS

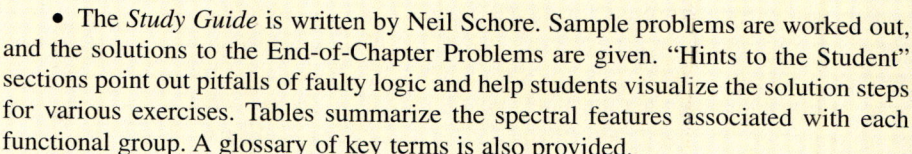

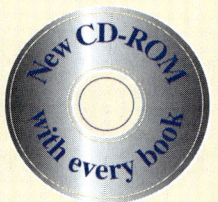

The CD-ROM found in the back of this text is a multimedia learning tool developed by W. H. Freeman and Company in conjunction with Sumanas, Inc. All the features of the CD function within the context of the book's coverage. Many of the structures mentioned in the book are depicted as three-dimensional animations with multiple-display options through a molecular-modeling program. These animations and many other molecular-level simulations bring the concepts of the book to life. Practice tools, such as interactive quizzes in every chapter and a preprofessional examination, help students review for exams. Interactive sample problems and solutions help guide students through difficult concepts. WebNotes provides direct links to relevant chemistry sites on the World Wide Web. Presentation software for instructors allows them to prepare series of illustrations and animations for lecture.

• The *Study Guide* is written by Neil Schore. Sample problems are worked out, and the solutions to the End-of-Chapter Problems are given. "Hints to the Student" sections point out pitfalls of faulty logic and help students visualize the solution steps for various exercises. Tables summarize the spectral features associated with each functional group. A glossary of key terms is also provided.

• The *Test Bank,* by Charles M. Garner and Kevin Pinney of Baylor University, is new to this edition. With the Windows and Macintosh software of the computerized versions, instructors can easily change and add questions as well as import their own electronic drawings.

• The *Maruzen Molecular Structure Model Set* and *Space-Filling Model Set* are also available for student purchase. These essential tools can be used to present orbitals; single, double, and triple bonds; and locations of atoms.

• *Experimental Organic Chemistry: Macroscale and Microscale.* With these texts, the laboratory becomes a place of discovery and critical thinking. Instead of simply following directions, students immerse themselves in the experimental process. Instructors will appreciate the versatility of the manuals' balanced approach, with enough experiments to use macroscale glassware, microscale glassware, or a combination of both. Innovative discovery-based experiments and multiweek projects encourage students in scientific investigation. A CD-ROM of techniques accompanies both texts.

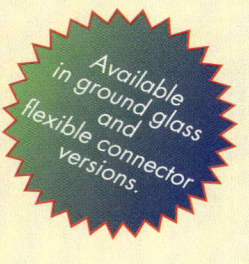

Available in ground glass and flexible connector versions.

A KEY TO THE FUNCTIONAL USE OF COLOR

We use color consistently and functionally to help students master basic principles, including nomenclature, orbitals, sequence rules in stereochemistry, the relation of spectral lines to functional groups, topological changes in molecular transformations, and the reactivity of functional groups. Color is suspended in exercises, chapter reviews, and problems, however, because it is important to learn how not to rely on it. In this edition, we have carefully reevaluated the application of color in reaction schemes and simplified its use.

For example, wherever possible, *s* orbitals are shown in red, 2*p* orbitals in blue, *spⁿ* hybrids in purple, and *p* orbitals in green.

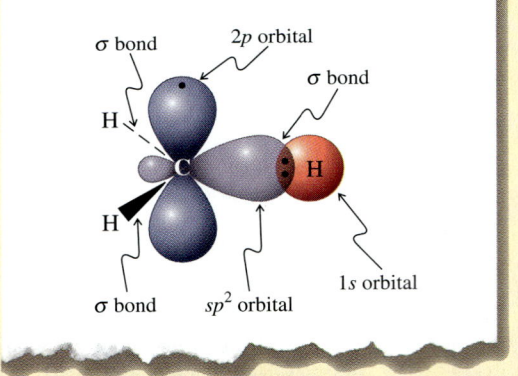

Color shows the relation of the names of organic molecules to their structures. In the illustration shown here, which is from Chapter 11, the functional group that gives the molecule its unique chemical behavior and other substituents are clearly differentiated from the stem.

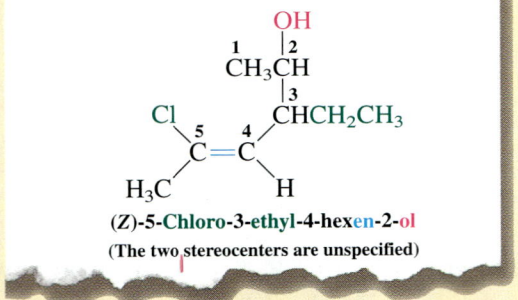

(Z)-5-Chloro-3-ethyl-4-hexen-2-ol

(The two stereocenters are unspecified)

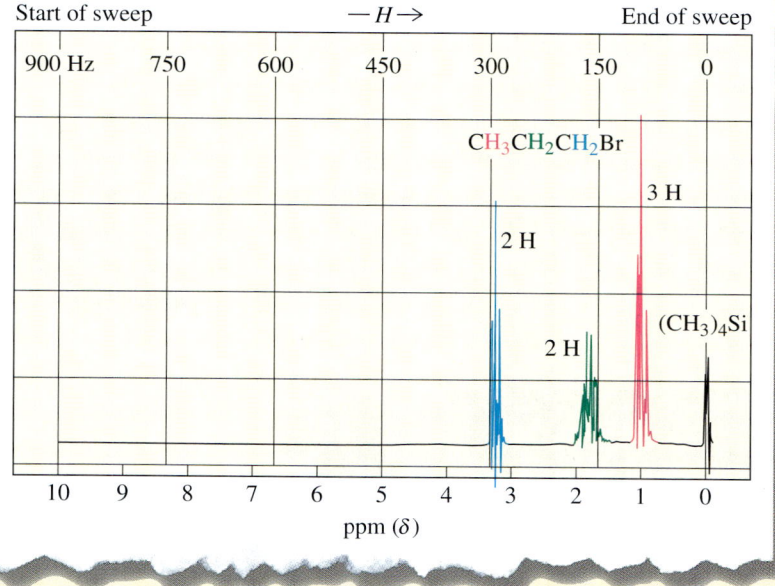

Start of sweep $-H\rightarrow$ End of sweep

900 Hz   750   600   450   300   150   0

$CH_3CH_2CH_2Br$

3 H

2 H

2 H

$(CH_3)_4Si$

10   9   8   7   6   5   4   3   2   1   0

ppm (δ)

Color is used to *associate* spectral features with certain molecular units. For example, in the adjoining spectrum, the three colors show how the three nonequivalent hydrogens give rise to three distinct "peaks"—an observation that will help the students identify a molecule when they know its spectrum.

Color offers clues to a molecule's stereochemistry, or the arrangement of its atoms in space. The student will see in Chapter 5 that substituents in three dimensions can be assigned a priority according to certain "sequence rules," and this assignment has been indicated, in diminishing order of priority, by red, blue, green, and black.

Remember the use of color to denote group priorities:
Highest—red
Second highest—blue
Third highest—green
Lowest—black

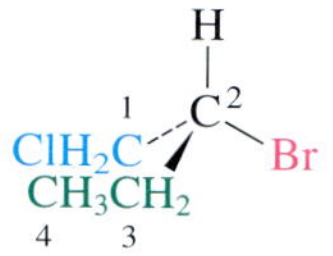

$$\underset{4\quad 3}{\overset{1}{ClH_2C}}\overset{H}{\underset{CH_3CH_2}{\overset{|}{C^2}}}Br$$

Optically active
2R

Most importantly, color frequently shows how the functional groups transform in the reaction mechanism. Electron-rich, or "nucleophilic," parts are shown in red; electron-deficient, or "electrophilic," fragments are blue; and radicals and leaving groups are green. Red arrows in these transformations indicate the movements of electrons.

STEP 4. Trapping by bromide

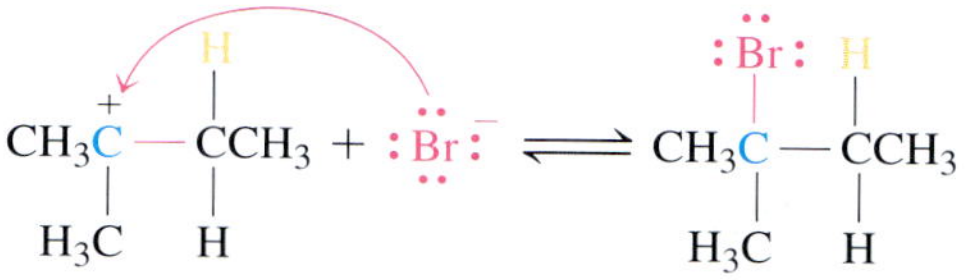

$$CH_3\overset{+}{\underset{\underset{H_3C}{|}}{C}}-\underset{\underset{H}{|}}{C}CH_3 + :\overset{..}{\underset{..}{Br}}:^- \rightleftharpoons CH_3\overset{:\overset{..}{Br}:}{\underset{\underset{H_3C}{|}}{C}}-\underset{\underset{H}{|}}{C}CH_3$$

ACKNOWLEDGMENTS

We are grateful to the following professors who reviewed the manuscript for the third edition:

Steven Angle, University of California at Riverside
Jeffrey Arterburn, New Mexico State University
Ronald Blankespoor, Calvin College
Frances Blase, Haverford College
Richard Broene, Bowdoin College
Patrick Buick, University of Toledo
Dee Ann Casteel, Bucknell University
Dana Chatellier, University of Delaware
James Deyrup, University of Florida
Morris Fishman, New York University
Thomas Flechtner, Cleveland State University
Francis Flores, California State Polytechnic University
Marcia France, Washington & Lee University
Andrew French, Radford University
Charles Garner, Baylor University
Rainer Glaser, University of Missouri at Columbia
Frank Guziec, Southwestern University
William Hagan, College of St. Rose
Eamonn Healy, St. Edward's University
Steven Kass, University of Minnesota at Minneapolis
Robert Kulawiec, Georgetown University
Mark Kurth, University of California at Davis
David Lemal, Dartmouth College
Ronald Magid, University of Tennessee at Knoxville
Roger Murray, University of Delaware
Thomas Newton, University of Southern Maine
Patrick O'Bannon, Kenyon College
Daniel O'Leary, Pomona College
Kenneth Piers, Calvin College
Michael Rathke, Michigan State University
Gretchen Rehburg, Bucknell University
John Richard, State University of New York at Buffalo
Adrian Schwan, University of Guelph
Larry Scott, Boston College
Raymond Shelden, La Sierra University
Jan Shepard, Millersville University of Pennsylvania
Sam Stevenson, Northeast State Technical Community College
Chaim Sukenik, Case Western Reserve University
Julie Tan, Cumberland College
Peter Trumper, University of Northern Maine at Orono
Jeffrey Ward, Georgetown University
Kraig Wheeler, Delaware State University
James White, Pepperdine University
John Williams, Temple University
John Wood, Indiana University of Pennsylvania

Peter Vollhardt thanks Administrative Assistants Kim Steele and Bonnie Kirk, for typing, photocopying, and coordinating manuscripts; graduate students Adam Matzger and Dan Holmes, for Spartan and artistic ideas; Kevin Cammack and Michael Eichberg, for running 300 MHz NMR and other spectra; and all four plus Dr. Christoph Erben, Sriram Kumaraswamy, Jennifer Moore, and Ian Wasser, for their assistance in checking page proofs.

We are indebted to Professor Richard Bozak for contributing the Preprofessional Problems and to Nancy Cox-Konopelski for contributing the Team Problems.

We express special gratitude to Professor Ronald Magid, his colleagues, and his students at the University of Tennessee at Knoxville, for the time and effort they have devoted to uncovering numerous typographical and factual errors. They deserve a great deal of credit for whatever improvements in accuracy and clarity we have achieved in this revision.

ABOUT THE AUTHORS

K. PETER C. VOLLHARDT was born in Madrid in 1946, raised in Buenos Aires and Munich, studied at the University of Munich, received his Ph.D. with Professor Peter Garratt at the University College London, and was a postdoctoral fellow with Professor Bob Bergman (then) at the California Institute of Technology. He moved to Berkeley in 1974, when he began his efforts toward the development of organocobalt reagents in organic synthesis, the preparation of theoretically interesting hydrocarbons, the assembly of novel transition metal arrays with potential in catalysis, and the discovery of a parking space. Among other pleasant experiences, he was a Studienstiftler, Adolf Windaus medalist,

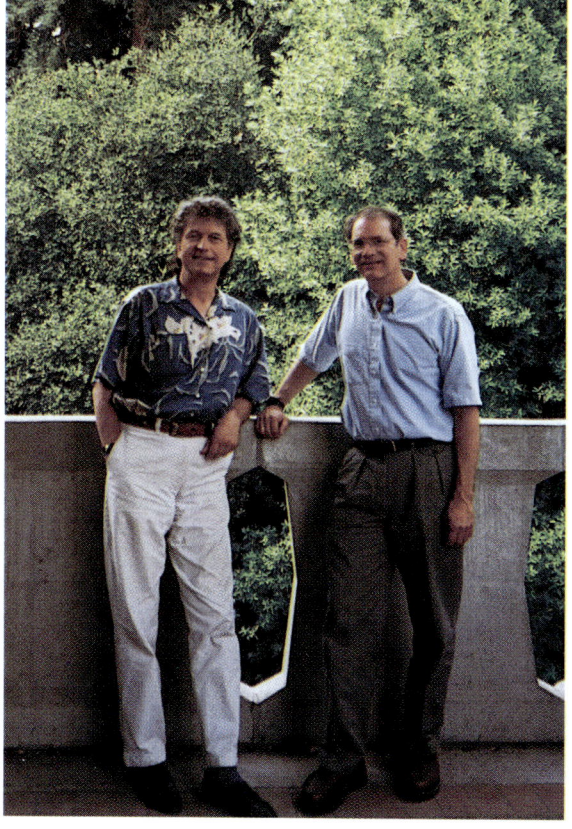

Humboldt Senior Scientist, ACS Organometallic Awardee, Otto Bayer Prize Awardee, and A. C. Cope Scholar. He is the current Chief Editor of SYNLETT. Among his more than 230 publications, he especially treasures this textbook in organic chemistry, translated into six languages. Peter is married to Marie-José Sat, a French artist, and they have two children, Paloma (b. 1994), whose picture you can admire in Chapter 5, and Julien (b. 1997), who refused to pose.

NEIL SCHORE was born in Newark, New Jersey, in 1948. His education took him through the public schools of the Bronx, New York, and Ridgefield, New Jersey, after which he completed a B.A. with honors in chemistry at the University of Pennsylvania in 1969. Moving back to New York, he worked with Professor Nicholas Turro at Columbia University, studying photochemical and photophysical processes of organic compounds for his Ph.D. thesis. He first met Peter Vollhardt when he and Peter were doing postdoctoral work in Professor Robert Bergman's laboratory at Cal Tech in the 1970s. Since joining the U. C. Davis faculty in 1976, he has taught organic chemistry to over 10,000 nonchemistry majors, winning three teaching awards, and published over 70 papers in various areas related to organic synthesis. Neil is married to Carrie Erickson, a microbiologist at the U. C. Davis School of Veterinary Medicine. They have two children, Michael (b. 1981) and Stefanie (b. 1983), both of whom carried out experiments for this book.

ORGANIC CHEMISTRY

Structure and Function

Structure and Bonding in Organic Molecules

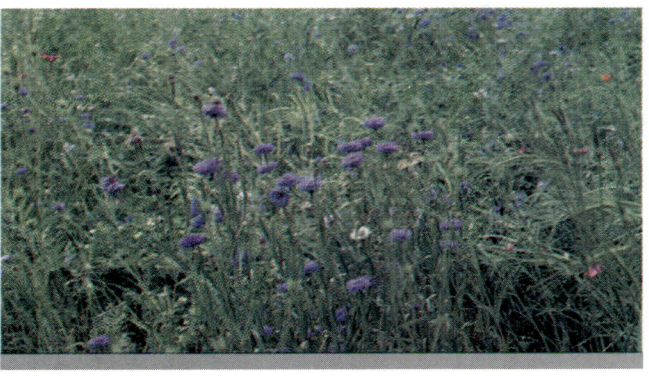

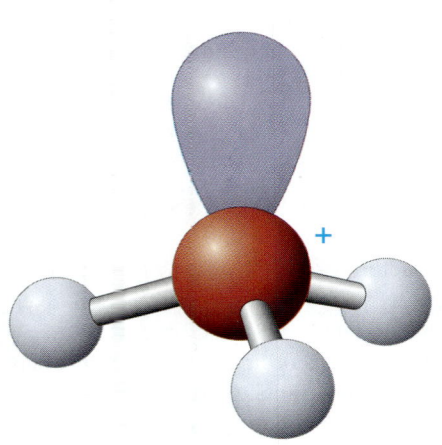

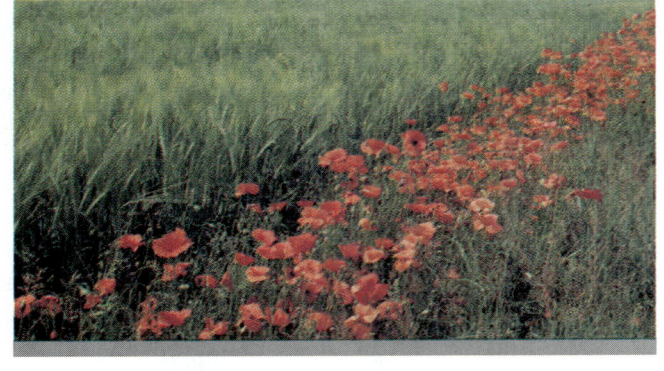

The hydronium ion, derived by protonation of water, is responsible for the acidity of aqueous solutions below pH 7. Similar protonation of an oxygen atom of the dye molecule responsible for the blue color of cornflowers gives rise to the color of the red poppy.

How does your body function? Why did your muscles ache this morning after last night's long jog? What is in the pill you took to get rid of that headache you got after studying all night? What happens to the gasoline you pour into the gas tank of your car? What is the molecular composition of the things you wear? What is the difference between a cotton shirt and one made of silk? What is the origin of the odor of garlic? You will find the answers to these questions, and many others that you may have asked yourself, in this book on organic chemistry.

Chemistry is the study of the structure of molecules and the rules that govern their interactions. As such, it interfaces closely with the fields of biology, physics, and mathematics. What, then, is organic chemistry? What distinguishes it from other chemical disciplines, such as physical, inorganic, or nuclear chemistry? A common definition provides a partial answer: *Organic chemistry is the chemistry of carbon and its compounds.* These compounds are called **organic molecules.**

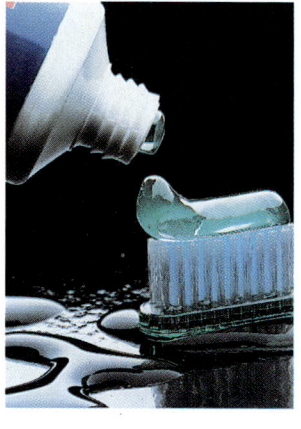

Toothbrushes and toothpaste consist of a mixture of many organic molecules.

Organic molecules constitute the chemical bricks of life. Fats, sugars, proteins, and the nucleic acids are compounds in which the principal component is carbon. So are countless substances that we take for granted in everyday use. Virtually all the clothes that we wear are made of organic molecules—some of natural fibers, such as cotton and silk; others artificial, such as polyester. Toothbrushes, toothpaste, soaps, shampoos, deodorants, perfumes—all contain organic compounds, as do furniture, carpets, the plastic in light fixtures and cooking utensils, paintings, food, and countless other items.

Organic substances such as gasoline, medicines, pesticides, and polymers have improved the quality of our lives. Yet the uncontrolled disposal of organic chemicals has polluted the environment, causing deterioration of animal and plant life, as well as injury and disease to humans. If we are to create useful molecules—*and* learn to control their effects—we need a knowledge of their properties and an understanding of their behavior. We must be able to apply the principles of organic chemistry. This chapter explains how the basic ideas of chemical structure and bonding apply to organic molecules.

1-1 The Scope of Organic Chemistry: An Overview

A goal of organic chemistry is to relate the structure of a molecule to the reactions that it can undergo. We can then study the steps by which each type of reaction takes place, and we can learn to create new molecules by applying those processes.

Thus, it makes sense to classify organic molecules according to the subunits and bonds that determine their chemical reactivity: These determinants are groups of atoms called **functional groups.** The study of the various functional groups and their respective reactions provides the structure of this book.

Functional groups determine the reactivity of organic molecules

$$H_3C—CH_3$$
Ethane

We begin with the **alkanes,** which contain the basic carbon framework of organic molecules. The alkanes are simple **hydrocarbons,** organic compounds composed of only hydrogen and carbon and lacking functional groups. As with other classes of molecules, we discuss the systematic rules for naming them, their structures, and their physical properties. An example of an alkane is ethane. Its structural mobility will be the starting point for a review of thermodynamics and kinetics. This review is then followed by a discussion of the strength of alkane bonds, which can be broken by heat, light, or chemical reagents. We shall illustrate these processes with the chlorination of alkanes (Chapter 3).

A Chlorination Reaction

$$CH_4 + Cl_2 \xrightarrow{\text{Energy}} CH_3—Cl + HCl$$

Next we shall look at cyclic alkanes (Chapter 4), which contain carbon atoms in a ring. This arrangement can lead to new properties and changes in reactivity. The recognition of a new type of isomerism in cycloalkanes bearing two or more substituents—either on the same side or on opposite sides of the ring plane—sets the stage for a general discussion of **stereoisomerism,** exhibited by compounds with the

Cyclohexane

same connectivity but differing in the relative positioning of their component atoms in space (Chapter 5).

We shall then study the haloalkanes, our first example of compounds containing a functional group—the carbon–halogen bond. The haloalkanes participate in two types of organic reactions: substitution and elimination (Chapters 6 and 7). In a **substitution** reaction, one halogen atom may be replaced by another; in an **elimination** process, adjacent atoms may be removed from a molecule to generate a double bond.

A Substitution Reaction

$$CH_3-Cl + K^+I^- \longrightarrow CH_3-I + K^+Cl^-$$

An Elimination Reaction

$$CH_2-CH_2 + K^{+\ -}OH \longrightarrow H_2C=CH_2 + HOH + K^+\ I^-$$
$$\ \ |\ \ \ \ \ \ |$$
$$\ \ H\ \ \ \ \ \ I$$

Like the haloalkanes, each of the major classes of organic compounds is characterized by a particular functional group. For example, the carbon–carbon triple bond is the functional group of alkynes; ethyne, a well-known alkyne, is the chemical burned in a welder's torch (Chapter 13). A carbon–oxygen double bond fulfills this role for aldehydes and ketones, the starting materials in many industrial processes (Chapters 16 and 17 ; and the amines, which include drugs such as nasal decongestants and amphetamines, contain nitrogen in their functional group (Chapter 21). We shall study a number of tools for identifying these molecular subunits, including various forms of spectroscopy (Chapters 10, 11, 14, and 20).

Subsequently, we shall encounter several important classes of organic molecules that are especially crucial in biology and industry. Many of these classes, such as the carbohydrates (Chapter 24) and amino acids (Chapter 26), contain multiple functional groups. However, in *every* class of organic compounds, the principle remains the same: *The structure of the molecule is related to the reactions that it can undergo.*

$$HC\equiv CH$$
An alkyne

$$H_2C=O$$
An aldehyde

$$\overset{\displaystyle O}{\overset{\|}{H_3C-C-CH_3}}$$
A ketone

$$H_3C-NH_2$$
An amine

Synthesis is the making of new molecules

Carbon compounds are called "organic" because it was originally thought that they could be produced only from living organisms. In 1828, Friedrich Wöhler* proved this idea to be false when he converted the inorganic salt lead cyanate into urea, an organic product of protein metabolism in mammals. [The average human excretes 30 g (grams) of urea each day.]

Wöhler's Synthesis of Urea

$$\text{Pb(OCN)}_2 + 2\ H_2O + 2\ NH_3 \longrightarrow 2\ \overset{\displaystyle O}{\overset{\|}{H_2NCNH_2}} + \text{Pb(OH)}_2$$

| Lead cyanate | Water | Ammonia | Urea | Lead hydroxide |

---

*Professor Friedrich Wöhler (1800–1882), University of Göttingen, Germany. In this and subsequent biographical notes, only the scientist's last known location of activity will be mentioned, even though much of his or her career may have been spent elsewhere.

Saccharin: One of the Oldest Synthetic Organic Compounds in Commercial Use

Saccharin was synthesized in the course of a study of the oxidation of organic chemicals containing sulfur and nitrogen. Its sweetness was discovered by Ira Remsen[*] in 1879, a time when chemists routinely *tasted* every new compound they made. This was an extremely dangerous practice, one that you should not

*Professor Ira Remsen (1846–1927), Johns Hopkins University, Baltimore.

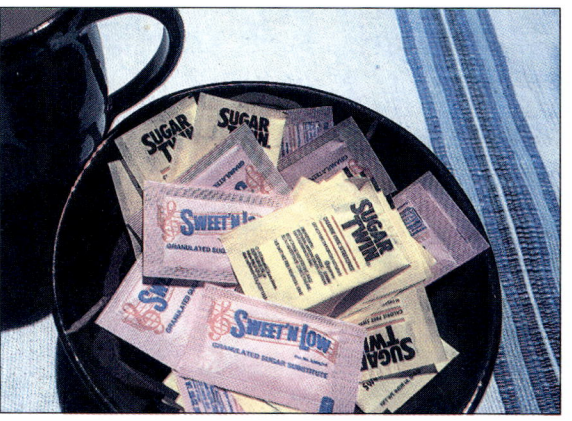

Familiar saccharin-containing packets.

observe under any circumstances, even with supposedly "safe" compounds that you may encounter in your laboratory. For example, had Remsen tasted brevetoxin B (illustrated on the cover of this book), he would have immediately felt a prickly sensation in his mouth and fingers, rapidly followed by hot and cold sensations, breathing problems, paralysis, and death.

Saccharin is 300 times as sweet as sugar and virtually nontoxic. It has proved to be a lifesaver for countless diabetics and of great value to people who need to control their caloric intake. The possibility that saccharin may be *carcinogenic*—that is, capable of causing cancer—was raised in the 1960s. In the 1970s, a connection was found between high doses of saccharin and bladder tumors in rats. Experiments completed in 1990 demonstrated that saccharin does not cause cancer directly, but at very high doses it promotes accelerated cell division, which may increase the likelihood of cell mutation and tumor formation. Warning labels are required on saccharin-containing products sold in the United States. These studies illustrate how society must balance the benefits that synthetic substances bring to our daily lives with the possible risks associated with their use.

Synthesis, or the making of molecules, is a very important part of organic chemistry (Chapter 8). Since Wöhler's time, more than 10 million organic substances have been synthesized from simpler materials, both organic and inorganic. These substances include many that also occur in nature, such as the penicillin antibiotics, as well as entirely new compounds. Some, like cubane, which gave chemists the opportunity to study special kinds of bonding and reactivity, are of largely theoretical interest. Others, like the artificial sweetener saccharin, have become a part of everyday life.

Typically, the goal of synthesis is to construct complex organic chemicals from simpler, more readily available ones. To be able to convert one molecule into another, chemists must know organic reactions. They must also know the physical conditions that govern such processes, such as temperature, pressure, solvent, and molecular structure. This knowledge is equally valuable in analyzing biological transformations.

As we study the chemistry of each functional group, we shall develop the tools both for planning effective syntheses and for predicting the processes that take place in nature. But how? The answer lies in looking at reactions step by step.

Benzylpenicillin

Cubane

Saccharin

Reactions are the vocabulary and mechanisms are the grammar of organic chemistry

When we introduce a chemical reaction, we will first show just the starting compounds, or **reactants** (also called **substrates**), and the **products.** In the chlorination process mentioned earlier, the substrates—methane, CH_4, and chlorine, Cl_2—may undergo a reaction to give chloromethane, CH_3Cl, and hydrogen chloride, HCl. The overall transformation was described as $CH_4 + Cl_2 \rightarrow CH_3Cl + HCl$. However, even a simple reaction like this one may proceed through a complex sequence of steps. The reactants could have first formed one or more *unobserved* substances—call these X—that rapidly changed into the observed products. These underlying details of the reaction constitute the **reaction mechanism.** In our example, the mechanism consists of a two-step sequence: $CH_4 + Cl_2 \rightarrow X$ followed by $X \rightarrow CH_3Cl + HCl$. Each step may have a part in determining whether the overall reaction will proceed.

Substance X in our chlorination reaction is an example of a **reaction intermediate,** a species formed on the pathway between reactants and products. We shall learn the mechanism of this chlorination process and the true nature of the reaction intermediates in Chapter 3.

How can we determine reaction mechanisms? The strict answer to this question is, We cannot. All we can do is amass circumstantial evidence that is consistent with (or points to) a certain sequence of molecular events that connect starting materials and products ("the postulated mechanism"). To do so, we exploit the fact that organic molecules are no more than collections of bonded atoms. We can, therefore, study how, when, and how fast bonds break and form, in which way they do so in three dimensions, and how changes in substrate structure affect the outcome of reactions. Thus, although we cannot strictly prove a mechanism, we can certainly rule out many (or even all) reasonable alternatives and propose a most likely pathway.

In a way, the "learning" and "using" of organic chemistry is much like learning and using a language. You need the vocabulary (i.e., the reactions) to be able to use the right words, but you also need the grammar (i.e., the mechanisms) to be able to converse intelligently. Neither one on its own gives complete knowledge and understanding, but together they form a powerful means of communication, rationalization, and predictive analysis. To highlight the interplay between reaction and mechanism, icons are displayed in the margin at appropriate places throughout the text.

Before we begin our study of the principles of organic chemistry, let us review some of the elementary principles of bonding. We shall find these concepts useful in understanding and predicting the chemical reactivity and the physical properties of organic molecules.

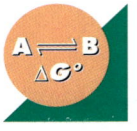

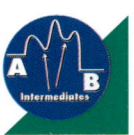

1-2 Coulomb Forces: A Simplified View of Bonding

The bonds between atoms hold a molecule together. But why are there bonds? Two atoms form a bond only if their interaction is energetically favorable; that is, if energy—heat, for example—is released when the bond is formed. Conversely, breaking that bond requires the input of the same amount of energy.

The two main causes of the energy release associated with bonding are based on fundamental laws of physics:

1. Opposite charges attract each other.
2. Electrons spread out in space.

Bonds are made by simultaneous Coulombic attraction and electron exchange

Each atom consists of a nucleus, containing electrically neutral particles, or neutrons, and positively charged protons. Surrounding the nucleus are negatively charged electrons, equal in number to the protons so that the net charge is zero. As two atoms approach each other, the positively charged nucleus of the first attracts the electrons of the second; similarly, the nucleus of the second attracts the electrons of the first. This sort of bonding is described by **Coulomb's* law:** Opposite charges attract each other with a force inversely proportional to the square of the distance between the centers of the charges.

Coulomb's Law

$$\text{Attracting force} = \text{constant} \times \frac{(+)\,\text{charge} \times (-)\,\text{charge}}{\text{distance}^2}$$

This attractive force causes energy to be released as the atoms are brought together. This energy is called the **bond strength.**

When the atoms reach a certain closeness, no more energy is released. The distance between the two nuclei at this point is called the **bond length** (Figure 1-1).

*Lieutenant-Colonel Charles Augustin de Coulomb (1736–1806), Inspecteur Général of the University of Paris, France.

FIGURE 1-1 _____

The changes in energy, *E,* that result when two atoms are brought into close proximity. At the separation defined as bond length, maximum bonding is achieved.

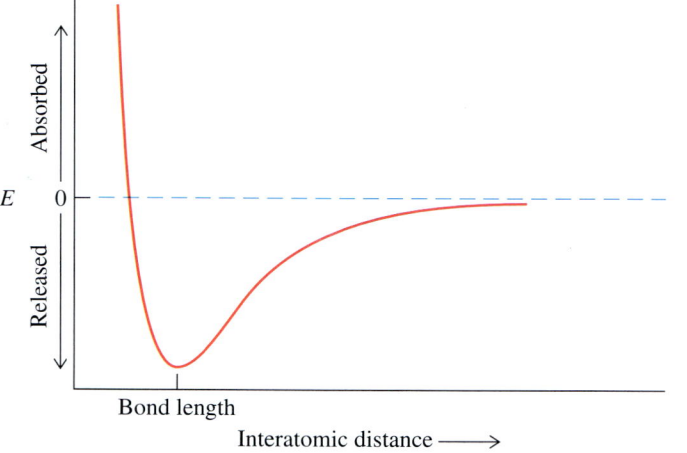

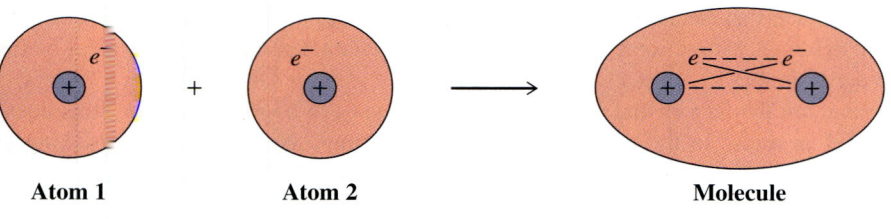

FIGURE 1-2

Covalent bonding. Attractive (solid line) and repulsive (dashed line) forces in the bonding between two atoms. The large circles represent areas in space in which the electrons are found around the nucleus. The small circle around the plus sign stands for the nucleus.

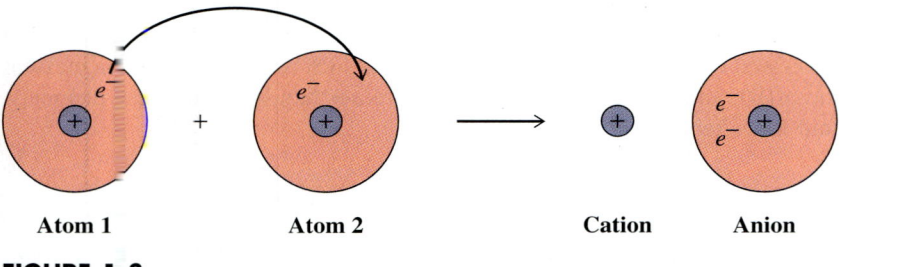

FIGURE 1-3

Ionic bonding. An alternative mode of bonding results from the complete transfer of an electron from atom 1 to atom 2, thereby generating two ions whose opposite charges attract each other.

Bringing the atoms closer together than this distance results in a sharp *increase* in energy. Why? Just as opposite charges attract, like charges repel. If the atoms are too close, the electron–electron and nuclear–nuclear repulsions become stronger than the attractive forces. When the nuclei are the appropriate bond length apart, the electrons are spread out around both nuclei, and attractive and repulsive forces balance for maximum bonding. The energy content of the two-atom system is then at a minimum, the most stable situation (Figure 1-2).

An alternative to this type of bonding results from the complete transfer of an electron from one atom to the other. The result is two charged *ions:* one positively charged, a *cation,* and one negatively charged, an *anion* (Figure 1-3). Again, the bonding is based on Coulombic attraction, this time between two ions.

The Coulombic bonding models of attracting and repelling charges shown in Figures 1-2 and 1-3 are highly simplified views of the interactions that take place in the bonding of atoms. Nevertheless, these models explain many of the properties of organic molecules.

We have seen that attraction between negatively and positively charged particles is a basis for bonding. How does this concept work in real molecules?

1-3 Ionic and Covalent Bonds: The Octet Rule

Two extreme types of bonding explain the interactions between atoms in organic molecules:

1. A **covalent bond** is formed by the sharing of electrons (as shown in Figure 1-2).
2. An **ionic bond** is based on the electrostatic attraction of two ions with opposite charges (as shown in Figure 1-3).

We shall see that many atoms bind to carbon in a way that is intermediate between these extremes: Some ionic bonds have covalent character, and some covalent bonds are partly ionic.

What are the factors that account for the two types of bonds? To answer this question, let us return to the atoms and their compositions. We shall start by looking at the periodic table and at how the electronic makeup of the elements changes as the atomic number increases.

The periodic table underlies the octet rule

The partial periodic table depicted in Table 1-1 includes those elements most widely found in organic molecules: carbon (C), hydrogen (H), oxygen (O), nitrogen (N), sulfur (S), chlorine (Cl), bromine (Br), and iodine (I). Certain reagents, indispensable for synthesis and commonly used, contain elements such as lithium (Li), magnesium (Mg), boron (B), and phosphorus (P). (If you are not familiar with these elements, you should learn Table 1-1.)

EXERCISE 1-1

(a) Redraw Figure 1-1 for a weaker bond than the one depicted. (b) Write Table 1-1 from memory.

The elements in the periodic table are listed according to their atomic number, or nuclear charge (number of protons), which also equals their number of electrons. This number increases by one with each element listed. The electrons occupy energy levels, or "shells," each with a fixed capacity. For example, the first shell has room for two electrons; the second, eight; and the third, eighteen. Helium, with two electrons in its shell, and the other noble gases, with eight electrons (called **octets**) in their outermost shells, are especially stable. These elements show very little chemical reactivity. All other elements lack octets in their outermost electron shells. *They will tend to form molecules in such a way as to reach an octet in the outer electron shell and attain a noble-gas configuration.* In the next two sections, we describe two extreme ways in which this goal may be accomplished: by the formation of pure ionic or pure covalent bonds.

TABLE 1-1	Partial Periodic Table							Halogens	Noble gases
Period									
First	H^1								He^2
Second	$Li^{2,1}$	$Be^{2,2}$	$B^{2,3}$	$C^{2,4}$	$N^{2,5}$	$O^{2,6}$		$F^{2,7}$	$Ne^{2,8}$
Third	$Na^{2,8,1}$	$Mg^{2,8,2}$	$Al^{2,8,3}$	$Si^{2,8,4}$	$P^{2,8,5}$	$S^{2,8,6}$		$Cl^{2,8,7}$	$Ar^{2,8,8}$
Fourth	$K^{2,8,8,1}$							$Br^{2,8,18,7}$	$Kr^{2,8,18,8}$
Fifth								$I^{2,8,18,18,7}$	$Xe^{2,8,18,18,8}$

Note: The superscripts indicate the number of electrons in each principal shell of the atom.

In pure ionic bonds, electron octets are formed by transfer of electrons

Sodium (Na), a reactive metal, interacts with chlorine, a reactive gas, in a violent manner to produce a stable substance: sodium chloride. Similarly, sodium reacts with fluorine (F), bromine, or iodine to give the respective salts. Other alkali metals, such as lithium and potassium (K), undergo the same reactions. These transformations succeed because both reaction partners attain noble-gas character by the *transfer of outer-shell electrons,* called **valence electrons,** from the alkali metals on the left side of the periodic table to the halogens on the right.

Let us see how this works for the ionic bond in sodium chloride. Why is the interaction energetically favorable? First, it takes energy to remove an electron from an atom. This energy is the **ionization potential (IP)** of the atom. For sodium gas, the ionization energy amounts to 119 kcal mol^{-1}.* Conversely, energy may be released when an electron attaches itself to an atom. For chlorine, this energy, called its **electron affinity (EA),** is -83 kcal mol^{-1}. These two processes result in the transfer of an electron from sodium to chlorine. Together, they require a net energy *input* of $119 - 83 = 36$ kcal mol^{-1}.

$$Na^{2,8,1} \xrightarrow{-1\,e} [Na^{2,8}]^+ \qquad IP = 119 \text{ kcal mol}^{-1}$$

Sodium cation
(Neon configuration)   Energy input required

$$Cl^{2,8,7} \xrightarrow{+1\,e} [Cl^{2,8,8}]^- \qquad EA = -83 \text{ kcal mol}^{-1}$$

Chloride anion
(Argon configuration)   Energy released

$$Na + Cl \longrightarrow Na^+Cl^- \qquad Total = 36 \text{ kcal mol}^{-1}$$

Why, then, do the atoms readily form NaCl? The reason is their electrostatic attraction, which pulls them together in an ionic bond. At the most favorable interatomic distance [about 2.8 Å (angstroms) in the gas phase], this attraction releases about 120 kcal mol^{-1} (see Figure 1-1). This energy release is enough to make the reaction of sodium with chlorine energetically highly favorable ($+36 - 120 = -84$ kcal mol^{-1}).

More than one electron may be donated (or accepted) to achieve favorable electronic configurations. Magnesium, for example, has two valence electrons. Donation to an appropriate acceptor produces the corresponding doubly charged cation with the electronic structure of neon. In this way, the ionic bonds of typical salts are formed.

Formation of Ionic Bonds by Electron Transfer

$$Na^{2,8,1} + Cl^{2,8,7} \longrightarrow [Na^{2,8}]^+ [Cl^{2,8,8}]^-, \text{ or } NaCl$$

A more convenient way of depicting valence electrons is by means of dots around the symbol for the element. In this case, the letters represent the nucleus and all the electrons in the inner shells, together called the **core configuration.**

---

*This book will cite energy values in the traditional units of kcal mol^{-1}, in which mol is the abbreviation for mole and a kilocalorie (kcal) is the energy required to raise the temperature of 1 kg (kilogram) of water by 1°C. In SI units, energy is expressed in joules (kg m^{2}s^{-2}, or kilogram-meter^2 per second^2). A joule (J) is the energy required to raise a mass of 1 kg every second by 1 m s^{-1}. The conversion factor is 1 kcal $= 4184$ J $= 4.184$ kJ (kilojoule).

Valence Electrons as Electron Dots

$$Li\cdot \quad \cdot Be \quad \cdot \overset{\cdot}{B}\cdot \quad \cdot \overset{\cdot}{\underset{\cdot}{C}}\cdot \quad \cdot \overset{\cdot}{\underset{\cdot}{N}}\cdot \quad :\overset{\cdot}{\underset{\cdot}{O}}\cdot \quad :\overset{\cdot}{\underset{\cdot}{F}}\cdot$$

$$Na\cdot \quad \cdot Mg \quad \cdot \overset{\cdot}{Al}\cdot \quad \cdot \overset{\cdot}{\underset{\cdot}{Si}}\cdot \quad \cdot \overset{\cdot}{\underset{\cdot}{P}}\cdot \quad :\overset{\cdot}{\underset{\cdot}{S}}\cdot \quad :\overset{\cdot}{\underset{\cdot}{Cl}}\cdot$$

Electron-Dot Picture of Salts

$$Na\cdot + \cdot \overset{\cdot\cdot}{\underset{\cdot\cdot}{Cl}}: \xrightarrow{1\,e \text{ transfer}} Na^+ : \overset{\cdot\cdot}{\underset{\cdot\cdot}{Cl}}:^-$$

$$\cdot Mg + 2\cdot \overset{\cdot\cdot}{\underset{\cdot\cdot}{Cl}}: \xrightarrow{2\,e \text{ transfer}} Mg^{2+}\,[:\overset{\cdot\cdot}{\underset{\cdot\cdot}{Cl}}:]_2^-$$

The hydrogen atom may either lose an electron to become a bare nucleus, the **proton,** or accept an electron to form the **hydride ion,** $[H:]^-$, which possesses the helium configuration. Indeed, the hydrides of lithium, sodium, and potassium (Li^+H^-, Na^+H^-, and K^+H^-) are commonly used reagents.

$$H\cdot \xrightarrow{-1\,e} [H]^+ \qquad \text{Bare nucleus} \qquad IP = 314 \text{ kcal mol}^{-1}$$
$$\textbf{Proton}$$

$$H\cdot \xrightarrow{+1\,e} [H:]^- \qquad \text{Helium configuration} \qquad EA = -18 \text{ kcal mol}^{-1}$$
$$\textbf{Hydride ion}$$

EXERCISE 1-2

Draw electron-dot pictures for ionic LiBr, Na_2O, BeF_2, $AlCl_3$, and MgS.

In covalent bonds, electron octets are formed by sharing electrons

Formation of ionic bonds between two identical elements is difficult because the electron transfer is usually very unfavorable. For example, in H_2, formation of H^+H^- would require an energy input of nearly 300 kcal mol^{-1}. For the same reason, none of the halogens, F_2, Cl_2, Br_2, and I_2, has an ionic bond. The high IP of hydrogen also prevents the bonds in the hydrogen halides from being ionic. For elements nearer the center of the periodic table, the formation of ionic bonds is unfeasible, because it becomes more and more difficult to donate or accept enough electrons to attain the noble-gas configuration. Such is the case for carbon. This element would have to shed four electrons to reach the helium electronic structure or add four electrons for a neon-like arrangement. The large amount of charge that would develop makes these processes very energetically unfavorable.

$$C^{4+} \quad \xleftarrow{-4\,e} \quad \cdot \overset{\cdot}{\underset{\cdot}{C}}\cdot \quad \xrightarrow{+4\,e} \quad :\overset{\cdot\cdot}{\underset{\cdot\cdot}{C}}:^{4-}$$

$$\underset{\textbf{configuration}}{\textbf{Helium}} \qquad\qquad\qquad \underset{\textbf{configuration}}{\textbf{Neon}}$$

Instead, **covalent bonding** takes place: The elements *share* electrons so that each attains a noble-gas configuration. Typical products of such sharing are H_2 and HCl. In HCl, the chlorine atom assumes an octet structure by sharing one of its valence electrons with that of hydrogen. Similarly, the chlorine molecule, Cl_2, is diatomic be-

cause both component atoms gain octets by sharing two electrons. Such bonds are called **covalent single bonds.**

Electron-Dot Picture of Covalent Single Bonds

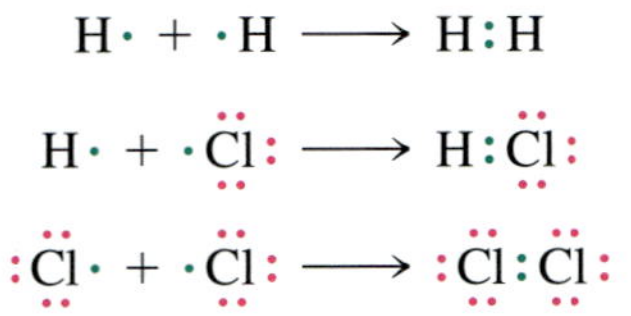

$$H \cdot + \cdot H \longrightarrow H : H$$

$$H \cdot + \cdot \ddot{\underset{..}{Cl}} : \longrightarrow H : \ddot{\underset{..}{Cl}} :$$

$$: \ddot{\underset{..}{Cl}} \cdot + \cdot \ddot{\underset{..}{Cl}} : \longrightarrow : \ddot{\underset{..}{Cl}} : \ddot{\underset{..}{Cl}} :$$

Because carbon has four valence electrons, it must acquire a share of four electrons to gain the neon configuration, as in methane. Nitrogen has five valence electrons and needs three to share, as in ammonia; and oxygen, with six valence electrons, requires only two to share, as in water.

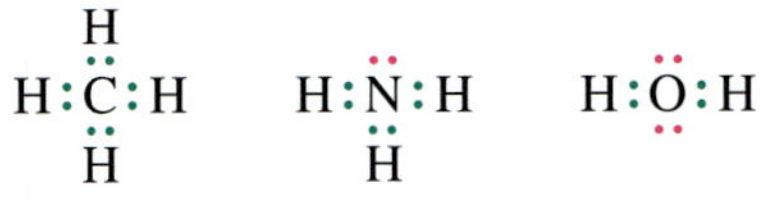

$$
\begin{array}{ccc}
H & & \\
H : \ddot{C} : H & H : \ddot{N} : H & H : \ddot{\underset{..}{O}} : H \\
H & H & \\
\textbf{Methane} & \textbf{Ammonia} & \textbf{Water}
\end{array}
$$

It is possible for one atom to supply both of the electrons required for covalent bonding. This occurs upon addition of a proton to ammonia, thereby forming NH_4^+, or to water, thereby forming H_3O^+.

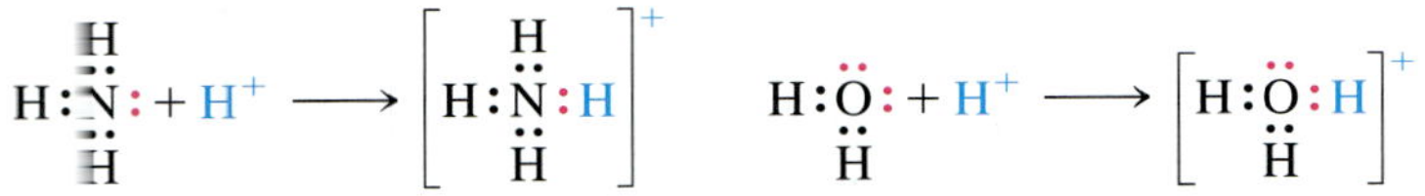

$$
H : \ddot{N} : + H^+ \longrightarrow \left[H : \ddot{N} : H \right]^+ \qquad H : \ddot{\underset{..}{O}} : + H^+ \longrightarrow \left[H : \ddot{\underset{..}{O}} : H \right]^+
$$

$$
\begin{array}{cc}
\textbf{Ammonium} & \textbf{Hydronium} \\
\textbf{ion} & \textbf{ion}
\end{array}
$$

Besides two-electron (**single**) bonds, atoms may form four-electron (**double**) and six-electron (**triple**) bonds to gain noble-gas configurations. Atoms that share more than one electron pair are found in ethene and ethyne.

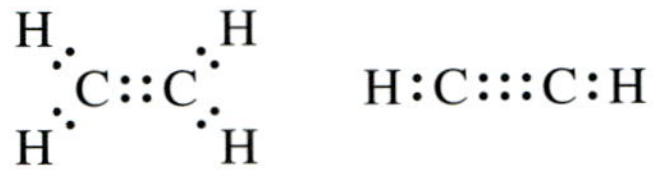

$$
\begin{array}{cc}
\underset{H}{\overset{H}{}} \ddot{C} :: \ddot{C} \underset{H}{\overset{H}{}} & H : C ::: C : H \\
\textbf{Ethene} & \textbf{Ethyne} \\
\textbf{(Ethylene)*} & \textbf{(Acetylene)*}
\end{array}
$$

EXERCISE 1-3

Draw electron-dot structures for F_2, CF_4, CH_2Cl_2, PH_3, BrI, OH^-, NH_2^-, and CH_3^-. (Where applicable, the first element is at the center of the molecule.) Make sure that all atoms have noble-gas electron configurations.

---

*In labels of molecules, systematic names (introduced in Section 2-3) will be given first, followed in parentheses by so-called common names that are still in frequent usage.

In most organic bonds, the electrons are not shared equally: polar covalent bonds

The preceding two sections presented two extreme ways in which atoms attain noble-gas configurations by entering into bonding: pure ionic and pure covalent. In reality, most bonds are of a nature that lies between these two extremes: **polar covalent.** Thus, the ionic bonds in most salts have some covalent character; conversely, the covalent bonds to carbon have some ionic or polar character. Recall (Section 1-2) that both sharing of electrons *and* Coulombic attraction contribute to the stability of a bond. How polar are polar covalent bonds and what is the direction of the polarity? We can answer these questions by considering the periodic table and keeping in mind that the positive nuclear charge of the elements increases from left to right. Thus, the elements on the left of the periodic table are often called **electropositive,** electron donating, or "electron pushing," because their electrons are held by the nucleus less tightly than are those of elements to the right. The latter are therefore described as being **electronegative,** electron accepting, or "electron pulling." Table 1-2 lists the relative electronegativity of some elements. On this scale, fluorine, the most electronegative of them all, is assigned the value 4.

Consideration of Table 1-2 readily explains why the most ionic (least covalent) bonds occur between elements at the two extremes (e.g., the alkali metal salts, such as sodium chloride). On the other hand, the purest covalent bonds are formed between atoms of equal electronegativity (i.e., identical elements, as in H_2, N_2, O_2, F_2, and so on) or in carbon–carbon bonds. However, most covalent bonds are between atoms of differing electronegativity, resulting in their **polarization.** The polarization of a bond is the consequence of a shift of the center of electron density in the bond toward the more electronegative atom. It is indicated in a very qualitative manner by designating a partial positive charge, δ^+, and partial negative charge, δ^-, to the respective less or more electronegative atom. The larger the difference in electronegativity, the bigger the charge separation.

Polar Bonds

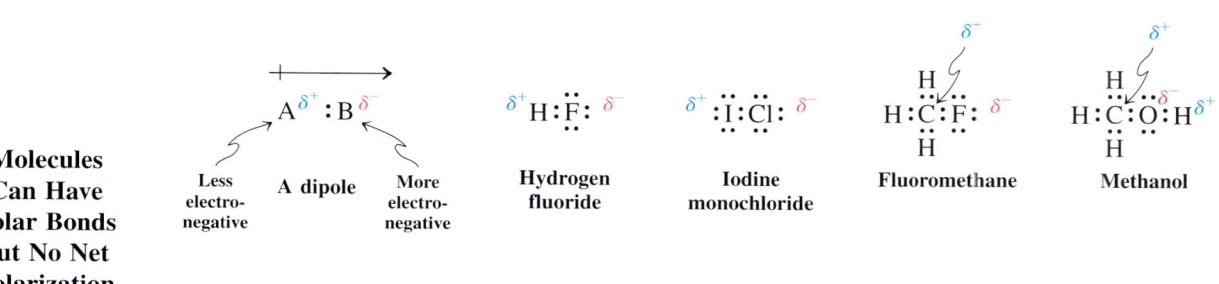

Molecules Can Have Polar Bonds but No Net Polarization

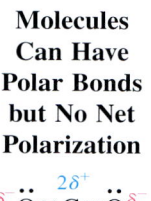

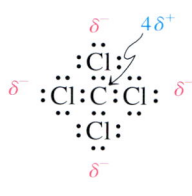

The separation of opposite charges is called an electric **dipole,** symbolized by an arrow crossed at its tail and pointing from positive to negative. A polarized bond can impart polarity to a molecule as a whole, as in HF, HCl, and CH_3F. In symmetrical structures, however, the polarizations of the individual bonds may cancel, thus leading to molecules with no net polarization, such as CO_2 and CCl_4. To know whether a molecule is polar, we have to know its shape, because the net polarity is the vector sum of the bond dipoles.

TABLE 1-2	Electronegativities of Selected Elements					
H 2.2						
Li 1.0	Be 1.6	B 2.0	C 2.6	N 3.0	O 3.4	F 4.0
Na 0.9	Mg 1.3	Al 1.6	Si 1.9	P 2.2	S 2.6	Cl 3.2
K 0.8						Br 3.0
						I 2.7

Note: Values established by L. Pauling and updated by A. L. Allred (see *Journal of Inorganic and Nuclear Chemistry,* 1961, *17,* 215).

Electron repulsion controls the shapes of molecules

Molecules adopt shapes in which electron repulsion is minimized. In diatomic species such as H_2 or LiH, there is only one bonding electron pair and one possible arrangement of the two atoms. However, beryllium fluoride, BeF_2, is a triatomic species. Will it be bent or linear? Electron repulsion is at a minimum in a **linear** structure, because the bonding and nonbonding electrons are placed as far from each other as possible, at 180°.* Linearity is also expected for other derivatives of beryllium, as well as of other elements in the same column of the periodic table.

BeF_2 Is Linear   **BCl_3 Is Trigonal**

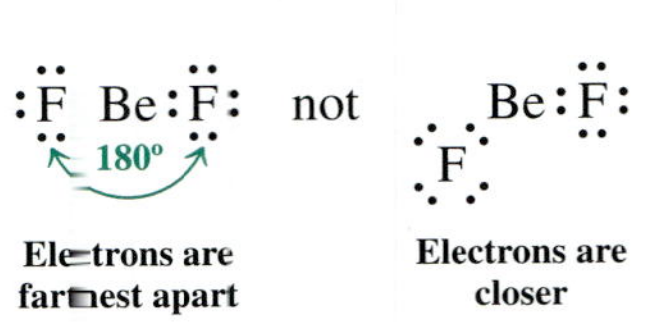

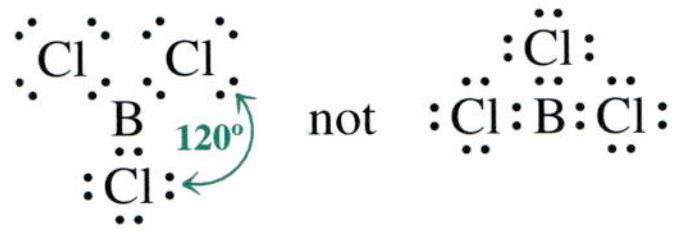

Electrons are farthest apart

Electrons are closer

In boron trichloride, the three valence electrons of boron allow it to form covalent bonds with three chlorine atoms. Electron repulsion enforces a regular **trigonal** arrangement—that is, the three halogens are at the corners of an equilateral triangle, the center of which is occupied by boron, and the bonding (and nonbonding) electron pairs of the respective chlorine atoms are at maximum distance from each other, that is, 120°. Other derivatives of boron, and the analogous compounds with other elements in the same column of the periodic table, are again expected to adopt trigonal structures.

*This is true only in the gas phase. At room temperature, BeF_2 is a solid (it is used in nuclear reactors) that exists as a complex network of linked Be and F atoms, not as a distinct linear triatomic structure.

Applying this principle to carbon, we can see that methane, CH_4, has to be **tetrahedral.** Pointing its four valences toward the vertices of a tetrahedron is the best arrangement for minimizing electron repulsion.

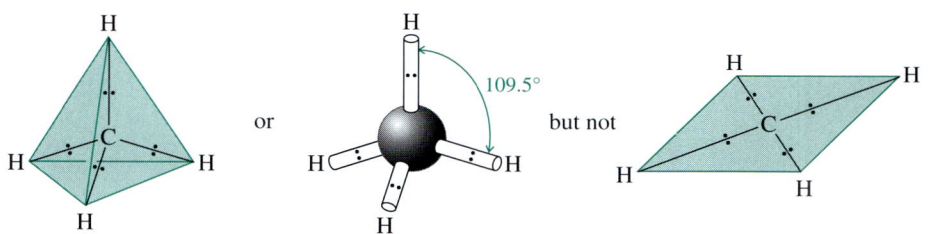

This method for determining molecular shape by minimizing electron repulsion is called the *valence shell electron pair repulsion (VSEPR)* method. Note that we often draw molecules such as BCl_3 and CH_4 as if they were flat and had 90° angles. *This depiction is for ease of drawing only.* Do *not* confuse such drawings with the true molecular shapes (trigonal for BCl_3 and tetrahedral for CH_4).

EXERCISE 1-4

Show the bond polarization in H_2O, SCO, SO, IBr, CH_4, $CHCl_3$, CH_2Cl_2, and CH_3Cl by using dipole arrows to indicate separation of charge. (In the last four examples, place the carbon in the center of the molecule.)

EXERCISE 1-5

Ammonia, $:NH_3$, is not trigonal but pyramidal, with bond angles of 107.3°. Water, $H_2\overset{..}{\underset{..}{O}}$, is not linear but bent (104.5°). Why? (**Hint:** Consider the effect of the nonbonding electron pairs.)

To summarize, there are two extreme types of bonding, ionic and covalent. Both derive favorable energetics from Coulomb forces and the attainment of noble-gas electronic structures. Most bonds are better described as something between the two types: the polar covalent (or covalent ionic) bonds. Polarity in bonds may give rise to polar molecules. The outcome depends on the shape of the molecule, which is determined in a simple manner by arrangement of its bonds and nonbonding electrons to minimize electron repulsion.

1-4 Electron-Dot Model of Bonding: Lewis Structures

The drawings in the preceding section, with pairs of electron dots representing bonds, are also called **Lewis* structures.** In this section, rules are given for writing such structures correctly and for keeping track of valence electrons.

Lewis structures are drawn by following simple rules

The procedure for drawing correct electron-dot structures is straightforward, as long as the following rules are observed.

---

*Professor Gilbert N. Lewis (1875–1946), University of California at Berkeley.

RULE 1. *Draw the molecular skeleton.* As an example, consider methane. The molecule has four hydrogen atoms bonded to one central carbon atom.

H
H C H   H H C H H
H

Correct   **Incorrect**

RULE 2. *Count the number of available valence electrons.* Add up all the valence electrons of the component atoms. Special care has to be taken with charged structures (anions or cations), in which case the appropriate number of electrons has to be added or subtracted to account for extra charges.

CH_4	4 H	4×1 electron =	4 electrons		HBr	1 H	1×1 electron =	1 electron
	1 C	1×4 electrons =	4 electrons			1 Br	1×7 electrons =	7 electrons
		Total	8 electrons				Total	8 electrons
H_3O^+	3 H	3×1 electron =	3 electrons		NH_2^-	2 H	2×1 electron =	2 electrons
	1 O	1×6 electrons =	6 electrons			1 N	1×5 electrons =	5 electrons
	Charge	+1 =	−1 electron			Charge	−1 =	+1 electron
		Total	8 electrons				Total	8 electrons

RULE 3. (The **octet rule**) *Depict all covalent bonds by two shared electrons, giving as many atoms as possible a surrounding electron octet, except for H, which requires a duet.* Make sure that the number of electrons used is *exactly* the number counted according to rule 2. Elements at the right in the periodic table may contain pairs of valence electrons not used for bonding, called **lone electron pairs** or just **lone pairs.**

Consider, for example, hydrogen bromide. The shared electron pair supplies the hydrogen atom with a duet, the bromine with an octet, because the bromine carries three lone-electron pairs. Conversely, in methane, the four C–H bonds satisfy the requirement of the hydrogens and, at the same time, furnish the octet for carbon. Examples of correct and incorrect Lewis structures for HBr are shown below.

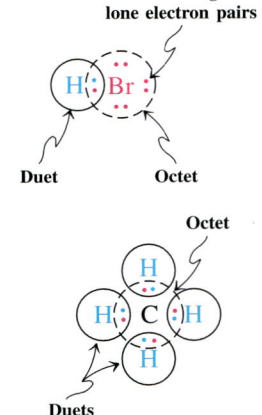

Correct Lewis Structure   **Incorrect Lewis Structures**

H : Br :

3 electrons around H   No octet

· H : Br :

4 electrons around H   Wrong number of electrons

H : : Br :   H : Br :

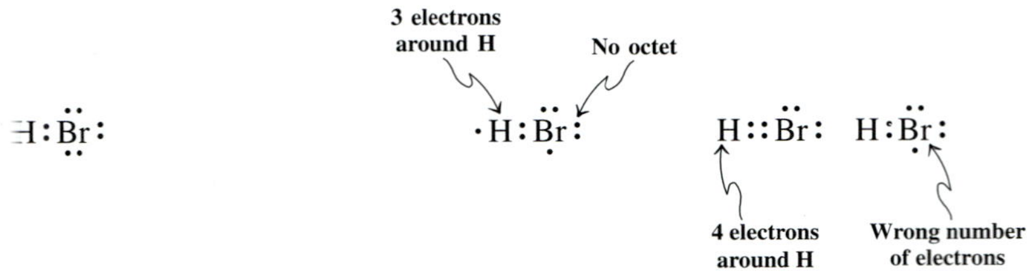

Frequently, the number of valence electrons is not sufficient to satisfy the octet rule only with single bonds. In this event, double bonds (two shared electron pairs) and even triple bonds (three shared pairs) are necessary to obtain octets. An example is the nitrogen molecule, N_2, which has ten valence electrons. An N–N single bond would leave both atoms with electron sextets, and a double bond provides only one nitrogen atom with an octet. It is the molecule with a triple bond that satisfies both atoms.

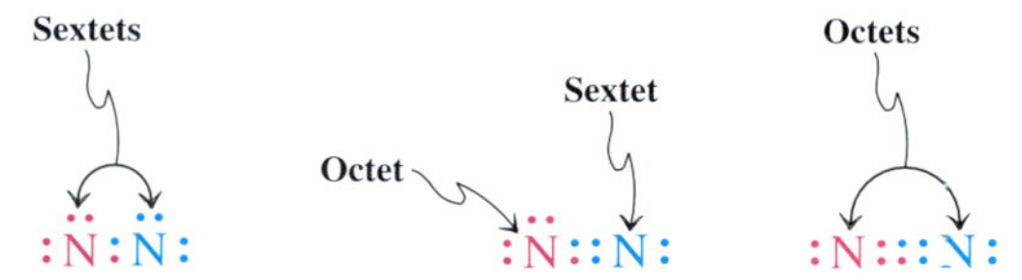

Sextets

Octet

Sextet

Octets

$:N:N:$

Single bond

$:N::N:$

Double bond

$:N:::N:$

Triple bond

Further examples of molecules with double and triple bonds are shown below.

Correct Lewis Structures

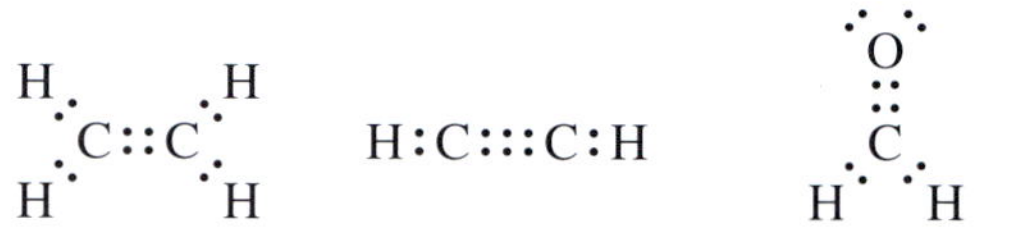

$H:C::C:H$ (with H's at each corner)

**Ethene
(Ethylene)**

$H:C:::C:H$

**Ethyne
(Acetylene)**

Formaldehyde structure with O double bonded to C and two H's

Formaldehyde

In practice, you may find a simple sequence useful: First, connect all mutually bonded atoms in your structure by single bonds (i.e., shared electron pairs); second, if there are any electrons left, distribute them as lone electron pairs to maximize the number of octets; and finally, if some of the atoms lack octet structures, change as many lone electron pairs into shared electron pairs as required to complete the octet shells (see also the Chapter Integration Problem on p. 40).

EXERCISE 1-6

Draw Lewis structures for the following molecules: HI, $CH_3CH_2CH_3$, CH_3OH, HSSH, SiO_2 (OSiO), O_2, CS_2 (SCS).

RULE 4. *Assign charges to atoms in the molecule.* Each lone pair contributes two electrons to the valence electron count of an atom in a molecule, and each bonding (shared) pair contributes one. An atom is charged if this total is different from the outer-shell electron count in the free, nonbonded atom. Thus we have the formula

$$\text{Charge} = \left(\begin{array}{c}\text{number of outer-shell} \\ \text{electrons on the} \\ \text{free, neutral atom}\end{array}\right) - \left(\begin{array}{c}\text{number of unshared} \\ \text{electrons on the atom} \\ \text{in the molecule}\end{array}\right) - \frac{1}{2}\left(\begin{array}{c}\text{number of bonding} \\ \text{electrons surrounding the} \\ \text{atom in the molecule}\end{array}\right)$$

$\overset{+}{H:\overset{..}{O}:H}$
$\underset{H}{}$

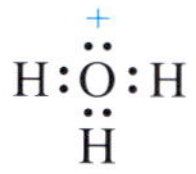

Hydronium ion

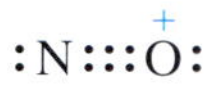

As an example, which atom bears the positive charge in the hydronium ion? Each hydrogen has a valence electron count of 1 from the shared pair in its bond to oxygen. Because this value is the same as the electron count in the free atom, the charge on each hydrogen is zero. The electron count on the oxygen in the hydronium ion is 2 (the lone pair) + 3 (half of 6 bonding electrons) = 5. This value is one short of the number of outer-shell electrons in the free atom, thus giving the oxygen a charge of +1. Hence the positive charge is assigned to oxygen.

$:N:::\overset{+}{O}:$

Nitrosyl cation

Another example is the nitrosyl cation, NO^+. The molecule bears a lone pair on nitrogen, in addition to the triple bond connecting the nitrogen to the oxygen atom. This gives nitrogen five valence electrons, a value that matches the count in the free atom; therefore the nitrogen atom has no charge. The same number of valence electrons (5) is found on oxygen. Because the free oxygen atom requires six valence electrons to be neutral, the oxygen in NO^+ possesses the +1 charge. Other examples follow on the next page.

H:C̈:H
    H
Methyl anion

    H
H:C̈:S̈:⁻
    H
Methanethiolate ion

         ⁺Ö—H
    Ḃ:C̈.
H   H
Protonated formaldehyde

Sometimes the octet rule leads to charges on atoms even in neutral molecules. The Lewis structure is then said to be **charge separated.** An example is carbon monoxide, CO. Some compounds containing nitrogen–oxygen bonds, such as nitric acid, HNO_3, also exhibit this behavior.

:C⁻≡O⁺:
Carbon monoxide

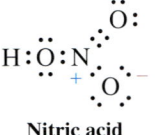
Nitric acid

The octet rule does not always hold

The octet rule strictly holds only for the elements of the second row and then only if there is a sufficient number of valence electrons to satisfy it. Thus, there are three exceptions to be considered.

EXCEPTION 1. You will have noticed that all our examples of "correct" Lewis structures contain an even number of electrons; that is, all are distributed as bonding or lone pairs. This distribution is not possible in species having an odd number of electrons, such as nitrogen oxide (NO) and neutral methyl (methyl radical, ·CH_3; see Section 3-1).

:N̈::Ö
Nitrogen oxide

H:Ċ:H
    H
Methyl radical

H:Be:H
Beryllium hydride

H. .H
   B
   H
Borane

EXCEPTION 2. Some compounds of the early second-row elements, such as BeH_2 and BH_3, have a deficiency of valence electrons.

Compounds falling under exceptions 1 and 2 reveal the consequences of being denied octet configurations: they are unusually reactive and transform readily in reactions that lead to octet structures. For example, ·CH_3 dimerizes spontaneously to ethane, $CH_3–CH_3$, and BH_3 reacts with hydride, H^-, to give borohydride, BH_4^-.

         H   H   H H
H:Ċ· + ·Ċ:H ⟶ H:C̈:C̈:H
         H   H   H H
                                    Ethane

H. .H   H
   B   + :H⁻ ⟶ H:B:H⁻
   H   H
                         Borohydride

EXCEPTION 3. Beyond the second row, the simple Lewis model is not strictly applied, and elements may be surrounded by more than eight valence electrons, a feature referred to as **valence shell expansion.** For example, not only are phosphorus

and sulfur (as relatives of nitrogen and oxygen) trivalent and divalent, respectively, and Lewis octet structures readily formulated for their derivatives, but they form stable compounds of higher valency, among them the familiar phosphoric and sulfuric acids. Some examples of octet and expanded-octet molecules containing these elements are shown below.

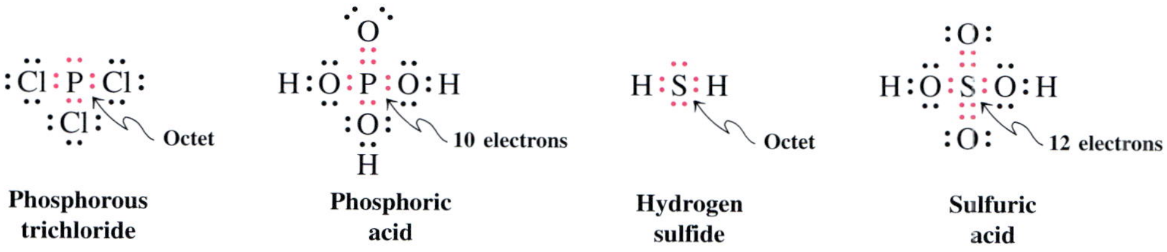

$$:\overset{\cdot\cdot}{\underset{\cdot\cdot}{Cl}}:\overset{\cdot\cdot}{P}:\overset{\cdot\cdot}{\underset{\cdot\cdot}{Cl}}: \qquad H:\overset{\cdot\cdot}{\underset{\cdot\cdot}{O}}:\overset{\overset{\overset{\cdot\cdot}{O}}{|}}{P}:\overset{\cdot\cdot}{\underset{\cdot\cdot}{O}}:H \qquad H:\overset{\cdot\cdot}{S}:H \qquad H:\overset{\cdot\cdot}{\underset{\cdot\cdot}{O}}:\overset{\overset{:O:}{|}}{S}:\overset{\cdot\cdot}{\underset{\cdot\cdot}{O}}:H$$

:Cl: Octet   :O: 10 electrons   Octet   :O: 12 electrons
      H

Phosphorous   **Phosphoric**   **Hydrogen**   **Sulfuric**
trichloride   **acid**   **sulfide**   **acid**

An explanation for this apparent violation of the octet rule is found in a more sophisticated description of atomic structure by quantum mechanics (Section 1-6). However, you will notice that, even in these cases, you can construct dipolar forms in which the Lewis octet rule is preserved (see Section 1-5).

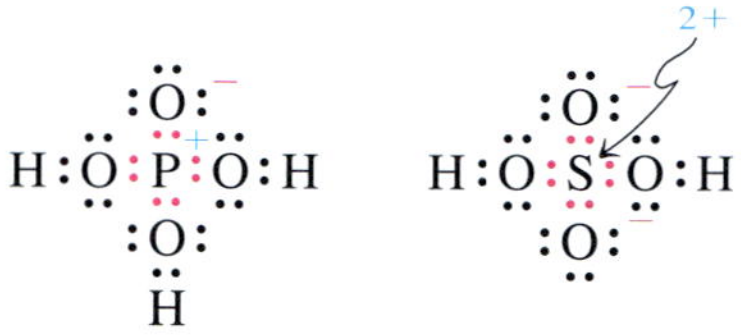

$$H:\overset{\cdot\cdot}{\underset{\cdot\cdot}{O}}:\overset{\overset{:\overset{\cdot\cdot}{O}:^-}{|}}{\overset{+}{P}}:\overset{\cdot\cdot}{\underset{\cdot\cdot}{O}}:H \qquad H:\overset{\cdot\cdot}{\underset{\cdot\cdot}{O}}:\overset{\overset{:\overset{\cdot\cdot}{O}:}{|}}{\overset{2+}{S}}:\overset{\cdot\cdot}{\underset{\cdot\cdot}{O}}:H$$

Covalent bonds can be depicted as straight lines

Electron-dot structures can be cumbersome, particularly for larger molecules. It is simpler to represent covalent single bonds by single straight lines; double bonds are represented by two lines and triple bonds by three. Lone electron pairs can either be shown as dots or simply omitted. The use of such notation was first suggested by the German chemist August Kekulé,* long before electrons were discovered; structures of this type are often called **Kekulé structures.**

Straight-Line Notation for the Covalent Bond

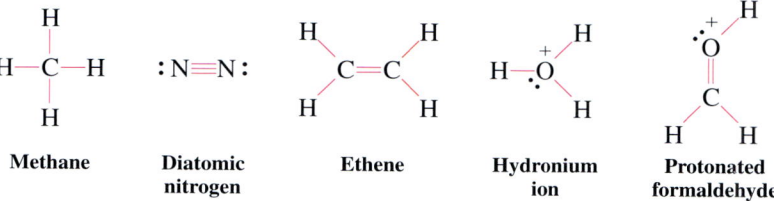

Methane   Diatomic   Ethene   Hydronium   Protonated
              nitrogen   ion   formaldehyde

EXERCISE 1-7

Draw Lewis structures of the following molecules, including the assignment of any charges to atoms (the order in which the atoms are attached is given in parentheses when it may not be obvious from the formula as it is commonly written): SO, F_2O (FOF), $HClO_2$ (HOClO), BF_3NH_3 (F_3BNH_3), $CH_3OH_2^+$ ($H_3COH_2^+$), $Cl_2C=O$, CN^-, C_2^{2-}.

---

*Professor F. August Kekulé von Stradonitz (1829–1896), University of Bonn, Germany.

In summary, Lewis structures describe bonding by the use of electron dots or straight lines. Whenever possible, they are drawn so as to give hydrogen an electron duet and other atoms an electron octet. Charges are assigned to each atom by evaluating its electron count.

1-5 Resonance Forms

In organic chemistry, we also encounter molecules for which there are *several* correct Lewis structures.

The carbonate ion has several correct Lewis structures

Let us consider the carbonate ion, CO_3^{2-}. Following our rules, we can easily draw a Lewis structure (A) in which every atom is surrounded by an octet. The two negative charges are located on the bottom two oxygen atoms; the third oxygen is neutral, connected to the central carbon by a double bond and bearing two lone pairs. But why choose the bottom two oxygen atoms as the charge carriers? There is no reason at all—it is a completely arbitrary choice. We could equally well have drawn structures B or C to describe the carbonate ion. The three correct Lewis pictures are called **resonance forms.**

Resonance Forms of the Carbonate Ion

The individual resonance forms are connected by double-headed arrows and all placed within one set of square brackets. They have the characteristic property of being interconvertible by *electron-pair movement only,* the nuclear positions in the molecule remaining *unchanged.* Note that, to turn A into B and then into C, we have to shift two electron pairs in each case. Such movement of electrons can be depicted by curved arrows, a procedure informally called "electron pushing."

The use of curved arrows to depict electron-pair movement is a useful technique that will prevent us from making the common mistake of changing the total number of electrons when we draw resonance forms. It is also advantageous in keeping track of electrons when formulating mechanisms (Section 6-4).

But what is its true structure?

Does the carbonate ion have one uncharged oxygen atom bound to carbon through a double bond and two other oxygen atoms bound through a single bond each, both bearing a negative charge, as suggested by the Lewis structures? *The answer is no.* If that were true, the carbon–oxygen bonds would be of different lengths, because double bonds are normally shorter than single bonds. But the carbonate ion is *perfectly symmetrical* and contains a trigonal central carbon, all C–O bonds being of equal length—between the length of a double and that of a single bond. The negative charge is evenly distributed over all three oxygens: It is said to be **delocalized.**

In other words, none of the individual Lewis representations of this molecule is correct on its own. Rather, *the true structure is a composite of A, B, and C.* The resulting picture is called a **resonance hybrid.** Because A, B, and C are equivalent (i.e., each is composed of the same number of atoms, bonds, and electron pairs), they contribute equally to the true structure of the molecule, but none of them by itself accurately represents it.

The word *resonance* may imply to you that the molecule vibrates or equilibrates from one form to another. This inference is incorrect. The molecule *never* looks like any of the individual resonance forms; it has only one structure, the resonance hybrid. Unlike substances in ordinary chemical equilibria, resonance forms are *not* real, although each makes a partial contribution to reality.

Dotted-Line Notation of Carbonate as a Resonance Hybrid

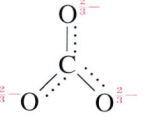

An alternative convention used to describe resonance hybrids such as carbonate is to represent the bonds as a combination of solid and dotted lines. The $\frac{2}{3}-$ sign here indicates that a partial charge ($\frac{2}{3}$ of a negative charge) resides on each oxygen atom. The equivalence of all three carbon–oxygen bonds and all three oxygens is clearly indicated by this convention. Other examples of resonance hybrids are the acetate anion and the 2-propenyl (allyl) cation.

Acetate anion

2-Propenyl (allyl) cation

When drawing resonance forms, keep in mind that (1) pushing one electron pair toward one atom and away from another results in a movement of charge; (2) the relative positions of all the atoms stay unchanged—only electrons are moved; (3) equivalent resonance forms contribute equally to the resonance hybrid; and (4) the arrows connecting resonance forms are double headed (↔).

EXERCISE 1-8

Draw two resonance forms for nitrite ion, NO_2^-. What can you say about the geometry of this molecule (linear or bent)? (**Hint:** Consider the effect of electron repulsion exerted by the lone pair on nitrogen.)

Not all resonance forms are equivalent

The carbonate and acetate anions and the 2-propenyl cation all have equivalent octet resonance forms. However, many molecules are described by resonance forms that are not equivalent. An example is the enolate anion. The two resonance forms differ in the locations of both the double bond and the charge.

The Two Nonequivalent Resonance Forms of the Enolate Ion

Although both forms are contributors to the true structure of the anion, we shall see that one contributes more than the other. The question is, which one? If we (greatly) extend our consideration of nonequivalent resonance forms to those devoid of octets, the question becomes more general.

[Octet ⟷ Nonoctet] Resonance Forms

Formaldehyde   Sulfuric acid

Such an extension requires that we relax our definitions of "correct" and "incorrect" Lewis structures and broadly regard *all* resonance forms as potential contributors to the true picture of a molecule. The task is then to recognize which resonance form is the most important one. In other words, which one is the **major resonance contributor?** Here are some guidelines.

GUIDELINE 1. *Structures with a maximum of octets are most important.* In the enolate ion, all component atoms in either structure are surrounded by octets. Consider, however, the nitrosyl cation, NO^+: The better resonance form has a positive charge on oxygen with electron octets around both atoms; the other form places the positive charge on nitrogen, thereby resulting in an electron sextet on this atom. Because of the octet rule, the second structure contributes less to the hybrid. Thus, the N–O linkage is closer to being a triple than a double bond, and more of the positive charge is on oxygen than on nitrogen. Similarly, the dipolar resonance form for formaldehyde (shown earlier) generates an electron sextet around carbon, rendering it a minor resonance contributor. The possibility of valence shell expansion for third-row elements (Section 1-4) makes the non-charge-separated picture of sulfuric acid with 12 electrons around sulfur a feasible resonance form, but the dipolar octet structure is better.

Major   Minor
resonance   resonance
contributor   contributor
Nitrosyl cation

GUIDELINE 2. *Charges should be preferentially located on atoms with compatible electronegativity.* Consider again the enolate ion. Which is the major contributing resonance form? Guideline 2 requires it to be the first, in which the negative charge resides on the more electronegative oxygen atom.

Looking again at NO^+, you might find guideline 2 confusing. The major resonance contributor to NO^+ has the positive charge on the more electronegative oxygen. In cases such as this, *the octet rule overrides the electronegativity criterion;* that is, guideline 1 takes precedence over guideline 2.

GUIDELINE 3. *Structures with less (opposite) charge separation are greater resonance contributors than those with more charge separation.* This rule is a simple consequence of Coulomb's law: Separating opposite charges requires energy; hence neutral structures are better than dipolar ones.

$$
\left[\quad \text{H}-\overset{:\ddot{\text{O}}:}{\underset{}{\text{C}}}-\ddot{\text{O}}-\text{H} \quad \longleftrightarrow \quad \text{H}-\overset{:\ddot{\text{O}}:^{-}}{\underset{}{\text{C}}}=\overset{+}{\ddot{\text{O}}}-\text{H} \quad \right]
$$

Major   Minor

Formic acid

$$
\left[:\text{C}=\ddot{\text{O}}: \longleftrightarrow :\bar{\text{C}}\equiv\overset{+}{\text{O}}: \right]
$$

Minor   Major

Carbon monoxide

In some cases, to ensure octet Lewis structures, charge separation is necessary; that is, guideline 1 takes precedence over guideline 3. An example is carbon monoxide. Other examples are phosphoric and sulfuric acids, although valence shell expansion allows the formulation of expanded octet structures (see also Section 1-4 and guideline 1).

When there are several charge-separated resonance forms that comply with the octet rule, the most favorable is the one in which the charge distribution best accommodates the relative electronegativities of the component atoms (guideline 2). In diazomethane, for example, nitrogen is more electronegative than carbon, thus allowing a clear choice between the two resonance contributors.

$$
\left[\quad \overset{\text{H}}{\underset{\text{H}}{\text{C}}}=\overset{+}{\text{N}}=\overset{-}{\ddot{\text{N}}}: \quad \longleftrightarrow \quad \overset{\text{H}}{\underset{\text{H}}{\ddot{\text{C}}}}-\overset{+}{\text{N}}\equiv\text{N}: \quad \right]
$$

Major   Minor

Diazomethane

EXERCISE 1-9

Draw resonance forms for the following two molecules. Indicate the more favorable resonance contributor in each case. **(a)** CNO^{-}; **(b)** NO^{-}.

In summary, there are molecules that cannot be described accurately by one Lewis structure but exist as hybrids of several extreme resonance forms. To find the most important resonance contributor, consider the octet rule, make sure that there is a minimum of charge separation, and place on the relatively more electronegative atoms as much negative and as little positive charge as possible.

1-6 Atomic Orbitals: A Quantum Mechanical Description of Electrons Around the Nucleus

So far, we have considered bonds in terms of electron pairs arranged around the component atoms in such a way as to maximize noble-gas configurations (e.g., Lewis octets) and minimize electron repulsion. This approach is useful as a descriptive and predictive tool with regard to the number and location of electrons in molecules. However, it does not answer some simple questions that you may have asked yourself while dealing with this material. For example, why are some Lewis structures "incorrect" or, ultimately, why are noble gases relatively stable? Why are some bonds stronger than others, and how can we tell? What is so good about the two-electron

bond, and what do multiple bonds look like? To get some answers, we will start by learning more about the way in which the electrons are distributed around the nucleus, both spatially and energetically. The simplified treatment presented here has as its basis the theory of quantum mechanics developed independently in the 1920s by Heisenberg, Schrödinger, and Dirac.[*] In this theory, the movement of an electron around a nucleus is expressed in the form of equations that are very similar to those characteristic of waves. The solutions to these equations, called **atomic orbitals,** allow us to describe the probability of finding the electron in a certain region in space. The shape of these domains depends on the energy of the electron.

The electron is described by wave equations

The classical description of the atom (Bohr[†] theory) assumed that electrons move on more or less defined trajectories around the nucleus. Their energy was thought to relate to their distance from the nucleus. This view is intuitively appealing because it coincides with our physical understanding of classical mechanics. Yet it is incorrect for several reasons.

First, the classical picture of an electron moving in an orbit requires (as does any moving charge) the emission of electromagnetic radiation. The resulting energy loss from the system would cause the electron to spiral toward the nucleus, a prediction that is completely at odds with reality.

Second, in the classical picture, an electron can have any energy, so it can have any of an infinite number of orbits of differing radii. This, again, is not what is observed. Rather, only certain defined energies, called **energy states,** are possible for an electron around a nucleus. Thus, classical mechanics does not satisfactorily explain atomic structure and, ultimately, bonding.

A better model is afforded by considering the wave nature of moving particles. Matter of mass m that moves with velocity v has a wavelength λ.

> **de Broglie[‡] Wavelength**
>
> $$\lambda = \frac{h}{mv}$$

in which h is Planck's[§] constant. As a result, an orbiting electron can be described by equations that are the same as those used in classical mechanics to describe waves (Figure 1-4; see page 24). The latter have amplitudes with alternating positive and negative signs. Points at which the sign changes are called **nodes.** Waves that interact in phase reinforce each other, as shown in Figure 1-4B. Those out of phase interfere with each other to make smaller waves (and possibly even cancel each other), as shown in Figure 1-4C.

[*]Professor Werner Heisenberg (1901–1976), University of Munich, Germany, Nobel Prize 1932 (physics); Professor Erwin Schrödinger (1887–1961), University of Dublin, Ireland, Nobel Prize 1933 (physics); Professor Paul Dirac (1902–1984), Florida State University, Tallahassee, Nobel Prize 1933 (physics).
[†]Professor Niels Bohr (1885–1962), University of Copenhagen, Denmark, Nobel Prize 1922 (physics).
[‡]Prince Louis-Victor de Broglie (1892–1987), Nobel Prize 1929 (physics).
[§]Professor Max K. E. L. Planck (1858–1947), University of Berlin, Germany, Nobel Prize 1918 (physics).

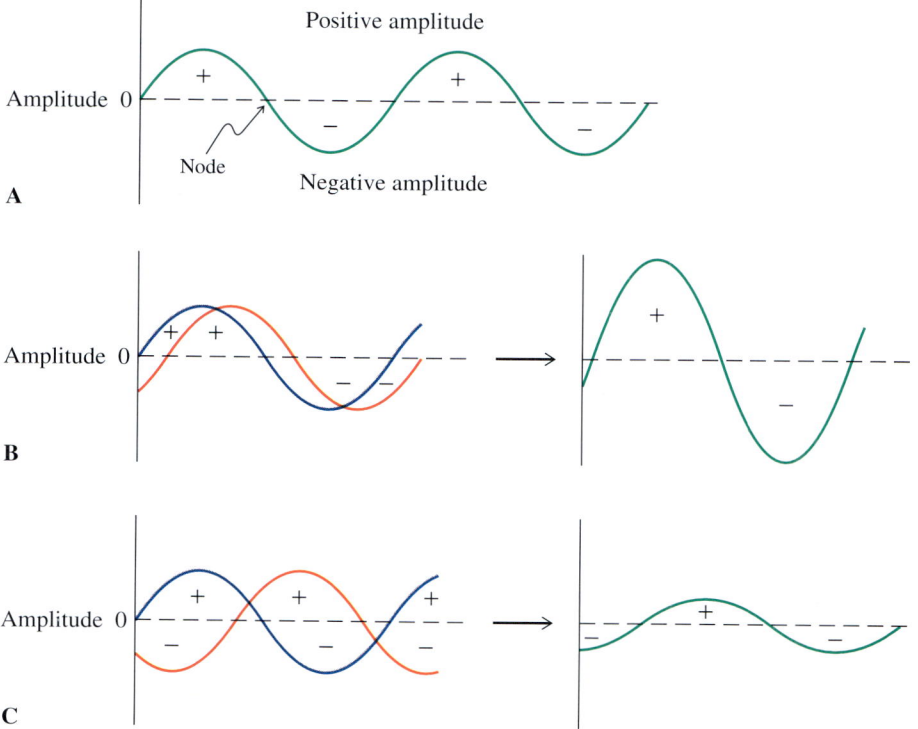

FIGURE 1-4
(A) A wave. The signs of the amplitude are assigned arbitrarily. At points of zero amplitude, called nodes, the wave changes sign.
(B) Waves with amplitudes of like sign (in phase) reinforce each other to make a larger wave. (C) Waves out of phase subtract from each other to make a smaller wave.

Note: The + and − signs in Figure 1-4 refer to signs of the mathematical functions describing the wave amplitudes and have nothing to do with electrical charges.

This theory of electron motion is called **quantum mechanics.** The equations developed in this theory, the **wave equations,** have a series of solutions called **wave functions,** usually described by the Greek letter psi, ψ. Their values around the nucleus are not directly identifiable with any observable property of the atom. However, *the squares (ψ^2) of their values at each point in space describe the probability of finding an electron at that point.* The physical realities of the atom make solutions attainable only for certain *specific energies.* The system is said to be **quantized.**

EXERCISE 1-10

Draw a picture similar to Figure 1-4 of two waves overlapping such that their amplitudes cancel each other.

Atomic orbitals have characteristic shapes

Plots of wave functions in three dimensions typically have the appearance of spheres or dumbbells with flattened or teardrop-shaped lobes. For simplicity, we may regard artistic renditions of **atomic orbitals** as indicating the regions in space in which the electron is likely to be found. Nodes separate portions of the wave function with opposite mathematical signs. The value of the wave function at a node is zero; therefore the probability of finding electron density there is zero. Higher energy wave functions have more nodes than do those of low energy.

Let us consider the shapes of the atomic orbitals for the simplest case, that of the hydrogen atom, consisting of a proton surrounded by an electron. The single lowest energy solution of the wave equation is called the 1*s* orbital, the number one referring to the first (lowest) energy level. An orbital label also denotes the shape and num-

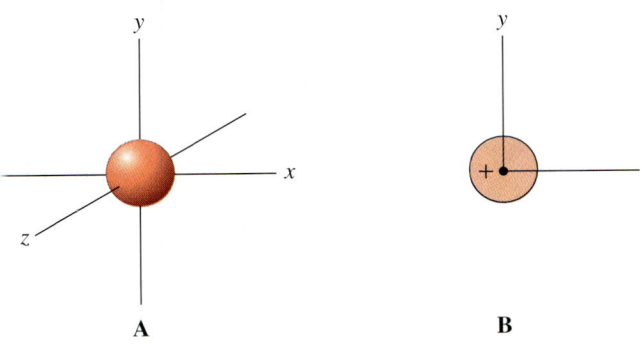

FIGURE 1-5

Representations of a 1s orbital. (A) The orbital is spherically symmetric in three dimensions. (B) A simplified two-dimensional view. The plus sign denotes the mathematical sign of the wave function and is *not a charge.*

ber of nodes of the orbital. The 1s orbital is *spherically symmetric* (Figure 1-5) and has no nodes. This orbital can be represented pictorially as a sphere (Figure 1-5A) or simply as a circle (Figure 1-5B).

The next higher energy wave function, the 2s orbital, also is unique and, again, spherical. The 2s orbital is larger than the 1s orbital; the higher energy 2s electron is on the average farther from the positive nucleus. In addition, the 2s orbital has one node, a spherical surface of zero electron density separating regions of the wave function of opposite sign (Figure 1-6). Like that of classical waves, the sign of the wave function on either side of the node is arbitrary, as long as it changes at the node. Remember that the sign of the wave function is not related to "where the electron is." As mentioned earlier, the probability of electron occupancy at any point of the orbital is given by the square of the value of the wave function. Moreover, the node does not constitute a barrier of any sort to the electron, which, in this description, is regarded not as a particle but as a wave.

After the 2s orbital, the wave equations for the electron around a hydrogen atom have three energetically equivalent solutions, the $2p_x$, $2p_y$, and $2p_z$ orbitals. Solutions of equal energy of this type are called **degenerate** (*degenus,* Latin, without genus or

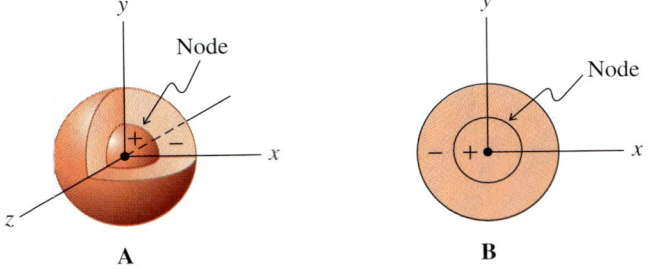

FIGURE 1-6

Representations of a 2s orbital. Notice that it is larger than the 1s orbital and that a node is present. The + and − denote the sign of the wave function. (A) The orbital in three dimensions with a section removed to allow the visualization of the node. (B) The more conventional two-dimensional representation of the orbital.

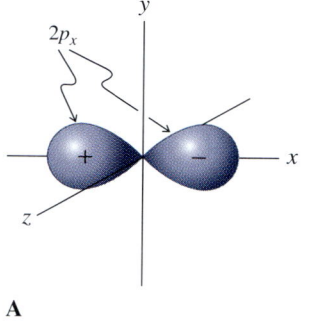

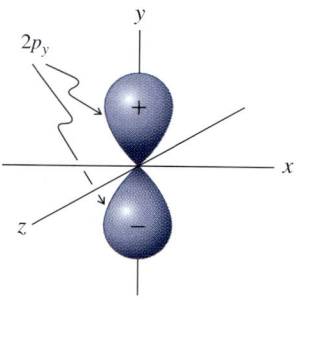

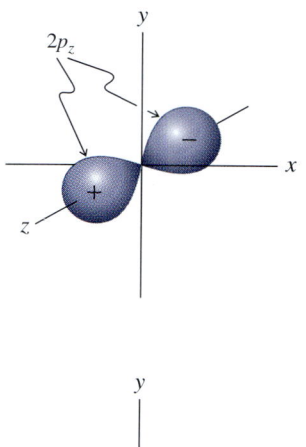

A

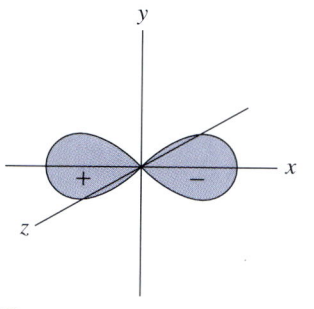

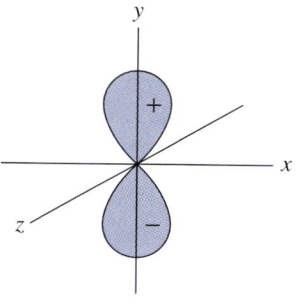

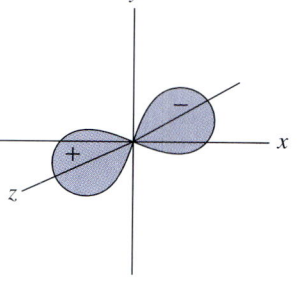

B

FIGURE 1-7

Representations of 2p orbitals (A) in three dimensions and (B) in two dimensions. Remember that the + and − signs refer to the wave functions and *not* to electrical charges. Lobes of opposite sign are separated by a nodal plane that is perpendicular to the axis of the orbital. For example, the p_x orbital is divided by a node in the *yz* plane.

kind). As shown in Figure 1-7, *p* orbitals consist of two lobes that resemble a solid figure eight. A *p* orbital is characterized by its directionality in space. The orbital axis can be aligned with any one of the *x, y,* and *z* axes, hence the labels p_x, p_y, and p_z. The two lobes of opposite sign of each orbital are separated by a nodal plane through the atom's nucleus and perpendicular to the orbital axis.

The third set of solutions furnishes the 3*s* and 3*p* atomic orbitals. They are similar in shape to, but more diffuse than, their lower energy counterparts and have two nodes. Still higher energy orbitals (3*d*, 4*s*, 4*p*, etc.) are characterized by an increasing number of nodes and a variety of shapes. They are of much less importance in organic chemistry than are the lower orbitals. To a first approximation, the shapes and nodal properties of the atomic orbitals of other elements are very similar to those of hydrogen. Therefore, we may use *s* and *p* orbitals in a description of the electronic configurations of helium, lithium, and so forth.

The Aufbau principle assigns electrons to orbitals

Approximate relative energies of the atomic orbitals up to the 5*s* level are shown in Figure 1-8. With its help, we can give an electronic configuration to every atom in the periodic table. To do so, we follow three rules for assigning electrons to atomic orbitals:

1. Lower energy orbitals are filled before those with higher energy.
2. No orbital may be occupied by more than two electrons, according to the **Pauli***
exclusion principle. Furthermore, these two electrons must differ in the orientation of their intrinsic angular momentum, their **spin.** There are two possible directions of the electron spin, usually depicted by vertical arrows pointing in opposite directions. An orbital is filled when it is occupied by two electrons of opposing spin, frequently referred to as **paired electrons.**
3. Degenerate orbitals, such as the *p* orbitals, are first occupied by one electron each, all of these electrons having the same spin. Subsequently, three more, each of opposite spin, are added to the first set. This assignment is based on **Hund's†** **rule.**

With these rules in hand, the determination of electronic configuration becomes simple. Helium has two electrons in the 1*s* orbital and its electronic structure is abbreviated $(1s)^2$. Lithium $[(1s)^2(2s)^1)]$ has one and beryllium $[(1s)^2(2s)^2]$ two additional electrons in the 2*s* orbital. In boron $[(1s)^2(2s)^2(2p)^1]$, we begin the filling of

*Professor Wolfgang Pauli (1900–1958), Swiss Federal Institute of Technology (ETH) Zurich,
Switzerland, Nobel Prize 1945 (physics).
†Professor Friedrich Hund (1896–1997), University of Göttingen, Germany.

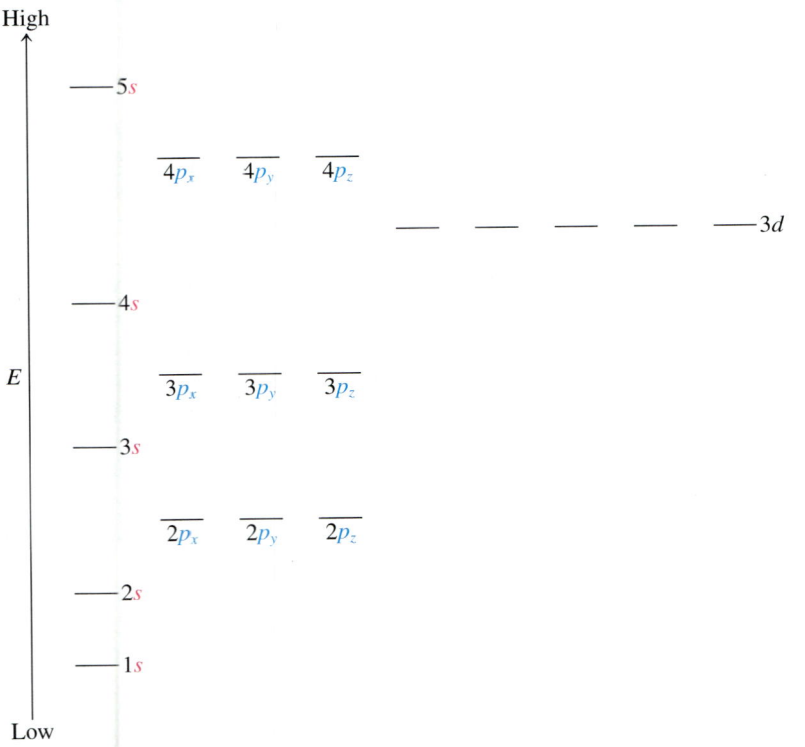

FIGURE 1-8
Approximate relative energies of atomic orbitals, corresponding roughly to the order in which they are filled in atoms. Orbitals of lowest energy are filled first; degenerate orbitals are filled according to Hund's rule.

That the description of electrons as waves is not simply a mathematical construct but is "visibly real" was demonstrated by researchers at IBM in 1993. Using a device called a scanning tunneling microscope, which allows pictures to be taken at the atomic level, they generated this computer-enhanced view of a circle of iron atoms deposited on a copper surface. The image, which they called a "quantum corral," reveals the electrons moving in waves over the surface, the maxima defining the "corral" hovering over the individual iron atoms. [*Photograph courtesy of Dr. Donald Eigler, IBM, San Jose, California.*]

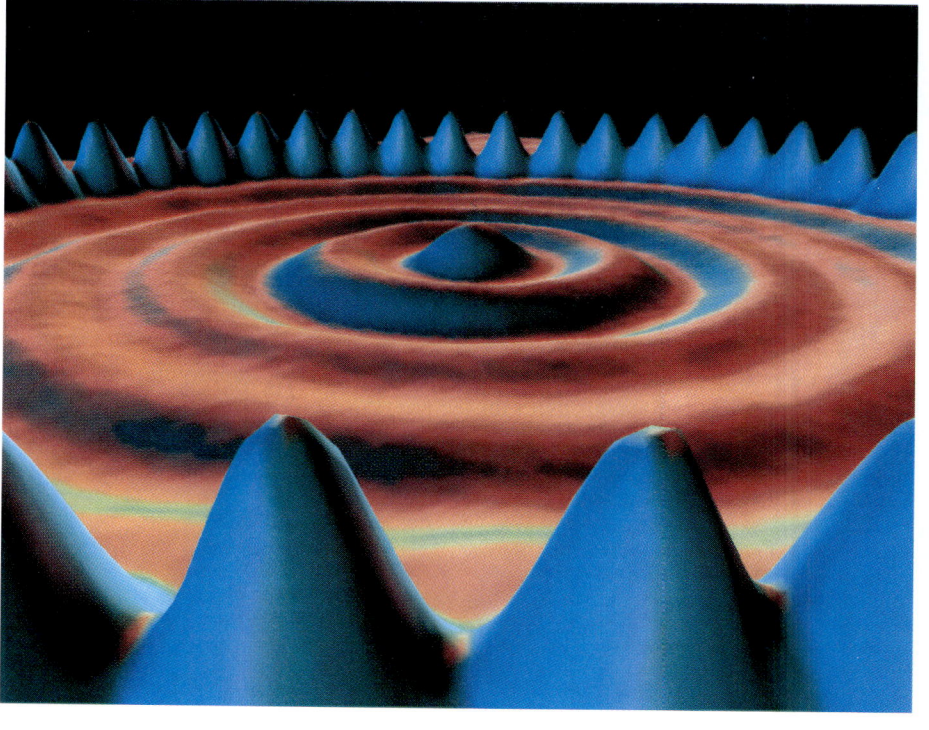

the three degenerate $2p$ orbitals. This pattern continues with carbon and nitrogen, and then the addition of electrons of opposite spin for oxygen, fluorine, and neon fills all p levels. The electronic configurations of four of the elements are depicted in Figure 1-9. Atoms with completely filled sets of atomic orbitals are said to have a **closed-shell configuration.** For example, helium, neon, and argon have this attribute (Figure 1-10). Carbon, in contrast, has an **open-shell configuration.**

The process of adding electrons one by one to the orbital sequence shown in Figure 1-8 is called the **Aufbau principle** (*Aufbau,* German, build up). It is easy to see that the Aufbau principle affords a rationale for the stability of the electron octet and duet. These numbers are required for closed-shell configurations. For helium, the closed-shell configuration is a $1s$ orbital filled with two electrons of opposite spin. In neon, the $2s$ and $2p$ orbitals are occupied by an additional eight electrons; in argon, the $3s$ and $3p$ levels accommodate eight more (Figure 1-10). The availability of $3d$ orbitals for the third-row elements provides an explanation for the phenomenon of valence shell expansion (Section 1-4) and the loosening up of the strict application of the octet rule beyond neon.

EXERCISE 1-11

Using Figure 1-8, draw the electronic configurations of sulfur and phosphorus.

To summarize, the motion of electrons around the nucleus is described by wave equations. Their solutions, atomic orbitals, can be symbolically represented as regions in space, with each point given a positive, negative, or zero (at the node) numerical value, the square of which represents the probability of finding the electron there. The Aufbau principle allows us to assign electronic configurations to all atoms.

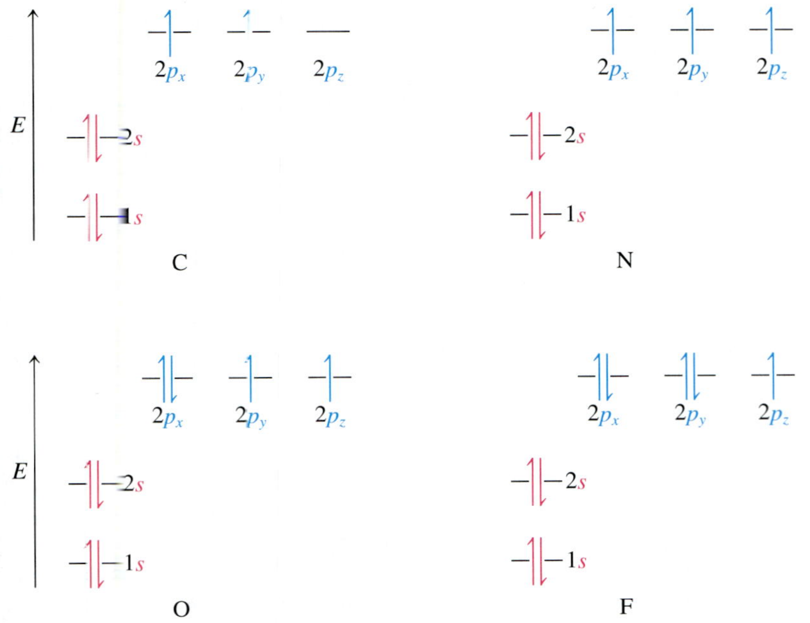

FIGURE 1-9
The most stable electronic configurations of carbon, $(1s)^2(2s)^2(2p)^2$; nitrogen, $(1s)^2(2s)^2(2p)^3$; oxygen, $(1s)^2(2s)^2(2p)^4$; and fluorine $(1s)^2(2s)^2(2p)^5$. Notice that the unpaired electron spins in the p orbitals are in accord with Hund's rule, and the paired electron spins in the filled 1s and 2s orbitals are in accord with the Pauli principle and Hund's rule. The order of filling the p orbitals has been arbitrarily chosen as p_x, p_y, and then p_z. Any other order would have been equally good.

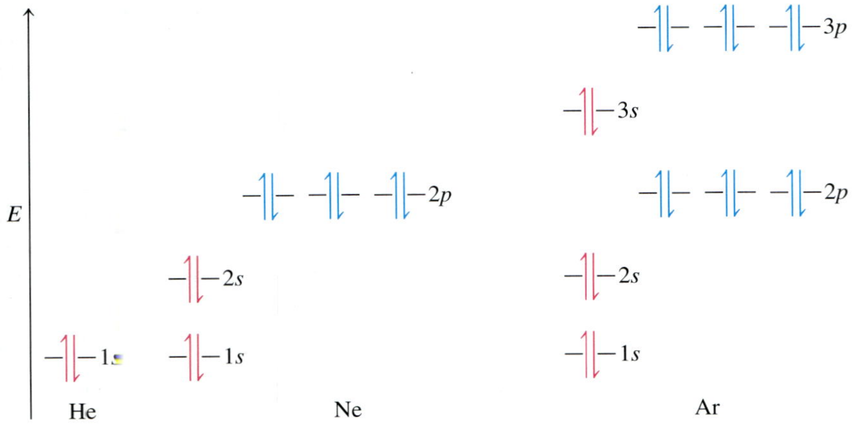

FIGURE 1-10
Closed-shell configurations of the noble gases helium, neon, and argon.

1-7 Molecular Orbitals and Covalent Bonding

We shall now see how covalent bonds are constructed from the overlap of atomic orbitals.

The bond in the hydrogen molecule is formed by the overlap of 1s atomic orbitals

Let us begin by looking at the simplest case: the bond between the two hydrogen atoms in H_2. In a Lewis structure of the hydrogen molecule, we would write the bond as an electron pair shared by both atoms to give each a helium configuration. How do we construct H_2 by using atomic orbitals? An answer to this question was developed by Pauling*: *Bonds are made by the in-phase overlap of atomic orbitals.* What is meant by that? Recall that atomic orbitals are solutions of wave equations. Like waves, they may interact in a reinforcing way (Figure 1-4B) if the overlap is between areas of the wave function of the same sign, or *in phase*. They may also interact in a destructive way if the overlap is between areas of opposite sign, or *out of phase* (Figure 1-4C).

The in-phase overlap of the two 1s orbitals results in a new orbital of lower energy called a **bonding molecular orbital** (Figure 1-11). In the bonding combination, the wave function in the space between the nuclei is strongly reinforced. Thus, the probability of finding the electrons occupying this molecular orbital in that region is very high: a condition for bonding between the two atoms. This picture is strongly reminiscent of that shown in Figure 1-2. The use of two wave functions with *positive* signs for representing the in-phase combination of the two 1s orbitals in Figure 1-11 is arbitrary. Overlap between two *negative* orbitals would give identical results. In other words, it is overlap between *like* lobes that makes a bond, regardless of the sign of the wave function.

On the other hand, out-of-phase overlap between the same two atomic orbitals results in a destabilizing interaction and formation of an **antibonding molecular orbital.** In the antibonding molecular orbital, the amplitude of the wave function is canceled in the space between the two atoms, thereby giving rise to a node.

Thus, the net result of the interaction of the two 1s atomic orbitals of hydrogen is the generation of two molecular orbitals. One is bonding and lower in energy; the

*Professor Linus Pauling (1901–1994), Stanford University, Nobel Prizes 1954 (chemistry) and 1963 (peace).

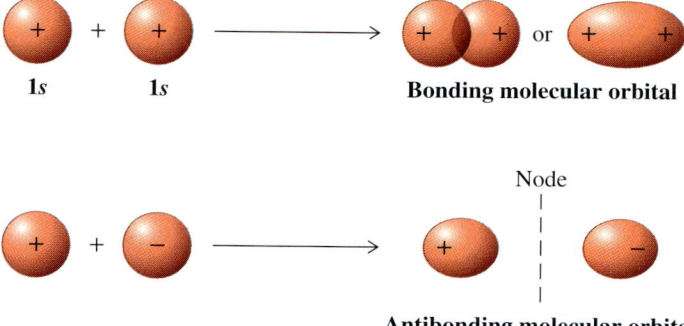

FIGURE 1-11

In-phase (bonding) and out-of-phase (antibonding) combinations of 1s atomic orbitals. The + and − signs denote the *sign* of the wave function, not charges. Electrons in bonding molecular orbitals have a high probability of occupying the space *between* the atomic nuclei, as required for good bonding (compare Figure 1-2). The antibonding molecular orbital has a nodal plane, where the probability of finding electrons is zero. Electrons in antibonding molecular orbitals are most likely to be found *outside* the space between the nuclei and therefore do not contribute to bonding.

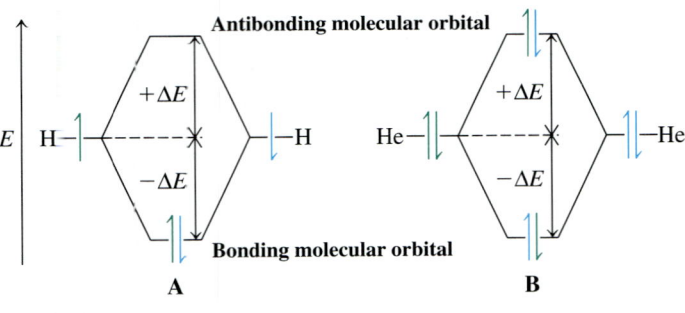

FIGURE 1-12

Schematic representation of the interaction of two (A) singly (as in H_2) and (B) doubly (as in He_2) occupied atomic orbitals to give two molecular orbitals (MO). (Not drawn to scale.) Formation of an H–H bond is favorable because it stabilizes two electrons. Formation of an He–He bond stabilizes two electrons (in the bonding MO) but destabilizes two others (in the antibonding MO). Bonding between He and He thus results in no net stabilization. Therefore, helium is monatomic.

other is antibonding and higher in energy. Because the total number of electrons available to the system is only two, they are placed in the lower energy molecular orbital: the two-electron bond. The result is a decrease in total energy, thereby making H_2 more stable than two free hydrogen atoms. This difference in energy levels corresponds to the strength of the H–H bond. The interaction can be depicted schematically in an energy diagram (Figure 1-12A).

It is now readily understandable why hydrogen exists as H_2, whereas helium is monatomic. The overlap of two filled atomic orbitals, as in helium, leads to bonding and antibonding orbitals, *both of which are filled* (Figure 1-12B). Therefore, making a He–He bond does not decrease the total energy.

The overlap of atomic orbitals gives rise to sigma and pi bonds

The formation of molecular orbitals by an overlap of atomic orbitals applies not only to the $1s$ orbitals of hydrogen, but also to other atomic orbitals. The amount of energy by which the bonding level drops and the antibonding level is raised is called the **energy splitting.** It indicates the strength of the bond being made and depends on a variety of factors. For example, overlap is best between orbitals of similar size and energy. Therefore, two $1s$ orbitals will interact with each other more effectively than a $1s$ and a $3s$.

Geometric factors also affect the degree of overlap. This consideration is important for orbitals with directionality in space, such as p orbitals. Such orbitals give rise to two types of bonds: one in which the atomic orbitals are aligned along the internuclear axis (parts A, B, C, and D in Figure 1-13) and the other in which they are perpendicular to it (part E). The first type is called a **sigma (σ) bond,** the second a **pi (π) bond.** All carbon–carbon single bonds are of the σ type; however, we shall find that double and triple bonds also have π components (Section 1-9).

EXERCISE 1-12

Construct a molecular-orbital and energy-splitting diagram of the bonding in He_2^+. Is it favorable?

FIGURE 1-13

Bonding between atomic orbitals. (A) 1s and 1s (e.g., H_2), (B) 1s and 2p (e.g., HF), (C) 2p and 2p (e.g., F_2), (D) 2p and 3p (e.g., FCl) aligned along internuclear axes, σ bonds; (E) 2p and 2p perpendicular to internuclear axis, a π bond. Note the arbitrary use of + and − signs to indicate in-phase interactions of the wave functions.

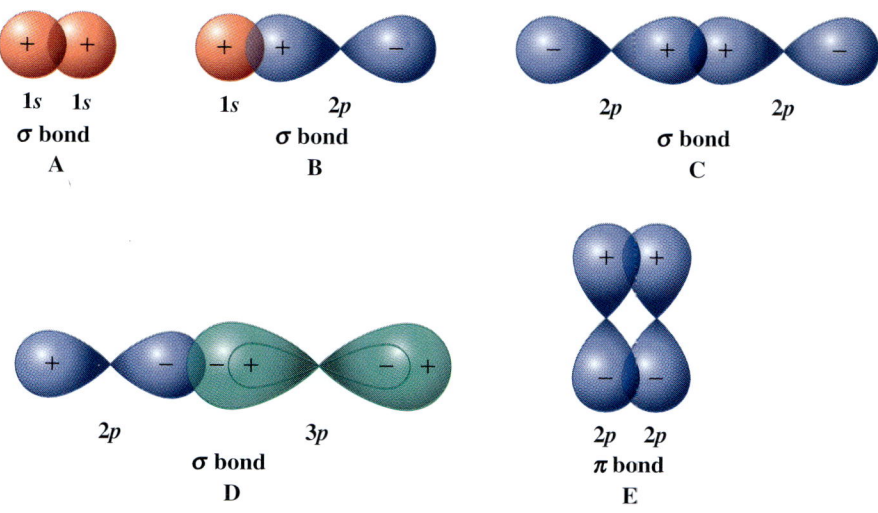

We have come a long way in our description of bonding. First, we thought of bonds in terms of Coulomb forces, then in terms of covalency and shared electron pairs, and now we have a quantum mechanical picture. Bonds are a result of the overlap of atomic orbitals. The two bonding electrons are placed in the bonding molecular orbital. Because it is stabilized relative to the two initial atomic orbitals, energy is given off during bond formation. This decrease in energy represents the bond strength.

1-8 Hybrid Orbitals: Bonding in Complex Molecules

Let us now construct bonding schemes for more complex molecules by using quantum mechanics. How can we use atomic orbitals to build linear (as in BeH_2), trigonal (as in BH_3), and tetrahedral molecules (as in CH_4)?

Mixing orbitals in a single atom gives hybrid orbitals

Consider the molecule beryllium hydride, BeH_2. Beryllium has two electrons in the 1s orbital and two electrons in the 2s orbital. Without unpaired electrons, this arrangement does not appear to allow for bonding.

However, it takes a relatively small amount of energy to promote one electron from the 2s orbital to one of the 2p levels (Figure 1-14), energy to be readily regained by bond formation. Thus, in the $1s^2 2s^1 2p^1$ configuration, there are now two singly filled atomic orbitals available for bonding overlap. One could propose bond formation by overlap of the Be 2s orbital with the 1s orbital of one H, on the one hand, and the Be 2p orbital with the second H, on the other (Figure 1-15). This scheme predicts two different bonds of unequal length, probably at an angle. However, the theory of electron repulsion predicts that compounds such as BeH_2 should have *linear* structures (Section 1-3). Experiments on related compounds confirm this prediction and also show that the bonds to beryllium are of *equal* length.*

*These predictions cannot be tested for BeH_2 itself, which exists as a complex network of Be and H atoms. However, both BeF_2 and $Be(CH_3)_2$ exist as individual molecules in the gas phase and possess the predicted structures.

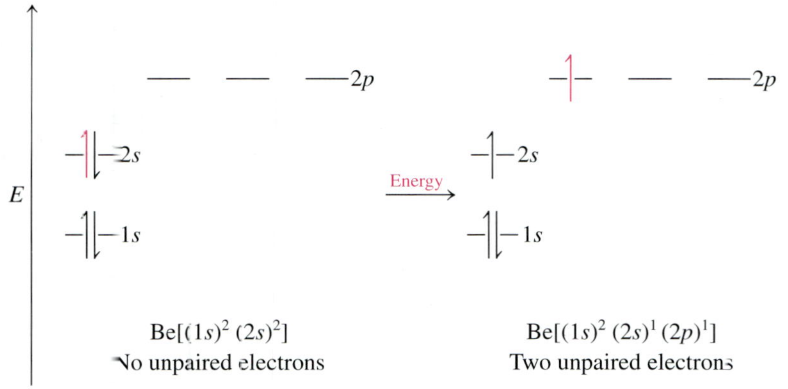

FIGURE 1-14
Promotion of an electron in beryllium to allow the use of both valence electrons in bonding.

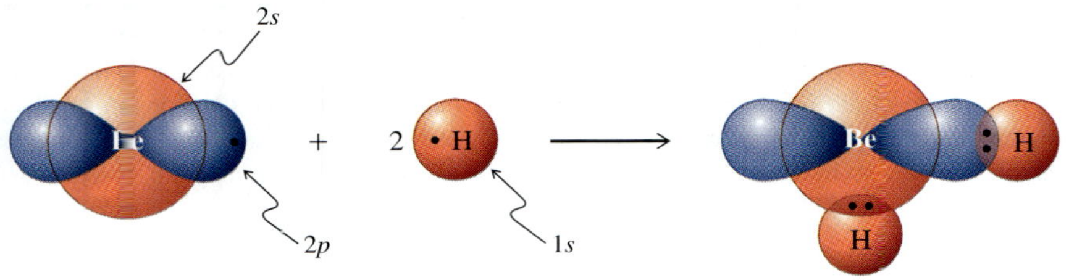

FIGURE 1-15
Possible but incorrect bonding in BeH_2 by separate use of a 2s and a 2p orbital on beryllium. The node in the former is not shown. Moreover, the other two empty p orbitals and the lower energy filled 1s orbital are omitted for clarity. The dots indicate the valence electrons.

sp Hybrids produce linear structures

How can we explain this geometry in orbital terms? To answer this question, we use a quantum mechanical approach called **orbital hybridization.** Like the mixing of atomic orbitals on different atoms to form molecular orbitals, the mixing of atomic orbitals on the same atom forms new **hybrid orbitals.**

When we mix the 2s and one of the 2p wave functions on beryllium, we obtain two new hybrids, called *sp* orbitals, made up of 50% s and 50% p character. This treatment rearranges the orbital lobes in space, as shown in Figure 1-16 (see page 34). The major parts of the orbitals, also called front lobes, point away from each other at an angle of 180°. There are two additional minor back lobes (one for each *sp* hybrid) with opposite sign. The remaining two p orbitals are left unchanged.

Overlap of the *sp* front lobes with two hydrogen 1s orbitals yields the bonds in BeH_2. The 180° angle that results from this hybridization scheme minimizes electron repulsion. The oversized front lobes of the hybrid orbitals also overlap better than do lobes of unhybridized orbitals; the result is improved bonding.

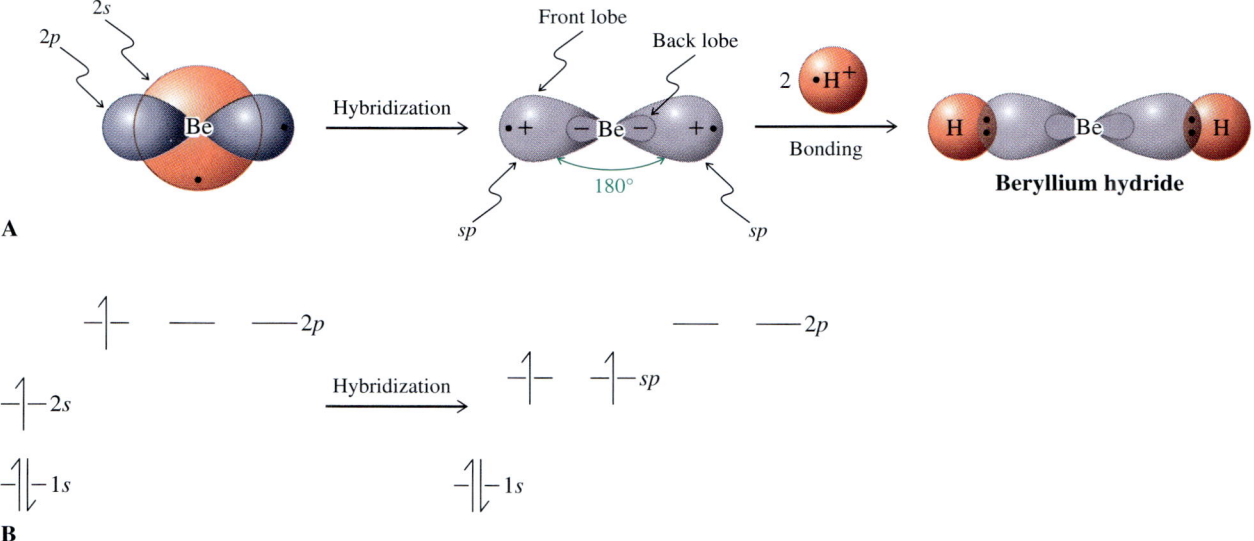

A

B

FIGURE 1-16

Hybridization in beryllium to create two *sp* hybrids. (A) The resulting bonding gives BeH_2 a linear structure. Again, both remaining *p* orbitals and the 1*s* orbital have been omitted for clarity. The sign of the wave function for the large *sp* lobes is opposite that for the small lobes. (B) The energy changes occurring on hybridization. The 2*s* orbital and one 2*p* orbital combine into two *sp* hybrids of intermediate energy. The 1*s* and remaining 2*p* energies remain the same.

Note that hybridization does not change the overall number of orbitals available for bonding. Hybridization of the four orbitals in beryllium gives a new set of four: two *sp* hybrids and two essentially unchanged 2*p* orbitals. We shall see shortly that carbon uses *sp* hybrids when it forms triple bonds.

*sp*² Hybrids create trigonal structures

Now let us consider the group of elements in the periodic table with three valence electrons. What bonding scheme can be derived for borane, BH_3? Promotion of a 2*s* electron in boron to one of the 2*p* levels gives the three singly filled atomic orbitals (one 2*s*, two 2*p*) needed for forming three bonds. Mixing these atomic orbitals creates *three* new hybrid orbitals, which are designated *sp*² to indicate the component atomic orbitals (Figure 1-17). The third *p* orbital is left unchanged, so the total number of orbitals stays the same—namely, four.

The front lobes of the three *sp*² orbitals of boron overlap the respective 1*s* orbitals of the hydrogen atoms to give trigonal planar BH_3. Again, hybridization minimizes electron repulsion and improves overlap, conditions giving stronger bonds. The remaining unchanged *p* orbital is perpendicular to the plane incorporating the *sp*² hybrids. It is empty and does not enter significantly into bonding.

The molecule BH_3 is **isoelectronic** with the methyl cation, CH_3^+; that is, they have the same number of electrons. Bonding in CH_3^+ requires three *sp*² hybrid orbitals, and we shall see shortly that carbon uses *sp*² hybrids in double-bond formation.

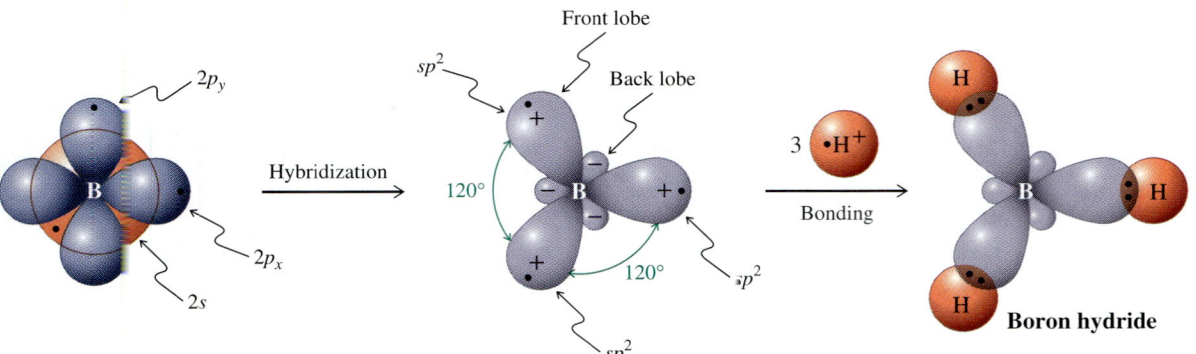

FIGURE 1-17
Hybridization in boron to create three sp^2 hybrids. The resulting bonding gives BH_3 a trigonal planar structure. There are three front lobes of one sign and three back lobes of opposite sign. The remaining p orbital (p_z) is perpendicular to the molecular plane (the plane of the page; one p_z lobe is above, the other below that plane) and has been omitted. In analogy to Figure 1-16B, the energy diagram for the hybridized boron features three singly occupied, equal-energy sp^2 levels and one remaining empty $2p$ level, in addition to the filled $1s$ orbital.

sp^3 Hybridization explains the shape of tetrahedral carbon compounds

Consider the element whose bonding is of most interest to us: carbon. Its electronic configuration is $(1s)^2(2s)^2(2p)^2$, with two unpaired electrons residing in two $2p$ orbitals. Promotion of one electron from $2s$ to $2p$ results in four singly filled orbitals for bonding. We have learned that the arrangement of the four C–H bonds of methane in space that would minimize electron repulsion is tetrahedral (Section 1-3). To be able to achieve this geometry, the $2s$ orbital on carbon is hybridized with *all three* $2p$ orbitals to make *four* equivalent sp^3 orbitals with tetrahedral symmetry, each occupied by one electron. Overlap with four hydrogen $1s$ orbitals furnishes methane with four equal C–H bonds. The HCH bond angles are typical of a tetrahedron: $109.5°$ (Figure 1-18).

Any combination of atomic and hybrid orbitals may overlap to form bonds. For example, the four sp^3 orbitals of carbon combine with four chlorine $2p$ orbitals to result in tetrachloromethane, CCl_4. Carbon–carbon bonds are generated by overlap of

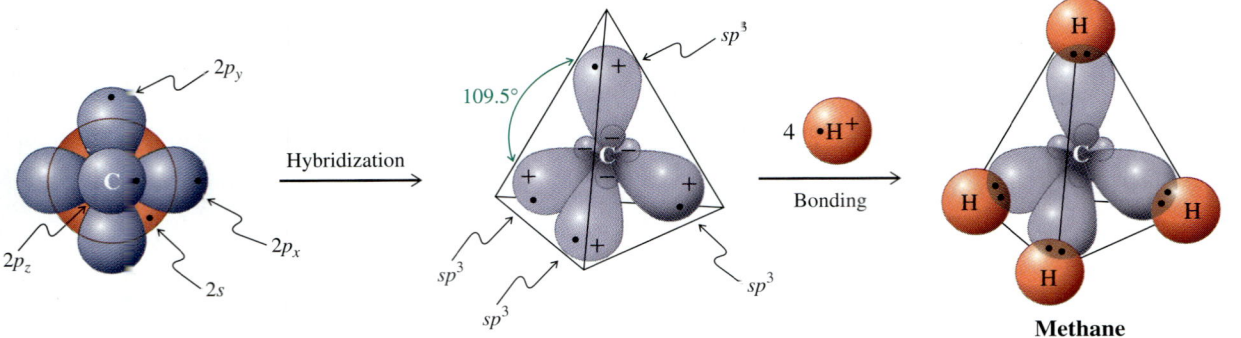

FIGURE 1-18
Hybridization in carbon to create four sp^3 hybrids. The resulting bonding gives CH_4 and other carbon compounds tetrahedral structures. The sp^3 hybrids contain small back lobes of sign opposite that of the front lobes. In analogy to Figure 1-16B, the energy diagram of sp^3-hybridized carbon contains four singly occupied, equal-energy sp^3 levels, in addition to the filled $1s$ orbital.

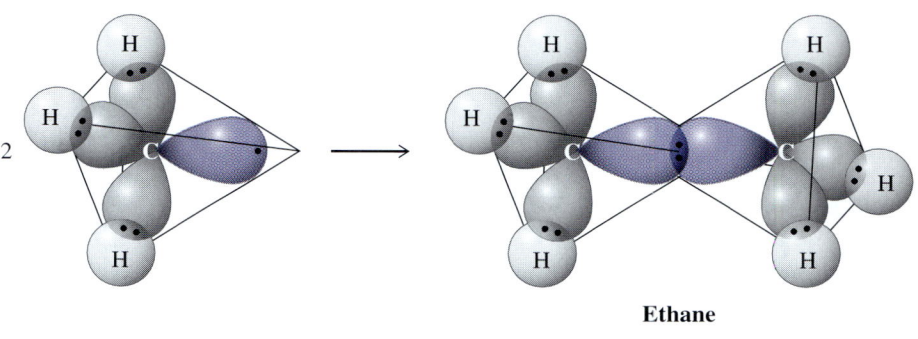

Ethane

FIGURE 1-19

Overlap of two sp^3 orbitals to form the carbon–carbon bond in ethane.

hybrid orbitals. In ethane, CH_3–CH_3 (Figure 1-19), this bond consists of two sp^3 hybrids, one from each of two CH_3 units. Any hydrogen atom in methane and ethane may be replaced by CH_3 or other groups to give new combinations.

In all of these molecules, and countless more, *carbon is approximately tetrahedral*. It is this ability of carbon to form chains of atoms bearing a variety of additional substituents that gives rise to the extraordinary diversity of organic chemistry.

Hybrid orbitals may contain lone electron pairs: ammonia and water

What sort of orbitals describe the bonding in ammonia and water (see Exercise 1-5)? Let us begin with ammonia. The electronic configuration of nitrogen, $(1s)^2(2s)^2(2p)^3$, explains why nitrogen is trivalent, three covalent bonds being needed for octet formation. We could use p orbitals for overlap, leaving the nonbonding electron pair in the $2s$ level. However, this arrangement does not minimize electron repulsion. The best solution is again sp^3 hybridization. Three of the sp^3 orbitals are used to bond to the hydrogen atoms, and the fourth contains the lone electron pair. The HNH bond angles (107.3°) in ammonia are almost tetrahedral (Figure 1-20).

Similarly, the bonding in water is best described by sp^3 hybridization on oxygen. The HOH bond angle is 104.5°, again close to tetrahedral.

The effect of the lone electron pairs explains why the bond angles in NH_3 and H_2O are reduced below the tetrahedral value of 109.5°. Because they are not shared, the lone pairs are relatively close to the nitrogen or oxygen. As a result, they exert increased repulsion on the electrons in the bonds to hydrogen, thereby leading to the observed bond-angle compression.

FIGURE 1-20

Bonding and electron repulsion in ammonia and water. The arcs indicate increased electron repulsion by the lone pairs located close to the central nucleus.

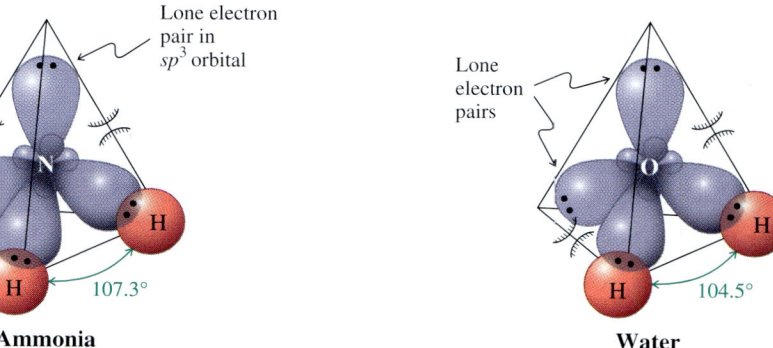

Lone electron pair in sp^3 orbital

107.3°

Ammonia

Lone electron pairs

104.5°

Water

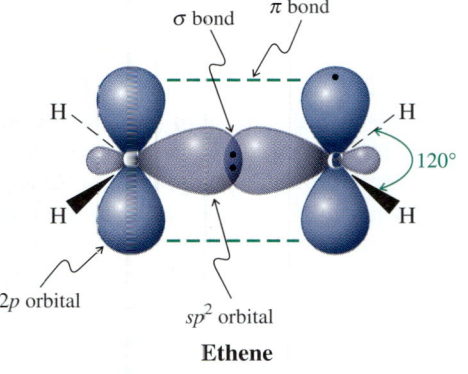

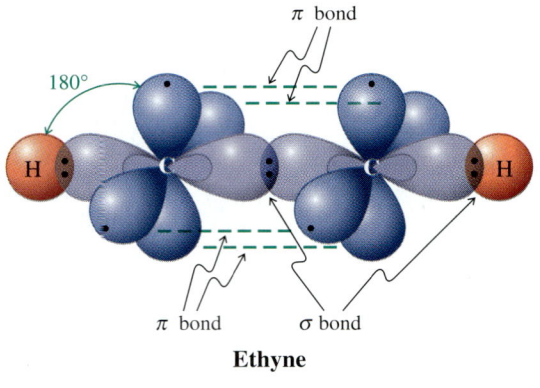

Ethene

Ethyne

FIGURE 1-21

The double bond in ethene (ethylene) and the triple bond in ethyne (acetylene).

Pi bonds are present in ethene (ethylene) and ethyne (acetylene)

The double bond in alkenes, such as ethene (ethylene), and the triple bond in alkynes, such as ethyne (acetylene), are the result of the ability of the atomic orbitals of carbon to adopt sp^2 and sp hybridization, respectively. Thus, the σ bonds in ethene are derived entirely from carbon-based sp^2 hybrid orbitals: Csp^2–Csp^2 for the C–C bond, and Csp^2–H1s for holding the four hydrogens (Figure 1-21). In contrast with BH_3, with an empty p orbital on boron, the leftover unhybridized p orbitals on the ethene carbons are occupied by one electron each, overlapping to form a π bond (recall Figure 1-13E). In ethyne, the σ frame is made up of bonds consisting of Csp hybrid orbitals. The arrangement leaves *two* singly occupied p orbitals on each carbon and allows the formation of two π bonds (Figure 1-21).

EXERCISE -13

Draw a scheme for the hybridization and bonding in methyl cation, CH_3^+, and methyl anion, CH_3^-.

In summary, to minimize electron repulsion and maximize bonding in triatomic and larger molecules, we apply the concept of atomic-orbital hybridization to construct orbitals of appropriate shape. Combinations of s and p atomic orbitals create hybrids. Thus, a $2s$ and a $2p$ orbital mix to furnish two linear sp hybrids, the remaining two p orbitals being unchanged. Combination of the $2s$ with two p orbitals gives three sp^2 hybrids used in trigonal molecules. Finally, mixing the $2s$ with all three p levels results in the four sp^3 hybrids that produce the geometry around tetrahedral carbon.

1-9 Structures and Formulas of Organic Molecules

A good understanding of the nature of bonding allows us to learn how chemists determine the identity of organic molecules and depict their structures. Do not underestimate the importance of the latter task. Sloppiness in drawing molecules had been the source of many errors in the literature and, perhaps of more immediate concern, in organic chemistry examinations.

Ethanol and Methoxymethane: Two Isomers

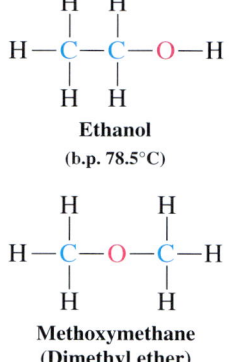

Ethanol
(b.p. 78.5°C)

Methoxymethane
(Dimethyl ether)
(b.p. −23°C)

To establish the identity of a molecule, we determine its structure

Organic chemists have many diverse techniques at their disposal with which to determine molecular structure. **Elemental analysis** reveals the **empirical formula,** which summarizes the kinds and ratios of the elements present. However, other procedures are usually needed to determine the molecular formula and to distinguish between structural alternatives. For example, the molecular formula C_2H_6O corresponds to *two* known substances: ethanol and methoxymethane (dimethyl ether). But we can tell them apart on the basis of their physical properties—for example, their melting points, boiling points (b.p.), refractive indices, specific gravities, and so forth. Thus ethanol is a liquid (b.p. 78.5°C) commonly used as a laboratory and industrial solvent and present in alcoholic beverages. In contrast, methoxymethane is a gas (b.p. −23°C) used as a refrigerant in place of Freon. Their other physical and chemical properties differ as well. Molecules such as these, which have the same molecular formula but differ in the sequence (**connectivity**) in which the atoms are held together, are called **constitutional** or **structural isomers.**

EXERCISE 1-14

Construct as many isomers with the molecular formula C_4H_{10} as you can.

Two naturally occurring substances illustrate the biological consequences of such structural differences. Prostacyclin I_2 prevents blood inside the circulatory system from clotting. Thromboxane A_2, which is released when bleeding occurs, *induces* platelet aggregation, causing clots to form over wounds. Incredibly, these compounds are constitutional isomers (both have the molecular formula $C_{20}H_{32}O_5$) with only relatively minor connectivity differences. Indeed, they are so closely related that they are synthesized in the body from a common starting material (see Section 19-14 for details).

When a compound is isolated in nature or from a reaction, a chemist may attempt to identify it by matching its properties with those of known materials. Suppose, however, that the chemical under investigation is new. In this case, structural elucidation requires the use of other methods, most of which are various forms of spectroscopy. These methods will be dealt with and applied often in later chapters.

The most complete methods for structure determination are X-ray diffraction of single crystals and electron diffraction or microwave spectroscopy of gases. These techniques reveal the exact position of every atom, as if viewed under very powerful magnification. The structural details that emerge in this way for the two isomers ethanol and methoxymethane are depicted in the form of ball-and-stick models in Figure 1-22A and B. Note the tetrahedral bonding around the carbon atoms and the bent arrangement of the bonds to oxygen, which is hybridized as in water. A more

FIGURE 1-22

Three-dimensional representations of (A) ethanol and (B) methoxymethane, depicted by ball-and-stick molecular models. Bond lengths are given in angstrom units, bond angles in degrees. (C) Space-filling rendition of methoxymethane, taking into account the effective size of the electron "clouds" around the component nuclei.

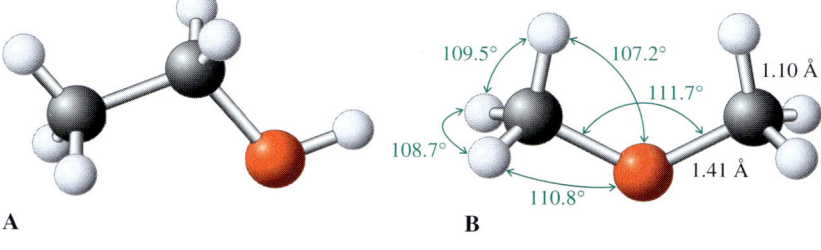

A   B

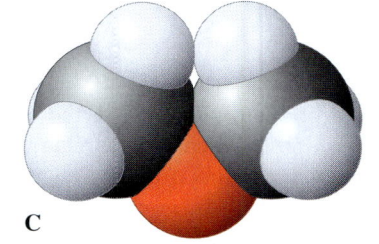

C

accurate picture of the actual size of methoxymethane and its component units is given in Figure 1-22C, a space-filling model.

The perception of organic molecules in three dimensions is essential for understanding their structures and, frequently, their reactivities. You may find it difficult to visualize the spatial arrangements of the atoms in even very simple systems. A good aid is a molecular model kit. You should acquire one and practice the assembly of organic structures. To encourage you in this practice and to indicate particularly good examples where building a molecular model can help you, the icon displayed in the margin will appear at the appropriate places in the text.

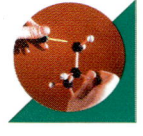

EXERCISE 1-15

Repeat Exercise 1-14, using your molecular model kit to construct as many isomers with the molecular formula C_4H_{10} as you can. Draw each isomer.

Several types of drawings are used to represent molecular structures

The representation of molecular structures is not new to us. It was first addressed in Section 1-4, which outlined rules for drawing Lewis structures. We learned that bonding and nonbonding electrons are depicted as dots. A simplification is the straight-line notation (Kekulé structure), with lone pairs (if present) added again as dots. To simplify even further, chemists use **condensed formulas** in which most single bonds and lone pairs have been omitted. The main carbon chain is written horizontally, the attached hydrogens usually to the right of the associated carbon atom. Other groups (the **substituents** on the main stem) are added through connecting vertical lines.

The most economical notation of all is the **bond-line formula.** It portrays the carbon frame by zigzag straight lines, omitting all hydrogen atoms. Each terminus represents a methyl group and each apex a carbon atom.

Kekulé	**Condensed**	**Bond-Line Formulas**

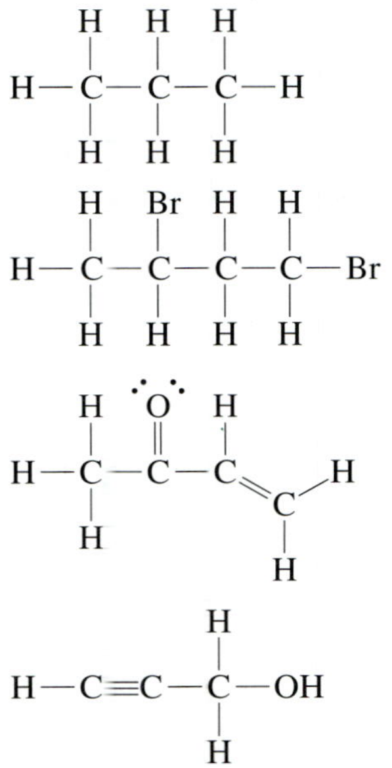

$CH_3CH_2CH_3$

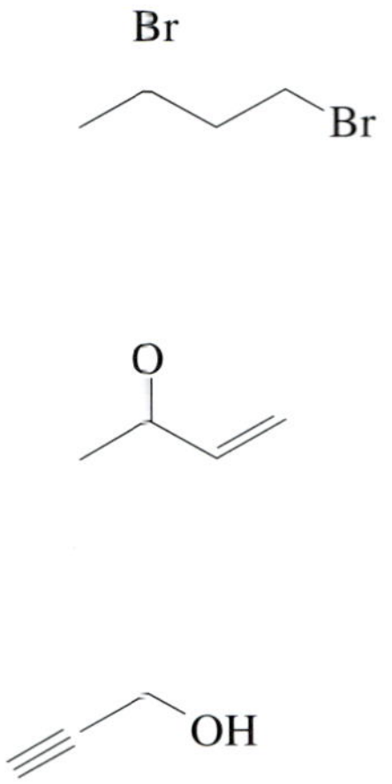

$CH_3CHCH_2CH_2Br$ with Br above

$CH_3CCH=CH_2$ with O above

$HC\equiv CCH_2OH$

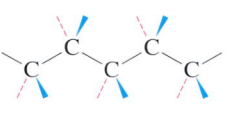 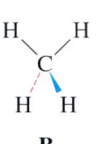 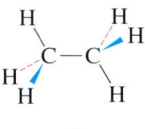 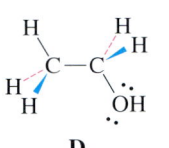

A   B   C   D   E

FIGURE 1-23 _____

Dashed (red) and wedged (blue) line notation for (A) a carbon chain; (B) methane; (C) ethane; (D) ethanol; and (E) methoxymethane. Atoms attached by ordinary straight lines lie in the plane of the page. Groups at the ends of dashed lines lie below that plane; groups at the ends of wedges lie above it.

EXERCISE 1-16

Draw condensed and bond-line formulas for each C_4H_{10} isomer.

Figure 1-22 calls attention to a problem: How can we draw the three-dimensional structures of organic molecules accurately, efficiently, and in accord with generally accepted conventions? For tetrahedral carbon, this problem is solved by the **dashed-wedged line notation.** It uses a zigzag convention to depict the main carbon chain, now defined to lie *in the plane* of the page. Each apex (carbon atom) is then connected to two additional lines, one dashed and one wedged, both pointing away from the chain. These represent the remaining two bonds to carbon; the dashed line corresponds to the bond that lies *below the plane* of the page and the wedged line to that lying *above that plane* (Figure 1-23). Substituents are placed at the appropriate termini. This convention is applied to molecules of all sizes, even methane (see Figure 1-23B–E).

EXERCISE 1-17

Draw dashed-wedged line formulas for each C_4H_{10} isomer.

In summary, determination of organic structures relies on the use of several experimental techniques, including elemental analysis and various forms of spectroscopy. Molecular models are useful aids for the visualization of the spatial arrangements of the atoms in structures. Condensed and bond-line notations are useful shorthand approaches to drawing two-dimensional representations of molecules, whereas dashed-wedged line formulas provide a means of depicting the atoms and bonds in three dimensions.

CHAPTER INTEGRATION PROBLEM

Propyne can be deprotonated twice with very strong base (i.e., the base removes two protons) to give a dianion.

$$H-C\equiv C-\underset{\underset{H}{|}}{\overset{\overset{H}{|}}{C}}-H \xrightarrow[\text{(}-2H^+\text{)}]{\text{strong base}} \left[CCCH_2\right]^{2-}$$

Propyne   **Propyne dianion**

Two resonance forms can be constructed in which all three carbons have Lewis octets.

a. Draw both structures and indicate which is the more important one.

SOLUTION

Let us analyze the problem one step at a time:

Step 1. What structural information is embedded in the picture given for propyne dianion? *Answer (Section 1-4, Rule 1):* The picture shows the connectivity of the atoms: a chain of three carbons, one of the terminal atoms bearing two hydrogens.

Step 2. How many valence electrons are available? *Answer (Section 1-4, Rule 2):*

$$
\begin{array}{llll}
2\text{ H} & = & 2 \times 1\text{ electron} & = & 2\text{ electrons} \\
3\text{ C} & = & 3 \times 4\text{ electrons} & = & 12\text{ electrons} \\
\text{Charge} = -2 & & & = & 2\text{ electrons} \\
\hline
\text{Total} & & & & 16\text{ electrons}
\end{array}
$$

Step 3. How do we get a Lewis octet structure for this ion? *Answer (Section 1-4, Rule 3):* Using the connectivity given in the structure for propyne dianion, we can immediately dispose of eight of the available electrons:

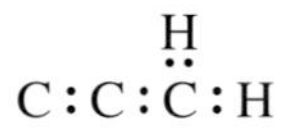

$$
\begin{array}{c}
\text{H} \\
\ddot{} \\
\text{C:C:C:H}
\end{array}
$$

Now, let us use the remaining eight electrons in the form of lone electron pairs to give as many carbons as possible octet surroundings. A good place to start is at the right, because that carbon requires only two electrons for this purpose, the center carbon needs two lone pairs, and, finally, the carbon at the left has to make do (for the time being) with one additional pair of electrons:

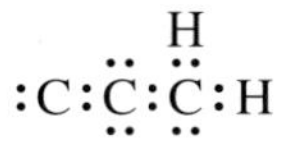

$$
\begin{array}{c}
\text{H} \\
\cdot\cdot \quad \cdot\cdot \\
\text{:C:}\ddot{\text{C}}\text{:C:H} \\
\cdot\cdot \quad \cdot\cdot
\end{array}
$$

This structure leaves the carbon at the left with only four electrons. Thus, we have to change the two lone pairs at the center into two shared pairs, furnishing the following dot structure:

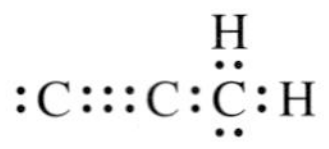

$$
\begin{array}{c}
\text{H} \\
\cdot\cdot \\
\text{:C:::C:}\ddot{\text{C}}\text{:H} \\
\cdot\cdot
\end{array}
$$

Step 4. Now every atom has its duet or octet satisfied, but we still have to concern ourselves with charges. What are the charges on each atom? *Answer (Section 1-4, Rule 4):* Starting again at the right, we can quickly ascertain that the hydrogens are charge neutral. Each is attached to carbon through a shared electron pair, giving it an effective electron count of one, the same as in a free, neutral hydrogen atom. On the other hand, the carbon atom bears three shared electron pairs and one lone pair, thus having an effective electron count of five, one more electron than the number associated with the neutral nucleus. Hence one of the negative charges is located on the carbon at the right. The central carbon nucleus is surrounded by four shared electron pairs and is therefore neutral. Finally, the carbon at the left is attached to its neigh-

bor by three shared pairs and has, in addition, two unbound electrons, giving it the other negative charge. The result is

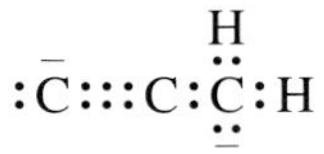

$$:\overset{-}{\underset{}{C}}:::C:\overset{..}{\underset{..}{C}}:H$$

with H above the third carbon

Step 5. We can now address the question of resonance forms. Is it possible to move pairs of electrons in such a way as to generate another Lewis octet form? *Answer (Section 1-5):* Yes, by shifting the lone pair on the carbon at the right into a sharing position and at the same time moving one of the three shared pairs to the left into an unshared location:

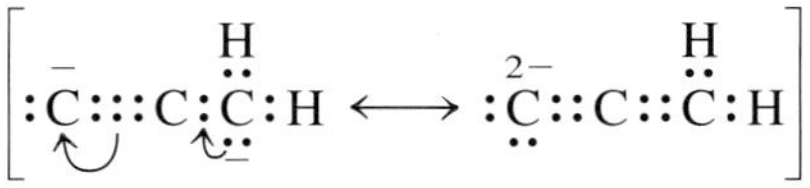

$$\left[:\overset{-}{C}:::C:\overset{H}{\underset{..}{C}}:H \longleftrightarrow :\overset{2-}{C}::C::\overset{H}{\underset{..}{C}}:H \right]$$

The consequence of this movement is the transfer of the negative charge from the carbon at the right to the carbon at the left, the latter therefore becoming doubly negative.

Step 6. Which one of the two resonance pictures is more important? *Answer (Section 1-5):* Electron repulsion makes the structure at the left a more important resonance contributor.

A final point: You could have derived the first resonance structure much more quickly by considering the information given in the reaction scheme leading to the dianion. Thus, the bond-line formula of propyne represents its Lewis structure, and the process of removing a proton each from the respective terminal carbons leaves these carbons with two lone electron pairs and the associated charges:

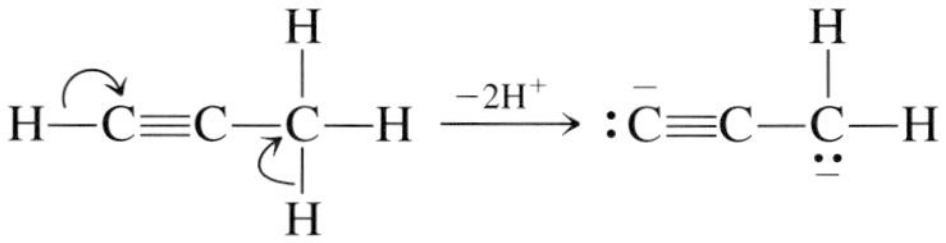

$$H-C\equiv C-\underset{\underset{H}{|}}{\overset{\overset{H}{|}}{C}}-H \xrightarrow{-2H^+} :\overset{-}{C}\equiv C-\overset{\overset{H}{|}}{\underset{..}{C}}-H$$

The important lesson to be learned from this final point is that, whenever you are confronted with a problem, you should take the time to complete an inventory (write it down) of all the information given explicitly or implicitly in stating the problem.

b. Provide an orbital drawing of propyne dianion, given the following hybridization: $[CspCspCsp^2H_2]^{2-}$.

SOLUTION

You can construct an orbital picture simply by attaching one half (the CH_2 group) of the rendition of ethene in Figure 1-21 to that of ethyne without its hydrogens.

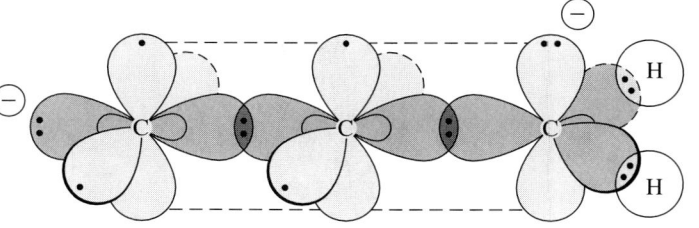

You can clearly see how the doubly filled *p* orbital of the CH_2 group enters into overlap with one of the π bonds of the alkyne unit, allowing the charge to delocalize as represented by the two resonance structures.

IMPORTANT CONCEPTS

1. Organic chemistry is the chemistry of **carbon** and its compounds.

2. **Coulomb's law** relates the attractive force between particles of opposite electrical charge to the distance between them.

3. **Ionic bonds** result from Coulombic attraction of oppositely charged ions. These ions are formed by the complete transfer of electrons from one atom to another, typically to achieve a noble-gas configuration

4. **Covalent bonds** result from electron sharing between two atoms. Electrons are shared to allow the atoms to attain noble-gas configurations.

5. **Bond length** is the average distance between two covalently bonded atoms. Bond formation releases energy bond breaking requires energy.

6. **Polar bonds** are formed between atoms of differing electronegativity (a measure of an atom's ability to attract electrons).

7. The **shape of molecules** is strongly influenced by electron repulsion.

8. **Lewis structures** describe bonding by the use of valence electron dots. They are drawn so as to give hydrogen an electron duet, the other atoms electron octets (octet rule). Charge separation should be minimized but may be enforced by the **octet rule.**

9. When two or more Lewis structures differing only in the positions of the electrons are needed to describe a molecule, they are called **resonance forms.** None correctly describes the molecule, its true representation being an average **(hybrid)** of all its Lewis structures. If the resonance forms of a molecule are unequal, those which best satisfy the rules for writing Lewis structures and the electronegativity requirements of the atoms are more important.

10. The **de Broglie relation** describes the wavelength of a moving particle in terms of its mass and velocity.

11. The motion of electrons around the nucleus can be described by **wave equations.** The solutions to these equations are **atomic orbitals,** which roughly delineate regions in space in which there is a high probability of finding electrons.

12. An *s* **orbital** is spherical; a *p* **orbital** looks like two touching teardrops or a "spherical figure eight." The mathematical sign of the orbital at any point can be positive, negative, or zero (node). With increasing energy, the number of nodes increases. Each orbital can be occupied by a maximum of two electrons of opposite spin **(Pauli exclusion principle, Hund's rule).**

13. The process of adding electrons one by one to the atomic orbitals, starting with those of lowest energy, is called the **Aufbau principle.**

14. A **molecular orbital** is formed when two atomic orbitals overlap to generate a bond. Atomic orbitals of the same sign overlap to give a **bonding molecular orbital** of lower energy. Atomic orbitals of opposite sign give rise to an **antibonding molecular orbital** of higher energy and containing a node. The number of molecular orbitals equals the number of atomic orbitals from which they derive.

15. Bonds made by overlap along the internuclear axis are called σ **bonds;** those made by overlap of *p* orbitals perpendicular to the internuclear axis are called π **bonds.**

16. The mixing of orbitals on the same atom results in new **hybrid orbitals** of different shape. One *s* and one *p* orbital mix to give two **linear *sp* hybrids,** used, for example, in the bonding of BeH_2. One *s* and two *p* orbitals result in three **trigonal *sp*^{2} hybrids,** used, for example, in BH_3. One *s* and three *p* orbitals furnish four **tetrahedral *sp*^{3} hybrids,** used, for example, in CH_4. The orbitals that are not hybridized stay unchanged. Hybrid orbitals may overlap with each other. Overlapping *sp*^{3} hybrid orbitals on different carbon atoms form the carbon–carbon bonds in ethane and other organic molecules. Hybrid orbitals may also be occupied by lone electron pairs, as in NH_3 and H_2O.

17. The composition (i.e., ratios of types of atoms) of organic molecules is revealed by **elemental analysis.** The **molecular formula** gives the number of atoms of each kind.

18. Molecules that have the same molecular formula but different connectivity order of their atoms are called **constitutional** or **structural isomers.** They have different properties.

19. **Condensed** and **bond-line formulas** are abbreviated representations of molecules. **Dashed-wedged line drawings** illustrate molecular structures in three dimensions.

PROBLEMS

18. Draw a Lewis structure for each of the following molecules and assign charges where appropriate. The order in which the atoms are connected is given in parentheses.

(a) ClF

(b) Br CN

(c) $SOCl_2$ (ClSCl) $\overset{O}{}$

(d) CH_3NH_2

(e) CH_3OCH_3

(f) N_2H_2 (HNNH)

(g) CH_2CO

(h) HN_3 (HNNN)

(i) N_2O (NNO)

19. Using electronegativity values from Table 1-2 (in Section 1-3), identify polar covalent bonds in several of the structures in Problem 18 and label the atoms δ^+ and δ^-, as appropriate.

20. Draw a Lewis structure for each of the following species. Again, assign charges where appropriate.

(a) H^-

(b) CH_3^-

(c) CH_3^+

(d) CH_3

(e) $CH_3NH_3^+$

(f) CH_3O^-

(g) CH_2

(h) HC_2^- (HCC)

(i) H_2O_2 (HOOH)

21. Each of the following structures contains at least one error, such as an incorrect number of electrons or bonds on one or more atoms or a violation of the octet rule. Identify the errors and suggest corrections for them.

(a) H_2OCH_3

(b) H_2COH

(c) $:CH_4$

(d) NH_3OH

(e) 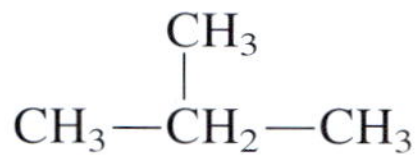 $CH_3-\overset{\overset{\displaystyle CH_3}{|}}{C}H_2-CH_3$

(f) $CH_3-CH_2=CH$

(g) 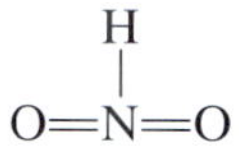 $O=N=O$ with H above N

(h) $H_2C\equiv CH_2$

(i) 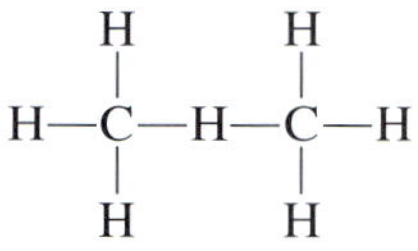 $H-\overset{\overset{\displaystyle H}{|}}{\underset{\underset{\displaystyle H}{|}}{C}}-H-\overset{\overset{\displaystyle H}{|}}{\underset{\underset{\displaystyle H}{|}}{C}}-H$

(j) $C=O=O$

22. (a) The structure of the bicarbonate (hydrogen carbonate) ion, HCO_3^-, is best described as a hybrid of several contributing resonance forms, two of which are shown here.

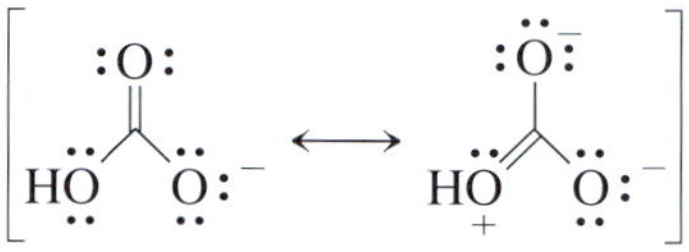

(i) Draw at least one additional resonance form. (ii) Using curved "electron pushing" arrows, show how these Lewis structures may be interconverted by movement of electron pairs. (iii) Determine which form or forms will be the major contributor(s) to the real structure of bicarbonate, explaining your answer on the basis of the criteria in Section 1-5.

(b) Draw two resonance forms for formaldehyde oxime, H_2CNOH. As in parts (ii) and (iii) of (a), use curved arrows to interconvert the resonance forms and determine which form is the major contributor.

(c) Repeat the exercises in (b) for formaldehyde oximate anion, $[H_2CNO]^-$.

23. Several of the compounds in Problem 18 can have resonance forms. Identify these molecules and write an additional resonance Lewis structure for each. In each case, indicate the major contributor to the resonance hybrid.

24. Draw two or three resonance forms for each of the following species. Indicate the major contributor or contributors to the hybrid in each case.

(a) OCN^- (b) CH_2CHNH^- (c) $HCONH_2$ ($HCNH_2$)

$$\overset{O}{\underset{}{}}$$

(d) O_3 (OOO) (e) $CH_2CHCH_2^-$ (f) SO_2 (OSO)

(g) $HOCHNH_2^+$ (h) CH_3CNO

25. Use a molecular-orbital analysis to predict which species in each of the following pair has the stronger bonding between atoms. (**Hint:** Refer to Figure 1-12).
(a) H_2 or H_2^+ (b) He_2 or He_2^+ (c) O_2 or O_2^+ (d) N_2 or N_2^+

26. Describe the hybridization of each carbon atom in each of the following structures. Base your answer on the geometry about the carbon atom.
(a) CH_3Cl (b) CH_3OH (c) $CH_3CH_2CH_3$ (d) $CH_2{=}CH_2$ (trigonal carbons)

(e) $HC{\equiv}CH$ (linear structure) (f) 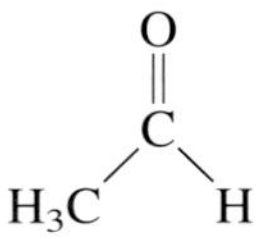 $\underset{H_3CH}{\overset{\overset{O}{\parallel}}{C}}$ (g) 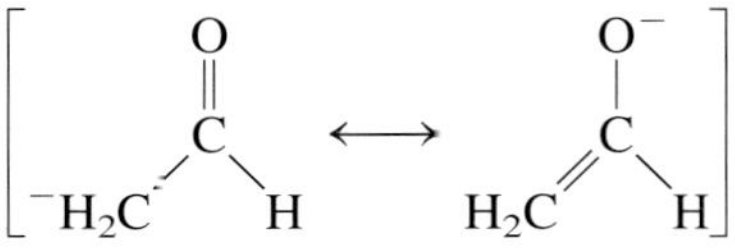 $\left[\underset{^-H_2CH}{\overset{\overset{O}{\parallel}}{C}} \longleftrightarrow \underset{H_2CH}{\overset{\overset{O^-}{|}}{C}}\right]$

27. Depict the following condensed formulas in Kekulé (straight-line) notation.

(a) CF_3CN (b) 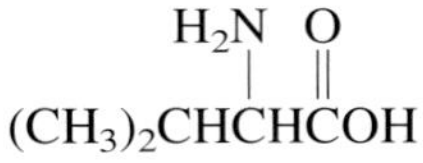 $(CH_3)_2CHCHCOH$ (c) $CH_3CHCH_2CH_3$
$$\overset{H_2NO}{\underset{}{|\|}}$$
$$\overset{}{\underset{OH}{|}}$$

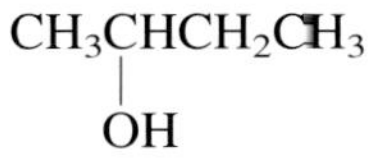

(d) $CF_2BrCHBr_2$ (e) 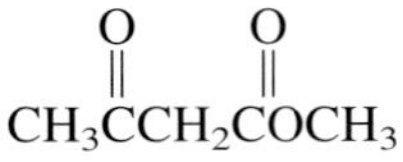 $CH_3CCH_2COCH_3$ (f) $HOCH_2CH_2OCH_2CH_2OH$
$$\overset{OO}{\underset{}{\|\|}}$$

28. Convert the following bond-line formulas into Kekulé (straight-line) structures.

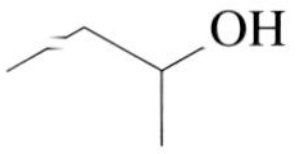

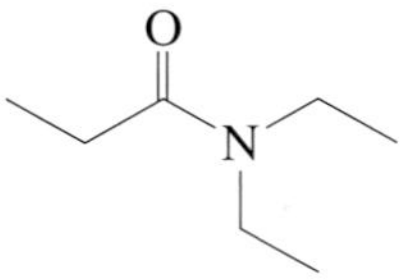

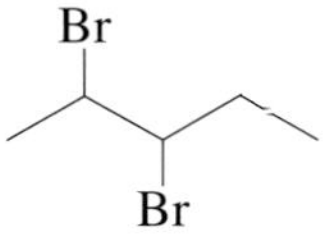

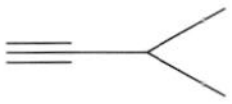

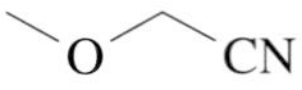

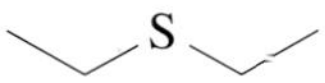

29. Convert the following dashed-wedged line formulas into condensed formulas.

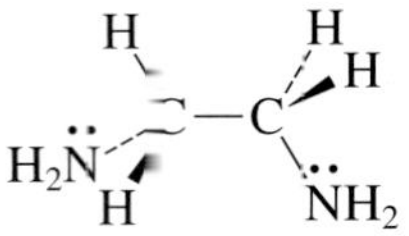

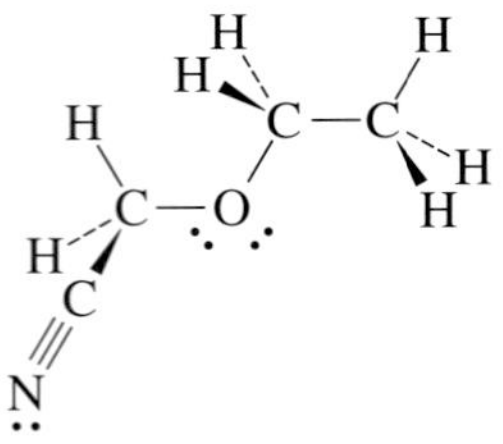

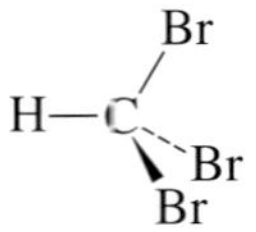

30. Depict the following straight-line formulas in their condensed form.

(a)   **(b)**

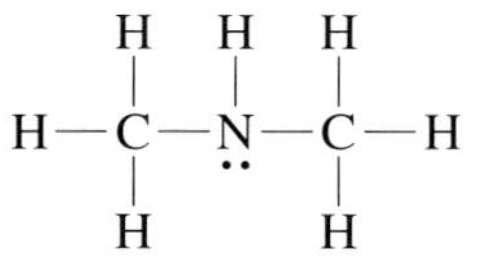

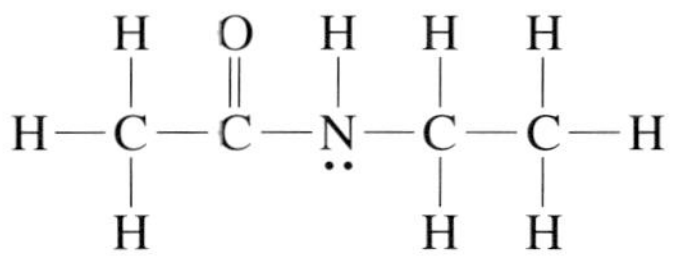

(c) 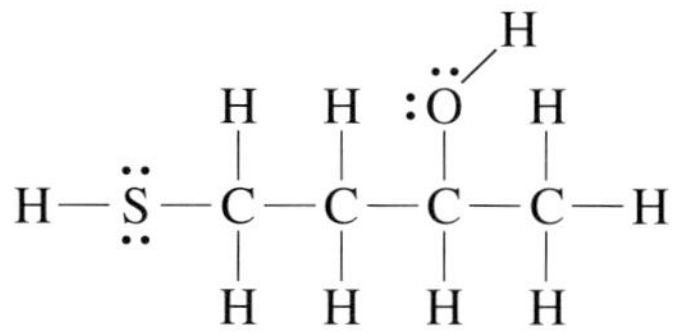   **(d)**

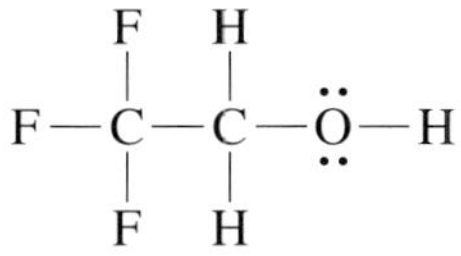

(e) 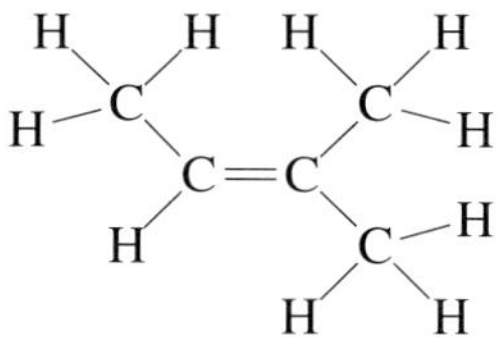   **(f)**

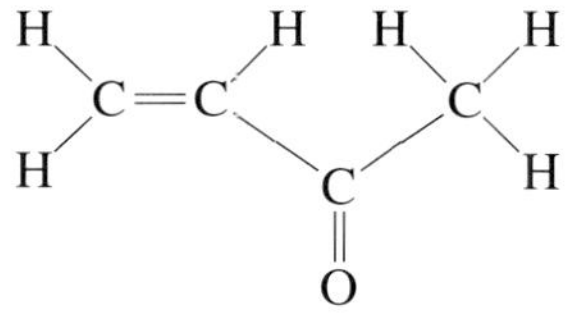

31. Redraw the structures depicted in Problems 27 and 30 using bond-line formulas.

32. Convert the following condensed formulas into dashed-wedged line structures.

(a) $CH_3\overset{CN}{\underset{|}{C}}HOCH_3$ **(b)** $CHCl_3$ **(c)** $(CH_3)_2NH$ **(d)** $CH_3\overset{SH}{\underset{|}{C}}HCH_2CH_3$

33. Construct as many isomers of each molecular formula as you can for **(a)** C_5H_{12}; **(b)** C_3H_8O. Draw both condensed and bond-line formulas for each isomer.

34. Draw condensed formulas showing the multiple bonds, charges, and lone electron pairs (if any) for each molecule in the following pairs of constitutional isomers. (**Hint:** First make sure that you can draw a proper Lewis structure for each molecule.) Do any of these pairs consist of resonance forms?

(a) $HCCCH_3$ and H_2CCCH_2 **(b)** CH_3CN and CH_3NC
(c) $CH_3\overset{O}{\overset{\|}{C}}H$ and H_2CCHOH

35. Two resonance forms can be written for a bond between trivalent boron and an atom with a lone pair of electrons. **(a)** Formulate them for (i) $(CH_3)_2BN(CH_3)_2$; (ii) $(CH_3)_2BOCH_3$; (iii) $(CH_3)_2BF$. **(b)** Using the guidelines in Section 1-5, determine which form in each pair of resonance forms is more important. **(c)** How do the electronegativity differences between N, O, and F affect the relative importance of the resonance forms in each case? **(d)** Predict the hybridization of N in (i) and O in (ii).

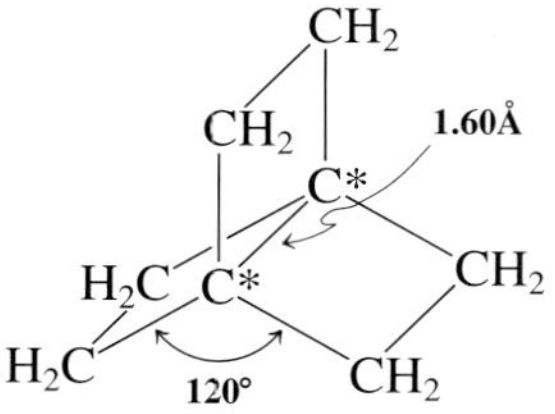

[2.2.2]Propellane

36. The unusual molecule [2.2.2]propellane is pictured in the margin. On the basis of the given structural parameters, what hybridization scheme best describes

the carbons marked by asterisks? (Make a model to help you visualize its shape.) What types of orbitals are used in the bond between them? Would you expect this bond to be stronger or weaker than an ordinary carbon–carbon single bond (which is usually 1.54 Å long)?

37. (a) On the basis of the information in Problem 26, give the likely hybridization of the orbital that contains the unshared pair of electrons (responsible for the negative charge) in each of the following species: $CH_3CH_2^-$; $CH_2{=}CH^-$; $HC{\equiv}C^-$. **(b)** Electrons in sp, sp^2, and sp^3 orbitals do not have identical energies. Because the $2s$ orbital is lower in energy than a $2p$, the more s character a hybrid orbital has, the lower its energy will be. Therefore the sp^3 ($\frac{1}{4}s$ and $\frac{3}{4}p$ in character) is highest in energy, and the sp ($\frac{1}{2}s$, $\frac{1}{2}p$) lowest. Use this information to determine the relative abilities of the three anions in (a) to accommodate the negative charge. **(c)** The strength of an acid HA is related to the ability of its conjugate base A^- to accommodate negative charge. In other words, the ionization $HA \rightleftharpoons H^+ + A^-$ is favored for a more stable A^-. Although CH_3CH_3, $CH_2{=}CH_2$, and $HC{\equiv}CH$ are all weak acids, they are not equally so. On the basis of your answer to (b), rank them in order of acid strength.

38. A number of substances containing positively polarized carbon atoms have been labeled as "cancer suspect agents" (i.e., suspected carcinogens or cancer-inducing compounds). It has been suggested that the presence of such carbon atoms is responsible for the carcinogenic properties of these molecules. Assuming that the extent of polarization is proportional to carcinogenic potential, how would you rank the following compounds with regard to cancer-causing potency?
(a) CH_3Cl **(b)** $(CH_3)_4Si$ **(c)** $ClCH_2OCH_2Cl$
(d) CH_3OCH_2Cl **(e)** $(CH_3)_3C^+$
(Note: Polarization is only one of the many factors known to be related to carcinogenicity. Moreover, none of them shows the type of straightforward correlation implied in this question.)

39. The structure of the substance lynestrenol, a component of certain oral contraceptives, is presented below. Locate an example of each of the following types of bonds or atoms: **(a)** a highly polarized covalent bond; **(b)** a nearly unpolarized covalent bond; **(c)** an sp-hybridized carbon atom; **(d)** an sp^2-hybridized carbon atom; **(e)** an sp^3-hybridized carbon atom; **(f)** a bond between atoms of different hybridization.

Lynestrenol

Team Problems

Team problems are meant to encourage discussion and collaborative learning. Try to solve team problems with a partner or small study group. Notice that the problems are divided into parts. Rather than tackling each part individually, discuss each section of the problem together. Try out the vocabulary that you learned in the chapter to question and convince yourselves that you are on the right track, before you move to the next part. In general, the more you use the terms and apply the concepts presented in the text, the better you will become at correlating molecular structure and reactivity and thus visualizing bond breaking and bond making. You will begin to see the elegant patterns of organic chemistry and will not be a slave to memorization. The collaborative process used in partner or group study will force you to articulate your ideas. Talking out a solution with an "audience" instead of to yourself builds in checks and balances. Your teammates will not let you get away with, "Well, you know what I mean," because they probably do not. You become responsible to others as well as yourself. By learning from and teaching others, you solidify your own understanding.

40. Consider the following reaction:

$$CH_3CH_2CH_2\mathbf{C}CH_3 + HCN \longrightarrow CH_3CH_2CH_2\mathbf{C}CH_3$$

A   B

(a) Draw these condensed formulas as Lewis dot structures. Label the geometry and hybridization of the bold carbons in compounds A and B. Did the hybridization change in the course of the reaction?

(b) Draw the condensed formulas as bond-line structures.

(c) Examine the components of the reaction in light of bond polarity. Using the notation for partial charge separation, δ^+ and δ^-, indicate, on the bond-line structures, any polar bonds.

(d) This reaction is actually a two-step process: cyanide attack followed by protonation. Depict these processes by using the same "electron pair pushing" technique that we employed for resonance structures in Section 1-5, but now show the flow of electrons for the two steps. Clearly position the beginning (an electron pair) and the end (a positively polarized or charged nucleus) of your arrows.

Preprofessional Problems

Preprofessional problems are included to give you practice solving the type of problems found on exams required for entry into professional schools, such as the MCAT, DAT, chemistry GRE, and ACS exams, as well as on many undergraduate tests. Do these multiple-choice questions as you go through this course and then return to them before you take a professional school exam. These questions are to be answered "closed-book"—that is, no periodic table, calculator, or the like.

41. A certain organic compound was found on combustion analysis to contain 84% carbon and 16% hydrogen (C = 12.0, H = 1.00). A molecular formula for the compound could be

(a) CH_4O (b) $C_6H_{14}O_2$ (c) C_7H_{16}

(d) C_6H_{10} (e) $C_{14}H_{22}$

42. The compound $\underset{\overset{|}{Br}\ \ \overset{|}{CH_3}}{\overset{\overset{Br}{|}\ \ \overset{CH_3}{|}}{Br-Al-N-CH_2CH_3}}$ has a formal charge of

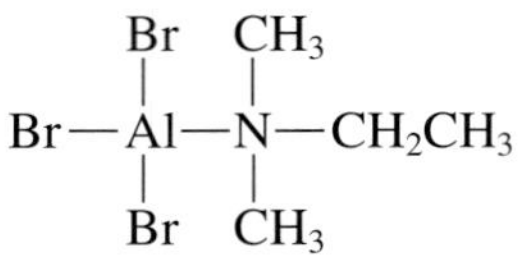

    (a) -1 on N   (b) $+2$ on N   (c) -1 on Al
    (d) $+1$ on Br   (e) none of the above

43. The arrow in the structure points to a bond that is formed by 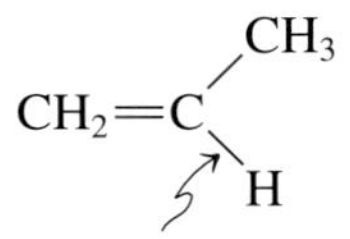
    (a) overlap of an s orbital on H and an sp^2 orbital on C
    (b) overlap of an s orbital on H and an sp orbital on C
    (c) overlap of an s orbital on H and an sp^3 orbital on C
    (d) none of the above

    $$CH_2{=}C\overset{\displaystyle CH_3}{\underset{\displaystyle H}{}}$$

44. Which compound has bond angles nearest to 120°?
    (a) $O{=}C{=}S$ (b) CHI_3 (c) $H_2C{=}O$
    (d) $H-C{\equiv}C-H$ (e) CH_4

45. The pair of structures that are resonance hybrids is
    (a) $H\overset{..}{\underset{..}{C}}{-}\overset{+}{C}HCH_3$ and $H\overset{+}{\underset{..}{O}}{=}CHCH_3$

    (b) 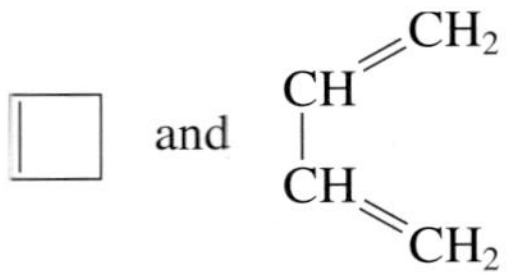 $\square$ and $\underset{\displaystyle CH}{\overset{\displaystyle CH}{\big|}}\Big\langle {\overset{\displaystyle {=}CH_2}{{=}CH_2}}$

    (c) 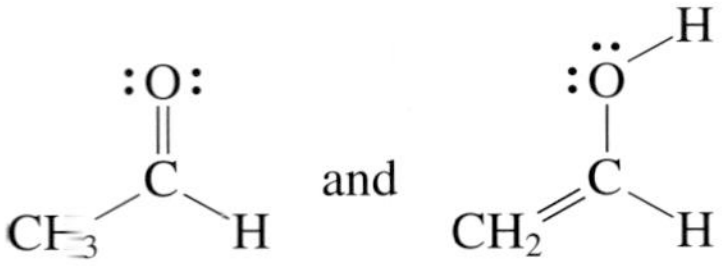 $\underset{\displaystyle CH_3}{\overset{\displaystyle :O:}{\underset{\displaystyle \overset{\|}{C}}{}}}\diagdown H$ and $\underset{\displaystyle CH_2}{\overset{\displaystyle :\overset{..}{O}}{\underset{\displaystyle \overset{}{C}}{}}}\diagup\overset{\displaystyle H}{}\diagdown H$

    (d) $CH_3\overset{+}{C}H_2$ and $\overset{+}{C}H_2CH_3$

Alkanes

Molecules Lacking Functional Groups

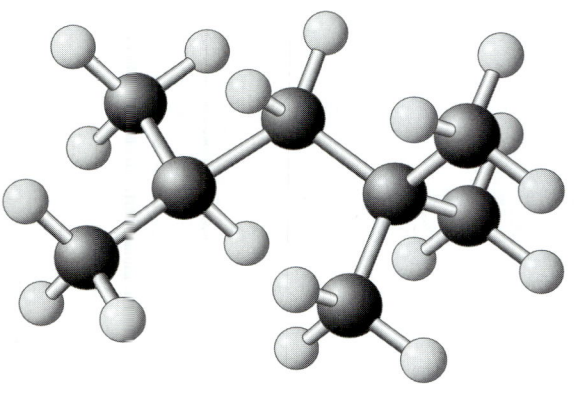

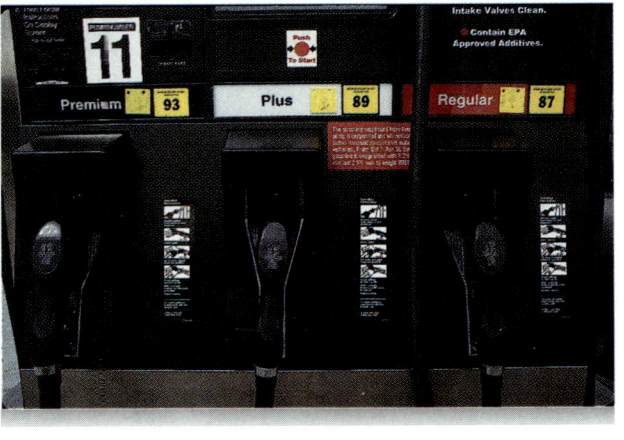

Different blends of alkanes and other additives give rise to gasolines with different octane number ratings.

Turn to page 90 of this book and look at the structures of the molecules illustrated in Problem 29. Each one contains a variety of types of bonds between various elements. Can we predict what kinds of chemical reactivity will be displayed by these substances? This chapter will begin to answer this question with a brief description of functional groups: the places in molecules where reactions tend to occur. Next we shall examine in depth the simplest class of organic molecules, the alkanes. If you have an appropriate kit, make a model of the structure shown at the top of this page. Does your model look exactly like the picture? Can it adopt other shapes by rotation of the atoms about bonds? This molecule is called 2,2,4-trimethylpentane, an alkane used in gasoline. As we proceed through this chapter, we shall explore the names, physical properties, and structural mobility of the members of the alkane family.

2-1 Functional Groups: Centers of Reactivity

Many organic molecules consist predominantly of a backbone of carbons linked by single bonds, with only hydrogen atoms attached. However, they may also contain doubly or triply bonded carbons, as well as other elements. These atoms or groups of atoms tend to be sites of comparatively high chemical reactivity and are referred to as **functional groups** or **functionalities.** Such groups have characteristic properties, and *they control the reactivity of the molecule as a whole.*

Hydrocarbons are molecules that contain only hydrogen and carbon

We begin our study with hydrocarbons, which have the general empirical formula C_xH_y. Those containing only single bonds, such as methane, ethane, and propane, are

called **alkanes.** Molecules such as cyclohexane, whose carbons form a ring, are called **cycloalkanes.** *Alkanes lack functional groups;* as a result, they are relatively nonpolar and unreactive. The properties and chemistry of the alkanes are described in this chapter and in Chapters 3 and 4.

Alkanes

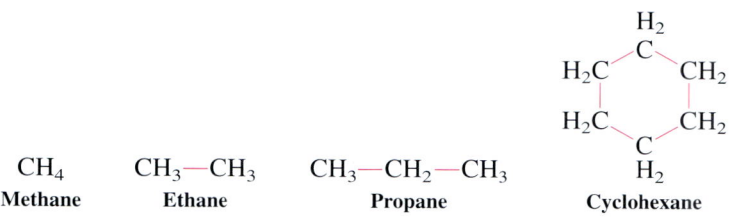

CH₄   CH₃—CH₃   CH₃—CH₂—CH₃

Methane   Ethane   Propane   Cyclohexane

Double and triple bonds are the functional groups of **alkenes** and **alkynes,** respectively. Their properties and chemistry are the topics of Chapters 11 through 13.

Alkenes and Alkynes

CH₂=CH₂   C=CH₂   HC≡CH   CH₃—C≡CH

Ethene   Propene   Ethyne   Propyne
(Ethylene)   (Acetylene)

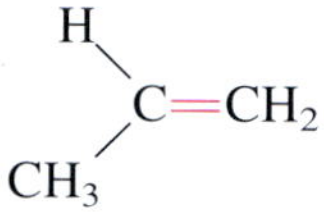

A special hydrocarbon is **benzene,** C_6H_6, in which three double bonds are incorporated into a six-membered ring. Benzene and its derivatives are traditionally called **aromatic,** because some substituted benzenes do have a strong fragrance. Aromatic compounds are discussed in Chapters 15, 16, 22, and 25.

Aromatic Compounds

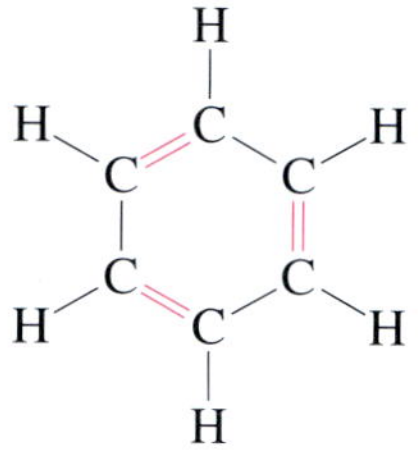

Benzene 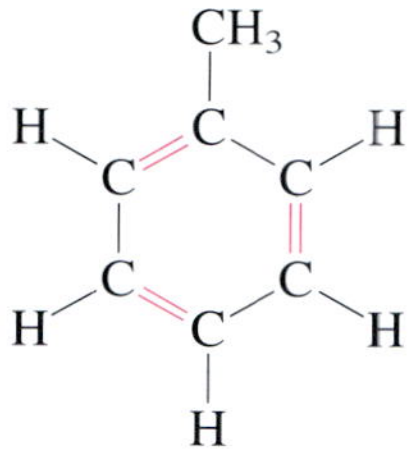   Methylbenzene
(Toluene)

Many functional groups contain polar bonds

Polar bonds determine the behavior of many classes of molecules. Recall that polarity is due to a difference in the electronegativity of two atoms bound to each other (Section 1-3). Chapters 6 and 7 will introduce the **haloalkanes,** which contain polar carbon–halogen bonds as their functional groups. Another example is the **hydroxy** group, –O–H, characteristic of **alcohols.** The characteristic functional unit of **ethers** is an oxygen atom bonded to two carbon atoms. The functional group in alcohols and

those in some ethers can be converted into a large variety of other functionalities and are therefore important in synthetic transformations. This chemistry is the subject of Chapters 8 and 9.

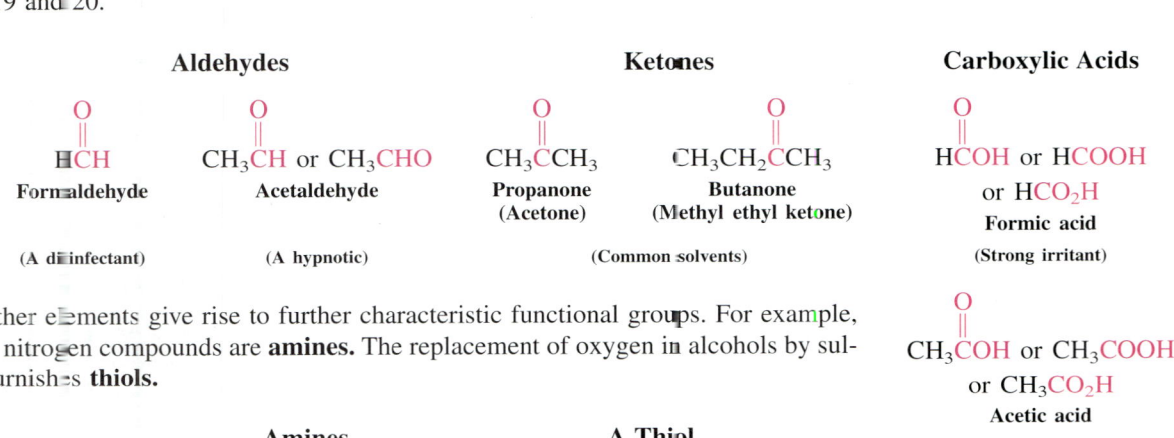

Haloalkanes

CH_3Cl
Chloromethane
(Methyl chloride)

CH_3CH_2Cl
Chloroethane
(Ethyl chloride)

Topical anesthetics

Alcohols

CH_3OH
Methanol

(Wood alcohol)

CH_3CH_2OH
Ethanol

(Grain alcohol)

Ethers

CH_3OCH_3
Methoxymethane
(Dimethyl ether)

(A refrigerant)

$CH_3CH_2OCH_2CH_3$
Ethoxyethane
(Diethyl ether)

(An inhalation anesthetic)

The **carbonyl** function, C=O, is found in **aldehydes,** in **ketones,** and, in conjunction with an attached –OH, in the **carboxylic acids.** Aldehydes and ketones are discussed in Chapters 17 and 18, the carboxylic acids and their derivatives in Chapters 19 and 20.

Aldehydes

O
‖
HCH
Formaldehyde

(A disinfectant)

O
‖
CH_3CH or CH_3CHO
Acetaldehyde

(A hypnotic)

Ketones

O
‖
CH_3CCH_3
Propanone
(Acetone)

O
‖
$CH_3CH_2CCH_3$
Butanone
(Methyl ethyl ketone)

(Common solvents)

Carboxylic Acids

O
‖
HCOH or HCOOH
or HCO_2H
Formic acid

(Strong irritant)

O
‖
CH_3COH or CH_3COOH
or CH_3CO_2H
Acetic acid

(In vinegar)

Other elements give rise to further characteristic functional groups. For example, alkyl nitrogen compounds are **amines.** The replacement of oxygen in alcohols by sulfur furnishes **thiols.**

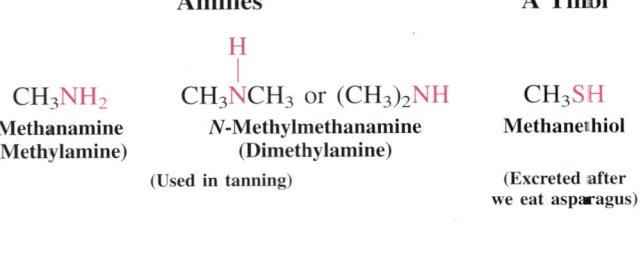

Amines

CH_3NH_2
Methanamine
(Methylamine)

H
|
CH_3NCH_3 or $(CH_3)_2NH$
N-Methylmethanamine
(Dimethylamine)

(Used in tanning)

A Thiol

CH_3SH
Methanethiol

(Excreted after we eat asparagus)

R represents a part of an alkane molecule

Table 2-1 depicts a selection of common functional groups, the class of compounds to which they give rise, a general structure, and an example. In the general structures, we commonly use the symbol **R** (for *radical* or *residue*) to represent an **alkyl group,** a molecular fragment derived by removal of one hydrogen atom from an alkane (Section 2-3). Therefore, a general formula for a haloalkane is R–X, in which R stands for any alkyl group and X for any halogen. Alcohols are similarly represented as R–O–H. In structures that contain multiple alkyl groups, we add a prime (′) or double prime ″) to R to distinguish groups that differ in structure from one another. Thus a general formula for an ether in which both alkyl groups are the same (a **symmetrical ether**) is R–O–R, whereas an ether with two dissimilar groups (an **unsymmetrical ether**) is represented by R–O–R′.

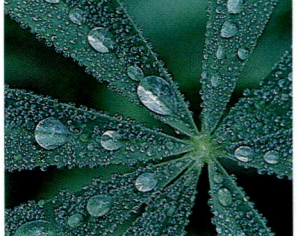

The alkanes $C_{29}H_{60}$ and $C_{31}H_{64}$ constitute the waxy, water-repellent coatings on these wild lupine leaves.

TABLE 2-1	Common Functional Groups

Compound class	General structure[a]	Functional group	Example
Alkanes	R—H	None	$CH_3CH_2CH_2CH_3$ **Butane**
Haloalkanes	R—X (X = F, Cl, Br, I)	—X	CH_3CH_2—Br **Bromoethane**
Alcohols	R—OH	—OH	$(CH_3)_2\overset{\overset{\text{H}}{\mid}}{C}$—OH **2-Propanol (Isopropyl alcohol)**
Ethers	R—O—R′	—O—	CH_3CH_2—O—CH_3 **Methoxyethane (Ethyl methyl ether)**
Thiols	R—SH	—SH	CH_3CH_2—SH **Ethanethiol**
Alkenes	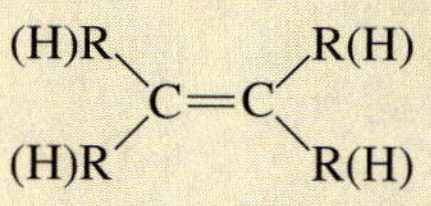 $\underset{\text{(H)R}}{\overset{\text{(H)R}}{}}C=C\underset{\text{R(H)}}{\overset{\text{R(H)}}{}}$	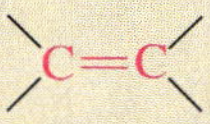 $C=C$	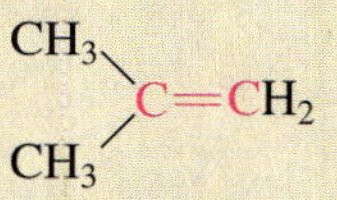 $\underset{CH_3}{\overset{CH_3}{}}C=CH_2$ **2-Methylpropene**
Alkynes	(H)R—C≡C—R(H)	—C≡C—	$CH_3C≡CCH_3$ **2-Butyne**
Aromatic compounds	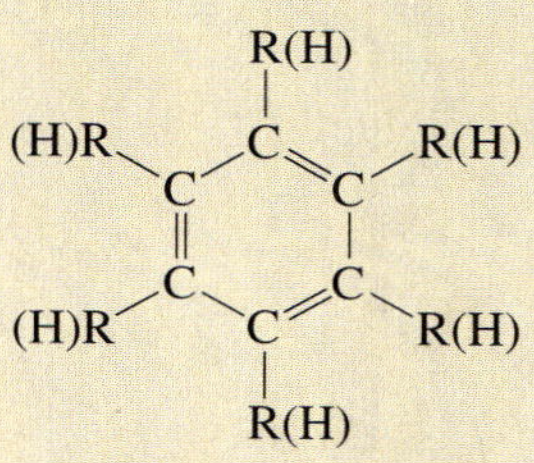 (see structure)	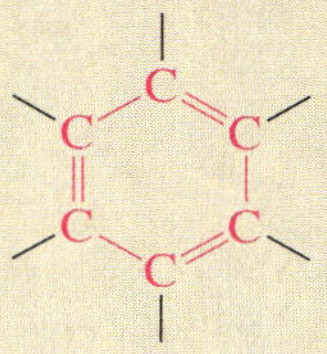 (see structure)	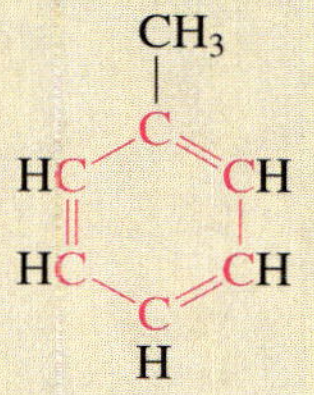 **Methylbenzene (Toluene)**
Aldehydes	$R-\overset{\overset{\text{O}}{\|}}{C}-H$	$-\overset{\overset{\text{O}}{\|}}{C}-H$	$CH_3CH_2\overset{\overset{\text{O}}{\|}}{C}H$ **Propanal**
Ketones	$R-\overset{\overset{\text{O}}{\|}}{C}-R'$	$-\overset{\overset{\text{O}}{\|}}{C}-$	$CH_3CH_2\overset{\overset{\text{O}}{\|}}{C}CH_2CH_2CH_3$ **3-Hexanone**
Carboxylic acids	$R-\overset{\overset{\text{O}}{\|}}{C}-O-H$	$-\overset{\overset{\text{O}}{\|}}{C}-OH$	$CH_3CH_2\overset{\overset{\text{O}}{\|}}{C}OH$ **Propanoic acid**
Anhydrides	$R-\overset{\overset{\text{O}}{\|}}{C}-O-\overset{\overset{\text{O}}{\|}}{C}-R'(H)$	$-\overset{\overset{\text{O}}{\|}}{C}-O-\overset{\overset{\text{O}}{\|}}{C}-$	$CH_3CH_2\overset{\overset{\text{O}}{\|}}{C}O\overset{\overset{\text{O}}{\|}}{C}CH_2CH_3$ **Propanoic anhydride**

[a]The letter R denotes an alkyl group. Different alkyl groups can be distinguished by adding primes to the letter R: R′, R″, and so forth.

TABLE 2-1 (continued)			
Compound class	**General structure**[a]	**Functional group**	**Example**
Esters	$$\overset{\displaystyle O}{\overset{\displaystyle \|}{(H)R-C-O-R'}}$$	$$\overset{O}{\overset{\|}{-C-O-}}$$	$\overset{\displaystyle O}{\overset{\displaystyle \|}{CH_3CH_2COCH_3}}$ **Methyl propanoate** **(Methyl propionate)**
Amides	$$\overset{\displaystyle O}{\overset{\displaystyle \|}{R-C-\underset{\underset{\displaystyle R''(H)}{\|}}{N}-R'(H)}}$$	$$\overset{C}{\overset{\|}{-C-\underset{\|}{N}-}}$$	$\overset{\displaystyle O}{\overset{\displaystyle \|}{CH_3CH_2CH_2CNH_2}}$ **Butanamide**
Nitriles	$R-C\equiv N$	$-C\equiv N$	$CH_3C\equiv N$ **Ethanenitrile** **(Acetonitrile)**
Amines	$$R-\underset{\underset{\displaystyle R''(H)}{\|}}{N}-R'(H)$$	$-\overset{\diagup}{\underset{\diagdown}{N}}$	$(CH_3)_3N$ **N,N-Dimethylmethanamine** **(Trimethylamine)**

2-2 Straight-Chain and Branched Alkanes

We begin with the alkanes, hydrocarbons that contain only single bonds. They are classified into several types according to structure: the linear **straight-chain alkanes;** the **branched alkanes,** in which the carbon chain contains one or several branching points; and the cyclic alkanes, or **cycloalkanes,** which will be covered in Chapter 4.

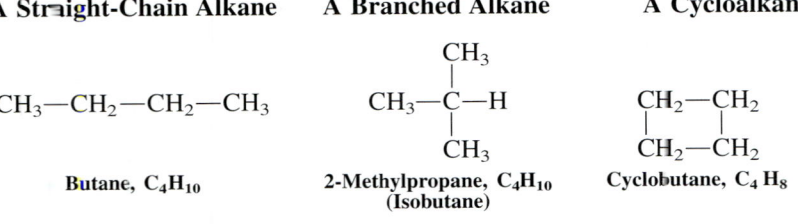

A Straight-Chain Alkane

$CH_3-CH_2-CH_2-CH_3$

Butane, C$_4$H$_{10}$

A Branched Alkane

$$CH_3-\overset{\overset{\displaystyle CH_3}{|}}{\underset{\underset{\displaystyle CH_3}{|}}{C}}-H$$

2-Methylpropane, C$_4$H$_{10}$
(Isobutane)

A Cycloalkane

$$\begin{array}{c} CH_2-CH_2 \\ |\qquad\quad| \\ CH_2-CH_2 \end{array}$$

Cyclobutane, C$_4$H$_8$

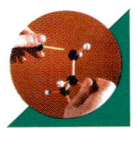

The alkanes form a homologous series

In the straight-chain alkanes, each carbon is bound to its two neighbors and to two hydrogen atoms. Exceptions are the two terminal carbon nuclei, which are bound to only one carbon atom and three hydrogen atoms. Several general formulas may be written for the straight-chain alkane series:

$$H-(CH_2)_n-H \quad CH_3-(CH_2)_{n-1}-H \quad CH_3-(CH_2)_{n-2}-CH_3$$

Each member of this series differs from the next lower one by the addition of a methylene group, $-CH_2-$. Molecules that are related in this way are **homologs** of each other (*homos,* Greek, same as), and the series is a **homologous series.** Methane ($n = 1$) is the first member of the homologous series of the alkanes, ethane ($n = 2$) the second, and so forth.

Branched alkanes are constitutional isomers of straight-chain alkanes

Branched alkanes are derived from the straight-chain systems by removal of a hydrogen from a methylene (CH_2) group and replacement with an alkyl group. Both branched and straight-chain alkanes have the same general formula, C_nH_{2n+2}. The smallest branched alkane is 2-methylpropane. It has the same molecular formula as that of butane (C_4H_{10}) but different connectivity; the two compounds therefore form a pair of constitutional isomers (Section 1-9).

For the higher alkane homologs ($n > 4$), more than two isomers are possible. There are three pentanes, C_5H_{12}, as shown below. There are five hexanes, C_6H_{14}; nine heptanes, C_7H_{16}; and eighteen octanes, C_8H_{18}.

The Isomeric Pentanes

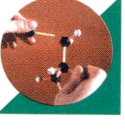

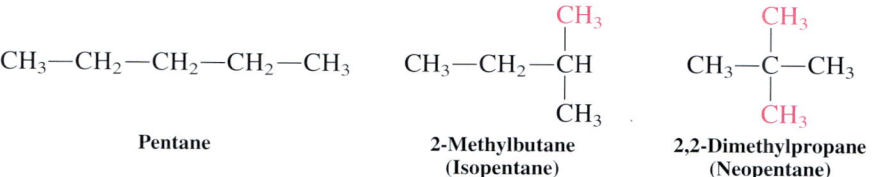

| Pentane | 2-Methylbutane (Isopentane) | 2,2-Dimethylpropane (Neopentane) |

The number of possibilities in connecting n carbon atoms to each other and to $2n + 2$ surrounding hydrogen atoms increases dramatically with the size of n (Table 2-2).

EXERCISE 2-1

(a) Draw the structures of the five isomeric hexanes. (b) Draw the structures of all the possible next higher and lower homologs of 2-methylbutane.

TABLE 2-2 | Number of Possible Isomeric Alkanes, C_nH_{2n+2}

n	Isomers
1	1
2	1
3	1
4	2
5	3
6	5
7	9
8	18
9	35
10	75
15	4,347
20	366,319

2-3 Naming the Alkanes

The multiplicity of ways of assembling carbon atoms and attaching various substituents accounts for the existence of the very large number of organic molecules. This diversity poses a problem: How can we systematically differentiate all these compounds by name? Is it possible, for example, to name all the C_6H_{14} isomers so that information on any of them (such as boiling points, melting points, reactions) might easily be found in the index of a handbook? And is there a way to name a compound that we have never seen in such a way as to be able to draw its structure?

This problem of naming organic molecules has been with organic chemistry from its very beginning, but the initial method was far from systematic. Compounds have been named after their discoverers ("Nenitzescu's hydrocarbon"), after localities ("sydnones"), after their shapes ("cubane," "basketane"), and after their natural sources ("vanillin"). Many of these **common** or **trivial names** are still widely used. However, there now exists a precise system for naming the alkanes. **Systematic nomenclature,** in which the name of a compound describes its structure, was first introduced by a chemical congress in Geneva, Switzerland, in 1892. It has continually been revised since then, mostly by the International Union of Pure and Applied Chemistry (IUPAC). Table 2-3 gives the systematic names of the first twenty straight-chain alkanes. Their stems, mainly of Latin or Greek origin, reveal the number of carbon atoms in the chain. For example, the name heptadecane is composed of the Greek

			Boiling point (°C)	Melting point (°C)	Density at 20°C (g ml⁻¹)

TABLE 2-3 — **Names and Physical Properties of Straight-Chain Alkanes, C_nH_{2n+2}**

n	Name	Formula	Boiling point (°C)	Melting point (°C)	Density at 20°C (g ml^{-1})
1	Methane	CH_4	−161.7	−182.5	0.466 (at −164°C)
2	Ethane	CH_3CH_3	−88.6	−183.3	0.572 (at −100°C)
3	Propane	$CH_3CH_2CH_3$	−42.1	−187.7	0.5853 (at −45°C)
4	Butane	$CH_3CH_2CH_2CH_3$	−0.5	−138.3	0.5787
5	Pentane	$CH_3(CH_2)_3CH_3$	36.1	−129.8	0.5262
6	Hexane	$CH_3(CH_2)_4CH_3$	68.7	−95.3	0.6603
7	Heptane	$CH_3(CH_2)_5CH_3$	98.4	−90.6	0.6837
8	Octane	$CH_3(CH_2)_6CH_3$	125.7	−56.8	0.7026
9	Nonane	$CH_3(CH_2)_7CH_3$	150.8	−53.5	0.7177
10	Decane	$CH_3(CH_2)_8CH_3$	174.0	−29.7	0.7299
11	Undecane	$CH_3(CH_2)_9CH_3$	195.8	−25.6	0.7402
12	Dodecane	$CH_3(CH_2)_{10}CH_3$	216.3	−9.6	0.7487
13	Tridecane	$CH_3(CH_2)_{11}CH_3$	235.4	−5.5	0.7564
14	Tetradecane	$CH_3(CH_2)_{12}CH_3$	253.7	5.9	0.7628
15	Pentadecane	$CH_3(CH_2)_{13}CH_3$	270.6	10	0.7685
16	Hexadecane	$CH_3(CH_2)_{14}CH_3$	287	18.2	0.7733
17	Heptadecane	$CH_3(CH_2)_{15}CH_3$	301.8	22	0.7780
18	Octadecane	$CH_3(CH_2)_{16}CH_3$	316.1	28.2	0.7768
19	Nonadecane	$CH_3(CH_2)_{17}CH_3$	329.7	32.1	0.7855
20	Icosane	$CH_3(CH_2)_{18}CH_3$	343	36.8	0.7886

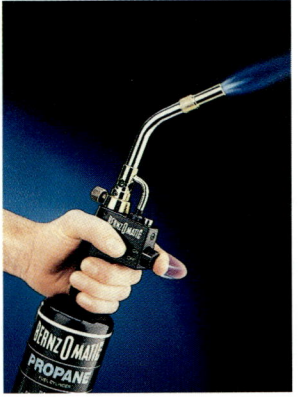

Propane, stored under pressure in liquefied form in canisters such as these, is a common fuel for torches, lanterns, and outdoor cooking stoves.

word *hepta*, seven, and the Latin word *decem*, ten. The first four alkanes have special names that have been accepted as part of the systematic nomenclature but also all end in **-ane**. It is important to know these names, because they serve as the basis for naming a large fraction of all organic molecules. A few smaller branched alkanes have common names that still have widespread use. They make use of the prefixes **iso-** and **neo-**, as in isobutane, isopentane, and neohexane.

$$CH_3 - \underset{\underset{\displaystyle H}{|}}{\overset{\overset{\displaystyle CH_3}{|}}{C}} - (CH_2)_n - CH_3$$

An isoalkane
(e.g., $n = 1$, Isopentane)

$$CH_3 - \underset{\underset{\displaystyle CH_3}{|}}{\overset{\overset{\displaystyle CH_3}{|}}{C}} - (CH_2)_n - H$$

A neoalkane
(e.g., $n = 2$, Neohexane)

EXERCISE 2-2

Draw the structures of isohexane and neopentane.

TABLE 2-4	Branched Alkyl Groups			
Structure	**Common name**	**Example of common name in use**	**Systematic name**	**Designation**
CH_3—$\overset{\overset{\displaystyle CH_3}{|}}{\underset{\underset{\displaystyle H}{|}}{C}}$—	Isopropyl	CH_3—$\overset{\overset{\displaystyle CH_3}{|}}{\underset{\underset{\displaystyle H}{|}}{C}}$—Cl (Isopropyl chloride)	1-Methylethyl	Secondary
CH_3—$\overset{\overset{\displaystyle CH_3}{|}}{\underset{\underset{\displaystyle H}{|}}{C}}$—$CH_2$—	Isobutyl	CH_3—$\overset{\overset{\displaystyle CH_3}{|}}{\underset{\underset{\displaystyle H}{|}}{C}}$—$CH_3$ (Isobutane)	2-Methylpropyl	Primary
CH_3—CH_2—$\overset{\overset{\displaystyle CH_3}{|}}{\underset{\underset{\displaystyle H}{|}}{C}}$—	sec-Butyl	CH_3—CH_2—$\overset{\overset{\displaystyle CH_3}{|}}{\underset{\underset{\displaystyle H}{|}}{C}}$—$NH_2$ (sec-Butyl amine)	1-Methylpropyl	Secondary
CH_3—$\overset{\overset{\displaystyle CH_3}{|}}{\underset{\underset{\displaystyle CH_3}{|}}{C}}$—	tert-Butyl	CH_3—$\overset{\overset{\displaystyle CH_3}{|}}{\underset{\underset{\displaystyle CH_3}{|}}{C}}$—Br (tert-Butyl bromide)	1,1-Dimethylethyl	Tertiary
CH_3—$\overset{\overset{\displaystyle CH_3}{|}}{\underset{\underset{\displaystyle CH_3}{|}}{C}}$—$CH_2$—	Neopentyl	CH_3—$\overset{\overset{\displaystyle CH_3}{|}}{\underset{\underset{\displaystyle CH_3}{|}}{C}}$—$CH_2$—OH (Neopentyl alcohol)	2,2-Dimethylpropyl	Primary

Alkyl groups

CH_3—

Methyl

CH_3—CH_2—

Ethyl

CH_3—CH_2—CH_2—

Propyl

As mentioned in Section 2-2, an **alkyl** group is formed by the removal of a hydrogen from an alkane. It is named by replacing the ending -ane in the corresponding alkane by **-yl,** as in methyl, ethyl, and propyl. Table 2-4 shows a few branched alkyl groups having common names. Note that some use the prefixes *sec-* (or *s-*), which stands for secondary, and *tert-* (or *t-*), for tertiary. To apply these prefixes, we must first see how to classify carbon atoms in organic molecules. A **primary** carbon is one attached to only one other carbon atom. For example, all carbon atoms at the ends of alkane chains are primary. The hydrogens attached to such carbons are designated primary hydrogens, and an alkyl group created by removing a primary hydrogen also is called primary. A **secondary** carbon is attached to two other carbon atoms, and a **tertiary** carbon to three others. Their hydrogens are labeled similarly. As shown in Table 2-4, removal of a secondary hydrogen results in a secondary alkyl group, and removal of a tertiary hydrogen in a tertiary alkyl group. Finally, a carbon bearing four alkyl groups is called **quaternary.**

Primary, Secondary, and Tertiary Carbons and Hydrogens

Primary C
Secondary C
Tertiary C
$CH_3CH_2CCH_2CH_3$ with CH_3 above and H below
Primary H
Secondary H
Tertiary H

3-Methylpentane

EXERCISE 2-3

Label the primary, secondary, and tertiary hydrogens in 2-methylpentane (isohexane).

The information in Table 2-3 enables us to name the first twenty straight-chain alkanes. How do we go about naming branched systems? A set of IUPAC rules makes this a relatively simple task, as long as they are followed carefully and in sequence.

IUPAC RULE 1. *Find the longest chain in the molecule and name it.* This task is not as easy as it seems. The problem is that, in the condensed formula, complex alkanes may be written in ways that mask the identity of the longest chain In the following examples, the longest chain, or **stem chain,** is clearly marked; the alkane stem gives the molecule its name. Groups other than hydrogen attached to the stem chain are called **substituents.**

Methyl $\longrightarrow$ CH$_3$

CH$_3$CHCH$_2$CH$_3$

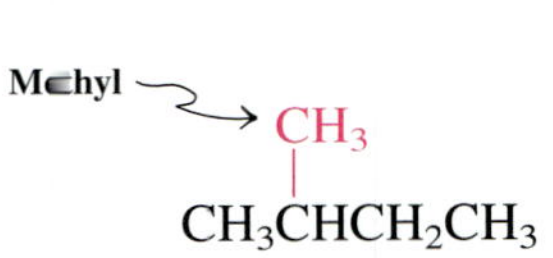

**A methyl-substituted butane
(A methylbutane)**

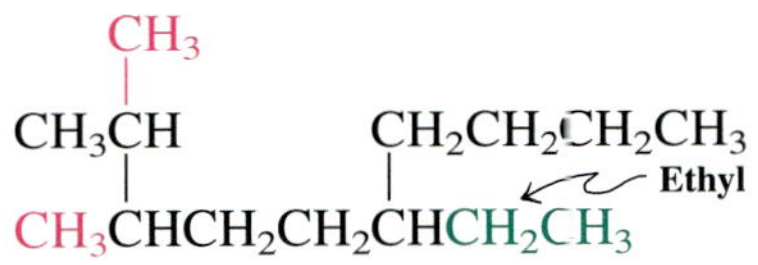

CH$_3$
|
CH$_3$CH   CH$_2$CH$_2$CH$_2$CH$_3$
| | $\longleftarrow$ Ethyl
CH$_3$CHCH$_2$CH$_2$CHCH$_2$CH$_3$

**An ethyl- and methyl-substituted decane
(An ethylmethyldecane)**

> **The stem chain is shown in black in the examples in this section.**

If a molecule has two or more chains of equal length, the chain with the largest number of substituents is the base stem chain.

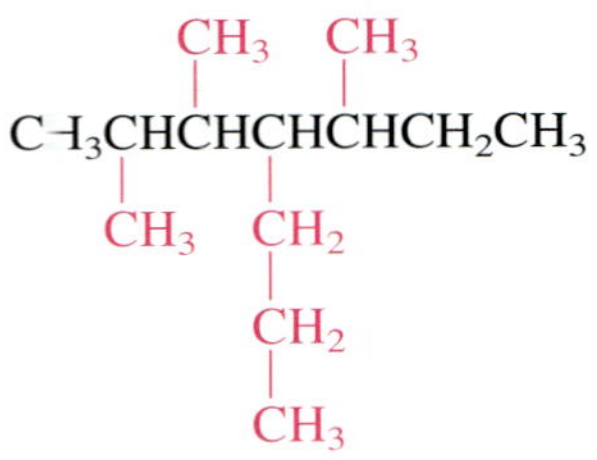

CH$_3$ CH$_3$
| |
CH$_3$CHCHCHCHCH$_2$CH$_3$
| |
CH$_3$ CH$_2$
|
CH$_2$
|
CH$_3$

**4 substituents
A heptane
Correct stem chain**

not

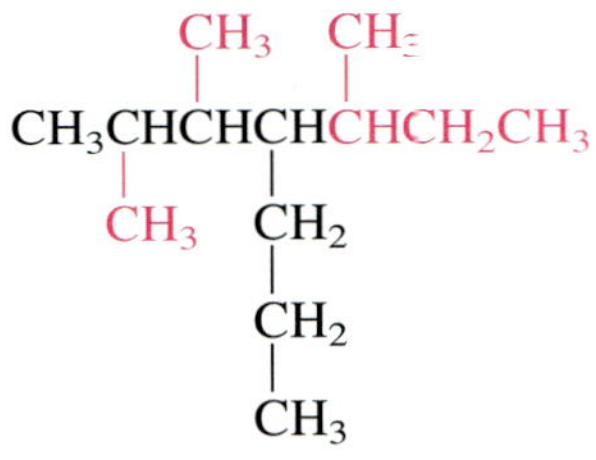

CH$_3$ CH$_3$
| |
CH$_3$CHCHCHCHCH$_2$CH$_3$
| |
CH$_3$ CH$_2$
|
CH$_2$
|
CH$_3$

**3 substituents
A heptane
Incorrect stem chain**

Here are two more examples, drawn with the use of bond-line notation:

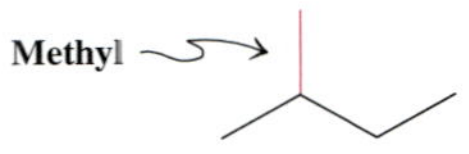

Methyl $\longrightarrow$

$\longleftarrow$ Ethyl

A methylbutane

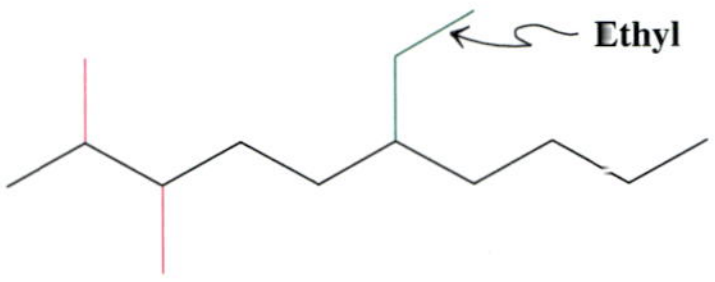

An ethylmethyldecane

IUPAC RULE 2. *Name all groups attached to the longest chain as alkyl substituents.* For straight-chain substituents, Table 2-3 can be used to derive the alkyl name. However, what if the substituent chain is branched? In this case, the same IUPAC rules apply to such complex substituents: First, find the longest chain in the substituent; next, name all *its* substituents.

IUPAC RULE 3. *Number the carbons of the longest chain beginning with the end that is closest to a substituent.*

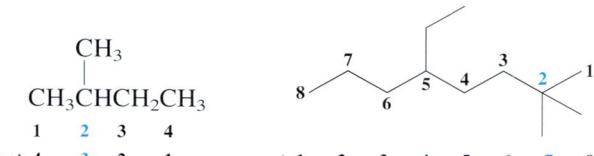

If there are two substituents at *equal* distance from the two ends of the chain, use the alphabet to decide how to number. The substituent to come first in alphabetical order is attached to the carbon with the lower number.

CH_3CH_2 CH_3

$CH_3CH_2CHCH_2CH_2CHCH_2CH_3$

| 1 | 2 | 3 | 4 | 5 | 6 | 7 | 8 |

Ethyl before methyl

16   14   12   10   8   6   4   2

17   15   13   11   9   7   5   3   1

Butyl before propyl

What if there are three or more substituents? Then number the chain in the direction that gives the lower number at the *first difference* between the two possible numbering schemes. This procedure follows the **first point of difference principle.**

CH_3 CH_3 CH_3

$CH_3CH_2CHCH_2CH_2CH_2CH_2CHCH_2CHCH_2CH_3$

| 1 | 2 | 3 | 4 | 5 | 6 | 7 | 8 | 9 | 10 | 11 | 12 |
| 12 | 11 | 10 | 9 | 8 | 7 | 6 | 5 | 4 | 3 | 2 | 1 |

Numbers for substituted carbons:
← 3, **8**, and 10 (incorrect)
← 3, **5**, and 10 (correct; 5 lower than 8)

3,5,10-Trimethyldodecane

Substituent groups are numbered outward from the main chain, with C1 of the group being the carbon attached to the main stem.

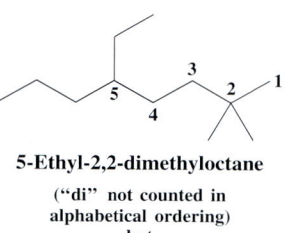

5-Ethyl-2,2-dimethyloctane

(**"di" not counted in alphabetical ordering**)
but

5-(1,1-Dimethylethyl)-3-ethyloctane

(**"di" counted: part of substituent name**)

IUPAC RULE 4. *Write the name of the alkane by first arranging all the substituents in alphabetical order (each preceded by the carbon number to which it is attached and a hyphen) and then add the name of the stem.* Should a molecule contain more than one of a particular substituent, its name is preceded by the prefix di, tri, tetra, penta, and so forth. The positions of attachment to the stem are given collectively before the substituent name and are separated by commas. These prefixes, as well as *sec-* and *tert-,* are not considered in the alphabetical ordering, except when they are part of a complex substituent name.

CH_3

$CH_3CHCH_2CH_3$

2-Methylbutane

CH_3

$CH_3CHCHCH_3$

      CH_3

2,3-Dimethylbutane

CH_3 CH_3

$CH_3CHCH_2CH_2CHCH_2CCH_3$

        CH_3CH_2 CH_3

4-Ethyl-2,2,7-trimethyloctane

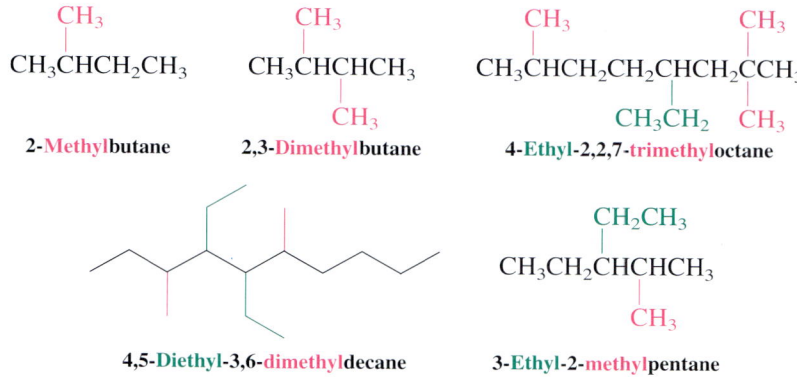

4,5-Diethyl-3,6-dimethyldecane

CH_2CH_3

$CH_3CH_2CHCHCH_3$

       CH_3

3-Ethyl-2-methylpentane

Although the common group names in Table 2-4 are permitted by IUPAC, it is preferable to use systematic names. Such complex names are usually enclosed in parentheses, to avoid possible ambiguities.

If a particular complex substituent is present more than once, its name is preceded by the prefix bis, tris, tetrakis, pentakis, and so on. In a substituent chain, the carbon numbered one (C1) is *always* the carbon atom bound to the principal chain.

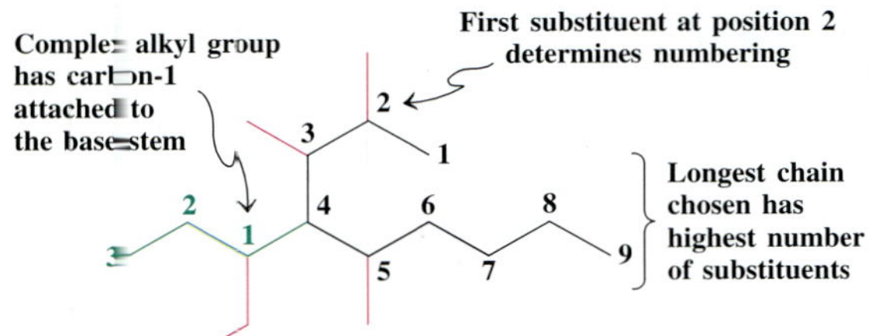

4-(1-Ethylpropyl)-2,3,5-trimethylnonane

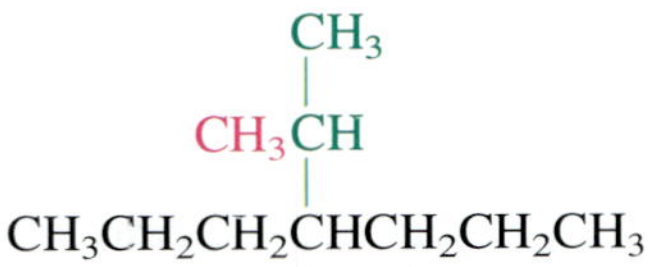

4-(1-Methylethyl)heptane
(4-Isopropylheptane)

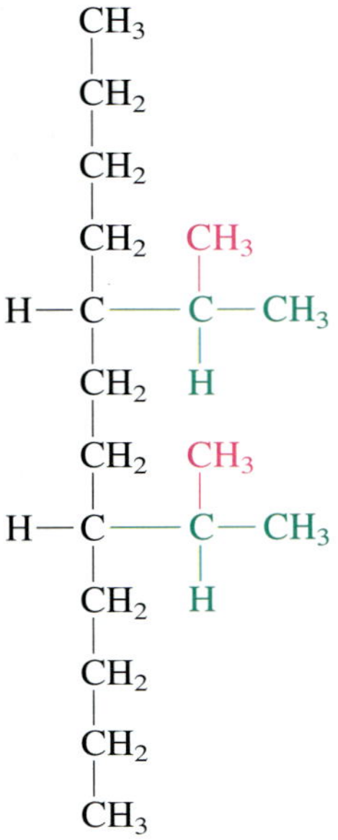

5,8-Bis(1-methylethyl)-codecane

Further instructions on nomenclature will be presented when new classes of compounds, such as the cycloalkanes and haloalkanes, are introduced.

Write down the names of the preceding eight branched alkanes, close the book, and reconstruct their structures from those names.

To summarize, four rules should be applied in sequence when naming a branched alkane: (1) find the longest chain; (2) find the names of all the alkyl groups attached to the stem; (3) number the chain; (4) name the alkane, with substituent names in alphabetical order and preceded by numbers to indicate their locations.

2-4 Structural and Physical Properties of Alkanes

What do the structures of alkanes look like in three dimensions? What are their physical appearances, and what are their physical properties? These questions will be addressed next.

Alkanes exhibit regular molecular structures and properties

The structural features of the alkanes are remarkably regular. The carbon atoms are tetrahedral, with bond angles close to 109° and with regular C–E ($\approx$ 1.10 Å) and

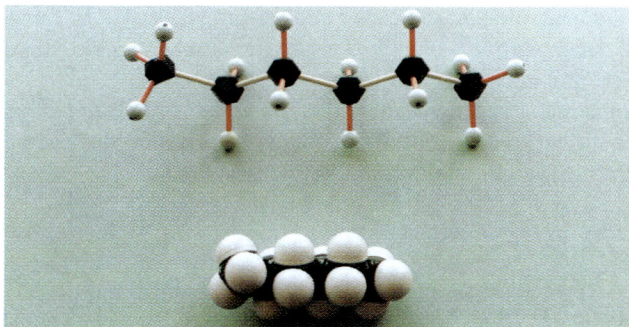

FIGURE 2-1

Ball-and-stick (top) and space-filling molecular models of hexane, showing the zigzag pattern of the carbon chain typical of the alkanes. [*Model sets courtesy of Maruzen Co., Ltd., Tokyo.*]

C–C ($\approx$ 1.54 Å) bond lengths. Alkane chains often adopt the zigzag patterns used in bond-line notation (Figure 2-1). To depict three-dimensional structures, we shall make use of the dashed-wedged line notation (see Figure 1-23). The main chain and a hydrogen at each end are drawn in the plane of the page (Figure 2-2).

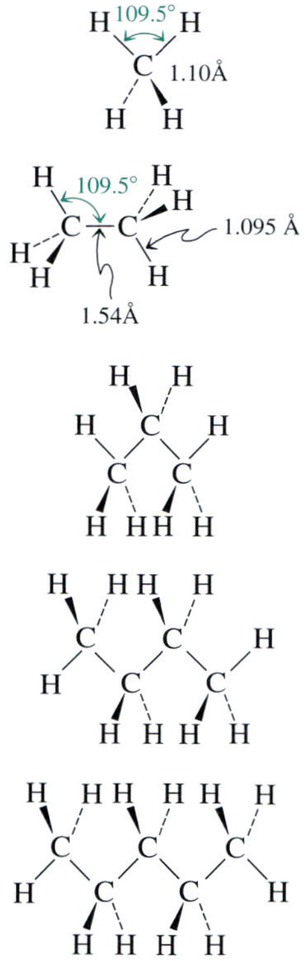

EXERCISE 2-5

Draw zigzag dashed-wedged line structures for 2-methylbutane and 2,3-dimethylbutane.

The regularity in alkane structures suggests that their physical constants would follow predictable trends. Indeed, inspection of the data presented in Table 2-3 reveals regular incremental increases along the homologous series. For example, at room temperature (25°C), the lower homologs of the alkanes are gases or colorless liquids, the higher homologs waxy solids. From pentane to pentadecane, each additional CH_2 group causes a 20°–30°C increase in boiling point (Figure 2-3).

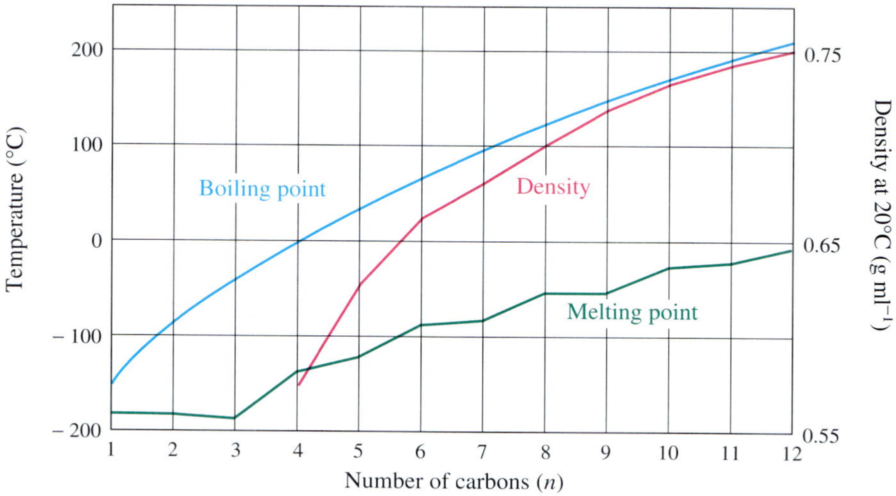

FIGURE 2-2

Dashed-wedged line structures of methane through pentane. Note the zigzag arrangement of the principal chain and two terminal hydrogens.

FIGURE 2-3

The physical constants of straight-chain alkanes. Their values increase with increasing size because molecular weights and London forces increase. Note that even-numbered systems have somewhat higher melting points than expected; these systems are more tightly packed in the solid state (notice their higher densities), thus allowing for stronger attractions between molecules.

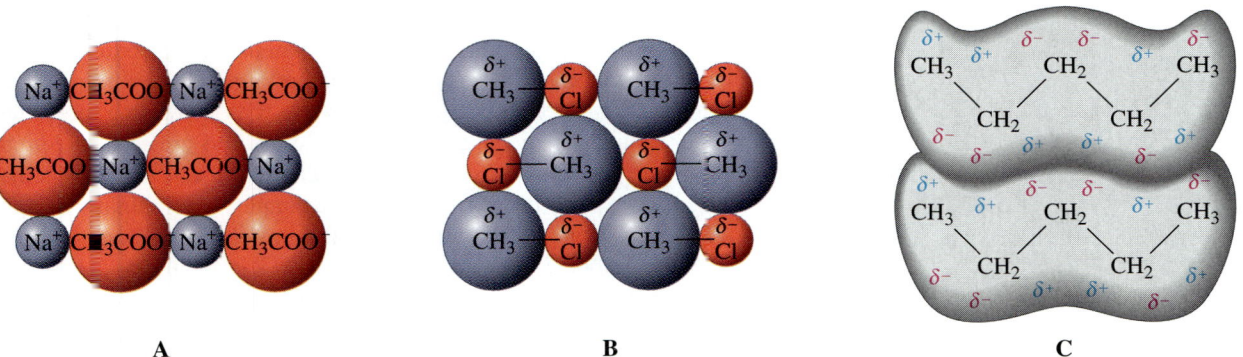

A B C

FIGURE 2-4

(A) Coulombic attraction in an ionic compound: crystalline sodium acetate, the sodium salt of acetic acid (in vinegar). (B) Dipole–dipole interactions in solid chloromethane. The polar molecules arrange to allow for favorable Coulombic attraction. (C) London forces in crystalline pentane. In this simplified picture, the electron clouds as a whole mutually interact to produce partial charges of opposite sign. The charge distributions in the two molecules change continually as the electrons continue to correlate their movements.

Attractive forces between molecules govern the physical properties of alkanes

Why are the physical properties of alkanes predictable? Such trends exist because of **intermolecular** or **van der Waals* forces.** Molecules exert several types of attractive forces on each other, causing them to aggregate into organized arrangements as solids and liquids. Most solid substances exist as highly ordered crystals. *Ionic* compounds, such as salts, are rigidly held in a crystal lattice, mainly by strong Coulomb forces. Nonionic but *polar* molecules, such as chloromethane (CH_3Cl), are attracted by weaker dipole–dipole interactions, also of Coulombic origin (Sections 1-2 and 6-2). Finally, the *nonpolar* alkanes attract each other by **London† forces,** which are due to **electron correlation.** When one alkane molecule approaches another, repulsion of the electrons in one molecule by those in the other results in correlation of their movement. Electron motion causes temporary bond polarization in one molecule; correlated electron motion in the bonds of the other induces polarization in the opposite direction, resulting in attraction between the molecules. Figure 2-4 is a simple picture comparing ionic, dipolar, and London attractions.

London forces are very weak. In contrast with Coulomb forces, which change with the square of the distance between charges, London forces fall off as the sixth power of the distance between molecules. There is also a limit to how close these forces can bring molecules together. At small distances, nucleus–nucleus and electron–electron repulsions outweigh these attractions.

How do these forces account for the physical constants of elements and compounds? The answer is that it takes energy, usually in the form of heat, to melt solids

*Professor Johannes D. van der Waals (1837–1923), University of Amsterdam, Netherlands, Nobel Prize 1910 (physics).

†Professor Fritz London (1900–1954), Duke University, North Carolina. *Note:* In older references the term "van der Waals forces" referred exclusively to what we now call *London forces; van der Waals force* now refers collectively to *all* intermolecular attractions.

and boil liquids. For example, to cause melting, the attractive forces responsible for the crystalline state must be overcome. In an ionic compound, such as sodium acetate (Figure 2-4A), the strong interionic forces require a rather high temperature (324°C) for the compound to melt. In alkanes, melting points rise with increasing molecular size: Molecules with relatively large surface areas are subject to greater London attractions. However, these forces are still relatively weak, and even high molecular weight alkanes have rather low melting points. For example, the straight-chain alkanes $C_{29}H_{60}$ and $C_{31}H_{64}$, waxy solids present in the protective coatings of plant leaves, have melting points below 70°C.

For a molecule to escape these same attractive forces in the liquid state and enter the gas phase, more heat has to be applied. When the vapor pressure of a liquid equals atmospheric pressure, boiling occurs. Like melting points, boiling points rise with increasing molecular weight: Heavy compounds require more kinetic energy to leave the liquid state. Boiling points of compounds are also relatively high if the intermolecular forces are relatively large. These effects lead to the smooth increase in boiling points seen in Figure 2-3.

Branched alkanes have smaller surface areas than do their straight-chain isomers. As a result, they are generally subject to smaller London attractions and are unable to pack as well in the crystalline state. The weaker attractions result in lower melting and boiling points. Branched molecules with highly compact shapes are exceptions. For example, 2,2,3,3-tetramethylbutane melts at +101°C because of highly favorable crystal packing (compare octane, m.p. –57°C). On the other hand, the greater surface area of octane, compared with that of the more spherical 2,2,3,3-tetramethylbutane, is clearly demonstrated in their boiling points (126°C and 106°C, respectively). Crystal packing differences also account for the slightly lower than expected melting points of odd-membered straight-chain alkanes relative to those of even-membered systems (Figure 2-3).

In summary, straight-chain alkanes have regular structures. Their melting points, boiling points, and densities increase with molecular weight because of increasing attraction between molecules.

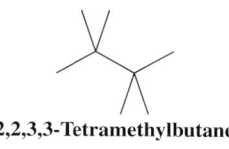

2,2,3,3-Tetramethylbutane

2-5 Rotation About Single Bonds: Conformations

We have considered how intermolecular forces can affect the physical properties of molecules. These forces act *between* molecules. In this section, we shall examine how the forces present *within* molecules (i.e., intramolecular forces) make some arrangements in space energetically more favorable than others.

Rotation interconverts the conformations of ethane

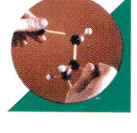

If we build a molecular model of ethane, we can see that the two methyl groups are readily rotated with respect to each other. The energy required to move the hydrogen atoms past each other, the *barrier to rotation,* is only 3 kcal mol^{-1}. This value turns out to be so low that chemists speak of "free rotation" of the methyl groups. In general, *there is free rotation about all single bonds.*

Figure 2-5 depicts the rotational movement in ethane by the use of dashed-wedged line structures (Section 1-9). There are two extreme ways of drawing ethane: the staggered conformation and the eclipsed one. If the **staggered conformation** is viewed along the C–C axis, each hydrogen atom on the first carbon is seen to be positioned perfectly between two hydrogen atoms on the second. The second extreme is derived

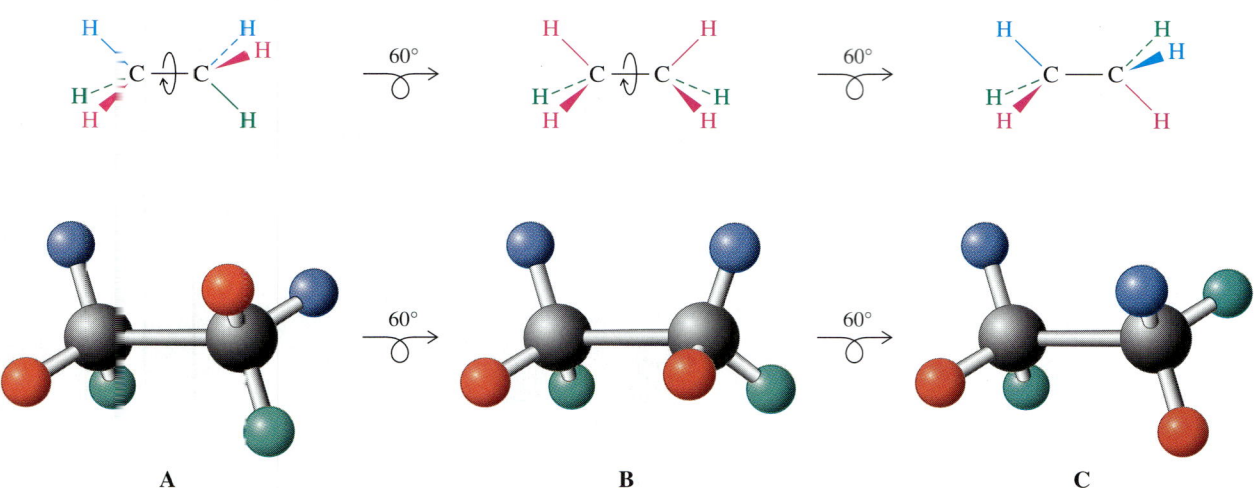

FIGURE 2-5
Rotation in ethane: (A and C) staggered conformations; (B) eclipsed. There is virtually "free rotation" between conformers.

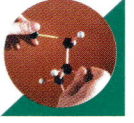

from the first by a 60° turn of one of the methyl groups about the C–C bond. Now, if this **eclipsed conformation** is viewed along the C–C axis, all hydrogen atoms on the first carbon are directly opposite those on the second—that is, those on the first eclipse those on the second. A further 60° turn converts the eclipsed form into a new but equivalent staggered arrangement. Between these two extremes, rotation of the methyl group results in numerous additional positions, referred to collectively as **skew conformations.**

The many forms of ethane (and, as we shall see, substituted analogs) created by such rotations are **conformations** (also called **conformers** and **rotamers**). All of them rapidly interconvert at room temperature. The study of their thermodynamic and kinetic behavior is **conformational analysis.**

Newman projections depict the conformations of ethane

A simple alternative to the dashed-wedged line structures for illustrating the conformers of ethane is the **Newman* projection.** We can arrive at a Newman projection from the dashed-wedged line picture by turning the molecule out of the plane of the page toward us and viewing it along the C–C axis (Figure 2-6A and B). In this notation, the front carbon obscures the back carbon, but the bonds emerging from both are clearly seen. The front carbon is depicted as the point of juncture of the three bonds attached to it, one of them usually drawn vertically and pointing up. The back carbon is a circle (Figure 2-6C). The bonds to this carbon project from the outer edge of the circle. The extreme conformational shapes of ethane are readily drawn in this way (Figure 2-7). To make the three rear hydrogen atoms more visible in eclipsed conformations, they are drawn somewhat rotated out of the perfectly eclipsing position.

*Professor Melvin S. Newman (1908–1993), Ohio State University.

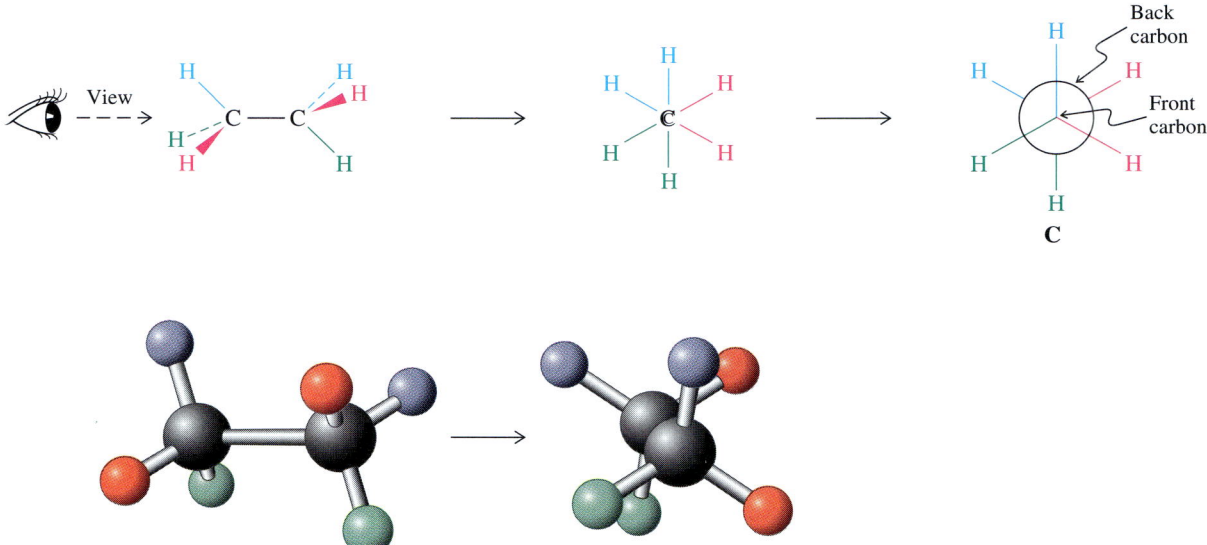

FIGURE 2-6
Representations of ethane. (A) Side-on views of the molecule. (B) End-on views of ethane, showing the carbon atoms directly in front of each other and the staggered positions of the hydrogens. (C) Newman projection of ethane derived from the view shown in (B). The "front" carbon is represented by the intersection of the bonds to its three attached hydrogens. The bonds from the remaining three hydrogens connect to the large circle, which represents the "back" carbon.

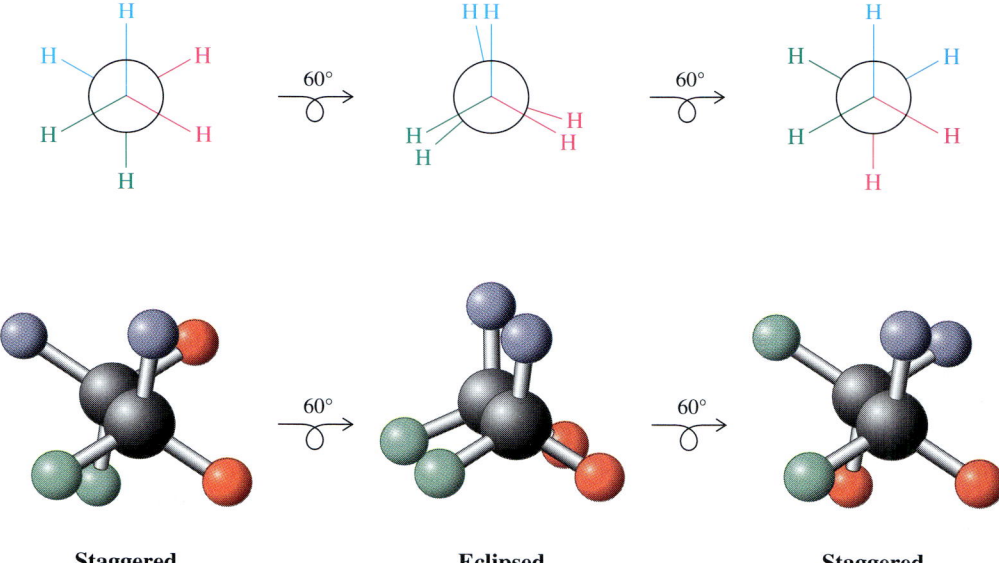

FIGURE 2-7
Newman projections and ball-and-stick models of staggered and eclipsed rotamers of ethane. In these representations, the back carbon is rotated clockwise in increments of 60°.

2-6 Potential-Energy Diagrams

As mentioned earlier, about 3 kcal mol^{-1} of heat is required to rotate the methyl groups in ethane. What is the reason for this requirement?

The rotamers of ethane have different potential energies

The various rotamers of ethane do not all have the same potential energies. A simple explanation is based on electron repulsion. As one methyl group turns about the C–C axis, starting from a staggered conformation, the distance between the hydrogen atoms of the respective methyl groups begins to diminish, resulting in increasing repulsion between the bonding pairs of electrons in the C–H bonds. Thus, the potential energy of the system rises steadily as the methyl group rotates from staggered through skew to eclipsed conformations. At the point of eclipsing, the molecule has its highest energy content, because at this stage the two sets of six bonding electrons are closest. This point is 3 kcal mol^{-1} above the lowest energy state of the molecule, the staggered rotamer. The change in energy resulting from bond rotation is called **rotational** or **torsional energy.**

Potential-energy diagrams are a convenient way to depict energy changes

The differences in potential energy between rotamers can be pictured by plotting the energy changes against the degree of rotation (Figure 2-8). Such a plot is called a **potential-energy diagram.** Potential-energy diagrams are useful in the description

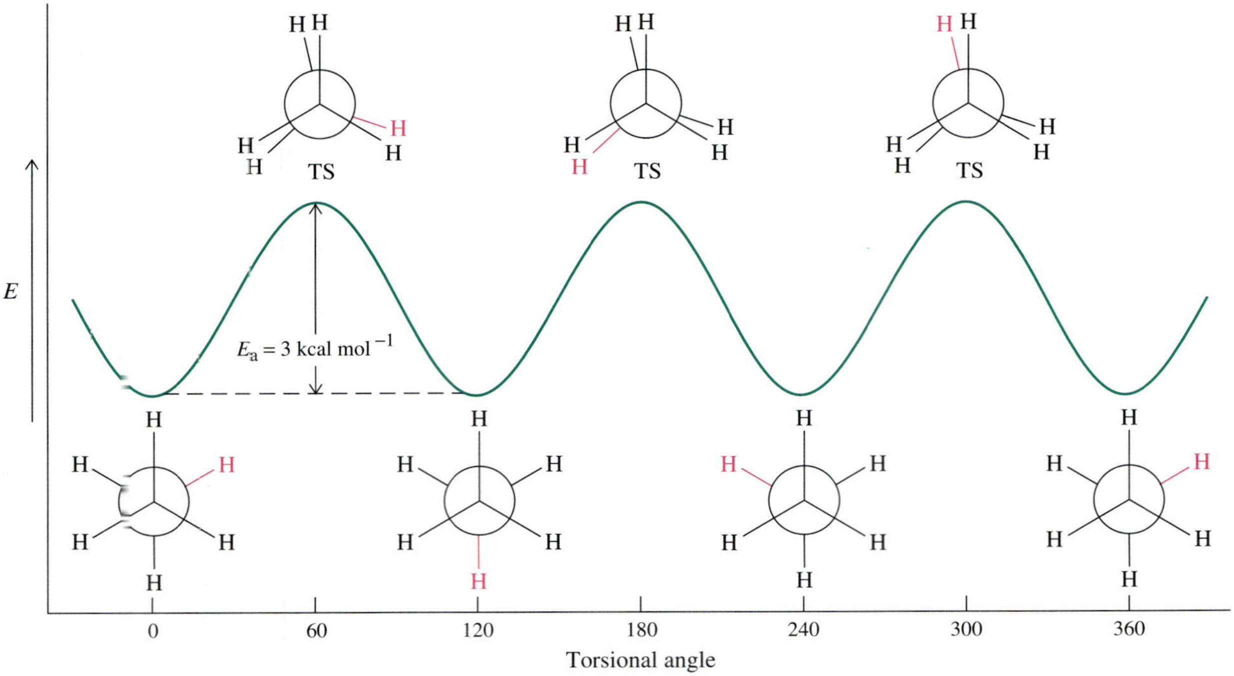

FIGURE 2-8 _____

Potential-energy diagram of the rotational isomerism in ethane. Because the eclipsed conformations have the highest energy, they correspond to peaks in the diagram. These maxima may be viewed as transition states (TS) between the more stable staggered rotamers. The activation energy (E_a) corresponds to the barrier to rotation.

$E_a = 3$ kcal mol^{-1}

Torsional angle

of other chemical processes as well. Changes in potential energy are plotted against a **reaction coordinate,** which describes the progress of the process or reaction. In the diagram for rotation of ethane, the reaction coordinate is degrees of rotation (Figure 2-8), usually called the **torsional angle.** Ethane is best described in its staggered conformation. In fact, the eclipsed rotamer has only a fleeting lifetime (of the order of 10^{-12} s) as the hydrogens rapidly move past each other, equilibrating one staggered arrangement with another. Because eclipsed conformations have the highest energy, they are maxima in potential-energy diagrams. Such points are called **transition states** (TS), marking the transition from one staggered rotamer to another. The energy of the transition state can be viewed as the barrier to be overcome when the molecule goes from one staggered arrangement to the next. This energy is called the **activation energy,** E_a, for the rotational process. The lower its value, the faster the rotation.

Collisions supply the energy to get past the activation-energy barrier

Where do organic molecules get the energy to overcome the barrier to rotation? Molecules have *kinetic energy* as a result of their motion, but at room temperature the average kinetic energy is only about 0.6 kcal mol^{-1}, far below the activation-energy barrier. To pick up enough energy, molecules must collide with each other or with the walls of the container. Each collision transfers energy from one molecule to another.

A graph called a **Boltzmann* distribution curve** depicts the distribution of kinetic energy. Figure 2-9 shows that, although most molecules have only average speed at any given temperature, some molecules have kinetic energies that are much higher. In ethane, part of this energy may be used to overcome the activation-energy barrier. Because continual collisions rapidly redistribute the kinetic energy, all molecules eventually get past this barrier. That is why we can speak of "free rotation."

The shape of the Boltzmann curve depends on the temperature. We can see that at higher temperatures, as the average kinetic energy increases, the curve flattens and shifts toward higher energies. More molecules now have energy higher than is required by the transition state, so the speed of rotation increases. Conversely, at lower temperatures, the rate of rotation decreases.

*Professor Ludwig Boltzmann (1844–1906), University of Vienna, Austria.

FIGURE 2-9

Boltzmann curves at two temperatures. At the higher temperature (green curve), there are more molecules of kinetic energy E than at the lower temperature (blue curve). Molecules with higher kinetic energy can more easily overcome the activation-energy barrier.

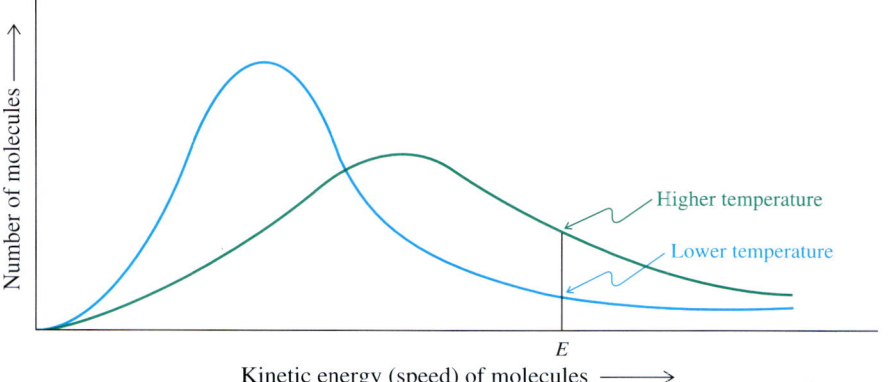

In summary, intramolecular forces control the arrangement of substituents on neighboring and bonded carbon atoms. In ethane, the relatively stable staggered conformations are interconverted through higher energy transition states in which substituents are eclipsed. To reach these transition states, molecules have to absorb the kinetic energy of others through collisions. The energy distribution of a collection of molecules at any given temperature is depicted by a Boltzmann curve. The energetics of rotation about the C–C bond is conveniently pictured in a potential-energy diagram.

2-7 Rotation in Substituted Ethanes

How does the potential-energy diagram change when a substituent is added to ethane? Consider, for example, propane, whose structure is similar to that of ethane, except that a methyl group replaces one of ethane's hydrogen atoms.

Steric hindrance raises the energy barrier to rotation

A potential-energy diagram for the rotation about a C–C bond in propane is shown in Figure 2-10. The Newman projections of propane differ from those of ethane only by the substituted methyl group. Again, the extreme conformations are staggered

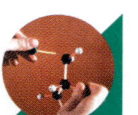

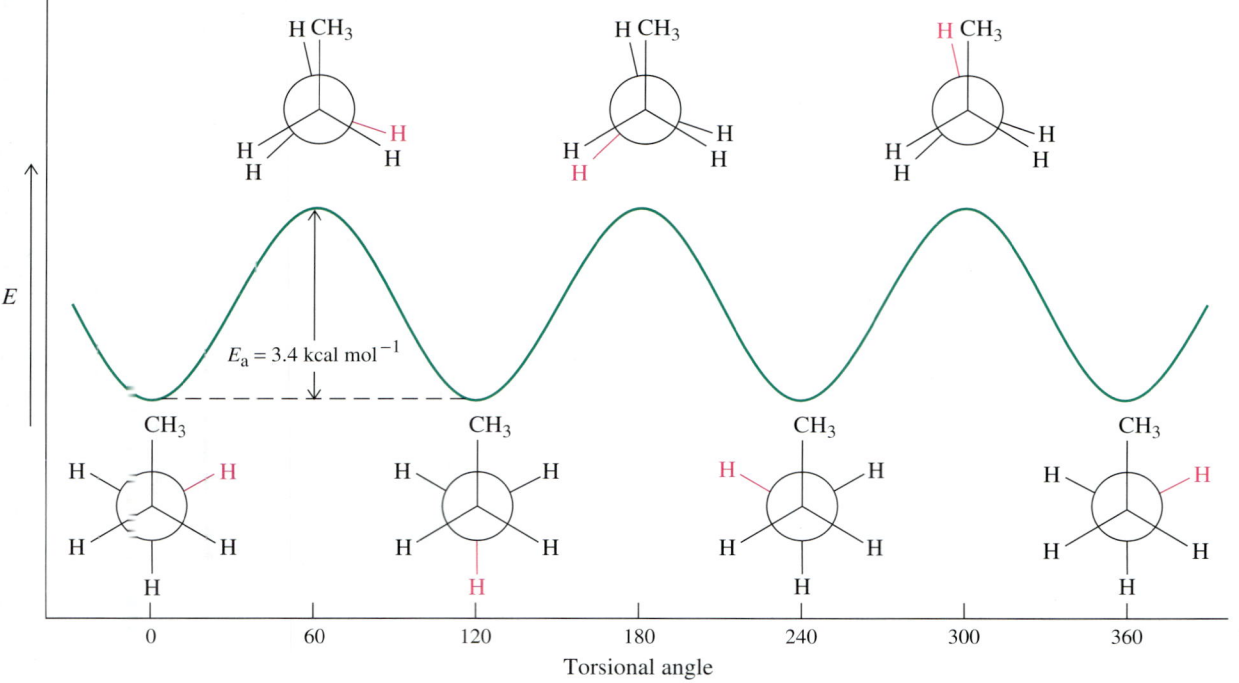

$E_a = 3.4$ kcal mol^{-1}

Torsional angle

FIGURE 2-10

Potential-energy diagram of the rotational isomerism in propane. Steric hindrance increases the relative energy of the eclipsed form.

and eclipsed. However, the activation barrier separating the two is 3.4 kcal mol^{-1}, slightly higher than that for ethane. This energy difference is due to unfavorable interference between the methyl substituent and the eclipsing hydrogen in the transition state, a phenomenon called **steric hindrance.** This effect arises from the fact that two molecular fragments cannot occupy the same region in space.

Steric hindrance in propane is actually worse than the E_a value for rotation indicates. Methyl substitution raises the energy not only of the eclipsed conformation, but also of the staggered (lowest energy, or *ground* state) one, the latter to a lesser extent because of less hindrance. Because the activation energy is equal to the *difference* in energy between ground and transition states, the net result is a small increase in E_a.

There can be more than one staggered and one eclipsed conformation: conformational analysis of butane

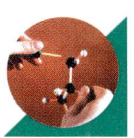

If we build a model and look at the rotation about the central C–C bond of butane, we find that there are more conformations than one staggered and one eclipsed (Figure 2-11). Consider the staggered conformer in which the two methyl groups are as far away from each other as possible. This arrangement, called *anti* (i.e., opposed), is the most stable because steric hindrance is minimized. Rotation of the rear carbon in the Newman projections in either direction (in Figure 2-11, the direction is clockwise) produces an eclipsed conformation with two CH$_3$–H interactions. This rotamer is 3.8 kcal mol^{-1} higher in energy than the *anti* precursor. Further rotation furnishes a *new* staggered structure in which the two methyl groups are closer than they are in the *anti* conformation. To distinguish this conformer from the others, it is named *gauche* (*gauche,* French, in the sense of awkward, clumsy). As a consequence of

Anti   *Gauche*   *Gauche*

FIGURE 2-11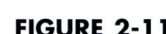
Clockwise rotation of the rear carbon along the C2–C3 bond in a Newman projection (top) and a ball-and-stick model (bottom) of butane.

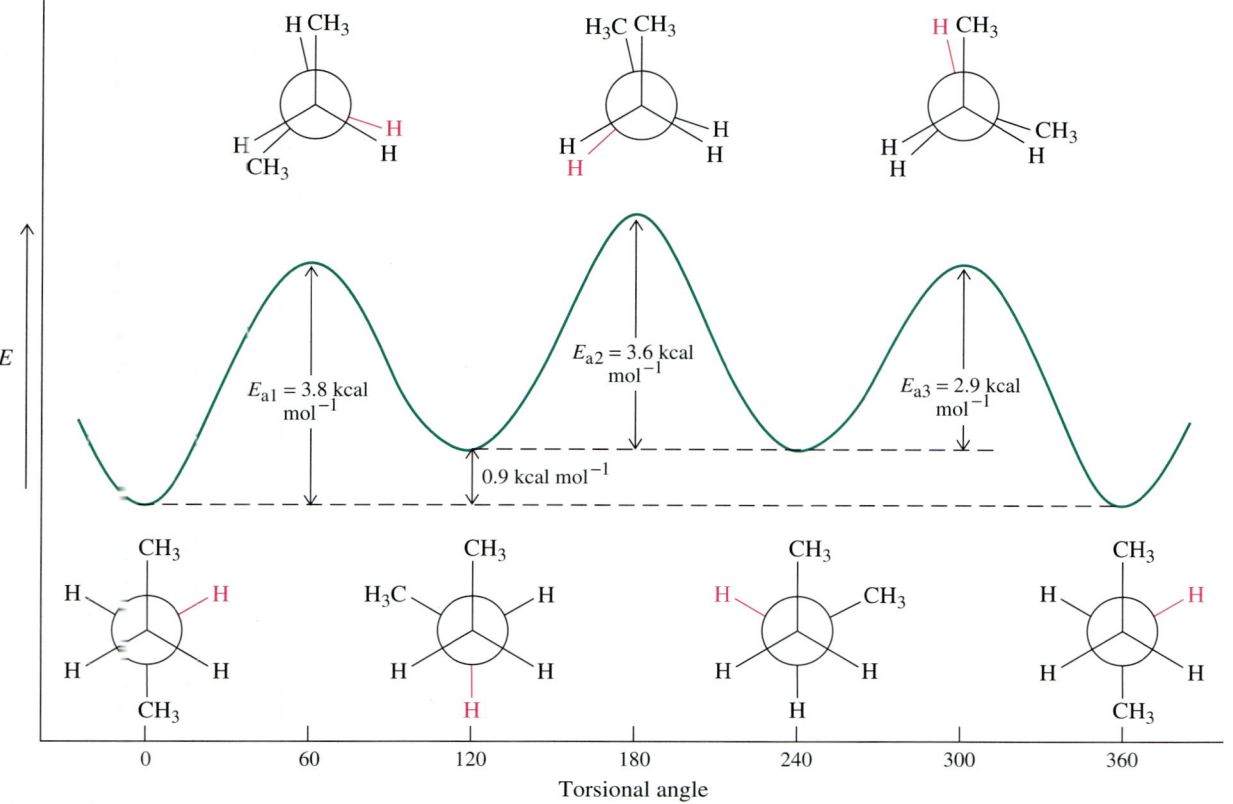

FIGURE 2-12

Potential-energy diagram of the rotation about the C2–C3 bond in butane. There are three processes: $anti \rightarrow gauche$ conversion with $E_{a1} = 3.8$ kcal mol^{-1}; $gauche \rightarrow gauche$ rotation with $E_{a2} = 3.6$ kcal mol^{-1}; and $gauche \rightarrow anti$ transformation with $E_{a3} = 2.9$ kcal mol^{-1}.

steric hindrance, the $gauche$ conformer is higher in energy than the $anti$ conformer by about 0.9 kcal mol^{-1}.

Further rotation (Figure 2-11) results in a *new* eclipsed arrangement in which the two methyl groups are superposed. Because the two bulkiest substituents eclipse in this rotamer it is energetically highest, 4.5 kcal mol^{-1} higher than the most stable $anti$ structure. Further rotation produces another $gauche$ conformer. The activation energy for $gauche \rightleftharpoons gauche$ interconversion is 3.6 kcal mol^{-1}. A potential-energy diagram summarizes the energetics of the rotation (Figure 2-12). The most stable $anti$ conformer is the most abundant in solution (about 72% at 25°C). Its less stable $gauche$ counterpart is present in lower concentration (28%).

We can see from Figure 2-12 that knowing the difference in thermodynamic stability of two conformers (e.g., 0.9 kcal mol^{-1} between the $anti$ and $gauche$ isomers) and the activation energy for proceeding from the first to the second (e.g., 3.8 kcal mol^{-1}) allows us to estimate the activation barrier of the reverse reaction. In this case, E_a for the $gauche$-to-$anti$ conversion is 3.8 − 0.9 = 2.9 kcal mol^{-1}.

EXERCISE 2-6

Draw the expected potential-energy diagram for the rotation about the C2–C3 bond in 2,3-dimethylbutane. Include the Newman projections of each staggered and eclipsed conformation.

In summary, if two substituents (one on each carbon atom) are 180° apart in a staggered Newman projection, they belong to an *anti* conformer. If they are 60° apart, the conformer is *gauche*. *Anti* conformations are usually more stable than their *gauche* counterparts. Conformational analysis is the study of the changes in potential energy that take place on rotation about single bonds.

2-8 Kinetics and Thermodynamics of Conformational Isomerism and of Simple Reactions

The *anti* $\rightleftharpoons$ *gauche* conformational isomerism is a typical example of an equilibrium between two distinct species. Although no bonds are broken or made, as in the usual chemical reaction, the process is controlled by the same physical criteria as are ordinary reactions.

We shall review two basic principles governing chemical reactions:

1. **Chemical thermodynamics** deals with the changes in energy that take place when processes such as conformational changes or chemical reactions occur, a feature that controls the *extent* to which a reaction will go to completion.
2. **Chemical kinetics** concerns the velocity or rate at which the concentrations of reactants and products change, in other words the *speed* at which a reaction will go to completion.

The two phenomena are frequently related. Reactions that are thermodynamically very favorable often proceed faster than do less favorable ones. Conversely, some reactions are faster than others even though they result in a comparatively less stable product. A transformation that yields the most stable products is said to be under **thermodynamic control.** Its outcome is determined by the net favorable change in energy in going from starting materials to products. A reaction in which the product obtained is the one formed fastest is defined as being under **kinetic control.** Let us put these statements on a more quantitative footing.

Equilibria are governed by the thermodynamics of chemical change

All chemical reactions are reversible, and reactants and products interconvert to various degrees. When the concentrations of reactants and products no longer change, the reaction is in a **state of equilibrium.** In many cases, equilibrium lies extensively (say, more than 99.9%) on the side of the products. When this occurs, the reaction is said to have *gone to completion.* (In such cases, the arrow indicating the reverse reaction is usually omitted.) Equilibria are described by equilibrium constants, K. To find an equilibrium constant, divide the arithmetic product of the concentrations of the components on the right side of the reaction by that of the components on the left, all given in units of moles per liter (mol L^{-1}). A large value for K indicates that a reaction will go to completion; it is said to have a large **driving force.**

Typical Chemical Equilibria

$$A \underset{}{\overset{K}{\rightleftharpoons}} B \qquad K = \frac{[B]}{[A]}$$

$$A + B \underset{}{\overset{K}{\rightleftharpoons}} C + D \qquad K = \frac{[C][D]}{[A][B]}$$

TABLE 2-5	Equilibria and Free Energy for A $\rightleftharpoons$ B; $K = $ [B]/[A]		
	Percentage		
K	B	A	$\Delta G°$ (kcal mol^{-1} at 25°C)
0.01	0.99	99.0	+2.73
0.1	9.1	90.9	+1.36
0.33	25	75	+0.65
1	50	50	0
2	67	33	−0.41
3	75	25	−0.65
4	80	20	−0.82
5	83	17	−0.95
10	90.9	9.1	−1.36
100	99.0	0.99	−2.73
1,000	99.9	0.1	−4.09
10,000	99.99	0.01	−5.46

If a reaction has gone to completion, a certain amount of energy has been released. The equilibrium constant can be related directly to the thermodynamic function of the **Gibbs*** **standard free energy change, $\Delta G°$**,† at equilibrium:

$$\Delta G° = -RT \ln K = -2.303 \, RT \log K \text{ (in kcal mol}^{-1})$$

in which R is the gas constant (1.986 cal deg^{-1} mol^{-1}) and T is the absolute temperature in kelvins‡ (K). A negative value for $\Delta G°$ signifies a release of energy. The equation shows that a large value for K indicates a large favorable free energy change. At room temperature (25°C, 298 K), the preceding equation becomes

$$\Delta G° = -1.36 \log K \text{ (in kcal mol}^{-1})$$

This expression tells us that an equilibrium constant of 10 would have a $\Delta G°$ of −1.36 kcal mol^{-1}, and, conversely, a K of 0.1 would have a $\Delta G° = +1.36$ kcal mol^{-1}. Because the relation is logarithmic, changing the $\Delta G°$ value affects the K value exponentially. When $K = 1$, starting materials and products are present in equal concentrations and $\Delta G°$ is zero (Table 2-5).

*Professor Josiah Willard Gibbs (1839–1903), Yale University, Connecticut.
†The descriptor $\Delta G°$ refers to the free energy of a reaction with the molecules in their standard states (e.g., ideal molar solutions) after it has reached equilibrium.
‡Temperature intervals in kelvins and degrees Celsius are identical. Temperature units are named after Lord Kelvin, Sir William Thomson (1824–1907), University of Glasgow, Scotland, and Anders Celsius (1701–1744), University of Uppsala, Sweden.

EXERCISE 2-7

Calculate the equilibrium concentration of *gauche* butane at 25°C and at 100°C. Use data from Figure 2-12. (**Hint:** Recall that $\Delta G°$ for a process in the thermodynamically favored direction—less stable to more stable—is negative.)

The free energy change is related to changes in bond strengths and the degree of order in the system

The Gibbs standard free energy change is related to two other thermodynamic quantities: the change in **enthalpy**, $\Delta H°$, and the change in **entropy**, $\Delta S°$.

Gibbs Standard Free Energy Change

$$\Delta G° = \Delta H° - T\Delta S°$$

In this equation, T is again in kelvins and $\Delta H°$ in kcal mol^{-1}, whereas $\Delta S°$ is in cal deg^{-1} mol^{-1}, also called entropy units (e.u.).

The **enthalpy change**, $\Delta H°$, is the heat of a reaction at constant pressure. Enthalpy changes in an organic chemical reaction relate mainly to changes in bond strengths in the course of the reaction. Thus, the value of $\Delta H°$ can be estimated by subtracting the sum of the strengths of the bonds formed from that of the bonds broken.

Enthalpy Change in a Reaction

$$\left(\begin{array}{c}\text{Sum of strengths}\\\text{of bonds broken}\end{array}\right) - \left(\begin{array}{c}\text{sum of strengths}\\\text{of bonds formed}\end{array}\right) = \Delta H°$$

If the bonds formed are stronger than those broken, the value of $\Delta H°$ is negative and the reaction is defined as **exothermic** (releasing heat). In contrast, a positive $\Delta H°$ is characteristic of an **endothermic** (heat-absorbing) process. An example of an exothermic reaction is the combustion of methane, the main component of natural gas, to carbon dioxide and liquid water. This process has a $\Delta H°$ value of -213 kcal mol^{-1}.

$$CH_4 + 2\ O_2 \rightarrow CO_2 + 2\ H_2O_{liq} \qquad \Delta H° = -213 \text{ kcal mol}^{-1}$$

The exothermic nature of this reaction is due to the very strong bonds formed in the products. Many hydrocarbons release a lot of energy on combustion and are therefore valuable fuels.

If the enthalpy of a reaction strongly depends on changes in bond strength, what is the significance of $\Delta S°$? The **entropy change**, $\Delta S°$, is a measure of the changes in the order of a system. The value of $S°$ increases with increasing disorder. Because of the negative sign in front of the $T\Delta S°$ term in the equation for $\Delta G°$, a positive value for $\Delta S°$ makes a negative contribution to the free energy of the system. In other words, going from order to disorder is thermodynamically favorable.

What is meant by disorder in a chemical reaction? Consider a transformation in which the number of reacting molecules differs from the number of product molecules formed. For example, upon strong heating, 1-pentene undergoes cleavage into ethene and propene. This process, in which two molecules are made from one, has a

relatively large positive $\Delta S°$. The increased number of particles present after bond cleavage means greater freedom of motion, thus representing an increase in disorder for the system.

$$CH_3CH_2CH_2CH=CH_2 \longrightarrow CH_2=CH_2 + CH_3CH=CH_2 \qquad \Delta H° = +22.4 \text{ kcal mol}^{-1}$$
$$\Delta S° = +33.3 \text{ e.u.}$$

1-Pentene	Ethene	Propene
	(Ethylene)	

EXERCISE 2-8

Calculate the $\Delta G°$ at 25°C for the preceding reaction. Is it thermodynamically feasible at 25°C? What is the effect of increasing T on $\Delta G°$? What is the temperature at which the reaction becomes favorable?

In contrast, disorder and entropy decrease when the number of product molecules is less than the number of molecules of starting materials. For example, the reaction of ethene (ethylene) with hydrogen chloride to give chloroethane is exothermic by $-15.5 \text{ kcal mol}^{-1}$, but the entropy makes an unfavorable contribution to the $\Delta G°$; $\Delta S° = -31.3$ e.u.

$$CH_2=CH_2 + HCl \longrightarrow CH_3CH_2Cl \qquad \Delta H° = -15.5 \text{ kcal mol}^{-1}$$
$$\Delta S° = -31.3 \text{ e.u.}$$

EXERCISE 2-9

Calculate the $\Delta G°$ at 25°C for the preceding reaction. In your own words, explain why a reaction that combines two molecules into one should have a large negative entropy change.

The rate of a chemical reaction depends on the activation energy

How fast is equilibrium established? The thermodynamic features of chemical reactions do not by themselves tell us anything about their rates. Consider the conversion of *gauche* butane into the *anti* rotamer (Figure 2-12). This process is thermodynamically favorable by only a small amount, and yet equilibrium is established exceedingly rapidly, even at very low temperatures. Now compare that with the combustion of methane considered earlier. This process releases 213 kcal mol^{-1}, a huge amount of energy, but we know that methane does not spontaneously ignite in air at room temperature. Why is the much more favorable combustion process so much slower? The answer is that rates of chemical processes are controlled by activation energies. We have already seen that E_a for bond rotation in butane is very low, which corresponds to a low-energy transition state, through which the molecule may pass very rapidly. Conversely, the transition state for methane combustion is very high in energy, corresponding to a high E_a and a very low rate (Figure 2-13).

How can there be such high activation energies for exothermic reactions? A simple answer is that partial bond-breaking usually precedes partial bond formation. Thus, before energy is released through bonding, energy must be expended to break bonds. The transition state is the point at which the initial energy input is compensated by a corresponding release of energy.

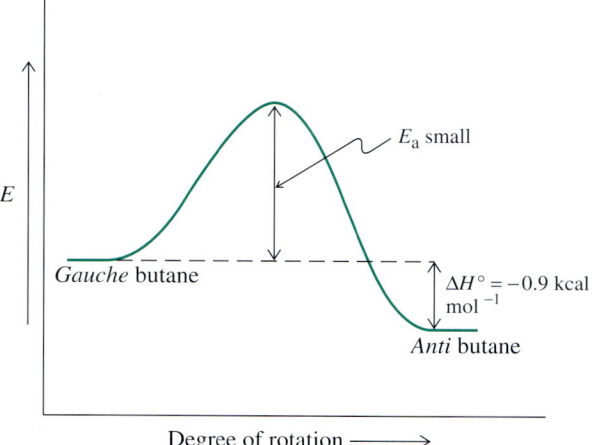

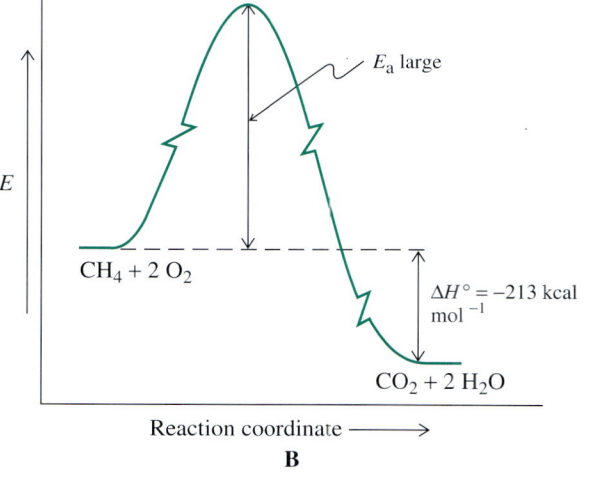

FIGURE 2-13

Comparison of potential-energy diagrams for (A) *gauche–anti* conversion in butane and (B) the combustion of methane. Comparison of the activation energies explains why bond rotation in butane is so much faster, even though the combustion reaction of methane is so much more thermodynamically favorable, as shown by its large negative $\Delta H°$. The diagrams are not drawn to scale.

The concentration of reactants can affect reaction rates

Consider the addition of reagent A to reagent B to give C:

$$A + B \longrightarrow C$$

In many transformations of this type, increasing the concentration of either reactant increases the rate of the reaction. In such cases, the transition-state structure incorporates both molecules A and B. The rate is expressed by

$$\text{Rate} = k[A][B] \text{ in units of mol } L^{-1} s^{-1}$$

in which the proportionality constant, k, is also called the **rate constant** of the reaction. A reaction for which the rate depends on the concentrations of two molecules in this way is said to be of **second order.**

There are processes whose rate depends on the concentration of only one reactant, such as the hypothetical reaction

$$A \longrightarrow B$$
$$\text{Rate} = k[A] \text{ in units of mol } L^{-1} s^{-1}$$

A reaction of this type is said to be of **first order.** Rotation about a carbon–carbon bond follows such a rate law.

EXERCISE 2-10

The dependence of reaction rates on reactant concentrations means that reactions slow down as starting materials are used up. For example, for a process following the first-order rate law, rate = $k[A]$, when half of A is consumed (i.e., after 50% conversion of the starting material), the reaction rate is reduced to half of its initial value. What would be the reduction in rate of a second-order reaction (in which rate = $k[A][B]$ after 50% conversion of the starting material?

The Arrhenius equation describes how temperature affects reaction rates

Temperature also greatly affects reaction rates. An increase in temperature leads to faster reactions. The kinetic energy of molecules increases when they are heated, which means that a larger fraction of them have sufficient energy to overcome the activation barrier (Figure 2-9). A useful rule of thumb applies to many reactions: Raising the reaction temperature by 10 degrees (Celsius) causes the rate to increase by a factor of 2 or 3. The Swedish chemist Arrhenius* noticed the dependence of reaction rate on temperature. He found that his measured data conformed to the equation

Arrhenius Equation

$$k = Ae^{-E_a/RT}$$

The A term is the maximum rate constant that the reaction would have if every molecule had sufficient collisional energy to overcome the activation barrier: At very high temperature, E_a/RT is small, $e^{-E_a/RT}$ approaches 1, and k nearly equals A. Each reaction has its own characteristic value for A. The Arrhenius equation describes how rates of reactions with different activation energies vary with temperature.

EXERCISE 2-11

(a) Calculate $\Delta G°$ at 25°C for the reaction $CH_3CH_2Cl \rightarrow CH_2{=}CH_2 + HCl$ (the reverse of the reaction in Exercise 2-9). **(b)** Calculate $\Delta G°$ at 500°C for the same reaction. (**Hint:** Apply $\Delta G° = \Delta H° - T\Delta S°$ and do not forget to first convert degrees Celsius into kelvins.)

EXERCISE 2-12

For the reaction in Exercise 2-11, $A = 10^{14}$ and $E_a = 58.4$ kcal mol^{-1}. Using the Arrhenius equation, calculate k at 500°C for this reaction. $R = 1.986$ cal deg^{-1} mol^{-1}.

This completes our review of the thermodynamic and kinetic relations governing many organic transformations. All reactions are described by equilibrating starting materials and products. On which side the equilibrium lies depends on the size of the equilibrium constant, in turn related to the Gibbs free energy changes, $\Delta G°$. An increase in the equilibrium constant by a factor of 10 is associated with a change in $\Delta G°$ of about -1.36 kcal mol^{-1} at 25°C. The free energy change of a reaction is

*Professor Svante Arrhenius (1859–1927), Technical Institute of Stockholm, Sweden, Nobel Prize 1903 (chemistry), director of the Nobel Institute from 1905 until shortly before his death.

composed of changes in enthalpy, $\Delta H°$, and entropy, $\Delta S°$. Contributions to the former stem mainly from variations in bond strengths, to the latter from the relative disorder in starting materials and products. Whereas these terms define the position of an equilibrium, the rate at which it is established depends on the concentration of starting materials, the activation barrier separating reactants and products, and the temperature. The relation between rate, E_a, and T is expressed by the Arrhenius equation.

2-9 Acids and Bases: A Review

In this chapter we have begun our study of organic compounds with the nonfunctional alkanes, and we have briefly examined the variety of functionalized compounds that may be derived from them. We have also looked into the energetic aspects of one of the physical processes common to virtually all organic molecules, rotation about single bonds. Now we turn to a fundamental application of thermodynamics, the chemistry of acids and bases. We shall see that acid-base processes provide models for the reactivity of polar organic functional groups. This section reviews the way in which acids and bases interact and how this process is quantified.

Acid and base strengths are measured by equilibrium constants

Brønsted and Lowry have given us a simple definition of acids and bases: An **acid** is a proton donor and a **base** is a proton acceptor. Acidity and basicity are commonly measured in water. An acid donates protons to water to give the hydronium ion, whereas a base removes them to give the hydroxide ion. Examples are hydrogen chloride for the former and sodium methoxide for the latter.

$$H-\ddot{\underset{..}{Cl}}: + H\ddot{O}H \rightleftharpoons H-\overset{H}{\underset{H}{\overset{|}{O}}}:^+ + :\ddot{\underset{..}{Cl}}:^-$$

Hydronium ion

$$CH_3\ddot{O}:^- Na^+ + H\ddot{O}H \rightleftharpoons CH_3OH + Na^+ + {}^-:\ddot{O}H$$

Hydroxide ion

Water itself is neutral. It forms an equal number of hydronium and hydroxide ions by self-dissociation. The process is described by the equilibrium constant K_w, the self-ionization constant of water. At 25°C,

$$H_2O + H_2O \xrightleftharpoons{K_w} H_3O^+ + OH^- \qquad K_w = [H_3O^+][OH^-] = 10^{-14} \text{ mol}^2 \text{ L}^{-2}$$

From the value for K_w, it follows that the concentration of H_3O^+ in pure water is 10^{-7} mol L^{-1}.

The pH is defined as the negative logarithm of the value for $[H_3O^+]$.

$$pH = -\log [H_3O^+]$$

Thus, for pure water, the pH is +7. An aqueous solution with a pH lower than 7 is acidic; that with a pH higher than 7 is basic.

The acidity of a general acid, HA, is expressed by the following general equation, together with its associated equilibrium constant.

$$HA + H_2O \xrightleftharpoons{K} H_3O^+ + A^- \qquad K = \frac{[H_3O^+][A^-]}{[HA][H_2O]}$$

Because, in aqueous solution, $[H_2O]$ is constant at 55 mol L^{-1}, that number may be incorporated into a new constant, the **acidity constant, K_a**.

$$K_a = K[H_2O] = \frac{[H_3O^+][A^-]}{[HA]} \text{ mol } L^{-1}$$

Like the concentration of H_3O^+ and its relation to pH, this measurement may be put on a logarithmic scale by the corresponding definition of pK_a.

$$pK_a = -\log K_a$$

The pK_a is the pH at which the acid is 50% dissociated. An acid with a pK_a lower than 1 is defined as strong, one with a pK_a higher than 4 as weak. The acidities of several common acids are compiled in Table 2-6 and compared with those of com-

TABLE 2-6	Relative Strengths of Common Acids (25°C)	
Acid	**K_a**	**pK_a**
Hydrogen iodide, HI (strongest acid)	1.6×10^5	-5.2
Sulfuric acid, H_2SO_4	1.0×10^5	-5.0^a
Hydrogen bromide, HBr	5.0×10^4	-4.7
Hydrogen chloride, HCl	160	-2.2
Hydronium ion, H_3O^+	50	-1.7
Methanesulfonic acid, CH_3SO_3H	16	-1.2
Hydrogen fluoride, HF	6.3×10^{-4}	3.2
Acetic acid, CH_3COOH	2.0×10^{-5}	4.7
Hydrogen cyanide, HCN	6.3×10^{-10}	9.2
Methanethiol, CH_3SH	1.0×10^{-10}	10.0
Methanol, CH_3OH	3.2×10^{-16}	15.5
Water, H_2O	2.0×10^{-16}	15.7
Ammonia, NH_3	1.0×10^{-35}	35
Methane, CH_4 (weakest acid)	$\sim 1.0 \times 10^{-50}$	~ 50

Note: $K_a = [H_3O^+][A^-]/[HA]$ mol L^{-1}.
^{a}First dissociation equilibrium.

pounds with higher pK_a values. Sulfuric acid and, with the exception of HF, the hydrogen halides, are very strong acids. Hydrogen cyanide, water, methanol, ammonia, and methane are decreasingly acidic, the last two being exceedingly weak.

Like acid dissociation, the protonation of bases and their basicity can be described by a corresponding set of equations. The basicity of a base, A^-, is governed by the equilibrium constant K'.

$$A^- + H_2O \xrightleftharpoons{K'} HO^- + HA \qquad K' = \frac{[HO^-][HA]}{[A^-][H_2O]}$$

By incorporation of the constant value for $[H_2O]$, this equilibrium constant transforms into the basicity constant, K_b, and gives rise to a set of pK_b values.

$$K_b = K'[H_2O] = \frac{[HO^-][HA]}{[A^-]} \text{ mol L}^{-1}$$

For an acid HA and its derived base A^-, K_a and K_b are related by simple multiplication.

$$K_a \times K_b = \frac{[H_3O^+][A^-]}{[HA]} \times \frac{[HO^-][HA]}{[A^-]} = [H_3O^+][HO^-] = K_w = 10^{-14}$$

We see that the product of the two is equal to the self-ionization constant of water. Hence,

$$pK_a + pK_b = 14$$

Therefore, if we know the pK_a of an acid HA, we automatically know the pK_b of A^-. Because of this relation, the species A^- derived from HA is frequently referred to as its **conjugate** base (*conjugatus*, Latin, joined). Conversely, a species HA would be the conjugate acid of base A^-. For example, Cl^- is the conjugate base of HCl, and CH_3OH is the conjugate acid of CH_3O^-. Or, HCl may be viewed as the conjugate acid of Cl^-, and CH_3O^- as the conjugate base of CH_3OH. It follows from this discussion that the conjugate base of a strong acid is weak, as is the conjugate acid of a strong base.

EXERCISE 2-13

Write the formula for the conjugate base of each of the following acids. **(a)** Sulfurous acid, H_2SO_3; **(b)** chloric acid, $HClO_3$; **(c)** hydrogen sulfide, H_2S; **(d)** dimethyloxonium, $(CH_3)_2OH^+$; **(e)** hydrogen sulfate, HSO_4^-.

EXERCISE 2-14

Write the formula for the conjugate acid of each of the following bases. **(a)** Dimethylamide, $(CH_3)_2N^-$; **(b)** sulfide, S^{2-}; **(c)** ammonia, NH_3; **(d)** propanone (acetone), $(CH_3)_2C{=}O$; **(e)** 2,2,2-trifluoroethoxide, $CF_3CH_2O^-$.

EXERCISE 2-15

Which is the stronger acid, nitrous (HNO_2, $pK_a = 3.3$) or phosphorous acid (H_3PO_3, $pK_a = 1.3$)? Calculate K_a for each.

We can predict relative acid and base strengths

Are there structural features that allow us to predict, at least qualitatively, the strength of an acid HA (and hence the weakness of its conjugate base)? Yes, there are several. Prominent among them are two:

1. The increasing *size* of A as we proceed down a column in the periodic table. This trend is seen in the ordering of the acid strengths of the hydrogen halides: HI > HBr > HCl > HF. Electron repulsion in the anionic conjugate base diminishes as the volume of space available for the charge to "spread out" increases.

2. The ability of the conjugate base A^- to accommodate the negative charge in either or both of two ways:

   (a) The increasing *electronegativity* of A as we proceed from left to right across a row in the periodic table. The more electronegative the atom to which the acidic proton is attached, the more acidic the latter will be. For example, the decreasing order of acidity in the series $HF > H_2O > H_3N > H_4C$ parallels the decreasing electronegativity of A (Table 1-2). In the hydrogen halides, this trend is outweighed by the size of A.

   (b) The *resonance* in A^- that allows delocalization of charge over several atoms. For example, acetic acid is more acidic than methanol. In both cases, an O–H bond dissociates into ions (heterolytic cleavage, Section 3-1). However, unlike methoxide, the acetate ion has two resonance structures to better accommodate the charge (Section 1-5) and is the weaker base.

Acetic Acid Is More Acidic Than Methanol Because of Resonance

$$CH_3\ddot{\text{O}}\!-\!H + H_2O \rightleftharpoons CH_3\!-\!\ddot{\text{O}}\!:^- + H_3O^+$$

Weaker acid   Stronger base

$$CH_3\overset{:O:}{\overset{\|}{C}}\!-\!\ddot{\text{O}}\!-\!H + H_2O \rightleftharpoons \left[CH_3\overset{:O:}{\overset{\|}{C}}\!-\!\ddot{\text{O}}\!:^- \longleftrightarrow CH_3C\!=\!\overset{:\ddot{\text{O}}:^-}{\ddot{\text{O}}} \right] + H_3O^+$$

Stronger acid   Weaker base

The effect of resonance is even more pronounced in sulfuric acid. The availability of *d* orbitals on sulfur enables us to write valence-shell-expanded Lewis structures containing as many as 12 electrons (Sections 1-4 and 1-5). Alternatively, charge-separated structures with one or two positive charges on sulfur can be used. Both representations indicate that the pK_a of H_2SO_4 should be low.

$$\left[\overset{:\ddot{\text{O}}:^-}{\underset{:\ddot{\text{O}}:^-}{HO\!-\!\overset{2+}{S}\!-\!\ddot{\text{O}}\!:^-}} \longleftrightarrow \overset{:O:}{\underset{:\ddot{\text{O}}:^-}{HO\!-\!\overset{\|+}{S}\!-\!\ddot{\text{O}}\!:^-}} \longleftrightarrow \overset{:O:}{\underset{:O:}{HO\!-\!\overset{\|}{S}\!-\!\ddot{\text{O}}\!:^-}} \longleftrightarrow \overset{:\ddot{\text{O}}:^-}{\underset{:O:}{HO\!-\!\overset{\|}{S}\!=\!\ddot{\text{O}}}} \longleftrightarrow \text{etc.} \right]$$

Hydrogen sulfate ion

The sulfonic acids also are quite strong, for the same reason. Consequently their conjugate bases, the sulfonates, are weak bases. As a rule, the acidity of HA increases to the right and down in the periodic table. Therefore, the basicity of A^- *decreases* in the same fashion.

The same molecule may act as an acid under one set of conditions and as a base under another. Water is the most familiar example of this behavior, but many other substances possess this capability as well. For instance, nitric acid acts as an acid in the presence of water but behaves as a base toward the more powerfully acidic H_2SO_4:

Nitric Acid Acting as an Acid

$$HNO_3 + H_2O \rightleftharpoons NO_3^- + H_3O^+$$

Nitric Acid Acting as a Base

$$H_2SO_4 + HNO_3 \rightleftharpoons HSO_4^- + H_2NO_3^+$$

Similarly, acetic acid protonates water, as shown earlier in this section, but is protonated by stronger acids such as HBr:

$$HBr + CH_3CO_2H \rightleftharpoons Br^- + CH_3CO_2H_2^+$$

EXERCISE 2-16

Suggest a structure for $CH_3CO_2H_2^+$. [**Hint:** Try placing the proton first on one, and then

$$\overset{\overset{\textstyle O}{\|}}{}$$

on the other of the two oxygen atoms in the molecule ($CH_3\overset{\overset{\textstyle O}{\|}}{C}-OH$), and consider which of the two resulting structures is better stabilized by resonance.]

Lewis acids and bases interact by sharing an electron pair

A more generalized description of acid-base interaction in terms of electron sharing was introduced by Lewis. A **Lewis acid** is a species that is at least two electrons short of a closed shell, whereas a **Lewis base** contains at least one lone pair of electrons.

Lewis Acids

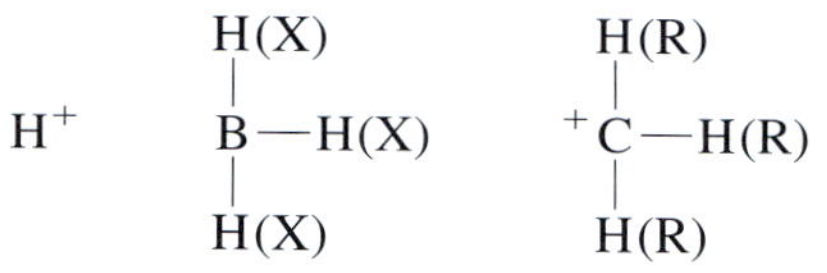

$$H^+ \qquad \overset{\textstyle H(X)}{\underset{\textstyle H(X)}{B-H(X)}} \qquad \overset{\textstyle H(R)}{\underset{\textstyle H(R)}{{}^+C-H(R)}} \qquad \begin{array}{c} MgX_2, AlX_3, \text{ many} \\ \text{transition metal} \\ \text{halide salts} \end{array}$$

Lewis Bases

$$^-:\!\ddot{O}H(R) \quad \overset{\textstyle}{\underset{\textstyle H(R)}{:\ddot{O}-H(R)}} \quad \overset{\textstyle}{\underset{\textstyle H(R)}{:\ddot{S}-H(R)}} \quad \overset{\textstyle H(R)}{\underset{\textstyle H(R)}{:N-H(R)}} \quad \overset{\textstyle H(R)}{\underset{\textstyle H(R)}{:P-H(R)}} \quad :\ddot{\ddot{X}}:$$

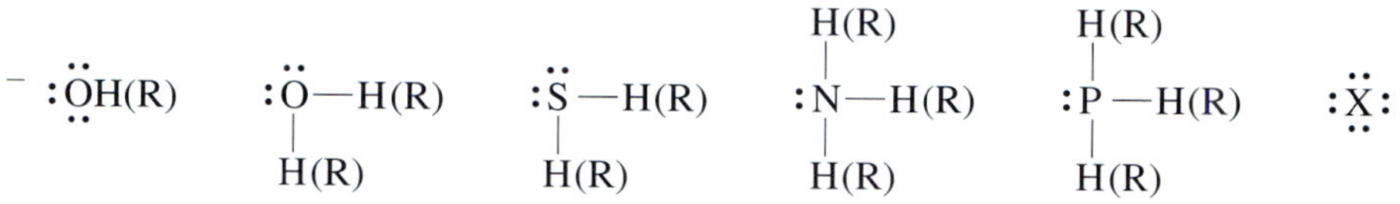

A Lewis base shares its lone pair with a Lewis acid to form a new covalent bond. From Section 1-5, we know that organic chemists routinely depict the movement of electron pairs through the use of curved arrows. A Lewis base–Lewis acid interaction may therefore be pictured by means of an arrow pointing in the direction that the

electron pair moves—from the base to the acid. The Brønsted acid-base reaction between hydroxide ion and a proton is an example of a Lewis acid-base process as well.

Lewis Acid-Base Reactions

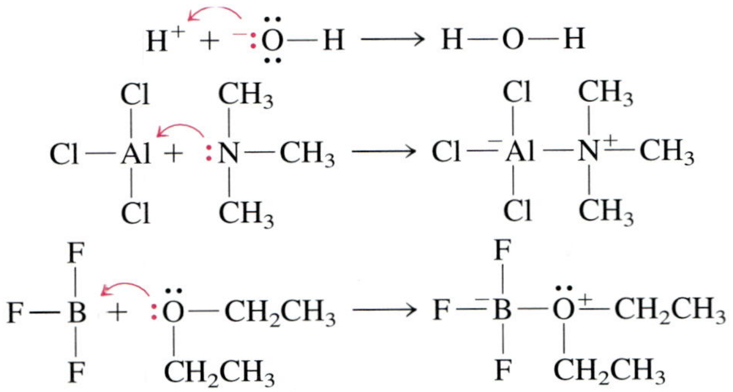

$$H^+ + :\overset{\displaystyle ..}{\underset{\displaystyle ..}{O}}-H \longrightarrow H-O-H$$

$$\underset{\underset{\displaystyle Cl}{|}}{\overset{\overset{\displaystyle Cl}{|}}{Cl-Al}} + :\underset{\underset{\displaystyle CH_3}{|}}{\overset{\overset{\displaystyle CH_3}{|}}{N}}-CH_3 \longrightarrow \underset{\underset{\displaystyle Cl}{|}}{\overset{\overset{\displaystyle Cl}{|}}{Cl-Al}}-\underset{\underset{\displaystyle CH_3}{|}}{\overset{\overset{\displaystyle CH_3}{|}}{N^+}}-CH_3$$

$$\underset{\underset{\displaystyle F}{|}}{\overset{\overset{\displaystyle F}{|}}{F-B}} + :\underset{\underset{\displaystyle CH_2CH_3}{|}}{\overset{..}{O}}-CH_2CH_3 \longrightarrow \underset{\underset{\displaystyle F}{|}}{\overset{\overset{\displaystyle F}{|}}{F-B}}-\underset{\underset{\displaystyle CH_2CH_3}{|}}{\overset{..}{O^+}}-CH_2CH_3$$

The dissociation of a Brønsted acid HA is just the reverse of the combination of the Lewis acid H^+ and the Lewis base A^-. We write it as follows:

Dissociation of a Brønsted Acid

$$H-A \longrightarrow H^+ + :A^-$$

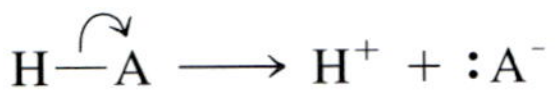

Notice that the curved arrow is from the bond *to A*, the direction in which the *pair of electrons* moves. *The curved arrow never points to the hydrogen atom in the dissociation of a Brønsted acid.*

Many processes in organic chemistry include either Lewis acid-base reactions or their reverse, dissociation of a covalent bond into ions. For example, certain haloalkanes are capable of ionization to give halide ions and alkyl cations, a process similar to dissociation of a Brønsted acid. As the curved arrow indicates, the electron pair originally constituting the C–Cl covalent bond shifts onto Cl, giving it a negative charge and leaving the carbon atom positive.

Dissociation of a Haloalkane into a Halide Ion and an Alkyl Cation

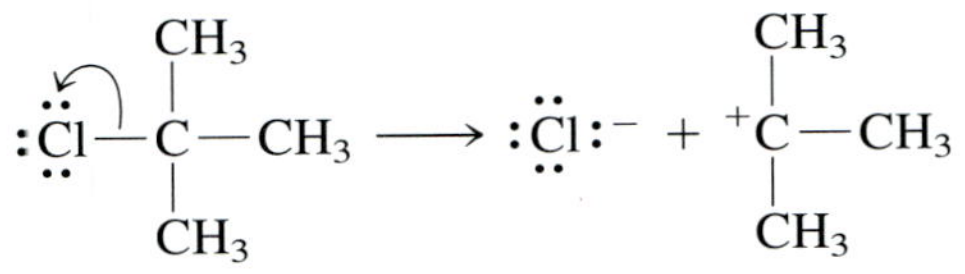

$$:\overset{..}{\underset{..}{Cl}}-\underset{\underset{\displaystyle CH_3}{|}}{\overset{\overset{\displaystyle CH_3}{|}}{C}}-CH_3 \longrightarrow :\overset{..}{\underset{..}{Cl}}:^- + \overset{+}{\underset{\underset{\displaystyle CH_3}{|}}{\overset{\overset{\displaystyle CH_3}{|}}{C}}}-CH_3$$

When the alkyl cation has formed, it may react with a Lewis base such as water. These processes are two of the steps in the conversion of haloalkanes into alcohols (Section 7-1).

Lewis Acid-Base Reaction of Water and an Alkyl Cation

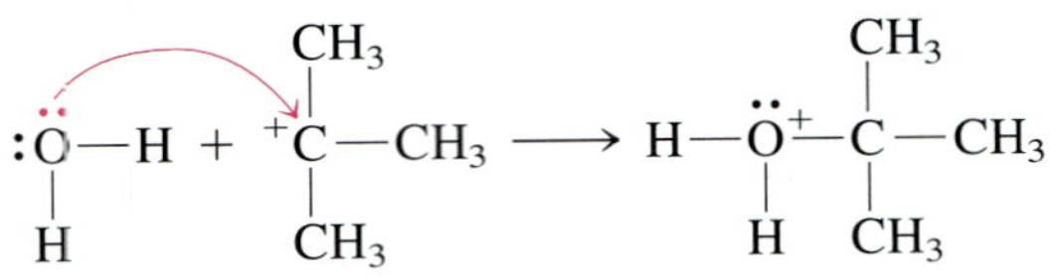

$$:\overset{..}{\underset{\underset{\displaystyle H}{|}}{O}}-H + \overset{+}{\underset{\underset{\displaystyle CH_3}{|}}{\overset{\overset{\displaystyle CH_3}{|}}{C}}}-CH_3 \longrightarrow H-\overset{..}{\underset{\underset{\displaystyle H}{|}}{O^+}}-\underset{\underset{\displaystyle CH_3}{|}}{\overset{\overset{\displaystyle CH_3}{|}}{C}}-CH_3$$

In summary, in Brønsted-Lowry terms, acids are proton donors and bases are proton acceptors. Acid-base interactions are governed by equilibria, which are quantitatively described by an acidity constant K_a. Removal of a proton from an acid generates its conjugate base; attachment of a proton to a base forms its conjugate acid. Lewis bases donate an electron pair to form a covalent bond with Lewis acids, a process depicted by a curved arrow pointing from the lone pair of the base toward the acid.

CHAPTER INTEGRATION PROBLEM

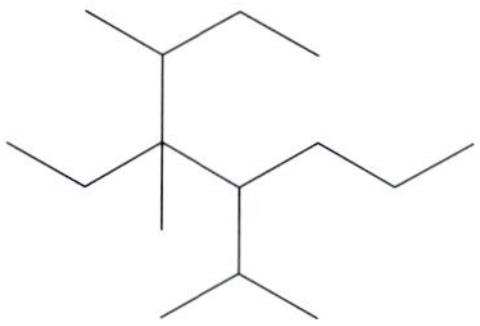

Consider the alkane shown in the margin.

a. Name this molecule according to the IUPAC system.

SOLUTION

Step 1. Locate the main, or stem, chain, the longest one in the molecule (shown in black in the illustration below). Do not be misled: The drawing of the stem chain can have almost any shape. The stem has eight carbons, so the base name is **octane.**
Step 2. Identify and name all substituents (shown in color): two **methyl** groups, an **ethyl** group, and a fourth, branched substituent. The branched substituent is named by first giving the number 1 (italicized in the illustration below) to the carbon that connects it to the main stem. By numbering away from the stem, we reach the number 2; therefore, the substituent is a derivative of the ethyl group (in green), onto which is attached a methyl group (red) at carbon 1. Thus, this substituent is called a **1-methylethyl** group.
Step 3. Number the stem chain, starting at the end closest to a carbon bearing a substituent. The numbering shown below gives a methyl-substituted carbon the number 3. Numbering the opposite way would have C4 as the lowest-numbered substituted carbon.
Step 4. Arrange the names of substituents alphabetically in the final name: *ethyl* comes first; then *methyl* comes before *methylethyl* (the "di" in *dimethyl,* denoting two methyl groups, is not considered in alphabetization because it is a multiplier of a substituent name and is therefore not considered part of the name).

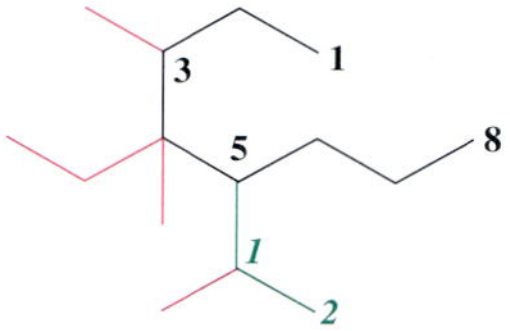

4-Ethyl-3,4-dimethyl-5-(1-methylethyl)octane

b. Draw structures to represent rotation about the C6–C7 bond. Correlate the structures that you draw with a qualitative potential energy diagram.

SOLUTION

Step 1. Identify the bond in question. Notice that much of the molecule can be treated simply as a large, complicated substituent on C6, the specific structure of which is

unimportant. For the purpose of this question, this large substituent may be replaced by "R." The "action" in this problem takes place between C6 and C7:

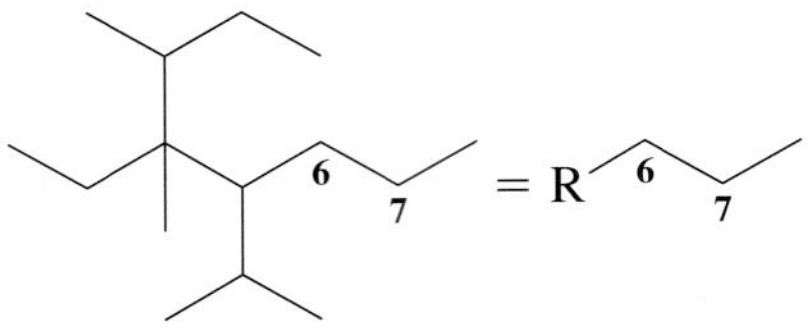

Step 2. Recognize that step 1 simplified the problem: rotation about the C6–C7 bond will give results very similar to rotation about the C2–C3 bond in butane. The only differences that a large "R" group has replaced one of the smaller methyl groups of butane.

Step 3. Draw conformations modeled after those of butane (Section 2-7) and superimpose them on an energy diagram similar to that in Figure 2-12. The only difference between this diagram and that for butane is that we do not know the exact heights of the energy maxima relative to the energy minima. However, we can expect them to be higher, qualitatively, because our "R" group is larger than a methyl group and thus can be expected to cause greater steric hindrance.

c. Two alcohols derived from this alkane are illustrated in the margin. Alcohols are categorized on the basis of the type of carbon atom that contains the –OH group (primary, secondary, or tertiary). Characterize the alcohols shown in the margin.

SOLUTION

In alcohol 1, the –OH group is located on a carbon atom that is directly attached to one other carbon, a primary carbon. Therefore, alcohol 1 is a primary alcohol. Similarly, the –OH group in alcohol 2 resides on a tertiary carbon (one attached to three other carbon atoms). It is a tertiary alcohol.

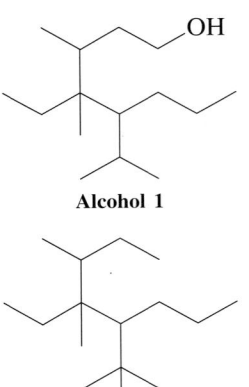

Alcohol 1

Alcohol 2

d. The –C–H bond in an alcohol is acidic to a similar degree to that in water. Primary alcohols have $K_a \approx 10^{-16}$; tertiary alcohols $K_a \approx 10^{-18}$. What are the approximate pK_a values for alcohols 1 and 2? Which is the stronger acid?

SOLUTION

The pK_a for alcohol 1 is approximately 16 ($-\log K_a$); that for alcohol 2 is about 18. Alcohol 1 with the lower pK_a value, is the stronger acid.

e. In which direction does the following equilibrium lie? Calculate K, the equilibrium constant, and $\Delta G°$, the free energy change, associated with the reaction as written in the left-to-right direction.

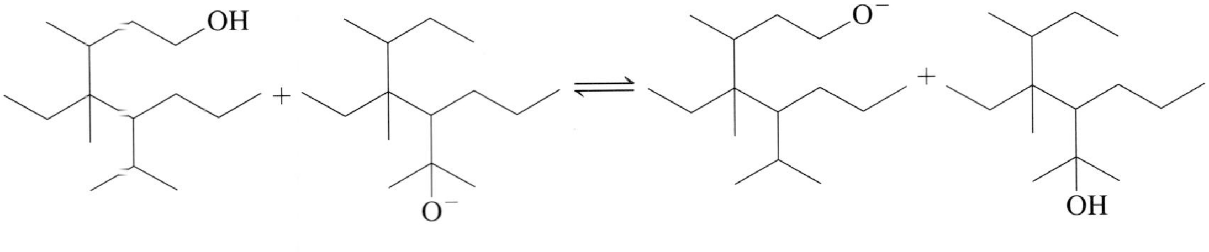

SOLUTION

The stronger acid (alcohol 1) is on the left; the weaker (alcohol 2) is on the right. Recall the relation between conjugate acids and bases: stronger acids have weaker conjugate bases, and vice versa. Relatively speaking, therefore, we have

		Conjugate base		Conjugate base		
Alcohol 1	+	of alcohol 2	⇌	of alcohol 1	+	Alcohol 2
(Stronger acid)		**(Stronger base)**		**(Weaker base)**		**(Weaker acid)**

The equilibrium lies to the right, on the side of the *weaker* acid-base pair. Recall that $K > 1$ and $\Delta G° < 0$ for a reaction that is thermodynamically favorable as written from left to right; *use* this information to be sure to get the magnitude of K and the sign of $\Delta G°$ correct. The equilibrium constant, K, for the process is the ratio of the K_a values, $(10^{-16}/10^{-18}) = 10^2$ (not 10^{-2}). With reference to Table 2-5, a K value of 100 corresponds to a $\Delta G°$ of -2.73 kcal mol^{-1} (not $+2.73$). If the reaction were written in the opposite direction, with the equilibrium lying to the left, the correct values would be those in parentheses.

IMPORTANT CONCEPTS

1. An organic molecule may be viewed as being composed of a carbon skeleton with attached **functional groups.**

2. **Hydrocarbons** are made up of carbon and hydrogen only. Hydrocarbons possessing only single bonds are also called **alkanes.** They do not contain functional groups. An alkane may exist as a single continuous chain or it may be branched or cyclic. The empirical formula for the **straight-chain** and **branched alkanes** is C_nH_{2n+2}.

3. Molecules that differ only in the number of methylene groups, CH_2, in the chain are called **homologs** and are said to belong to a homologous series.

4. A **primary carbon** is attached to only one other carbon. A **secondary carbon** is attached to two and a **tertiary** to three other carbon atoms. The hydrogen atoms bound to such carbon atoms are likewise designated primary, secondary, or tertiary.

5. The **IUPAC rules** for naming saturated hydrocarbons are (a) find the longest continuous chain in the molecule and name it; (b) name all groups attached to the longest chain as alkyl substituents; (c) number the carbon atoms of the longest chain; (d) write the name of the alkane, citing all substituents as prefixes arranged in alphabetical order and preceded by numbers designating their positions.

6. Alkanes attract each other through weak **London forces,** polar molecules through stronger dipole–dipole interactions, and salts mainly through very strong ionic interactions.

7. Rotation about carbon–carbon single bonds is rel-atively easy and gives rise to **conformations** (conformers, rotamers). Substituents on adjacent carbon atoms may be **staggered** or **eclipsed.** The eclipsed conformation is a transition state between staggered conformers. The energy required to reach the eclipsed state is called the activation energy for rotation. When both carbons bear alkyl or other groups, there may be additional conformers: Those in which the groups are in close proximity (60°) are *gauche;* those in which the groups are directly opposite (180°) each other are *anti.* Molecules tend to adopt conformations in which steric hindrance, as in *gauche* conformations, is minimized.

8. Chemical reactions can be described as equilibria controlled by **thermodynamic** and **kinetic** parameters. The change in the **Gibbs free energy,** $\Delta G°$, is related to the **equilibrium constant** by $\Delta G° = -RT \ln K = -1.36 \log K$ (at 25°C). The free energy has contributions from changes in **enthalpy,** $\Delta H°$, and **entropy,** $\Delta S°$: $\Delta G° = \Delta H° - T\Delta S°$. Changes in enthalpy are mainly due to differences between the strengths of the bonds made and those of the bonds broken. A reaction is **exothermic** when the former is larger than the latter. It is **endothermic** when there is a net loss in combined bond strengths. Changes in entropy are controlled by the relative degree of order in starting materials compared with that in products. The greater the increase in disorder, the larger a positive $\Delta S°$.

9. The rate of a chemical reaction depends mainly on the concentrations of starting material(s), the activation energy, and temperature. These correlations

are expressed in the **Arrhenius equation:** rate constant $k = Ae^{-E_a/RT}$.

10. If the rate depends on the concentration of only one starting material, the reaction is said to be of **first order.** If the rate depends on the concentrations of two reagents, the reaction is of **second order.**

11. **Brønsted acids** are proton donors; **bases** are proton acceptors. Acid strength is measured by the **acidity constant** K_a; $pK_a = -\log K_a$. Acids and their deprotonated forms have a **conjugate** relation. **Lewis acids** and **bases** are electron pair acceptors and donors, respectively.

PROBLEMS

17. For each example in Table 2-1, identify all polarized covalent bonds and label the appropriate atoms with partial positive or negative charges. (Do not consider carbon–hydrogen bonds.)

18. Circle and identify by name each functional group in the compounds pictured.

(a) 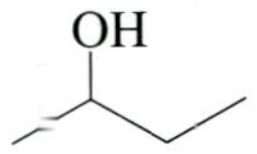   (b) 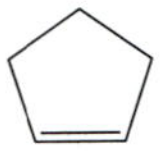   (c) 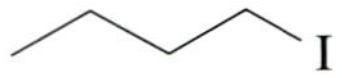   (d)

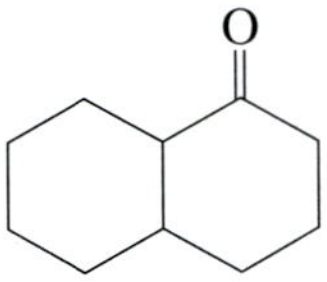

(e) 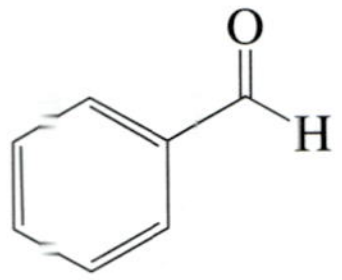   (f) 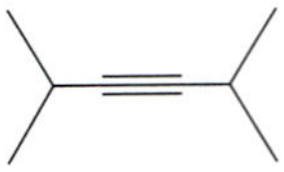   (g)

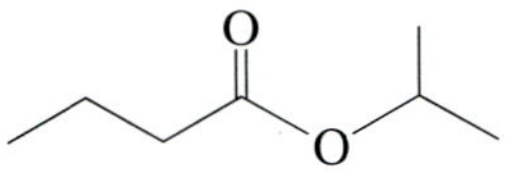

(h) 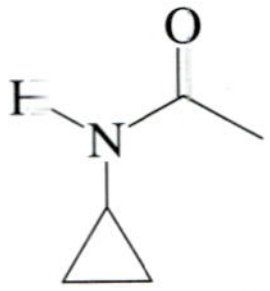   (i) 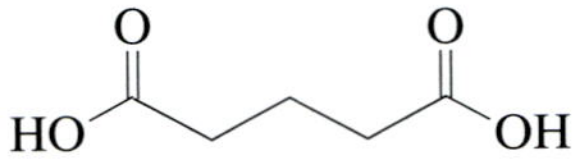   (j)

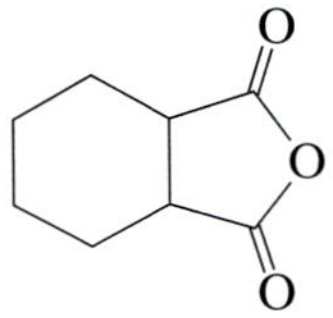

19. On the basis of electrostatics (Coulomb attraction), predict which atom in each of the following organic molecules is likely to react with the indicated reagent. Write "no reaction" if none seems likely. (See Table 2-1 for the structures of the organic molecules.) **(a)** Bromoethane, with the oxygen of HO^-; **(b)** propanal, with the nitrogen of NH_3; **(c)** methoxyethane, with H^+; **(d)** 3-pentanone, with the carbon of CH_3^-; **(e)** ethanenitrile (acetonitrile), with the carbon of CH_3^+; **(f)** butane, with HO^-.

20. Name the following molecules according to the IUPAC system of nomenclature.

(a) 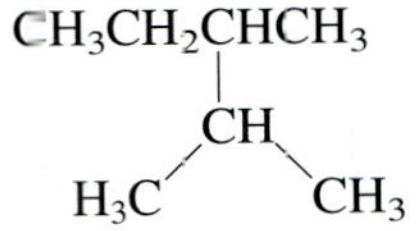 $CH_3CH_2CHCH_3$

(b)

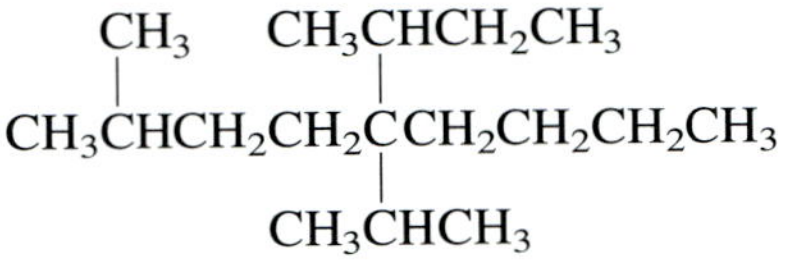

(c)

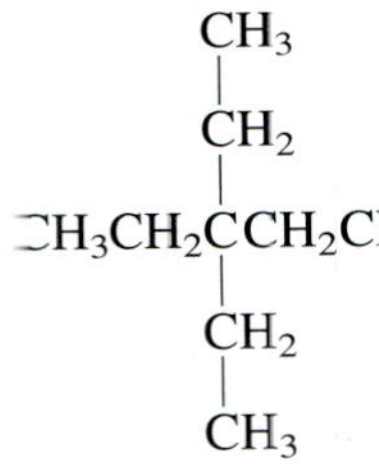

(d)

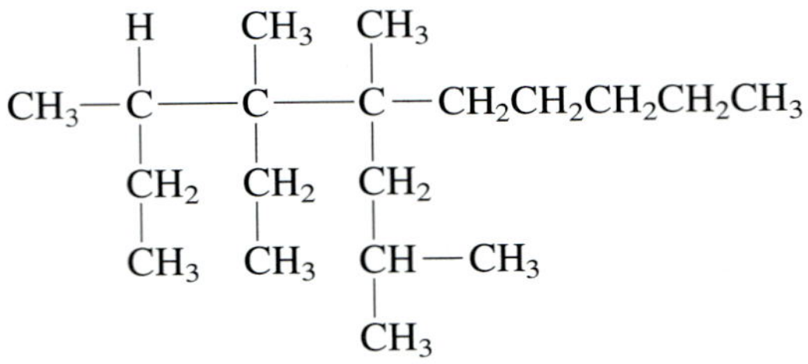

(e) $CH_3CH(CH_3)CH(CH_3)CH(CH_3)CH(CH_3)_2$ (f) 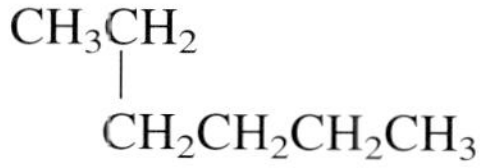 $\begin{array}{c} CH_3CH_2 \\ | \\ CH_2CH_2CH_2CH_3 \end{array}$

(g) 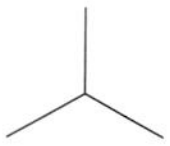 (h) 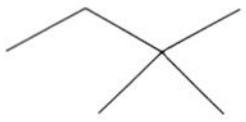 (i) 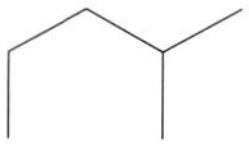 (j)

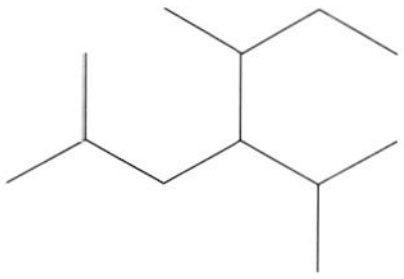

21. Convert the following names into the corresponding molecular structures. After doing so, check to see if the name of each molecule as given here is in accord with the IUPAC system of nomenclature. If not, name the molecule correctly. (a) 2-methyl-3-propylpentane; (b) 5-(1,1-dimethylpropyl)nonane; (c) 2,3,4-trimethyl-4-butylheptane; (d) 4-*tert*-butyl-5-isopropylhexane; (e) 4-(2-ethyl-butyl)decane; (f) 2,4,4-trimethylpentane; (g) 4-*sec*-butylheptane; (h) isohep-tane; (i) neoheptane.

22. Draw and name all possible isomers of C_7H_{16} (isomeric heptanes).

23. Identify the primary, secondary, and tertiary carbon atoms and the hydrogen atoms in each of the following molecules: (a) ethane; (b) pentane; (c) 2-methylbutane; (d) 3-ethyl-2,2,3,4-tetramethylpentane.

24. Identify each of the following alkyl groups as being primary, secondary, or tertiary, and give it a systematic IUPAC name.

(a) 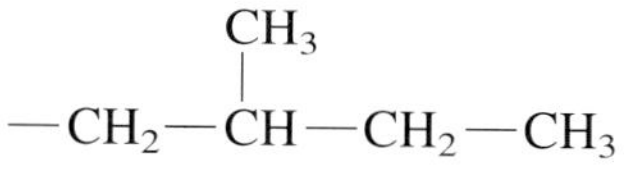 $\begin{array}{c} CH_3 \\ | \\ -CH_2-CH-CH_2-CH_3 \end{array}$
(b) $\begin{array}{c} CH_3 \\ | \\ CH_3-CH-CH_2-CH_2- \end{array}$
(c) $\begin{array}{c} CH_3 \quad CH_3 \\ | \qquad | \\ CH_3-CH \quad\quad CH- \end{array}$

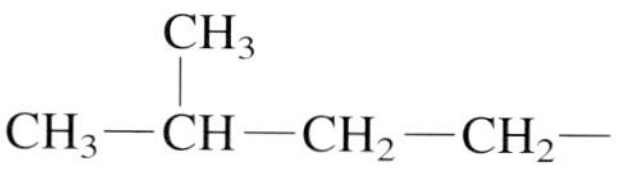

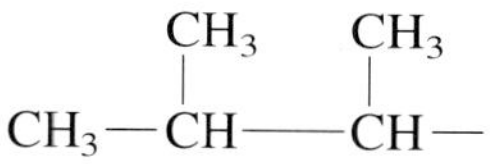

(d) 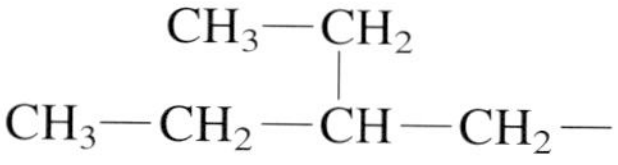 $\begin{array}{c} CH_3-CH_2 \\ | \\ CH_3-CH_2-CH-CH_2- \end{array}$
(e) 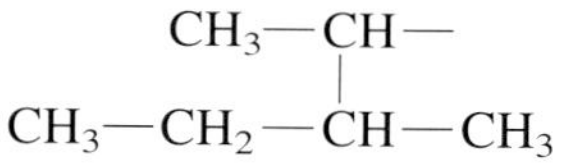 $\begin{array}{c} CH_3-CH- \\ | \\ CH_3-CH_2-CH-CH_3 \end{array}$
(f) 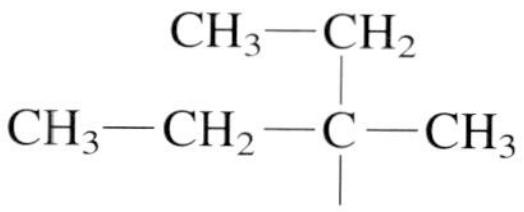 $\begin{array}{c} CH_3-CH_2 \\ | \\ CH_3-CH_2-C-CH_3 \\ | \end{array}$

25. Rank the following molecules in order of increasing boiling point (*without* looking up the real values): (a) 2-methylhexane; (b) heptane; (c) 2,2-dimethylpentane; (d) 2,2,3-trimethylbutane.

26. Draw dashed-wedged line structures for the following molecules in the conformations indicated: (a) staggered propane; (b) eclipsed propane; (c) *anti* butane; (d) *gauche* butane. (**Hint:** Refer to Figure 2-50.)

27. Using Newman projections, draw each of the following molecules in its most stable conformation with respect to the bond indicated: (a) 2-methylbutane, C2–C3 bond; (b) 2,2-dimethylbutane, C2–C3 bond; (c) 2,2-dimethylpentane, C3–C4 bond; (d) 2,2,4-trimethylpentane, C3–C4 bond.

28. At room temperature, 2-methylbutane exists primarily as two alternating conformations of rotation about the C2–C3 bond. About 90% of the molecules exist in the more favorable conformation and 10% in the less favorable one. (a) Calculate the free energy change ($\Delta G°$, more favorable conformation − less favorable conformation) between these conformations. (b) Draw a potential-energy diagram for rotation about the C2–C3 bond in 2-methylbutane. To the best of your ability, assign relative energy values to all the conformations on your diagram. (c) Draw Newman projections for all staggered and eclipsed rotamers in (b) and indicate the two most favorable ones.

29. For each of the following naturally occurring compounds, identify the compound class(es) to which it belongs, and circle all functional groups.

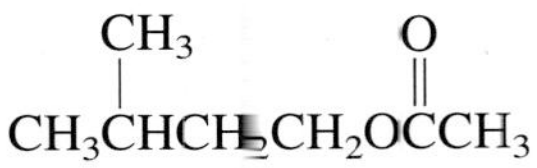

CH₃

$$CH_3CHCH_2CH_2OCCH_3$$

3-Methylbutyl acetate
(In banana oil)

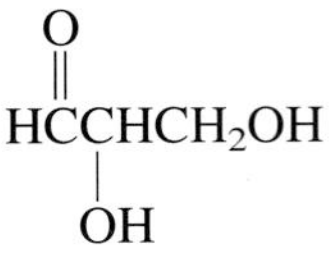

$$HCCHCH_2OH$$
$$\quad\;\; OH$$

2,3-Dihydroxypropanal
(The simplest sugar)

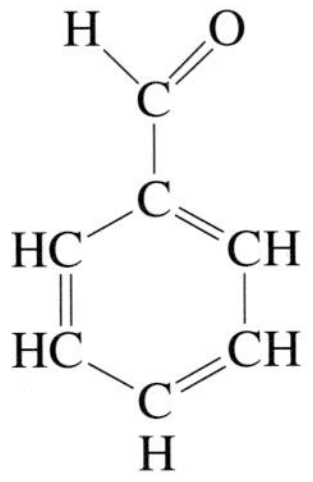

Benzaldehyde
(In fruit pits)

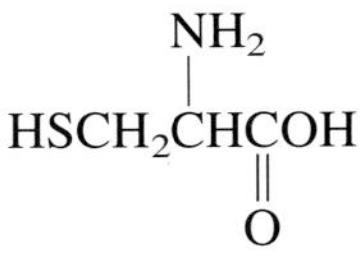

NH₂

$$HSCH_2CHCOH$$

Cysteine
(In proteins)

$$CH_3CH{=}CHC{\equiv}CC{\equiv}CCH{=}CHCH_2OH$$

Matricarianol
(From chamomile)

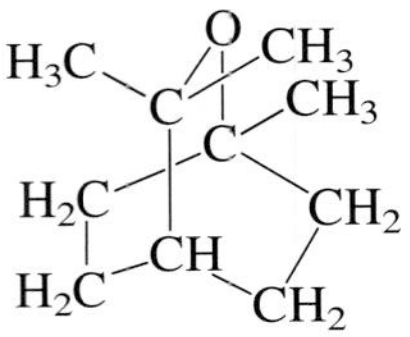

Cineole
(From eucalyptus)

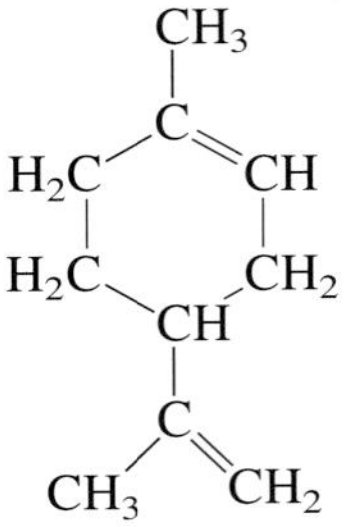

Limonene
(In lemons)

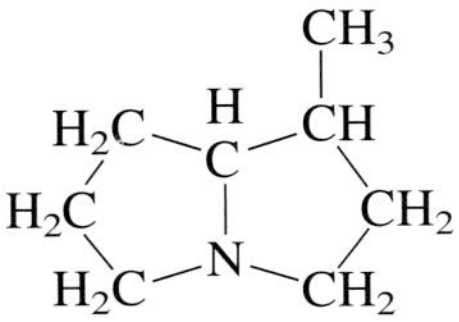

Heliotridane
(An alkaloid)

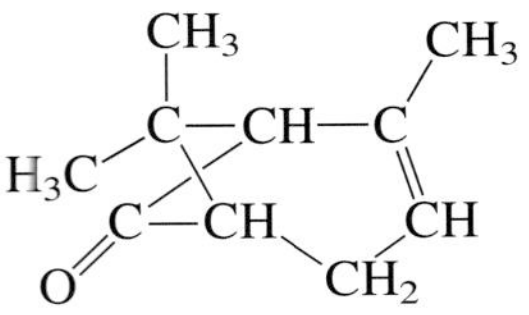

Chrysanthenone
(In chrysanthemums)

30. Give IUPAC names for all alkyl groups marked by dashed lines in each of the
following biologically important compounds. Identify each group as a primary,
secondary, or tertiary alkyl substituent.

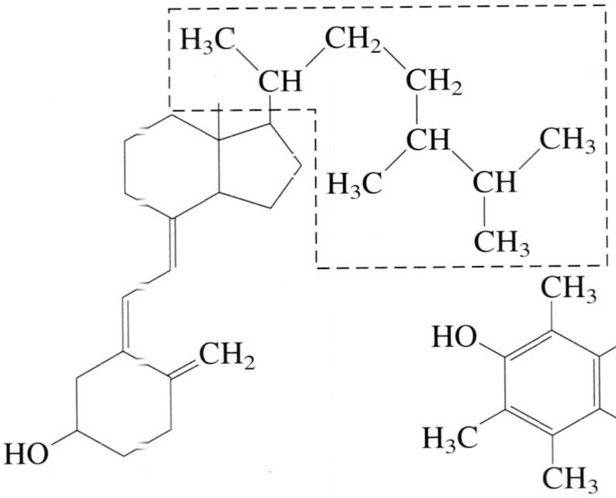

Vitamin D₄

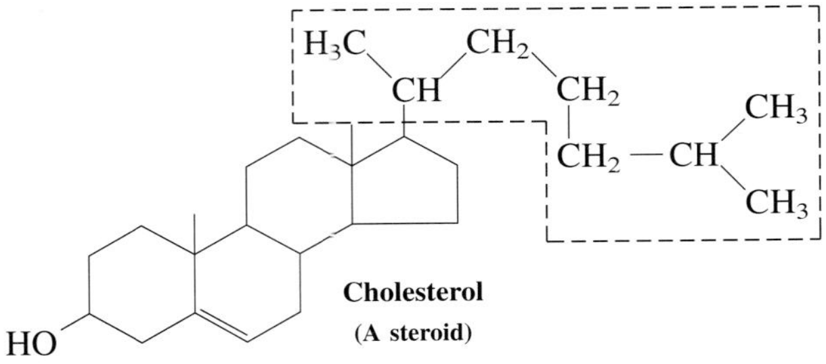

Cholesterol
(A steroid)

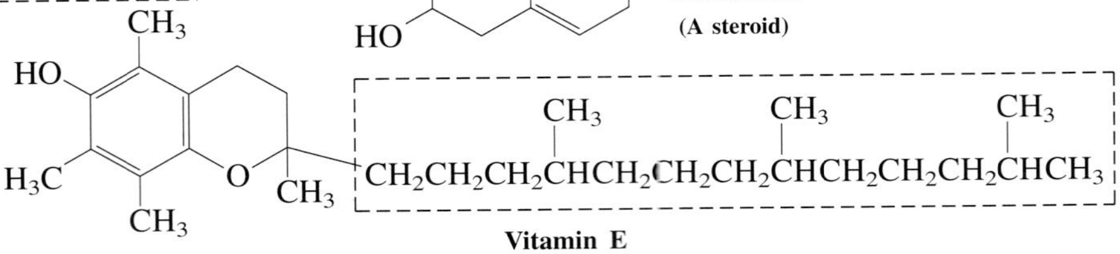

Vitamin E

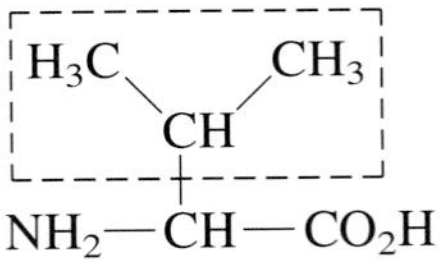

$$CH_2CH_2CH_2CHCH_2CH_2CH_2CHCH_2CH_2CH_2CHCH_3$$

Valine
(An amino acid)

$$NH_2{-}CH{-}CO_2H$$

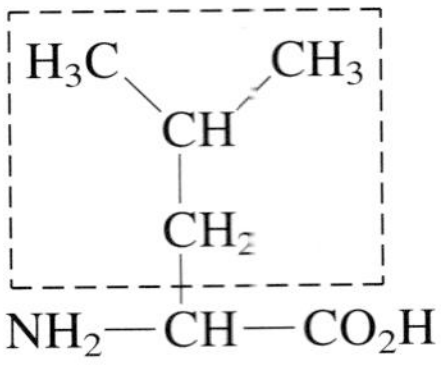

Leucine
(Another amino acid)

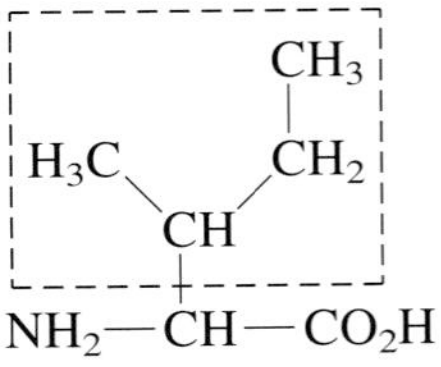

$$NH_2{-}CH{-}CO_2H$$

Isoleucine
(Still another amino acid)

$$NH_2{-}CH{-}CO_2H$$

31. The equation relating $\Delta G°$ to K contains a temperature term. Refer to your answer to Problem 28(a) to calculate the answers to the questions that follow. You will need to know that $\Delta S°$ for the formation of the more stable conformer of 2-methylbutane from the next most stable conformer is $+1.4$ cal deg^{-1} mol^{-1}. **(a)** Calculate the enthalpy difference ($\Delta H°$) between these two conformers from the equation $\Delta G° = \Delta H° - T\Delta S°$. How well does this agree with the $\Delta H°$ calculated from the number of *gauche* interactions in each conformer? **(b)** Assuming that $\Delta H°$ and $\Delta S°$ do not change with temperature, calculate $\Delta G°$ between these two conformations at the following three temperatures: $-250°C$; $-100°C$; $+500°C$. **(c)** Calculate K for these two conformations at the same three temperatures.

32. The hydrocarbon propene (CH_3–CH=CH_2) can react in two different ways with bromine (Chapters 12 and 14).

$$\text{(i)} \quad CH_3-CH=CH_2 + Br_2 \longrightarrow CH_3-\underset{\underset{Br}{|}}{CH}-\underset{\underset{Br}{|}}{CH_2}$$

$$\text{(ii)} \quad CH_3-CH=CH_2 + Br_2 \longrightarrow \underset{\underset{Br}{|}}{CH_2}-CH=CH_2 + HBr$$

Bond	Average Strength
C—C	83
C=C	146
C—H	99
Br—Br	46
H—Br	87
C—Br	68

(a) Using the bond strengths (kcal mol^{-1}) given in the margin, calculate $\Delta H°$ for each of these reactions. **(b)** $\Delta S° \approx 0$ cal deg^{-1} mol^{-1} for one of these reactions and -35 cal deg^{-1} mol^{-1} for the other. Which reaction has which $\Delta S°$? Briefly explain your answer. **(c)** Calculate $\Delta G°$ for each reaction at 25°C and at 600°C. Are both of these reactions thermodynamically favorable at 25°C? At 600°C?

33. Using the Arrhenius equation, calculate the effect on k of increases in temperature of 10, 30, and 50 degrees (Celsius) for the following activation energies. Use 300 K (approximately room temperature) as your initial T value, and assume that A is a constant. **(a)** $E_a = 15$ kcal mol^{-1}; **(b)** $E_a = 30$ kcal mol^{-1}; **(c)** $E_a = 45$ kcal mol^{-1}.

34. The Arrhenius equation can be reformulated in a way that permits the experimental determination of activation energies. For this purpose, we take the natural logarithm of both sides and convert into the base 10 logarithm.

$$\ln k = \ln (Ae^{-E_a/RT}) = \ln A - E_a/RT \quad \text{becomes} \quad \log k = \log A - \frac{E_a}{2.3RT}$$

The rate constant k is measured at several temperatures T and a plot of log k versus $1/T$ is prepared, a straight line. What is the slope of this line? What is its intercept (i.e., the value of log k at $1/T = 0$)? How is E_a calculated?

35. (i) Determine whether each species in the following equations is acting as a Brønsted acid or base, and label it. (ii) Indicate whether the equilibrium lies to the left or to the right. (iii) Estimate K for each equation if possible. (**Hint:** Use the data in Table 2-6.)
(a) $H_2O + HCN \rightleftharpoons H_3O^+ + CN^-$
(b) $CH_3O^- + NH_3 \rightleftharpoons CH_3OH + NH_2^-$
(c) $HF + CH_3COO^- \rightleftharpoons F^- + CH_3COOH$
(d) $CH_3^- + NH_3 \rightleftharpoons CH_4 + NH_2^-$
(e) $H_3O^+ + Cl^- \rightleftharpoons H_2O + HCl$
(f) $CH_3COOH + CH_3S^- \rightleftharpoons CH_3COO^- + CH_3SH$

36. Identify each of the following species as either a Lewis acid or a Lewis base, and write an equation illustrating a Lewis acid-base reaction for each one. Use curved arrows to depict electron-pair movement. Be sure that the product of each reaction is depicted by a complete, correct Lewis structure.

(a) Cl^-

(b) CH_3OH

(c) $(CH_3)_2CH^+$

(d) $MgBr_2$

(e) CH_3BH_2

(f) CH_3S^-

37. Reexamine your answers to Problem 19. Rewrite each one in the form of a complete equation describing a Lewis acid-base process, showing the product and using curved arrows to depict electron-pair movement. [**Hint:** For (b) and (d), start with a Lewis structure that represents a second resonance form of the starting organic molecule.]

Team Problem

38. Consider the difference in the rate between the following two second-order substitution reactions.

Reaction 1 The reaction of bromoethane and iodide ion to produce iodoethane and bromide ion is second order; that is, the rate of the reaction depends on the concentrations of both bromoethane and iodide ion:

$$\text{Rate} = k[CH_3CH_2Br][I^-] \text{ mol } L^{-1} \text{ s}^{-1}$$

Reaction 2 The reaction of 1-bromo-2,2-dimethylpropane (neopentyl bromide) with iodide ion to produce neopentyl iodide and bromide ion is more than 10,000 times slower than the reaction of bromoethane with iodide ion.

$$\text{Rate} = k[\text{neopentyl bromide}][I^-] \text{ mol } L^{-1} \text{ s}^{-1}$$

(a) Formulate each reaction by using bond-line structural drawings in your reaction scheme.

(b) Identify the reactive site of the starting haloalkane as primary, secondary, or tertiary.

(c) Discuss how the reaction might take place; that is, how would the species have to interact in order for the reaction to proceed. Remember that, because the reaction is second order, *both* reagents must be present in the transition state. Use your model kits to help you visualize the trajectory of approach of the iodide ion toward the bromoalkane that enables the simultaneous iodide bond-making and bromide bond-breaking required by the second-order kinetics of these two reactions. Of all the possibilities, which one best explains the experimentally determined difference in rate between the reactions?

(d) Use dashed-wedged line structures to make a three-dimensional drawing of the trajectory upon which you agree.

Preprofessional Problems

39. The compound 2-methylbutane has

(a) no secondary H's

(b) no tertiary H's

39. The compound 2-methylbutane has
  (a) no secondary H's
  (b) no tertiary H's
  (c) no primary H's
  (d) twice as many secondary H's as tertiary H's
  (e) twice as many primary H's as secondary H's

40.

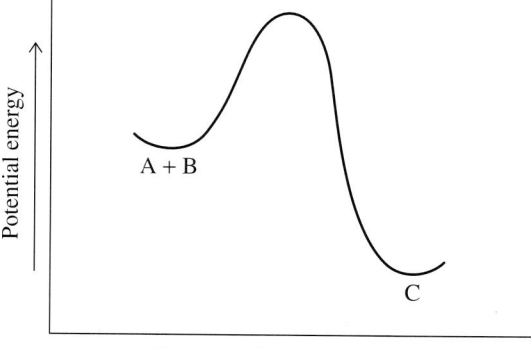

This energy profile diagram represents
(a) an endothermic reaction
(b) an exothermic reaction
(c) a fast reaction
(d) a thermolecular reaction

41. In 4-(1-methylethyl)heptane, any H–C–C angle has the value
  (a) 120°   (b) 109.5°   (c) 180°
  (d) 90°   (e) 360°

42. The structural representation shown in the margin is a Newman projection of the conformer of butane that is

(a) *gauche* eclipsed   (b) *anti* gauche   (c) *anti* staggered   (d) *anti* eclipsed

Reactions of Alkanes

Bond-Dissociation Energies, Radical Halogenation, and Relative Reactivity

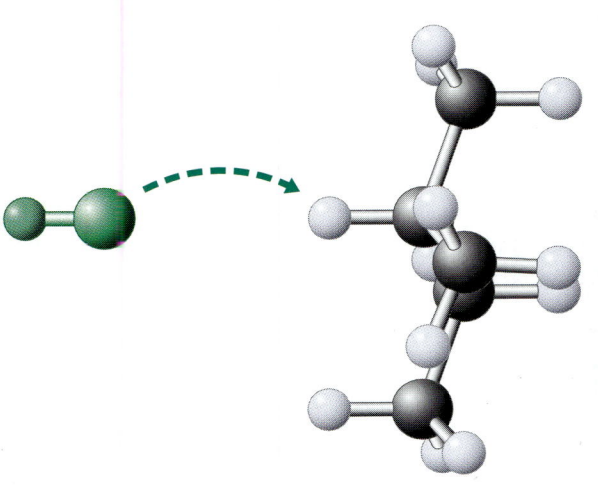

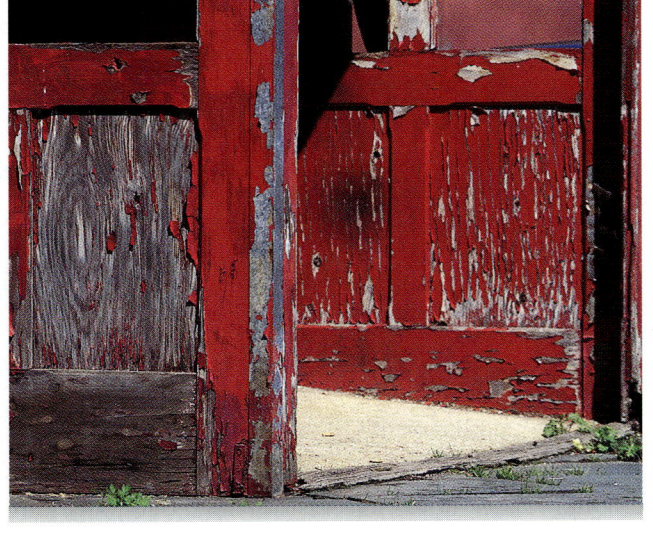

Combustion of alkanes releases most of the energy that powers modern industrialized society. How does this process occur? Is it related to the chemistry that converts alkanes into other organic molecules? We will find that both transformations rely on the same essential step: the breaking of a bond, or **bond dissociation.**

Many liquid and solid alkanes are obtained cheaply from petroleum (Section 3-3). Thus, nature has given us large quantities of hydrocarbons that can be used as "chemical feedstocks," or starting materials, for the synthesis of other organic molecules. Alkanes are also produced naturally by the slow decomposition of animal and vegetable matter in the presence of water but in the absence of oxygen, a process lasting millions of years. The smaller alkanes—methane, ethane, propane, and butane—are present in natural gas, methane being by far its major component. In the United States, natural gas is a major source of energy, with annual production in the hundreds of millions of tons.

As stated in Chapter 2, alkanes are organic chemicals that lack functional groups. Chapter 3 begins by explaining how bond dissociation in alkanes can be made to occur. Next, we learn how to introduce functional groups, to turn alkanes into compounds useful for synthesis, a process called **functionalization.** An important functionalization reaction of alkanes is **halogenation,** in which a hydrogen atom is replaced by a halogen. For each of these processes, we shall use a description of its **mechanism** to explain the conditions under which each takes place. You will see that these mechanistic concepts also explain the effects of halogen-containing chemicals on the stratospheric ozone layer. Finally, a discussion of alkane combustion leads to a description of the methods used to establish the heat contents and relative stabilities of molecules.

Reactive odd-electron species such as halogen atoms and hydroxy radicals (shown here) are involved in the oxidative degradation of organic materials and contribute to weathering.

3-1 Strength of Alkane Bonds: Radicals

Chapter 1 explained how bonds are formed and that energy is released on bond formation. For example, bringing two hydrogen atoms into bonding distance produces 104 kcal mol^{-1} of heat (refer to Figures 1-1 and 1-12).

$$H\cdot + H\cdot \longrightarrow H—H \qquad \Delta H° = -104 \text{ kcal mol}^{-1}$$

Consequently, breaking such a bond *requires* heat, in fact the same amount of heat that was released when the bond was made. This energy is called **bond-dissociation energy,** *DH°*, or **bond strength.**

$$H—H \longrightarrow H\cdot + H\cdot \qquad \Delta H° = DH° = 104 \text{ kcal mol}^{-1}$$

Radicals are formed by homolytic cleavage

In our example, the bond breaks in such a way that the two bonding electrons divide equally between the two participating atoms or fragments. This process is called **homolytic cleavage** or **bond homolysis.**

> **Homolytic Cleavage**
>
> $A—B \longrightarrow A\cdot + \cdot B$
> **Radicals**

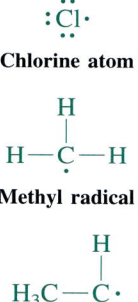

Chlorine atom

Methyl radical

Ethyl radical

The fragments that form have an unpaired electron, for example, H·, Cl·, CH$_3$·, and CH$_3$CH$_2$·. When these species are composed of more than one atom, they are called **radicals.** Because of the unpaired electron, radicals and free atoms are very reactive and usually cannot be isolated. However, radicals and free atoms are present in low concentration as unobserved *intermediates* in many reactions, such as the oxidation of fats that leads to the spoilage of perishable foods (Chapter 22).

In Section 2-9 we were introduced to an alternative way of breaking a bond, in which the entire bonding electron pair is donated to one of the atoms. This process is **heterolytic cleavage** and results in the formation of **ions.**

> **Heterolytic Cleavage**
>
> $A—B \longrightarrow A^+ + :B^-$
> **Ions**

Homolytic cleavage may be observed in nonpolar solvents or even in the gas phase. In contrast, heterolytic cleavage normally occurs in polar solvents, which are capable of stabilizing ions, and is restricted to situations in which the electronegativities of atoms A and B and the groups attached to them stabilize positive and negative charges, respectively.

Dissociation energies, DH°, refer only to homolytic cleavages. They have characteristic values for the various bonds that can be formed between the elements. Table 3-1 lists dissociation energies of some common bonds. Note the relatively strong bonds to hydrogen, as in H–F and H–OH. Even though these bonds have high *DH°* values, they readily undergo *heterolytic* cleavage in water to H$^+$ and F$^-$ or HO$^-$; *do not confuse homolytic with heterolytic processes.*

Bonds are strongest when made by overlapping orbitals that are closely matched in energy and size. For example, the strength of the bonds between hydrogen and the

TABLE 3-1	Bond-Dissociation Energies of Various A–B Bonds ($DH°$ in kcal mol^{-1})						

	B in A–B						
A in A–B	**–H**	**–F**	**–Cl**	**–Br**	**–I**	**–OH**	**–NH$_2$**
H—	104	135	103	87	71	119	107
CH$_3$—	105	110	85	71	57	93	80
CH$_3$CH$_2$—	98	107	80	68	53	92	77
CH$_3$CH$_2$CH$_2$—	98	107	81	68	53	91	78
(CH$_3$)$_2$CH—	94.5	106	81	68	53	92	93
(CH$_3$)$_3$C—	93	110	81	67	52	93	93

Note: These numbers are being revised continually because of improved methods for their measurement. Some of the values given here may be in (small) error.

halogens decreases in the order F > Cl > Br > I, because the *p* orbital of the halogen contributing to the bonding becomes larger and more diffuse along the series. Thus, the efficiency of its overlap with the relatively small 1*s* orbital on hydrogen diminishes. A similar trend holds for bonding between the halogens and carbon.

EXERCISE 3-1

Compare the bond-dissociation energies of CH$_3$–F, CH$_3$–OH, and CH$_3$–NH$_2$. Why do the bonds get weaker along this series even though the orbitals participating in overlap become better matched in size and energy? (**Hint:** Consider Figure 1-2 and Table 1-2 for a simple explanation.)

The stability of radicals determines the C–H bond strengths

How strong are the C–H and C–C bonds in alkanes? The bond-dissociation energies of various alkane bonds are given in Table 3-2. Note that bond energies generally

TABLE 3-2	Bond-Dissociation Energies for Some Alkanes		

Compound	$DH°$ (kcal mol^{-1})	Compound	$DH°$ (kcal mol^{-1})
CH$_3$–H	105	CH$_3$–CH$_3$	90
C$_2$H$_5$–H	98	C$_2$H$_5$–CH$_3$	86
C$_3$H$_7$–H	98	C$_3$H$_7$–CH$_3$	87
(CH$_3$)$_2$CHCH$_2$–H	98	C$_2$H$_5$–C$_2$H$_5$	82
(CH$_3$)$_2$CH–H	94.5	(CH$_3$)$_2$CH–CH$_3$	86
(CH$_3$)$_3$C—H	93	(CH$_3$)$_3$C–CH$_3$	84
		(CH$_3$)$_3$C–C(CH$_3$)$_3$	72

Note: See footnote for Table 3-1.

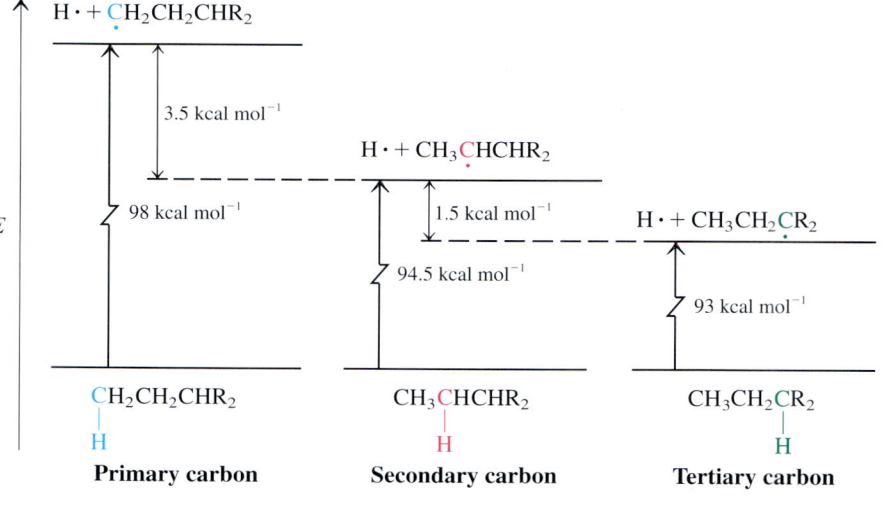

FIGURE 3-1

The different energies needed to form radicals from an alkane $CH_3CH_2CHR_2$. Radical stability increases from primary to secondary to tertiary.

decrease with the progression from methane to primary, secondary, and tertiary carbon. For example, the C–H bond in methane has a high $DH°$ value of 105 kcal mol^{-1}. In ethane, this bond energy is less: $DH° = 98$ kcal mol^{-1}. The latter number is typical for primary C–H bonds, as can be seen for the bond in propane. The secondary C–H bond is even weaker, with a $DH°$ of 94.5 kcal mol^{-1}, and a tertiary carbon atom bound to hydrogen has a $DH°$ of only 93 kcal mol^{-1}.

C–H bond is weaker and easier to break ↓	CH_3—H $\longrightarrow$ $CH_3\cdot$ + H·	$DH° = 105$ kcal mol^{-1}	Radical formed is more stable ↓
	R—H $\longrightarrow$ R· + H·		
	R primary	$DH° = 98$ kcal mol^{-1}	
	secondary	$DH° = 94.5$ kcal mol^{-1}	
	tertiary	$DH° = 93$ kcal mol^{-1}	

A similar trend is seen for C–C bonds. The extremes are the central linkages in ethane ($DH° = 90$ kcal mol^{-1}) and 2,2,3,3-tetramethylbutane ($DH° = 72$ kcal mol^{-1}).

Why do all of these dissociations exhibit different $DH°$ values? One explanation is that the radicals formed have different energies. Radical stability *increases* along the series from primary to secondary to tertiary; consequently, the energy required to create them *decreases* (Figure 3-1).

Stability of Alkyl Radicals

$CH_3\cdot$ < primary < secondary < tertiary

EXERCISE 3-2

Which C–C bond would break first, the bond in ethane or that in 2,2-dimethylpropane?

In summary, bond homolysis in alkanes yields radicals and free atoms. The heat required to do so is called the bond-dissociation energy, $DH°$. Its value is character-

istic only for the bond between the two participating elements. Bond-breaking that results in tertiary radicals demands less energy than that furnishing secondary radicals; the latter are in turn formed more readily than primary radicals. The methyl radical is the most difficult to obtain in this way.

3-2 Structure of Alkyl Radicals: Hyperconjugation

What is the reason for the ordering in stability of alkyl radicals? To answer this question, we need to inspect the alkyl radical structure more closely. Radicals are stabilized by electron delocalization in which the *p* orbital at the radical center overlaps with a neighboring C–H *σ* bond.

Consider the structure of the methyl radical, formed by removal of a hydrogen atom from methane. It could be described as an sp^3-hybridized carbon with three sp^3 C–H bonds, the odd electron occupying the fourth sp^3 molecular orbital. Spectral measurements, however, have shown that the methyl radical, and probably other alkyl radicals, adopt a *nearly planar* configuration, more accurately described by sp^2 hybridization (Figure 3-2). The unpaired electron occupies the remaining *p* orbital perpendicular to the molecular plane.

Let us see how the planar structures of alkyl radicals help explain their relative stabilities. Figure 3-3A (see page 98) shows that there is a conformer in the ethyl radical in which a C–H bond of the CH_3 group is aligned with and overlaps one of the lobes of the singly occupied *p* orbital on the radical center. This arrangement allows the bonding pair of electrons in the *σ* orbital to delocalize into the partly empty *p* lobe, a phenomenon called **hyperconjugation.** The interaction between a filled orbital and a singly occupied orbital has a net stabilizing effect (recall Exercise 1-12). Both hyperconjugation and resonance (Section 1-5) are forms of electron delocalization. They are distinguished by type of orbital: resonance normally refers to *π*-type overlap of *p* orbitals, whereas hyperconjugation incorporates overlap with the orbitals of *σ* bonds.

As further hydrogen atoms on the radical carbon are replaced successively by alkyl groups, the number of hyperconjugation interactions increases (Figure 3-3B). The order of stability of the radicals is a consequence of this effect. Notice in Figure 3-1 that the degree of stabilization arising from each hyperconjugation interaction is relatively small ($1.5–3.5$ kcal mol^{-1}); we shall see later that stabilization of radicals by resonance is considerably greater (Chapter 14). Another contribution to the relative

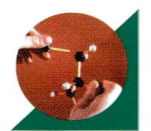

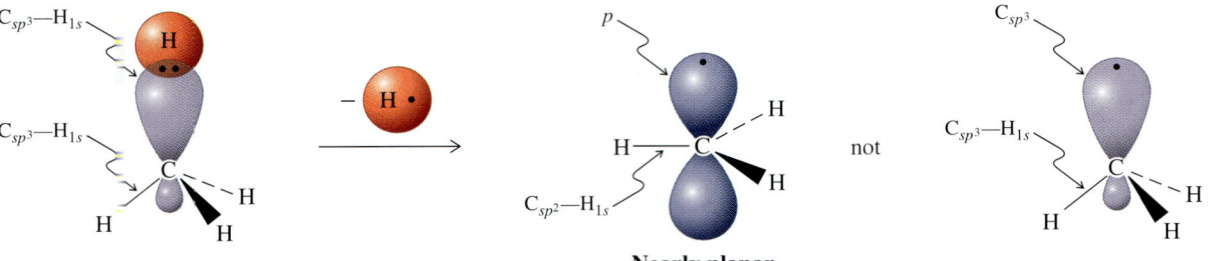

Nearly planar

FIGURE 3-2

The hybridization change upon formation of a methyl radical from methane. The nearly planar arrangement is reminiscent of the hybridization in BH_3 (Figure 1-17).

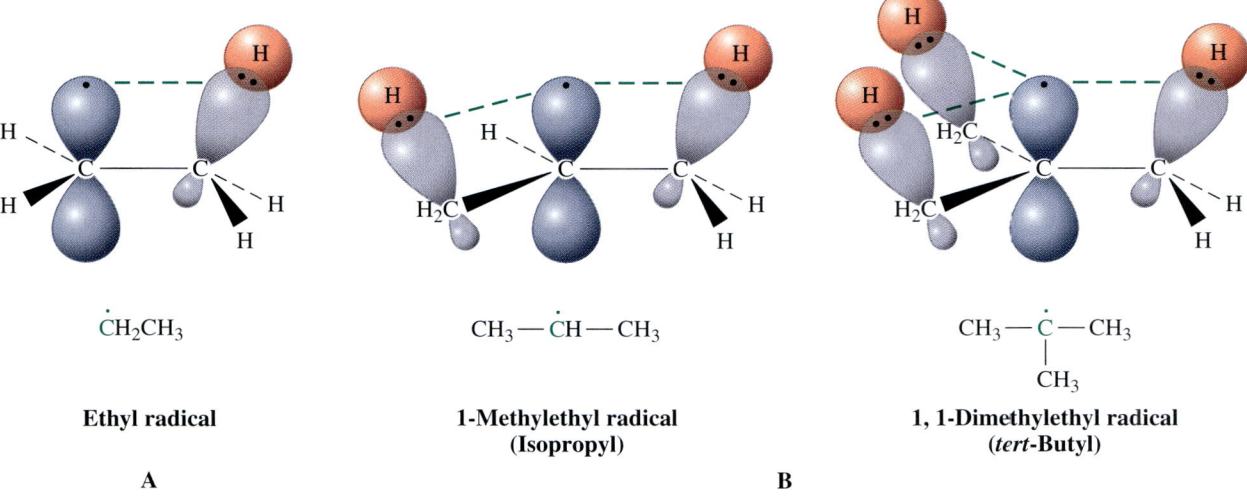

Ethyl radical — **A**

1-Methylethyl radical (Isopropyl)

1, 1-Dimethylethyl radical (*tert*-Butyl)

B

FIGURE 3-3
Hyperconjugation (green dashed lines) resulting from the donation of electrons in filled sp^3 hybrids to the partly filled p orbital in (A) ethyl and (B) 1-methylethyl and 1,1-dimethylethyl radicals. The resulting delocalization of electron density has a net stabilizing effect.

stability of secondary and tertiary radicals is the greater relief of steric crowding between the substituent groups as the geometry changes from tetrahedral in the alkane to planar in the radical.

3-3 Conversion of Petroleum: Pyrolysis

A knowledge of bond-dissociation energies helps us understand the high-temperature reactivity of hydrocarbons. Consider, for example, the conversion of crude petroleum into gasoline and other volatile materials. Distillation alone does not meet the demand for these desired lower molecular weight hydrocarbons. Additional heating is required to break up longer carbon chains into smaller fragments. How does this occur? Let us look first at the behavior of simple alkanes under these conditions and then move on to petroleum.

High temperatures cause bond homolysis

When alkanes are heated to a high temperature, both C–H bonds and C–C bonds rupture, a process called **pyrolysis.** In the absence of oxygen, the resulting radicals can combine to form new higher or lower alkanes. They can also remove hydrogen atoms from the carbon atom adjacent to another radical center to give alkenes, a process called *hydrogen abstraction*. Indeed, very complicated mixtures of alkanes and alkenes form in pyrolyses. Under special conditions, however, these transformations can be controlled to obtain a large proportion of hydrocarbons of a defined chain length.

Pyrolysis of Hexane

Examples of cleavage into radicals

$$\overset{1}{C}H_3\overset{2}{C}H_2\overset{3}{C}H_2\overset{4}{C}H_2CH_2CH_3$$

Hexane

Cl, C2 cleavage → $CH_3\cdot \;+\; \cdot CH_2CH_2CH_2CH_2CH_3$

C2, C3 cleavage → $CH_3CH_2\cdot \;+\; \cdot CH_2CH_2CH_2CH_3$

C3, C4 cleavage → $CH_3CH_2CH_2\cdot \;+\; \cdot CH_2CH_2CH_3$

Examples of radical combination reactions

$$CH_3\cdot + \cdot CH_2CH_3 \longrightarrow CH_3CH_2CH_3$$
Propane

$$CH_3CH_2CH_2CH_2CH_2\cdot + \cdot CH_2CH_2CH_3 \longrightarrow CH_3CH_2CH_2CH_2CH_2CH_2CH_2CH_3$$
Octane

Examples of hydrogen abstraction reactions

$$\underset{}{CH_3CH_2\cdot} + CH_3\overset{\overset{\text{H}}{|}}{CH}-CH_2\cdot \longrightarrow CH_3\overset{\overset{\text{H}}{|}}{CH_2} + CH_3CH\!=\!CH_2$$
Ethane   **Propene**

$$\overset{\overset{\text{H}}{|}}{CH_2}CH_2\cdot + CH_3CH_2CH_2\cdot \longrightarrow CH_2\!=\!CH_2 + CH_3CH_2\overset{\overset{\text{H}}{|}}{CH_2}$$
Ethene   **Propane**

Such control frequently requires the use of special catalysts, such as crystalline sodium aluminosilicates, also called zeolites. For example, zeolite-catalyzed pyrolysis of dodecane yields a mixture in which hydrocarbons containing from three to six carbons predominate.

$$\text{Dodecane} \xrightarrow{\text{Zeolite, 482°C, 2 min}} \underset{17\%}{C_3} + \underset{31\%}{C_4} + \underset{23\%}{C_5} + \underset{18\%}{C_6} + \underset{11\%}{\text{other products}}$$

Petroleum is an important source of alkanes

Breaking an alkane down into smaller fragments is also known as **cracking.** Such processes are important in the oil-refining industry for the production of gasoline and other liquid fuels from petroleum.

CHEMICAL HIGHLIGHT 3-1   **The Function of a Catalyst**

What is the function of the zeolite catalyst? As shown in the illustration, a *catalyst* is a substance that speeds up a reaction; that is, it increases the rate at which equilibrium is established. It does so by allowing reactants and products to be interconverted by a new pathway that has a lower activation energy (E_{cat}) than that of the reaction in the absence (E_a) of the catalyst. Apart from zeolites and other mineral-derived surfaces, many metals act as catalysts. In nature, *enzymes* usually fulfill this function (Chapter 26). The presence of catalysts allows many transformations to take place at lower temperatures and under generally milder conditions.

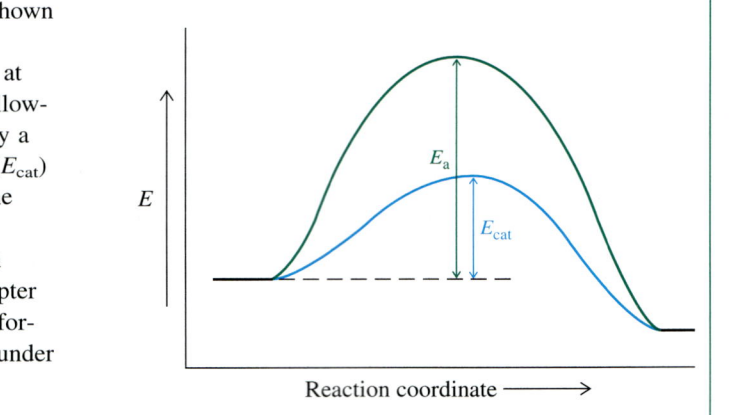

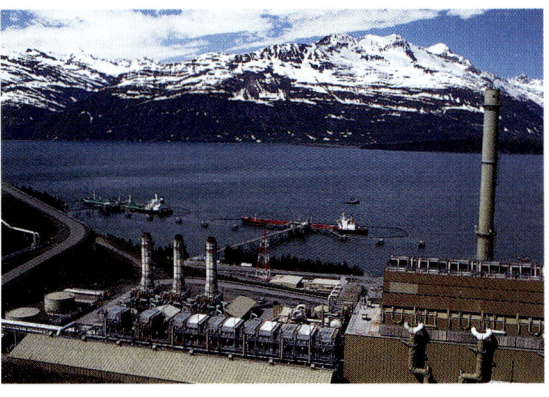

The Alyeska Pipeline Marine Terminal, Valdez, Alaska. Alaska is second only to Texas in oil production in the United States.

As mentioned in the introduction to this chapter, petroleum, or crude oil, is believed to be the product of microbial degradation of living organic matter that existed several hundred million years ago. Crude oil, a dark viscous liquid, is primarily a mixture of several hundred different hydrocarbons, particularly straight-chain alkanes, some branched alkanes, and varying quantities of aromatic hydrocarbons. Distillation yields several fractions with a typical product distribution, as shown in Table 3-3. The composition of petroleum varies widely, depending on the origin of the oil.

To increase the quantity of the much-needed gasoline fraction, the oils with higher boiling points are cracked by pyrolysis. Originally (in the 1920s), this process required high temperatures (800°–1000°C), but modern cracking processes use catalysts, such as zeolites, at relatively low temperatures (500°C). Cracking the residual oil from crude petroleum distillation gives approximately 30% gas, 50% gasoline, and 20% higher molecular weight oils and a residue called coke.

Another process converts alkanes into aromatic hydrocarbons with approximately the same number of carbon atoms. The aromatics are highly efficient fuels and are used as feedstocks for the chemical industry. Because the process reforms a new hydrocarbon from an old one, it is referred to as **reforming.** An example of reforming is the conversion of heptane into methylbenzene (toluene). Hundreds of millions of liters of reformate gasoline are produced in the United States alone.

$CH_3CH_2CH_2CH_2CH_2CH_2CH_3$

Heptane

Pt-SiO$_2$-Al$_2$O$_3$,
500°C,
20 atm H$_2$

CH$_3$

+ 4 H$_2$

**Methylbenzene
(Toluene)**

TABLE 3-3	Product Distribution in a Typical Distillation of Crude Petroleum		
Amount (% of volume)	Boiling point (°C)	Carbon atoms	Products
1–2	<30	$C_1–C_4$	Natural gas, methane, propane, butane, liquefied petroleum gas (LPG)
15–30	30–200	$C_4–C_{12}$	Petroleum ether ($C_{5,6}$), ligroin (C_7), naphtha, straight-run gasoline[a]
5–20	200–300	$C_{12}–C_{15}$	Kerosene, heater oil
10–40	300–400	$C_{15}–C_{25}$	Gas oil, diesel fuel, lubricating oil, waxes, asphalt
8–69	>400 (Nonvolatiles)	>C_{25}	Residual oil, paraffin waxes, asphalt (tar)

[a]This refers to gasoline straight from petroleum, without having been treated in any way.

CHEMICAL HIGHLIGHT **3-2**

Petroleum and Gasoline: Our Main Energy Sources

Oil and natural gas supply most of the U.S. energy requirement. Other industrialized nations have similar dependence on petroleum as a source of energy. Yearly production of natural gas in the United States approximates 500 million liters. In 1995, U.S. energy sources apart from gas (24%) and oil (38%) were coal (23%), nuclear power (8%), and hydroelectric power (4%). Domestic annual oil production peaked in 1971 at 4.2 billion barrels. In 1995, it had dropped below 3 billion barrels, with some 2 billion barrels imported. (One barrel equals 42 gallons, about 158 liters.)

The dependence of many countries on imported oil has had important economic and political consequences, as demonstrated by the wild swings in oil prices that accompanied the Iraqi invasion of Kuwait in 1990 and the subsequent Persian Gulf War. Renewed efforts are being undertaken to decrease economic dependence on imported oil and to develop new energy sources to satisfy demand when oil and gas reserves are depleted.

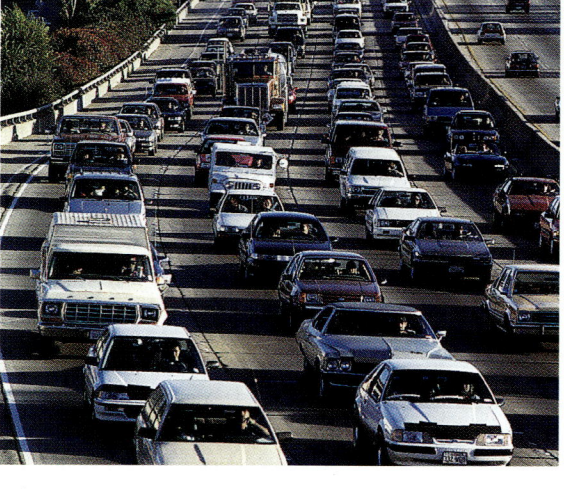

The United States, with more than 200 million motor vehicles, has the highest annual energy consumption of any nation in the world. More than 70% of that need is met by fossil fuels.

To summarize, the pyrolysis of alkanes often leads to complex mixtures of hydrocarbons, control being attained in the presence of special catalysts. Are there other reactions of alkanes? If so, can they be carried out with even greater control? Moreover, can they introduce functional groups into the chain? The sections that follow will answer these questions.

3-4 Chlorination of Methane: The Radical Chain Mechanism

We have seen that alkanes undergo chemical transformations when subjected to pyrolysis, and that these processes include the formation of radical intermediates. Do alkanes participate in other reactions? In this section, we shall consider the effect of exposing an alkane, methane, to a halogen, chlorine. A **chlorination** reaction, in which radicals again play a key role, takes place to produce chloromethane and hydrogen chloride. We shall analyze each step in this transformation to establish the *mechanism* of the reaction.

Chlorine converts methane into chloromethane

When methane and chlorine gas are mixed in the dark at room temperature, there is no reaction. The mixture must be heated to a temperature above 300°C (Δ) *or* irra-

diated with ultraviolet light ($h\nu$) before reaction occurs. One of the two initial products is chloromethane, derived from methane in which a hydrogen atom is removed and replaced by chlorine. The other product of this transformation is hydrogen chloride. Further substitution leads to dichloromethane (methylene chloride), CH_2Cl_2; trichloromethane (chloroform), $CHCl_3$; and tetrachloromethane (carbon tetrachloride), CCl_4.

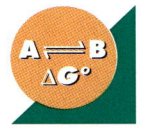

Why should this reaction proceed? Consider its $\Delta H°$. Note that a C–H bond in methane ($DH° = 105$ kcal mol^{-1}) and a Cl–Cl bond ($DH° = 58$ kcal mol^{-1}) are broken, whereas the C–Cl bond of chloromethane ($DH° = 85$ kcal mol^{-1}) and an H–Cl linkage ($DH° = 103$ kcal mol^{-1}) are formed. The net result is the release of 25 kcal mol^{-1} in forming stronger bonds: The reaction is substantially *exothermic*.

$$CH_3{-}H + \overset{..}{\underset{..}{:}}Cl{-}\overset{..}{\underset{..}{:}}Cl: \xrightarrow{\Delta \text{ or } h\nu} CH_3{-}\overset{..}{\underset{..}{:}}Cl: + H{-}\overset{..}{\underset{..}{:}}Cl:$$

$$105 \qquad 58 \qquad\qquad 85 \qquad 103$$

$DH°$ **(kcal mol^{-1})**   **Chloromethane**

$$\Delta H° = \text{energy input} - \text{energy output}$$
$$= \Sigma DH° \text{ (bonds broken)} - \Sigma DH° \text{ (bonds formed)}$$
$$= (105 + 58) - (85 + 103)$$
$$= -25 \text{ kcal mol}^{-1}$$

Why then does the thermal chlorination of methane not occur at room temperature? The fact that a reaction is exothermic does not necessarily mean that it should proceed rapidly and spontaneously. Remember (Section 2-8) that the rate of a chemical transformation depends on its activation energy, which in this case is evidently high. Why is this so? What is the function of irradiation when the reaction does proceed at room temperature? Answering these questions requires an investigation of the mechanism of the reaction.

The mechanism explains the experimental conditions required for reaction

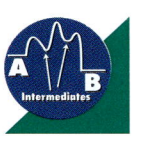

A **mechanism** is a detailed, step-by-step description of all the changes in bonding that occur in a chemical reaction (Section 1-1). Even simple reactions may consist of several separate steps. The mechanism shows the sequence in which bonds are broken and formed, as well as the energy changes associated with each step. This information is of great value in both analyzing possible transformations of complex molecules and understanding the experimental conditions required for reactions to occur.

The mechanism for the chlorination of methane consists of three stages: initiation, propagation, and termination. Let us look at these stages and the experimental evidence for each of them in more detail.

The chlorination of methane can be studied step by step

Experimental observation. Clorination occurs when a mixture of CH_4 and Cl_2 is either heated to 300°C or irradiated with light, as mentioned earlier. Under such conditions, methane by itself is completely stable, but Cl_2 undergoes homolysis to two atoms of chlorine.

Interpretation. The first step in the mechanism of chlorination of methane is the heat- or light-induced homolytic cleavage of the Cl–Cl bond (which happens to be the weakest bond in the starting mixture, with $DH° = 58$ kcal mol^{-1}). This event is required to start the chlorination process and is therefore called the **initiation** step.

As implied by its name, the initiation step generates reactive species (in this case, chlorine atoms) that permit the subsequent steps in the overall reaction to take place.

INITIATION

$$\overset{..}{\underset{..}{Cl}}-\overset{..}{\underset{..}{Cl}}: \xrightarrow{\Delta \text{ or } h\nu} 2 : \overset{..}{\underset{..}{Cl}} \cdot \qquad \Delta H° = DH°(Cl_2)$$
$$= +58 \text{ kcal mol}^{-1}$$

Chlorine atom

NOTE: In this scheme and in those that follow, all radicals and free atoms are in green.

Experimental observation. Only a relatively small number of initiation events are necessary to enable a great many methane and chlorine molecules to undergo conversion into products.

Interpretation. After initiation has taken place, the subsequent steps in the mechanism are self-sustaining, or self-propagating; that is, they can occur many times without the addition of further chlorine atoms from the homolysis of Cl_2. Two **propagation** steps fulfill this requirement. In the first step, a chlorine atom attacks methane by abstracting a hydrogen atom. The resulting products are hydrogen chloride and a methyl radical.

PROPAGATION STEP 1

$$: \overset{..}{\underset{..}{Cl}} \cdot + H-\overset{H}{\underset{H}{C}}-H \longrightarrow : \overset{..}{\underset{..}{Cl}}-H + \cdot \overset{H}{\underset{H}{C}}-H \qquad \Delta H° = DH°(CH_3-H)$$
$$- DH°(H-Cl)$$
$$= +2 \text{ kcal mol}^{-1}$$

$DH°$ (kcal mol^{-1})   05   103   Methyl radical

The $\Delta H°$ for this transformation is positive; its equilibrium is slightly unfavorable. What is its activation energy, E_a? Is there enough heat to overcome this barrier? In this case, the answer is yes. A molecular-orbital description of the transition state (Section 2-6) of hydrogen removal from methane (Figure 3-4) reveals the details of

FIGURE 3-4
Approximate molecular-orbital description of the abstraction of a hydrogen atom by a chlorine atom to give a methyl radical and hydrogen chloride. Notice the rehybridization at carbon in the planar methyl radical. The additional three nonbonded electron pairs on chlorine have been omitted. The orbitals are not drawn to scale. The symbol ‡ identifies the transition state.

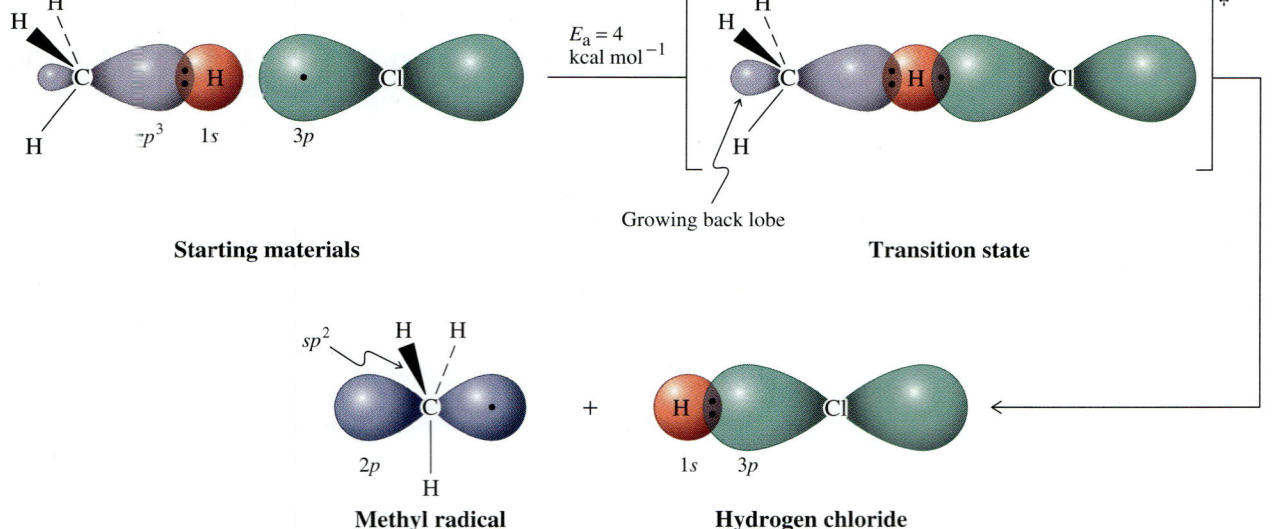

$E_a = 4$ kcal mol^{-1}

Starting materials · Growing back lobe · Transition state

Methyl radical · Hydrogen chloride

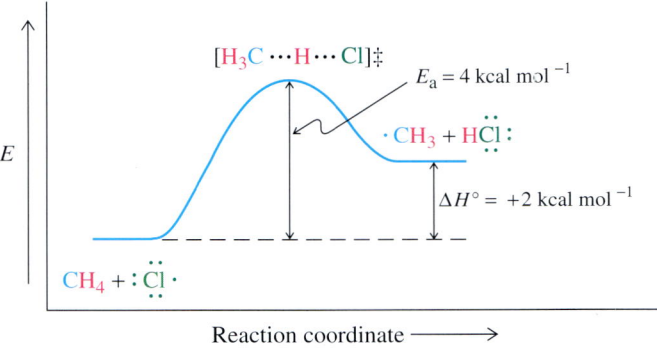

FIGURE 3-5

Potential-energy diagram of the reaction of methane with a chlorine atom. Partial bonds in the transition state are depicted by dotted lines. This process, propagation step 1 in the radical chain chlorination of methane, is slightly endothermic.

the process. The reacting hydrogen is positioned between the carbon and the chlorine, partly bound to both: H–Cl bond formation has occurred to about the same extent as C–H bond-breaking. The transition state, which is labeled by the symbol ‡, is located only about 4 kcal mol^{-1} above the starting materials. A potential-energy diagram describing this step is shown in Figure 3-5.

Propagation step 1 gives one of the products of the chlorination reaction: HCl. What about the organic product, CH_3Cl? Chloromethane is formed in the *second* propagation step. Here the methyl radical abstracts a chlorine atom from one of the starting Cl_2 molecules, thereby furnishing chloromethane *and a new chlorine atom*. The latter reenters propagation step 1 to react with a new molecule of methane. Thus one propagation cycle is closed, and a new one begins, *without the need for another initiation step to take place*. Note how exothermic propagation step 2 is, -27 kcal mol^{-1}. It supplies the overall driving force for the reaction of methane with chlorine.

PROPAGATION STEP 2

$$\Delta H° = DH°(Cl_2) - DH°(CH_3-Cl)$$
$$= -27 \text{ kcal mol}^{-1}$$

$DH°$ (kcal mol^{-1})   58   85

Because propagation step 2 is exothermic, the unfavorable equilibrium in the first propagation step is pushed toward the product side by the rapid depletion of its methyl radical product in the subsequent reaction.

$$CH_4 + Cl\cdot \rightleftharpoons CH_3\cdot + HCl \overset{Cl_2}{\rightleftharpoons} CH_3Cl + Cl\cdot + HCl$$

Slightly unfavorable   Very favorable; "drives" first equilibrium

The potential-energy diagram in Figure 3-6 illustrates this point by continuing the progress of the reaction begun in Figure 3-5. Propagation step 1 has the higher acti-

vation energy and is therefore slower than step 2. The diagram also shows that the overall $\Delta H^\ominus$ of the reaction is made up of the $\Delta H°$ values of the propagation steps: $+2 - 27 = -25$ kcal mol^{-1}. You can see that this should be so by adding the equations for the two.

$$\Delta H° \text{ (kcal mol}^{-1}\text{)}$$

$$:\ddot{\underset{..}{C}}l\cdot + CH_4 \longrightarrow CH_3\cdot + H\ddot{\underset{..}{C}}l: \qquad +2$$

$$CH_3\cdot + Cl_2 \longrightarrow CH_3\ddot{\underset{..}{C}}l: + :\ddot{\underset{..}{C}}l\cdot \qquad -27$$

$$CH_4 + Cl_2 \longrightarrow CH_3\ddot{\underset{..}{C}}l: + H\ddot{\underset{..}{C}}l: \qquad -25$$

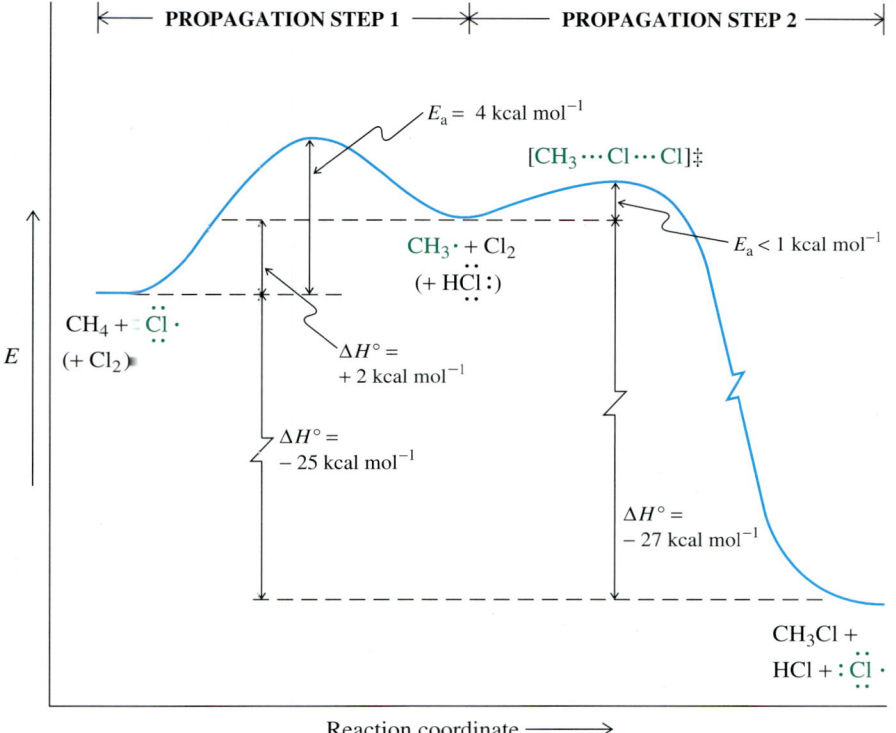

FIGURE 3-6 _____
Complete potential-energy diagram for the formation of CH_3Cl from methane and chlorine. Propagation step 1 has the higher transition-state energy and is therefore slower. The $\Delta H°$ of the overall reaction $CH_4 + Cl_2 \rightarrow CH_3Cl + HCl$ amounts to -25 kcal mol^{-1}, the sum of the $\Delta H°$ values of the two propagation steps.

Experimental observation. Small amounts of *ethane* are identified among the products of chlorination of methane.

Interpretation. Radicals and free atoms are capable of undergoing direct covalent bonding with one another. In the methane chlorination process, three such combination processes are possible, one of which, the reaction of two methyl groups, furnishes ethane. The concentrations of radicals and free atoms in the reaction mixture are very low, however, and hence the chance of one radical or free atom finding another is small. Such combinations are therefore relatively infrequent. When such an event does take place, the propagation of the chains giving rise to the radicals or atoms is terminated. We thus describe these combination processes as **termination** steps.

CHAIN TERMINATION

$$:\ddot{C}l\cdot + :\ddot{C}l\cdot \longrightarrow Cl_2$$

$$:\ddot{C}l\cdot + CH_3\cdot \longrightarrow CH_3\ddot{C}l:$$

$$CH_3\cdot + CH_3\cdot \longrightarrow CH_3—CH_3$$

The mechanism for the chlorination of methane is an example of a **radical chain mechanism.**

A Radical Chain Mechanism

Initiation	Propagation steps	Chain termination

$$X_2 \longrightarrow 2:\ddot{X}\cdot \qquad :\ddot{X}\cdot + RH \longrightarrow R\cdot + H\ddot{X}: \qquad :\ddot{X}\cdot + :\ddot{X}\cdot \longrightarrow X_2$$

$$X_2 + R\cdot \longrightarrow R\ddot{X}: + :\ddot{X}\cdot \qquad R\cdot + :\ddot{X}\cdot \longrightarrow RX$$

$$R\cdot + R\cdot \longrightarrow R_2$$

Only a few halogen atoms are necessary for initiating the reaction, because the propagation steps are self-sufficient in $:\ddot{X}\cdot$. The first propagation step consumes a halogen atom, the second produces one. The newly generated halogen atom then reenters the propagation cycle in the first propagation step. In this way, a *radical chain* is set in motion that can drive the reaction for many thousands of cycles.

EXERCISE 3-3

Chlorination of ethane furnishes chloroethane. Write a mechanism for this transformation and calculate $\Delta H°$ for each step (see Tables 3-1 and 3-2).

One of the practical problems in chlorinating methane is the control of product selectivity. As mentioned earlier, the reaction does not stop at the formation of chloromethane but continues to form di-, tri-, and tetrachloromethane by further substitution. A practical solution to this problem is the use of a large excess of methane in the reaction. Under such conditions, the reactive intermediate chlorine atom is at any given moment surrounded by many more methane molecules than product CH_3Cl. Thus, the chance of $Cl\cdot$ finding CH_3Cl to eventually make CH_2Cl_2 is greatly diminished, and product selectivity is achieved.

In summary, chlorine transforms methane into chloromethane. The reaction proceeds through a mechanism in which heat or light causes a small number of Cl_2 molecules to undergo homolysis to chlorine atoms (initiation). The latter induce and maintain a radical chain sequence consisting of two (propagation) steps: (1) hydrogen abstraction to generate the methyl radical and HCl and (2) conversion of $CH_3\cdot$ by Cl_2 into CH_3Cl and regenerated $Cl\cdot$. The chain is terminated by various combinations of radicals and free atoms. The heats of the individual steps are calculated by comparing the strengths of the bonds that are being broken with those of the bonds being formed.

3-5 Other Radical Halogenations of Methane

Fluorine and bromine, but not iodine, also react with methane by similar radical mechanisms to furnish the corresponding halomethanes. The dissociation energies of X_2

(X = F, Br, I) are lower than that of Cl_2, thus ensuring ready initiation of the radical chain (Table 3-4).

Fluorine is most reactive, iodine least reactive

Let us compare the enthalpies of the two propagation steps in the different halogenations of methane (Table 3-5). It is apparent that there are quite striking differences in the driving force for hydrogen abstraction. For fluorine, this step is exothermic by -30 kcal mol^{-1}. We have already seen that, for chlorine, the same step is slightly endothermic; for bromine, it is substantially so, and for iodine even more so. This trend has its origin in the decreasing bond strengths of the hydrogen halides in the procession from fluorine to iodine (Table 3-1). The strong hydrogen–fluorine bond is the cause of the high reactivity of fluorine atoms in hydrogen abstraction reactions. Fluorine is more reactive than chlorine, chlorine is more reactive than bromine, and the least reactive halogen atom is iodine.

The contrast between fluorine and iodine is illustrated by comparing potential-energy diagrams for their respective hydrogen abstractions from methane (Figure 3-7) (see page 108). The highly exothermic reaction of fluorine has a negligible activation barrier. Moreover, in its transition state, the fluorine atom is relatively far from the hydrogen that is being transferred, and the H–CH$_3$ distance is only slightly greater than that in CH$_4$ itself. Why should this be so? At the transition state the energy needed for (partial) bond-breaking exactly equals that gained by (partial) bond-making. In the present case, the full H–CH$_3$ bond is 30 kcal mol^{-1} weaker than that of H–F (Table 3-1). Only a small shift of the H toward the F· is necessary for bonding between the two to overcome that between hydrogen and carbon. If we view the reaction coordinate as a measure of the degree of hydrogen shift from C to F, the transition state is reached *early* and is much closer in appearance to the starting materials than to the products. *Early transition states are frequently characteristic of fast, exothermic processes.*

On the other hand, reaction of I· with CH$_4$ has a very high E_a (at least as large as its endothermicity, $+34$ kcal mol^{-1}; Table 3-1). Thus, the transition state is not reached until the H–C bond is nearly completely broken and the H–I bond is almost fully formed. The transition state is said to be *late:* It is substantially further along the reaction coordinate and is much closer in structure to the products of this process, CH$_3$· and HI. *Late transition states are frequently typical of relatively slow, endothermic transformations.* Together these rules are known as the **Hammond* postulate.**

*Professor George S. Hammond (b. 1921), Georgetown University, Washington, D.C.

TABLE 3-4	
DH° Values for the Elemental Halogens	

Halogen	DH° (kcal mol^{-1})
F$_2$	37
Cl$_2$	58
Br$_2$	46
I$_2$	36

Relative Reactivities of X· in Hydrogen Abstractions
F· > Cl· > Br· > I·

TABLE 3-5	Enthalpies of the Propagation Steps in the Halogenation of Methane (kcal mol^{-1})			
Reaction	**F**	**Cl**	**Br**	**I**
:X· + CH$_4$ ⟶ ·CH$_3$ + HX:	-30	$+2$	$+18$	$+34$
·CH$_3$ + X$_2$ ⟶ CH$_3$X: + :X·	-73	-27	-25	-21
CH$_4$ + X$_2$ ⟶ CH$_3$X: + HX:	-103	-25	-7	$+13$

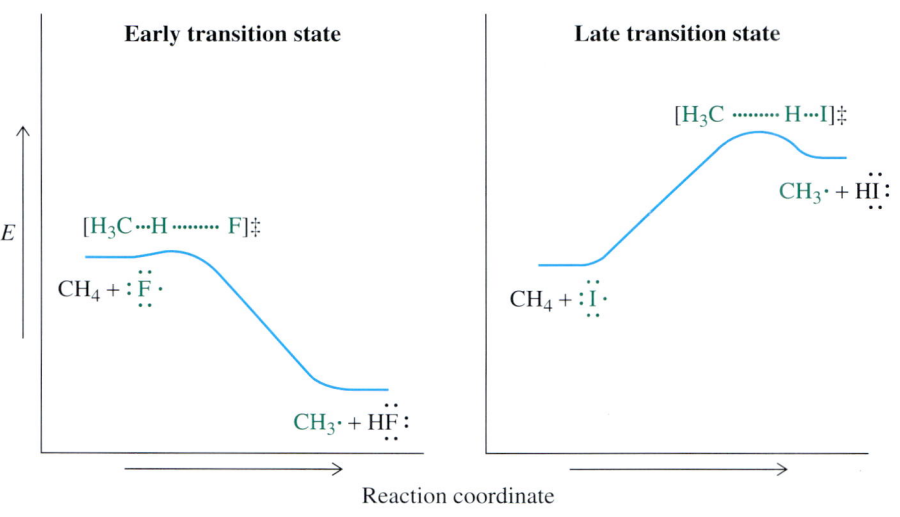

FIGURE 3-7

Potential-energy diagrams: *(left)* the reaction of a fluorine atom with CH_4, an exothermic process with an early transition state; and *(right)* the reaction of an iodine atom with CH_4, an endothermic transformation with a late transition state. Both are thus in accord with the Hammond postulate.

The second propagation step is exothermic

Let us now consider the second propagation step for each halogenation in Table 3-5. This process is exothermic for all the halogens. Again, the reaction is fastest and most exothermic for fluorine. The combined enthalpies of the two steps for the fluorination of methane result in a $\Delta H°$ of -103 kcal mol^{-1}. Indeed, this value is so large that, at sufficiently high concentrations of methane and fluorine gas, a violent reaction occurs. The formation of chloromethane is less exothermic, and that of bromomethane even less so. In the latter case, the appreciably endothermic nature of the first step ($\Delta H° = +18$ kcal mol^{-1}) is barely overcome by the enthalpy of the second ($\Delta H° = -25$ kcal mol^{-1}), resulting in an energy change of only -7 kcal mol^{-1} for the overall substitution. Finally, inspection of the thermodynamics of iodination reveals why iodine does not react with methane to furnish methyl iodide and hydrogen iodide. The first step costs so much energy that the second step, although exothermic, cannot drive the reaction.

EXERCISE 3-4

Predict the product distribution of the reaction of methane with an equimolar mixture of chlorine and bromine at low conversion.

In summary, fluorine, chlorine, and bromine react with methane to give halomethanes. All three reactions follow the radical chain mechanism described for chlorination. In these processes, the first propagation step is always the slower of the two. It becomes more exothermic and its activation energy decreases in the proces-

sion from bromine to chlorine to fluorine. This trend explains the relative reactivity of the halogens, fluorine being the most reactive. Iodination of methane is endothermic and does not occur. Strongly exothermic reaction steps are often characterized by early transition states. Conversely, endothermic or relatively less exothermic steps typically have late transition states.

3-6 Chlorination of Higher Alkanes: Relative Reactivity and Selectivity

What happens in the radical halogenation of other alkanes? Will the different types of R–H bonds—namely, primary, secondary, and tertiary—react in the same way as those in methane? Let us consider the chlorination of ethane, then propane, and finally 2-methylpropane.

The monochlorination of ethane gives chloroethane as the product.

Chlorination of Ethane

$$CH_3CH_3 + Cl_2 \xrightarrow{\Delta \text{ or } h\nu} CH_3CH_2Cl + HCl \qquad \Delta H° = -27 \text{ kcal mol}^{-1}$$
Chloroethane

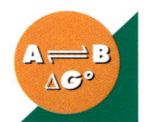

This reaction proceeds by a radical chain mechanism analogous to the one observed for methane. The propagation steps include formation of the ethyl radical from ethane and the chlorine atom, followed by generation of chloroethane and the release of another chlorine atom (Exercise 3-3). The greatest difference between the two mechanisms lies in the change in values for $\Delta H°$. Thus, the abstraction of a hydrogen from ethane is no longer endothermic, as in methane, but favorable by -5 kcal mol^{-1}. The reason is the weaker C–H bond of ethane ($DH° = 98$ kcal mol^{-1}).

Propagation Steps in the Mechanism of the Chlorination of Ethane

$$CH_3CH_3 + :\ddot{C}l\cdot \longrightarrow CH_3CH_2\cdot + H\ddot{C}l: \qquad \Delta H° = -5 \text{ kcal mol}^{-1}$$

$$CH_3CH_2\cdot - Cl_2 \longrightarrow CH_3CH_2\ddot{C}l: + :\ddot{C}l\cdot \qquad \Delta H° = -22 \text{ kcal mol}^{-1}$$

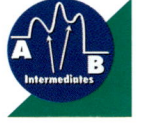

What can be expected for the next homolog, propane?

Secondary C–H bonds are more reactive than primary ones

The eight hydrogen atoms in propane fall into two groups: six primary and two secondary hydrogens. If chlorine atoms were to abstract and replace primary and secondary hydrogens at equal rates, we should expect to find a product mixture containing three times as much 1-chloropropane as 2-chloropropane. We call this outcome a **statistical product ratio,** because it derives from the statistical fact that in propane there are three times as many primary sites for reaction as secondary sites. In other words, it is three times as likely for a chlorine atom to collide with a primary

$CH_3CH_2CH_3$
Propane

Six primary hydrogens (blue)
Two secondary hydrogens (red)

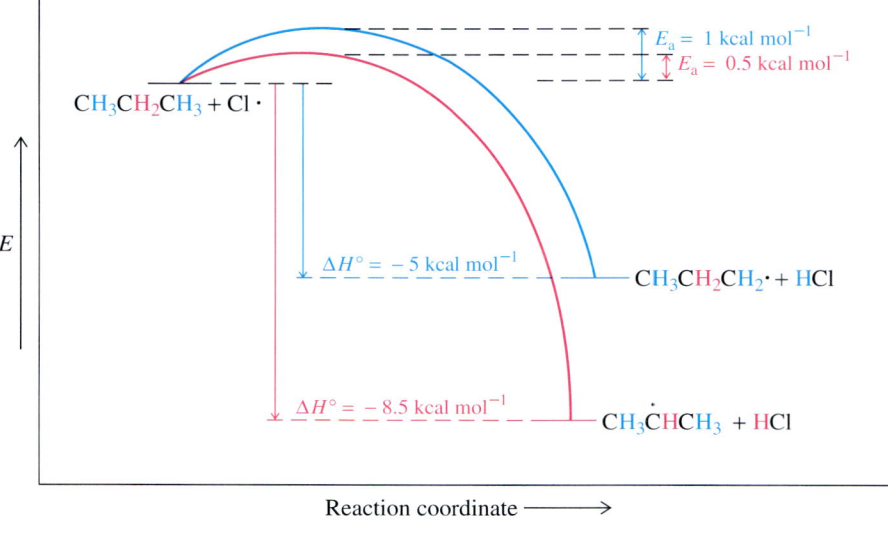

FIGURE 3-8

Hydrogen abstraction by a chlorine atom from the secondary carbon in propane is more exothermic and faster than that from the primary carbon.

hydrogen in propane, of which there are six, as with a secondary hydrogen atom, of which there are only two.

However, secondary C–H bonds are weaker than primary ones ($DH° = 94.5$ versus 98 kcal mol^{-1}). Abstraction of a secondary hydrogen is therefore more exothermic and proceeds with a smaller activation barrier (Figure 3-8). We might thus expect secondary hydrogens to react faster, leading to more 2-chloro- than 1-chloropropane. What is actually observed? At 25°C, the experimental ratio of 1-chloropropane : 2-chloropropane is found to be 43 : 57. This result indicates that both statistical and bond-energy factors determine the product formed.

Chlorination of Propane

$$Cl_2 + CH_3CH_2CH_3 \xrightarrow{h\nu} CH_3CH_2CH_2Cl + CH_3\overset{\overset{\displaystyle Cl}{|}}{C}HCH_3 + HCl$$

1-Chloropropane   **2-Chloropropane**

Expected statistical ratio	3 :	1
Expected C–H bond reactivity ratio	Less :	More
Experimental ratio (25°C)	43 :	57

We can calculate the *relative reactivity of secondary and primary hydrogens* in chlorination by factoring out the statistical contribution to the product ratio.

$$\frac{\text{Relative reactivity of a secondary hydrogen}}{\text{Relative reactivity of a primary hydrogen}} = \frac{\left(\begin{array}{c}\text{yield of product from}\\\text{secondary hydrogen abstraction}\end{array}\right) \Big/ \left(\begin{array}{c}\text{number of}\\\text{secondary hydrogens}\end{array}\right)}{\left(\begin{array}{c}\text{yield of product from}\\\text{primary hydrogen abstraction}\end{array}\right) \Big/ \left(\begin{array}{c}\text{number of}\\\text{primary hydrogens}\end{array}\right)} = \frac{57/2}{43/6} \approx 4$$

In other words, each secondary hydrogen in the chlorination of propane at 25°C is about four times as reactive as each primary one. We say that chlorine exhibits a **selectivity** of 4 : 1 in the removal of secondary versus primary hydrogens at 25°C.

We could predict from this analysis that *all* secondary hydrogens are four times as reactive as all primary hydrogens in *all* radical chain reactions. Is this prediction true? Unfortunately, no. Although secondary C–H bonds generally undergo dissociation faster than do their primary counterparts, their relative reactivity very much depends on the nature of the attacking species, X·, the strength of the resulting H–X bond, and the temperature. For example, at 600°C, the chlorination of propane results in the statistical distribution of products: roughly three times as much 1-chloropropane as 2-chloropropane. At such a high temperature, virtually every collision between a chlorine atom and any hydrogen in the propane molecule takes place with sufficient energy to lead to successful reaction. Chlorination is said to be **unselective** at this temperature, and the product ratio is governed by statistical factors.

EXERCISE 3-5

What do you expect the products of monochlorination of butane to be? In what ratio will they be formed at 25°C?

Tertiary C–H bonds are more reactive than secondary ones

Let us now find the relative reactivity of a tertiary hydrogen in the chlorination of alkanes. For this purpose, we expose 2-methylpropane, a molecule containing one tertiary and nine primary hydrogens, to chlorination conditions at 25°C. The resulting two products, 2-chloro-2-methylpropane (*tert*-butyl chloride) and 1-chloro-2-methylpropane (isobutyl chloride), are formed in the relative yields of 36 and 64%, respectively.

Chlorination of 2-Methylpropane

$$Cl_2 + CH_3{-}\overset{\overset{\displaystyle CH_3}{|}}{\underset{\underset{\displaystyle CH_3}{|}}{C}}{-}H \xrightarrow{h\nu} ClCH_2{-}\overset{\overset{\displaystyle CH_3}{|}}{\underset{\underset{\displaystyle CH_3}{|}}{C}}{-}H \;+\; CH_3{-}\overset{\overset{\displaystyle CH_3}{|}}{\underset{\underset{\displaystyle CH_3}{|}}{C}}{-}Cl \;+\; HCl$$

64%	36%
1-Chloro-2-methylpropane	**2-Chloro-2-methylpropane**
(Isobutyl chloride)	(***tert*-Butyl chloride)**

Expected statistical ratio	9 : 1	
Expected C–H bond reactivity ratio	Less : More	
Experimental ratio (25°C)	64 : 36	

We can determine the reactivity of tertiary relative to primary hydrogens as follows: we combine the experimental result of 64% primary chlorination and 36% tertiary chlorination in 2-methylpropane with the statistical presence of nine primary hydrogens and one tertiary hydrogen atom in the starting alkane. Dividing the proportionate amount observed of each product by the number of hydrogens that contribute to its formation gives a measure of reactivity per hydrogen atom:

$$CH_3{-}\overset{\overset{\displaystyle CH_3}{|}}{\underset{\underset{\displaystyle CH_3}{|}}{C}}{-}H$$

2-Methylpropane

Nine primary hydrogens (blue)
One tertiary hydrogen (red)

$$\frac{\text{Relative reactivity of a tertiary hydrogen}}{\text{Relative reactivity of a primary hydrogen}} = \frac{\left(\begin{array}{c}36\% \text{ tertiary}\\ \text{chlorination}\end{array}\right)\Big/\left(\begin{array}{c}1 \text{ tertiary}\\ \text{hydrogen}\end{array}\right)}{\left(\begin{array}{c}64\% \text{ primary}\\ \text{chlorination}\end{array}\right)\Big/\left(\begin{array}{c}9 \text{ primary}\\ \text{hydrogens}\end{array}\right)} = \frac{36/1}{64/9} \approx 5$$

Thus, tertiary hydrogen atoms are about five times as reactive as primary ones. This selectivity, again, decreases at higher temperatures. However, we can say that, at 25°C, the relative reactivities of the various C–H bonds in chlorinations are roughly

$$\text{Tertiary : secondary : primary} = 5:4:1$$

The result agrees well with the relative reactivity expected from consideration of bond strength: The tertiary C–H bond is weaker than the secondary, and the latter in turn is weaker than the primary.

We can verify this ordering by looking at the competition among all three types of hydrogens within a single substrate. 2-Methylbutane, which contains nine primary hydrogens, two secondary hydrogens, and one tertiary hydrogen, is an example. Because this molecule has two types of primary hydrogens, one set of six and one set of three, reaction with chlorine yields a total of four different monochlorination products.

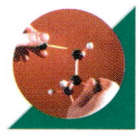

Chlorination of 2-Methylbutane

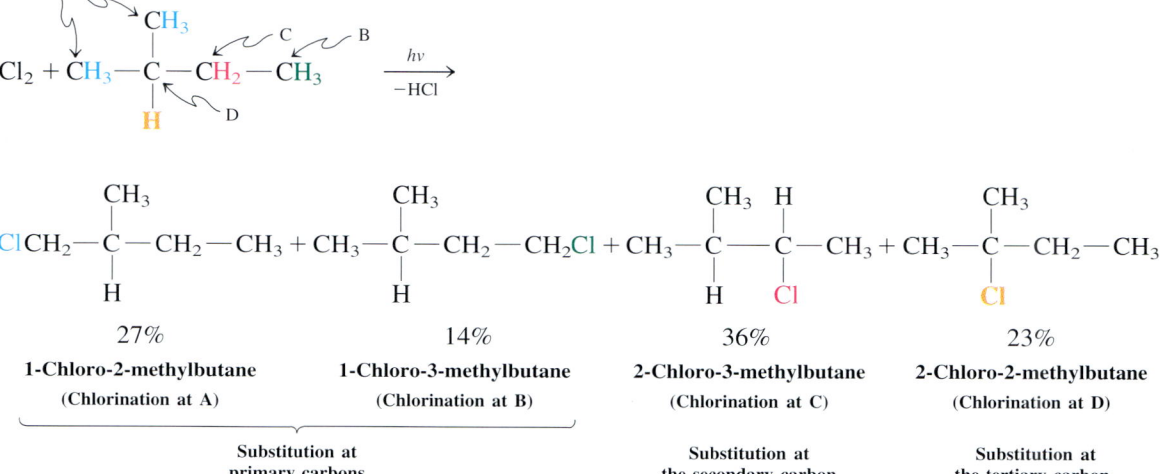

27%	14%	36%	23%
1-Chloro-2-methylbutane	**1-Chloro-3-methylbutane**	**2-Chloro-3-methylbutane**	**2-Chloro-2-methylbutane**
(Chlorination at A)	(Chlorination at B)	(Chlorination at C)	(Chlorination at D)
Substitution at primary carbons		**Substitution at the secondary carbon**	**Substitution at the tertiary carbon**

The combined yield of the two primary halide products is 41% (1-chloro-2-methylbutane plus 1-chloro-3-methylbutane), the secondary halide 2-chloro-3-methylbutane is formed in 36% yield, and the tertiary halide in 23%. Therefore,

$$\text{Primary : secondary : tertiary halide} = 41:36:23$$
$$\textit{Relative reactivity} \quad \text{primary : secondary : tertiary} = 41/9:36/2:23/1$$
$$= 1:4:5$$

as expected.

EXERCISE 3-6

Give products and the ratio in which they are expected to form for the monochlorination of 3-methylpentane at 25°C. Be careful to take into account the number of hydrogens in each distinct group in the starting alkane.

To summarize, the relative reactivity of primary, secondary, and tertiary hydrogens follows the trend expected on the basis of their relative C–H bond strengths. Relative reactivity ratios can be calculated by factoring out statistical considerations. These ratios are temperature dependent, with greater selectivity at lower temperatures.

3-7 Selectivity in Radical Halogenation with Fluorine and Bromine

How selectively do halogens other than chlorine halogenate the alkanes? Table 3-5 and Figure 3-7 show that fluorine is the most reactive halogen: Hydrogen abstraction is highly exothermic and has negligible activation energy. Conversely, bromine is much less reactive, because the same step has a large positive $\Delta H°$ and a high activation barrier. Does this difference affect their selectivity in halogenation of alkanes?

To answer this question, consider the reactions of fluorine and bromine with 2-methylpropane. Single fluorination at 25°C furnishes two possible products, in a ratio very close to that expected statistically.

2-Fluoro-2-methylpropane : 1-Fluoro-2-methylpropane
(*tert*-Butyl fluoride)   (**Isobutyl fluoride**)

Observed 14 : 86
Expected   1 : 9

Fluorine thus displays very little selectivity. Why? Because the transition states for the two competing processes are reached very early, their energies and structures are similar to each other, as well as similar to those of the starting material (Figure 3-9).

Fluorination of 2-Methylpropane

$$F_2 + (CH_3)_3CH \xrightarrow{h\nu} (CH_3)_3CF + FCH_2-\underset{\underset{CH_3}{|}}{\overset{\overset{CH_3}{|}}{C}}-H + HF$$

14%   86%
2-Fluoro-2-methylpropane   **1-Fluoro-2-methylpropane**
(*tert*-**Butyl fluoride**)   (**Isobutyl fluoride**)

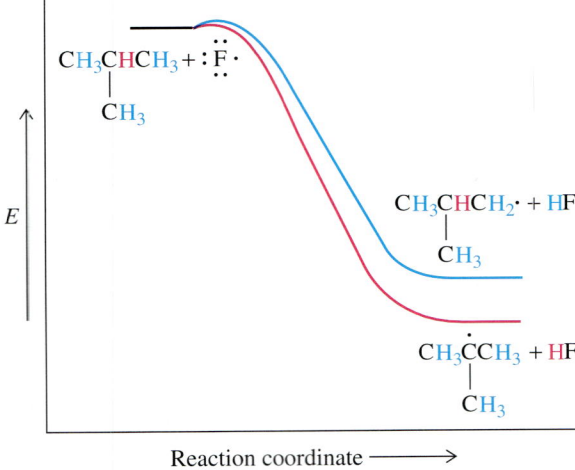

Reaction coordinate ⟶

FIGURE 3-9

Potential-energy diagram for the abstraction of a primary or a tertiary hydrogen by a fluorine atom from 2-methylpropane. The energies of the respective early transition states are almost the same and barely higher than that of starting material (i.e., both E_a values are close to zero), resulting in little selectivity.

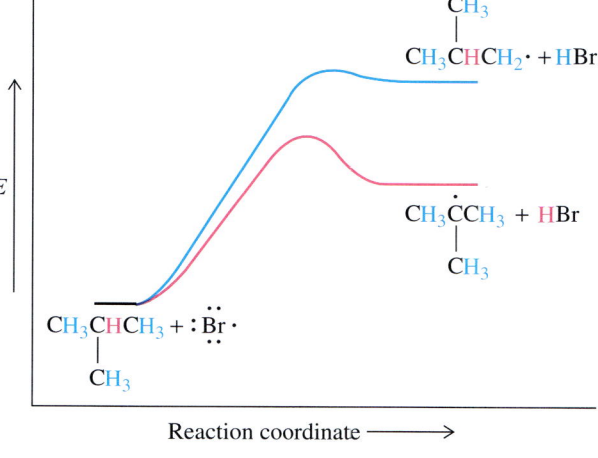

FIGURE 3-10
Potential-energy diagram for the abstraction of a primary or a tertiary hydrogen by a bromine atom from 2-methylpropane. The two late transition states are dissimilar in energy, indicative of the energy difference between the resulting primary and tertiary radicals, respectively, leading to the products with greater selectivity.

Conversely, *bromination of the same compound is highly selective,* giving the tertiary bromide almost exclusively. Hydrogen abstractions by bromine have *late* transition states in which extensive C–H bond-breaking and H–Br bond-making have occurred. Thus, their respective structures and energies resemble those of the corresponding radical products. As a result, the activation barriers for the reaction of bromine with primary and tertiary hydrogens, respectively, will differ by almost as much as the difference in stability between primary and tertiary radicals (Figure 3-10), a difference leading to the observed high selectivity (more than 1700:1).

Bromination of 2-Methylpropane

$$Br_2 + (CH_3)_3CH \xrightarrow{h\nu} (CH_3)_3CBr + BrCH_2{-}\underset{\underset{CH_3}{|}}{\overset{\overset{CH_3}{|}}{C}}{-}H + HBr$$

$>99\%$ $<1\%$
2-Bromo-2-methylpropane   **1-Bromo-2-methylpropane**
(*tert*-Butyl bromide)   (Isobutyl bromide)

In summary, increased reactivity goes hand in hand with reduced selectivity in radical substitution reactions. Fluorine and chlorine, the more reactive halogens, discriminate between the various types of C–H bonds much less than does the less reactive bromine (Table 3-6).

TABLE 3-6	Relative Reactivities of the Four Types of Alkane C–H Bonds in Halogenations		
C–H bond	**F·** (25°C, gas)	**Cl·** (25°C, gas)	**Br·** (150°C, gas)
$CH_3{-}H$	0.5	0.004	0.002
$RCH_2{-}H^a$	1	1	1
$R_2CH{-}H$	1.2	4	80
$R_3C{-}H$	1.4	5	1700

[a]For each halogen, reactivities with four types of alkane C–H bonds are normalized to the reactivity of the primary C–H bond.

3-8 Synthetic Radical Halogenation

How can we devise a successful and cost-effective alkane halogenation? We must take into account selectivity, convenience, efficiency, and price.

Fluorinations are unattractive, because fluorine is relatively expensive and corrosive; and, even worse, its reactions are often violently uncontrollable. Radical iodinations, at the other extreme, fail because of unfavorable thermodynamics.

Chlorinations are important, particularly in industry, simply because chlorine is cheap. (It is prepared by electrolysis of sodium chloride, ordinary table salt.) The drawback to chlorination is low selectivity, so the process results in mixtures of isomers that are difficult to separate. To circumvent the problem, an alkane that contains a single type of hydrogen can be used as a substrate, thus giving (at least initially) a single product. Cyclopentane is one such alkane.

Chlorination of a Molecule with Only One Type of Hydrogen

$$\text{Cyclopentane} + Cl_2 \xrightarrow{h\nu} \text{Chlorocyclopentane} + HCl$$

92.7%

Cyclopentane   **Chloro**cyclopentane
(Large excess)

To minimize overhalogenation, chlorine is used as the limiting reagent (Section 3-4). Even then, multiple substitution can complicate the reaction. Conveniently, the more highly chlorinated products have higher boiling points and can be separated by distillation.

On an industrial scale, alkanes are chlorinated in large vessels fitted with elaborate controls to ensure smooth, safe operation. In the research laboratory, the use of chlorine gas is often avoided, because it is toxic, corrosive, and difficult to measure accurately. Other chlorinating agents have been developed that can be handled more safely and accurately. These agents are usually liquids or solids, such as sulfuryl chloride (SO_2Cl_2) and N-chlorobutanimide (N-chlorosuccinimide, NCS).

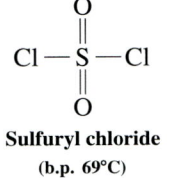

Sulfuryl chloride
(b.p. **69°C**)

Which of the following compounds will give a monochlorination product with reasonable selectivity: propane, 2,2-dimethylpropane, cyclohexane, methylcyclohexane?

Because bromination is selective (and bromine is a liquid), it is frequently the method of choice for halogenating an alkane on a relatively small scale in the research laboratory. Reaction occurs at the more substituted carbon, even in statistically unfavorable situations. Typical solvents are chlorinated methanes (CCl_4, $CHCl_3$, CH_2Cl_2), which are comparatively unreactive with bromine.

Bromine is obtained from aqueous sodium bromide solutions (found in natural brines) by treatment with chlorine. Bromine is used less in industry because of its relatively greater cost per mole compared with chlorine. A popular solid brominating agent is N-bromobutanimide (N-bromosuccinimide, NBS; see Section 14-2), an analog of NCS.

In summary, even though more expensive, bromine is the reagent of choice for selective radical halogenations. Chlorinations furnish product mixtures, a problem that can be minimized by choosing alkanes with only one type of hydrogen and treating them with a deficiency of chlorine. For ease in handling, research chemists often

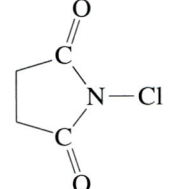

N-Chlorobutanimide
(**N-Chlorosuccinimide**)
(m.p. **148°C**)

CCl$_3$
Cl—⟨ ⟩—CH—⟨ ⟩—Cl

**1,1,1-Trichloro-2,2-bis(4-chlorophenyl)ethane
(DDT)**

Chlorination of ethanol in the production of trichloroacetaldehyde, CCl$_3$CHO, was first described in 1832. The hydrated form is commonly called *chloral* and is a powerful hypnotic with the nickname "knockout drops." Chloral is also a key reagent in the synthesis of the powerful insecticide DDT (an abbreviation derived from the nonsystematic name *di*chloro*di*phenyl*tri*chloroethane).

The use of DDT in the control of insect-borne diseases has saved many millions of lives in the past half-century, chiefly through the decimation of the *Anopheles* mosquito, the main carrier of the parasite that causes malaria. Although its toxicity toward mammals is low (the fatal human dose is about 500 mg kg^{-1} of body weight), DDT is very resistant

to biodegradation. Its accumulation in the food chain makes it a hazard to birds and fish, and consequently it has been banned by the U.S. Environmental Protection Agency since 1972.

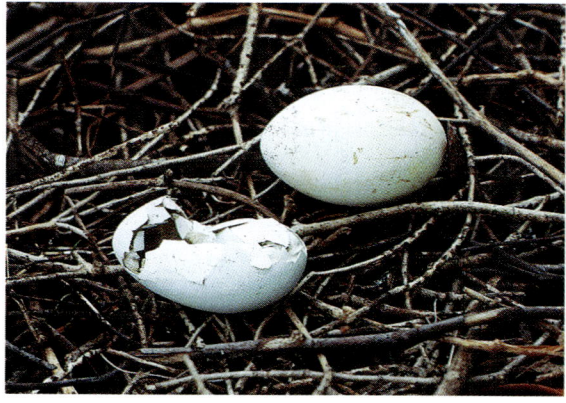

Eggshells damaged by high concentrations of pesticide residues.

prefer solid reagents, such as *N*-bromobutanimide, or liquids, such as sulfuryl chloride, to the respective liquid and gaseous reagents.

3-9 Synthetic Chlorine Compounds and the Stratospheric Ozone Layer

We have seen how bond homolysis can be caused by both heat and light. Such chemical events can occur on a grand scale in nature and may have substantial environmental consequences. This section will explore an example of radical chemistry that has had a significant effect on our lives and will continue to do so well into the next century.

The ozone layer shields Earth's surface from high-energy ultraviolet light

Earth's atmosphere consists of several distinct layers. The lowest layer, extending to about 15 km in altitude, is the troposphere, the region where weather occurs. Above the troposphere, the stratosphere extends upward to an altitude of some 50 km. Although the density of the stratosphere is too low to sustain terrestrial life, the stratosphere is the home of the **ozone layer,** which plays a critical role in the ability of life

to survive on Earth. Ozone (O_3) and ordinary molecular oxygen (O_2) equilibrate in the stratosphere by the action of ultraviolet light from the sun:

Interconversion of Ozone and Molecular Oxygen in the Stratosphere

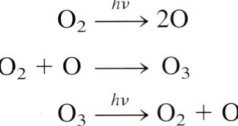

$$O_2 \xrightarrow{hv} 2O$$
$$O_2 + O \longrightarrow O_3$$
$$O_3 \xrightarrow{hv} O_2 + O$$

In the first two reactions high-energy solar radiation splits O_2 into oxygen atoms, which may combine with other molecules of O_2 to produce ozone, a bluish gas with a characteristically sharp and penetrating odor. Occasionally, ozone may be detected in the vicinity of high-voltage equipment, in which electrical discharges cause the conversion of O_2 into O_3. In urban areas, nitrogen dioxide (NO_2) is produced through oxidation of nitric oxide (NO), a product of high temperature combustion. Nitrogen dioxide, in turn, is split by sunlight to release oxygen atoms, which produce O_3 in the lower atmosphere. The presence of ozone as an air pollutant near Earth's surface causes severe irritation of the respiratory membranes and the eyes. However, in the upper atmosphere, the third reaction in the interconversion of ozone and molecular oxygen shown above occurs, in which ozone absorbs UV light in the 200 to 300 nm wavelength range. Irradiation at these wavelengths is capable of destroying the complex molecules that make up biochemical systems. Ozone serves as a natural atmospheric filter to prevent this light from reaching the surface, thereby protecting Earth's life from damage.

CFCs release chlorine atoms upon ultraviolet irradiation

Chlorofluorocarbons (CFCs), or **Freons,** are alkanes in which all of the hydrogens have been replaced by fluorine and chlorine. In general, chlorofluorocarbons are thermally stable, essentially odorless, and nontoxic gases. Among their many commercial applications, their use as refrigerants predominates, because they absorb large quantities of heat upon vaporization. Compression liquefies gaseous CFC, which flows through the cooling coils in refrigerators, freezers, and air conditioners. The liquid absorbs heat from the environment outside the coils and evaporates. The gas then reenters the compressor to be liquefied again, and the cycle continues.

CFCs are among the most effective and widely used synthetic organic compounds in modern society. Why, then, have the nations of the world, with almost unprecedented unanimity, agreed to phase them completely out of use? This remarkable event dates its origins to the late 1960s and early 1970s, when chemists Johnston, Crutzen, Rowland, and Molina* pointed to the existence of radical mechanisms that could convert several kinds of compounds, including CFCs, into reactive species capable of destroying ozone in the Earth's stratosphere.

Upon irradiation by UV light, the weaker C–Cl bonds in the CFC molecules undergo homolysis, giving rise to atomic chlorine.

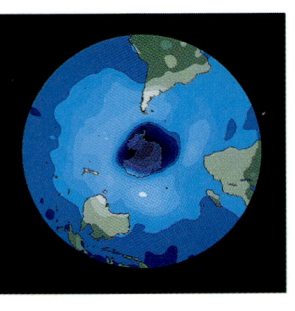

A color-enhanced view of the upper atmosphere above Antarctica during October 1996, showing a region in which the concentration of ozone has dropped below 35% of normal (dark gray).

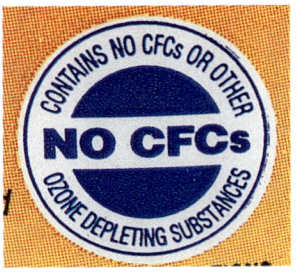

No CFCs!

Common CFCs

CCl_3F

CFC-11

CCl_2F_2

CFC-12

CCl_2FCClF_2

CFC-113

---

*Professor Harold S. Johnston (b. 1920), University of California at Berkeley; Professor Paul Crutzen (b. 1933), Max Planck Institute, Mainz, Germany, Nobel Prize 1995 (chemistry); Professor F. Sherwood Rowland (b. 1927), University of California at Irvine, Nobel Prize 1995 (chemistry); Professor Mario Molina (b. 1943), Massachusetts Institute of Technology, Nobel Prize 1995 (chemistry).

INITIATION STEP

$$F_3C-\overset{\cdot\cdot}{\underset{\cdot\cdot}{Cl}}: \xrightarrow{h\nu} F_3C\cdot + :\overset{\cdot\cdot}{\underset{\cdot\cdot}{Cl}}\cdot$$

Chlorine atoms, in turn, react efficiently with ozone in a radical chain sequence.

PROPAGATION STEPS

$$:\overset{\cdot\cdot}{\underset{\cdot\cdot}{Cl}}\cdot + O_3 \longrightarrow \cdot ClO + O_2$$

$$\cdot ClO + O \longrightarrow O_2 + :\overset{\cdot\cdot}{\underset{\cdot\cdot}{Cl}}\cdot$$

The net result of these two steps is the conversion of a molecule of ozone and an oxygen atom into two molecules of ordinary oxygen. As in the other radical chain processes that we have examined in this chapter, however, the reactive species consumed in one propagation step ($:\overset{\cdot\cdot}{\underset{\cdot\cdot}{Cl}}\cdot$) is regenerated in the other. As a consequence, a small concentration of chlorine atoms is capable of destroying many molecules of ozone. Does such a process actually occur in the atmosphere?

Stratospheric ozone has decreased by about 3% since 1978

Since measurements of atmospheric composition were first made, measurable decreases in stratospheric ozone have been recorded. These changes are seasonal, being most severe in the winter. They show extreme variations with latitude: Large reductions in the ozone layer over the Antarctic were noticeable as early as 1978. Satellite measurements confirmed that total ozone content in this part of the atmosphere in 1987 was less than half of its usual value, and by 1994 it had dropped further, to less than one-third of normal. Localized regions of the Antarctic had no ozone layer above them at all: An "ozone hole" was observed. Ozone amounts north of the Arctic Circle sank to 45% of normal, the lowest readings ever, in the winter of 1996. Stratospheric clouds, which can form only in the extreme cold of the polar regions, appeared to correlate with ozone-hole formation.

Reduction in total ozone above the temperate regions of the Northern Hemisphere currently averages 3%, approximately the worldwide mean, but 6% depletion is measured at 40°N latitude during the winter months. Epidemiological studies suggest that a reduction in stratospheric ozone density of 1% might be expected to give rise to a 1–3% increase in skin cancers. As a consequence, considerable effort has been made to identify the causes of ozone depletion. Are CFCs responsible? Or could natural sources of atmospheric chlorine or other substances be contributing significantly?

The answers to these questions were obtained in systematic studies of satellite observations made from 1987 to 1994: Chlorine monoxide (ClO), a critical component in the ozone-destroying chain reaction illustrated earlier, rises to more than 500 times normal levels in regions of the Antarctic ozone hole, where O_3 concentration is the lowest. Furthermore, this chlorine monoxide and at least 75% of stratospheric chlorine come from CFCs. This connection has been proved by the observation of corresponding concentrations of gaseous hydrogen fluoride (HF) in the stratosphere as well. Neither HF nor any other gaseous fluorine compound occurs naturally or is produced by natural chemical processes anywhere on (or in) this planet. However, CFC decomposition in the presence of hydrocarbons in the atmosphere is known to produce HF. The observations demonstrate that natural sources such as volcanic eruptions or sea spray are not major contributors to stratospheric chlorine, relative to CFCs.

Volcanic aerosols do, however, contribute to depletion of the ozone layer indirectly, by interfering with chemical processes that remove stratospheric chlorine.

The world is searching for CFC substitutes

The Montreal Protocol on Substances That Deplete the Ozone Layer was signed in 1987 and called for a 50% reduction in output of CFCs by 1998. Increasingly alarming evidence regarding ozone depletion led to amendments in 1990 and again in 1992, finally setting an advanced deadline for complete rather than partial phaseout of CFC production: December 31, 1995, marked the end of production of virtually all CFCs in the industrialized world. Meanwhile, hydrochlorofluorocarbons (HCFCs) and hydrofluorocarbons (HFCs) have been developed as replacements for CFCs in commercial applications. HCFCs are more chemically reactive than CFCs and are much more prone to decomposition at lower atmospheric altitudes. Their threat to stratospheric ozone is therefore less, because a smaller proportion survives the time necessary to diffuse to the upper atmosphere. Currently, HFC-134a is replacing CFC-12 in refrigerator and motor vehicle air conditioner compressors. HCFCs-22 and -141b have replaced CFC-11 in the manufacture of foam insulation.

Hydrofluorocarbons have been demonstrated to be safe for the ozone layer. However, hydrochlorofluorocarbons are potential ozone-destroying agents and are themselves scheduled for total phaseout by no later than 2030. Efforts to replace all HCFCs with HFCs are actively underway. It is hoped that this worldwide effort will finally bring the depletion of the ozone layer to a halt by the year 2000. Recovery to nearly normal levels is expected over the next century.

CFC Substitutes

CH_2FCF_3

HFC-134a

$CHClF_2$

HCFC-22

$CHCl_2CF_3$

HCFC-123

CH_3CCl_2F

HCFC-141b

CH_3CClF_2

HCFC-142b

3-10 Combustion and the Relative Stabilities of Alkanes

Let us review what we have learned in this chapter so far. We started by defining bond strength as the energy required to cleave a molecule homolytically. Some typical values were then presented in Tables 3-1 and 3-2 and explained through a discussion of relative radical stabilities, a major factor being the varying extent of hyperconjugation. We then used this information to calculate the $\Delta H°$ values of the steps making up the mechanism of radical halogenation, a discussion leading to an understanding of reactivity and selectivity. It is clear that knowing bond-dissociation energies is a great aid in the thermochemical analysis of organic transformations, an idea that we shall explore on numerous occasions later on. How are these numbers found experimentally?

Chemists determine bond strengths by first establishing the relative heat contents of entire molecules, or their relative positions along the energy axis in our potential-energy diagrams. The reaction chosen for this purpose is complete oxidation (literally, "burning"), or **combustion,** a process common to almost all organic structures, in which all carbon atoms are converted into CO_2 (gas) and all of the hydrogens into H_2O (liquid).

Both products in the combustion of alkanes have a very low energy content, and hence their formation is associated with a large negative $\Delta H°$, released as heat.

$$2\ C_nH_{2n+2} + (3n + 1)\ O_2 \longrightarrow 2n\ CO_2 + (2n + 2)\ H_2O + heat$$

The heat released in the burning of a molecule is called its **heat of combustion,** $\Delta H°_{comb}$, many of which have been measured with high accuracy, thus allowing com-

TABLE 3-7	Heats of Combustion (kcal mol^{-1}, normalized to 25°C) of Various Organic Compounds	
Compound (state)	**Name**	**$\Delta H°_{comb}$**
CH_4 (gas)	Methane	-212.8
C_2H_6 (gas)	Ethane	-372.8
$CH_3CH_2CH_3$ (gas)	Propane	-530.6
$CH_3(CH_2)_2CH_3$ (gas)	Butane	-687.4
$(CH_3)_3CH$ (gas)	2-Methylpropane	-685.4
$CH_3(CH_2)_3CH_3$ (gas)	Pentane	-845.2
$CH_3(CH_2)_3CH_3$ (liquid)	Pentane	-838.8
$CH_3(CH_2)_4CH_3$ (liquid)	Hexane	-995.0
(liquid)	Cyclohexane	-936.9
CH_3CH_2OH (gas)	Ethanol	-336.4
CH_3CH_2OH (liquid)	Ethanol	-326.7
$C_{12}H_{22}O_{11}$ (solid)	Cane sugar (sucrose)	-1348.2

Note: Combustion products are CO_2 (gas) and H_2O (liquid).

parisons of the relative energy content of the alkanes (Table 3-7) and other compounds. Such comparisons have to take into account the physical state of the compound undergoing combustion (gas, liquid, solid). For example, the difference between the heats of combustion of liquid and gaseous ethanol corresponds to its heat of vaporization, $\Delta H°_{vap} = 9.7$ kcal mol^{-1}.

It is not surprising that the $\Delta H°_{comb}$ of alkanes increases with chain length, simply because there is more carbon and hydrogen to burn along the homologous series. Con-

---

CHEMICAL HIGHLIGHT **3-4**

Enzymatic Oxidation of the Alkanes

Combustion is one way to activate the ordinarily quite unreactive alkane. Unfortunately, at high temperatures, oxidation is relatively unselective and destroys much of the starting molecule. A milder approach, called *enzymatic activation,* uses enzymes, the catalysts in living systems (see Chemical Highlight 3-1 and Chapter 26).

Enzymatic Alkane Activation

$$R-H \xrightarrow{\text{Enzyme, } O_2} R-OH$$

Alkanes   **Alcohols**
(C_1–C_8)

The monooxygenases, found in mammalian tissue, catalyze the oxidation of drugs, steroids (Chapter 4), and fatty acids (Chapter 19). In microbial systems, these same enzymes catalyze the oxidation of alkanes. Thus, an enzyme from *Methylococcus capsulatus* inserts oxygen into a number of hydrocarbons, to give alcohols.

Environmental pollutants can promote *non*enzymatic oxidations of molecules of biological importance, such as lipids, the building blocks of fatty tissue (Chapter 20). This chemistry will be discussed in Chapter 22.

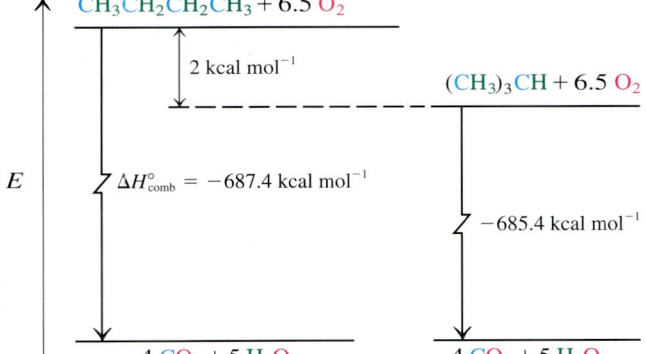

FIGURE 3-11
Butane has a higher energy content than does 2-methylpropane, as measured by the release of energy on combustion. Butane is therefore thermodynamically less stable than its isomer.

versely, isomeric alkanes contain the same number of carbons and hydrogens, and one might expect that their respective combustions would be equally exothermic. However, that is not the case.

A comparison of the heats of combustion of isomeric alkanes reveals that their values are usually *not* the same. Consider butane and 2-methylpropane. The combustion of butane has a ΔH°_{comb} of -687.4 kcal mol^{-1}, whereas its isomer releases $\Delta H^\circ_{comb} = -685.4$ kcal mol^{-1}, 2 kcal mol^{-1} less (Table 3-7). This finding shows that 2-methylpropane has a *smaller* energy content than does butane, because combustion yielding the identical kind and number of products produces less energy (Figure 3-11). Butane is said to be *thermodynamically less stable* than its isomer.

EXERCISE 3-8

The hypothetical thermal conversion of butane into 2-methylpropane should have a $\Delta H^\circ = -2.0$ kcal mol^{-1}. What value do you obtain by using the bond-dissociation data in Table 3-2?

To summarize, the heats of combustion values of alkanes and other organic molecules give quantitative estimates of their energy content and, therefore, their relative stabilities.

CHAPTER INTEGRATION PROBLEM

Iodomethane reacts with hydrogen iodide under free radical conditions ($h\nu$) to give methane and iodine. The overall equation of the reaction is

$$CH_3I + HI \xrightarrow{h\nu} CH_4 + I_2$$

a. Write a mechanism for this process including initiation, propagation, and at least one termination step. Use bond strength data from Tables 3-1 and 3-4.

SOLUTION

Step 1. Begin by proposing a likely **initiation** step. Recall—from Section 3-4, for example—that initiation steps of radical reactions include cleavage of the *weakest* bond in the starting compounds. According to Tables 3-1 and 3-4, that is the carbon–

iodine bond in CH_3I, with $DH° = 57$ kcal mole^{-1}. Therefore

Initiation Step

$$H_3C-I \xrightarrow{hv} H_3C\cdot + \overset{\cdot\cdot}{\underset{\cdot\cdot}{:I}}\cdot$$

Step 2. Again following the model in Section 3-4, propose a **propagation** step in which one of the species produced in the initiation step reacts with one of the molecules shown in the overall equation of the reaction. Try to design the step so that one of its products corresponds to a molecule formed in the overall reaction and the other is a species that can give rise to a second propagation step. The possibilities are

(i) $H_3C\cdot + HI \longrightarrow CH_4 + \overset{\cdot\cdot}{\underset{\cdot\cdot}{:I}}\cdot$

(iii) $\overset{\cdot\cdot}{\underset{\cdot\cdot}{:I}}\cdot + HI \longrightarrow I_2 + H\cdot$

(ii) $H_3C\cdot + CH_3I \longrightarrow CH_4 + \cdot CH_2I$

(iv) $\overset{\cdot\cdot}{\underset{\cdot\cdot}{:I}}\cdot + CH_3I \longrightarrow I_2 + \cdot CH_3$

Propagation steps (i) and (ii) both convert methyl radical into methane by removing a hydrogen atom from HI and CH_3I, respectively. Processes (iii) and (iv) show the removal of an iodine atom by another iodine atom from either HI or CH_3I, giving I_2. All four propagation steps convert a molecule of starting material in the overall equation of the reaction into a molecule of product. How do we choose the correct steps? *Look at the radical products of each hypothetical propagation step. The two correct steps are those whose product radicals are each other's reactants.* Propagation step (i) consumes a methyl radical and produces an iodine atom. Step (iv) consumes iodine, and produces methyl. Therefore, steps (i) and (iv) are the correct steps of a propagation cycle.

Propagation Steps

$$H_3C\cdot + HI \longrightarrow CH_4 + \overset{\cdot\cdot}{\underset{\cdot\cdot}{:I}}\cdot$$

$$\overset{\cdot\cdot}{\underset{\cdot\cdot}{:I}}\cdot + CH_3I \longrightarrow I_2 + \cdot CH_3$$

You can check this answer by adding up all species on the left and right sides of these equations, respectively, to see if they correspond to the reactants and products of the overall reaction. They do: the radicals cancel on both sides of the equations, leaving only the correct molecules.

Step 3. Finally, combination of *any pair of radicals* to give a single molecule constitutes a legitimate **termination** step. There are three:

$$2 \overset{\cdot\cdot}{\underset{\cdot\cdot}{:I}}\cdot \longrightarrow I_2$$

$$2 H_3C\cdot \longrightarrow H_3C-CH_3 \quad \text{(ethane)}$$

$$H_3C\cdot + \overset{\cdot\cdot}{\underset{\cdot\cdot}{:I}}\cdot \longrightarrow CH_3I$$

b. Calculate the enthalpy changes, $\Delta H°$, associated with the overall reaction and all of the mechanistic steps. Use Tables 3-1, 3-2, and 3-4, as appropriate.

Solution

Breaking a bond requires energy *input,* forming a bond gives rise to energy *output,* and $\Delta H° = $ (energy in) $-$ (energy out). For the overall reaction, we have the following bond strength values to consider:

$$CH_3-I + H-I \xrightarrow{hv} CH_3-H + I-I$$

$$DH°: \quad 57 \qquad 71 \qquad \qquad 105 \qquad 36$$

The answer is $\Delta H° = (57 + 71) - (105 + 36) = -13$ kcal mole^{-1} (see also Table 3-5).

For the mechanistic steps, the same principle applies. With one exception, the same four $DH°$ values just shown are all that you need, because they correspond to the only four bonds that are either made or broken in any of the steps in the mechanism.

Initiation step: $\Delta H° = DH° \, (CH_3–I) = +57$ kcal mole^{-1}

Propagation step (i): $\Delta H° = DH° \, (H–I) - DH° \, (CH_3–H) = -34$ kcal mole^{-1}

Propagation step (iv): $\Delta H° = DH° \, (CH_3–I) - DH° \, (I–I) = +21$ kcal mole^{-1}

Notice that the sum of the $\Delta H°$ values for the two propagation steps equals $\Delta H°$ for the overall reaction. *This is always true.*

Termination steps: $\Delta H° = -DH°$ for the bond formed; -36 kcal mole^{-1} for I_2, -57 kcal mole^{-1} for CH_3I, and -90 kcal mole^{-1} for the C–C bond in ethane.

IMPORTANT CONCEPTS

1. The $\Delta H°$ of **bond homolysis** is defined as the **bond-dissociation energy,** $DH°$. Bond homolysis gives radicals or free atoms.

2. The C–H bond strengths in the alkanes decrease in the order

$$CH_3–H > RCH_2–H > R–\underset{H}{\overset{R}{CH}} > R–\underset{R}{\overset{R}{C}}–H$$

Methyl (Strongest)   Primary   Secondary   Tertiary (Weakest)

because the order of stability of the corresponding alkyl radicals is

$$CH_3\cdot < RCH_2\cdot < R–CH\cdot < R–\underset{R}{\overset{R}{C}}\cdot$$

Methyl (Least stable)   Primary   Secondary   Tertiary (Most stable)

This is the order of increasing **hyperconjugative stabilization.**

3. **Catalysts** speed up the establishment of an equilibrium between starting materials and products.

4. Alkanes react with halogens (except iodine) by a **radical chain mechanism** to give haloalkanes. The mechanism consists of **initiation** to create a halogen atom, two **propagation steps,** and various **termination steps.**

5. In the first propagation step, the slower of the two, a hydrogen atom is abstracted from the alkane chain, a reaction resulting in an alkyl radical and

HX. Hence, **reactivity** increases from I_2 to F_2. **Selectivity** decreases along the same series, as well as with increasing temperature.

6. The **Hammond postulate** states that fast, exothermic reactions are typically characterized by **early transition states,** which are similar in structure to the starting materials. On the other hand, slow, endothermic processes usually have **late (productlike) transition states.**

7. The $\Delta H°$ for a reaction may be calculated from the $DH°$ values of the bonds affected in the process as follows: $\Delta H° = \Sigma \, DH°_{\text{bonds broken}} - \Sigma \, DH°_{\text{bonds formed}}$.

8. The $\Delta H°$ for a radical halogenation process equals the sum of the $\Delta H°$ values for the propagation steps.

9. The relative reactivities of the various types of alkane C–H bonds in halogenations can be estimated by factoring out statistical contributions. They are roughly constant under identical conditions and follow the order

$$CH_4 < \underset{CH}{\text{primary}} < \underset{CH}{\text{secondary}} < \underset{CH}{\text{tertiary}}$$

In radical chlorinations of alkanes at 25°C, the approximate relative reactivities of the tertiary : secondary : primary positions are 5 : 4 : 1. In fluorinations, these ratios are about 1.4 : 1.2 : 1, whereas, in brominations (150°), they are 1700 : 80 : 1.

10. Chemists often use halogenating agents other than the halogens. Examples are sulfuryl chloride, SO_2Cl_2, and N-chloro- and N-bromobutanimide.

11. The $\Delta H°$ of the combustion of an alkane is called the **heat of combustion,** $\Delta H°_{\text{comb}}$. The heats of combustion of isomeric compounds provide an experimental measure of their relative stabilities.

PROBLEMS

9. Label the primary, secondary, and tertiary hydrogens in each of the following compounds.

    (a) $CH_3CH_2CH_2CH_3$ (b) $CH_3CH_2CH_2CH_2CH_3$

    (c)

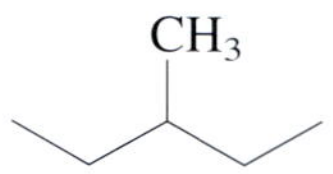

    (d)

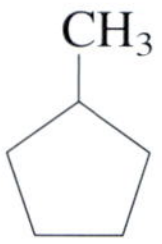

10. Within each of the following sets of alkyl radicals, name each radical; identify each as either primary, secondary, or tertiary; rank in order of decreasing stability; and sketch an orbital picture of the most stable radical, showing the hyperconjugative interaction(s).
    (a) $CH_3CH_2\dot{C}HCH_3$ and $CH_3CH_2CH_2CH_2\cdot$
    (b) $(CH_3CH_2)_2CHCH_2\cdot$ and $(CH_3CH_2)_2\dot{C}CH_3$
    (c) $(CH_3)_2CH\dot{C}HCH_3$, $(CH_3)_2\dot{C}CH_2CH_3$, and $(CH_3)_2CHCH_2CH_2\cdot$

11. Write as many products as you can think of that might result from the pyrolytic cracking of propane. Assume that the only initial process is C–C bond cleavage.

12. Answer the question posed in Problem 11 for (a) butane and (b) 2-methylpropane. Use the data in Table 3-2 to determine the bond most likely to cleave homolytically, and use that bond cleavage as your first step.

13. Calculate $\Delta H°$ values for the following reactions. (a) $H_2 + F_2 \rightarrow 2\ HF$; (b) $H_2 + Cl_2 \rightarrow 2\ HCl$; (c) $H_2 + Br_2 \rightarrow 2\ HBr$; (d) $H_2 + I_2 \rightarrow 2\ HI$; (e) $(CH_3)_3CH + F_2 \rightarrow (CH_3)_3CF + HF$; (f) $(CH_3)_3CH + Cl_2 \rightarrow (CH_3)_3CCl + HCl$; (g) $(CH_3)_3CH + Br_2 \rightarrow (CH_3)_3CBr + HBr$; (h) $(CH_3)_3CH + I_2 \rightarrow (CH_3)_3CI + HI$.

14. For each compound in Problem 9, determine how many constitutional isomers can form upon monohalogenation. (**Hint:** Identify all groups of hydrogens that reside in distinct structural environments in each molecule.)

15. (a) Using the information given in Sections 3-6 and 3-7, write the products of the radical monochlorination of (i) pentane and (ii) 3-methylpentane. (b) For each, estimate the ratio of the isomeric monochlorination products that would form at 25°C. (c) Using the bond-strength data from Table 3-1, determine the $\Delta H°$ values of the propagation steps for the chlorination of 3-methylpentane at C3. What is the overall $\Delta H°$ value for this reaction?

16. Write in full the mechanism for monobromination of methane. Be sure to include initiation, propagation, and termination steps.

17. Write a mechanism for the radical bromination of the hydrocarbon benzene, C_6H_6 (for structure, see Section 2-1). Use propagation steps similar to those in the halogenation of alkanes, as presented in Sections 3-4 through 3-6. Calculate $\Delta H°$ values for each step and for the reaction as a whole. How does this reaction compare thermodynamically with the bromination of other hydrocarbons? Data: $DH°_{C_6H_5-H} = 112\ kcal\ mol^{-1}$; $DH°_{C_6H_5-Br} = 81\ kcal\ mol^{-1}$.

18. Write the major organic product(s), if any, of each of the following reactions.

    (a) $CH_3CH_3 + I_2 \xrightarrow{\Delta}$ (b) $CH_3CH_2CH_3 + F_2 \longrightarrow$

(c) 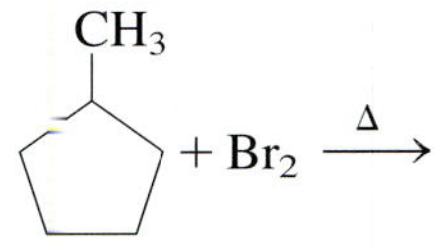 [structure: methylcyclopentane] $+ Br_2 \xrightarrow{\Delta}$

(d) $CH_3CH-CH_2-CCH_3 + Cl_2 \xrightarrow{hv}$ [with CH_3 groups]

(e) $CH_3CH-CH_2-CCH_3 + Br_2 \xrightarrow{hv}$ [with CH_3 groups]

19. Calculate product ratios in each of the reactions in Problem 18. Use relative reactivity data for F_2 and Cl_2 at 25°C and for Br_2 at 150°C (Table 3-6).

20. Which, if any, of the reactions in Problem 18 give the major product with reasonable selectivity (i.e., are useful "synthetic methods")?

21. Predict the major product(s) of radical monobromination of each of the following compounds (identified by their common names). Point out any reaction that gives the major product with reasonable selectivity. Except for twistane, all the hydrocarbons shown are derived from molecules representative of the class of naturally occurring compounds called terpenes (see Section 4-7).

(a) 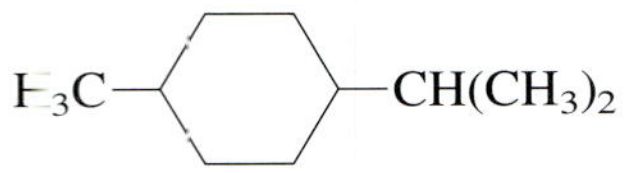 H_3C- [cyclohexane ring] $-CH(CH_3)_2$

Menthane

(b) 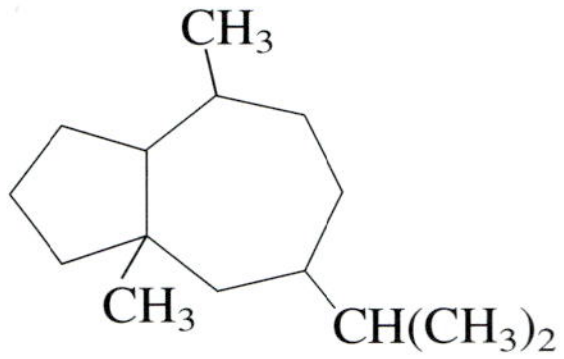 [bicyclic structure with CH_3, CH_3, and $CH(CH_3)_2$ groups]

Pseudoguaiane

(c) 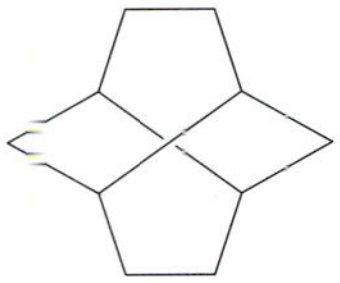 [cage structure]

Twistane

(d) 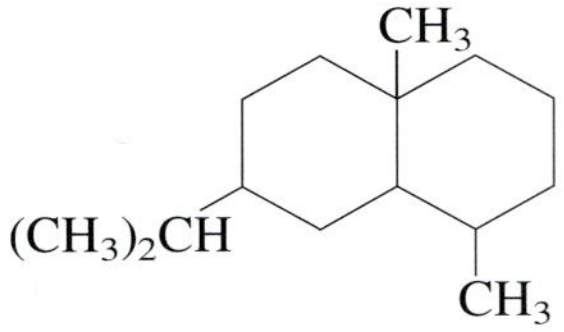 [bicyclic decalin structure with $(CH_3)_2CH$, CH_3, and CH_3 groups]

Eudesmane

22. Write balanced equations for the combustion of each of the following substances (molecular formulas may be obtained from Table 3-7): (a) methane; (b) propane; (c) cyclohexane; (d) ethanol; (e) sucrose.

23. Propanal ($CH_3CH_2\overset{O}{\overset{\|}{C}}H$) and propanone (acetone; $CH_3\overset{O}{\overset{\|}{C}}CH_3$) are isomers with the formula C_3H_6O. The heat of combustion of propanal is -434.1 kcal mol^{-1}, that of propanone -427.9 kcal mol^{-1}. (a) Write a balanced equation for the combustion of either compound. (b) What is the energy difference between propanal and propanone? Which has the lower energy content? (c) Which substance is more thermodynamically stable, propanal or propanone? (**Hint:** Draw a diagram similar to that in Figure 3-11.)

24. Propose a mechanism for chlorination of CH_4, using sulfuryl chloride, SO_2Cl_2. (**Hint:** Follow the usual model for a radical chain process, substituting SO_2Cl_2 for Cl_2 where appropriate.)

25. Use the Arrhenius equation (Section 2-8) to estimate the ratio of the rate constants k for the reactions of a C–H bond in methane with a chlorine atom and

with a bromine atom at 25°C. Assume that the A values for the two processes are equal, and use $E_a = 19$ kcal mol^{-1} for the reaction between Br· and CH_4.

26. Reexamine Exercise 3-4 regarding the reaction between methane and an equimolar mixture of Br_2 and Cl_2. As this reaction proceeds, CH_3Br is observed to eventually form in significantly greater quantities than would be expected, considering the very large difference in reactivity between Cl· and Br· toward C–H bonds (Problem 25). Suggest an explanation. (**Hint:** The mechanism for formation of CH_3Br in this situation differs in one important way from that of the reaction between CH_4 and Br_2 alone. In analyzing this problem, consider which step in the radical chain mechanism determines the structure of the *organic* product.)

27. When an alkane with different types of C–H bonds, such as propane, reacts with an equimolar mixture of Br_2 and Cl_2, the selectivity in the formation of the brominated products is much worse than that observed when reaction is carried out with Br_2 alone. (In fact, it is very similar to the selectivity for *chlorination*.) Explain. (**Hint:** Recall Problem 26, and consider which step in the radical chain mechanism is responsible for the *selectivity* of halogenation.)

28. A hypothetical alternative mechanism for the halogenation of methane has the following propagation steps.

$$\textbf{(i)} \;\; X· + CH_4 \longrightarrow CH_3X + H·$$
$$\textbf{(ii)} \;\; H· + X_2 \longrightarrow HX \;\; + X·$$

(**a**) Using $DH°$ values from appropriate tables, calculate $\Delta H°$ for these steps for any one of the halogens. (**b**) Compare your $\Delta H°$ values with those for the accepted mechanism (Table 3-5). Do you expect this alternative mechanism to compete successfully with the accepted one? (**Hint:** Consider activation energies.)

29. The addition of certain materials called radical inhibitors to halogenation reactions causes the reactions to come to a virtually complete stop. An example is the inhibition by I_2 of the chlorination of methane. Explain how this inhibition might come about. (**Hint:** Calculate $\Delta H°$ values for possible reactions of the various species present in the system with I_2, and evaluate the possible further reactivity of the products of these I_2 reactions.)

30. One additional piece of experimental evidence in support of the radical chain mechanism is the observation that traces of oxygen, O_2, strongly inhibit halogenation reactions of alkanes. (**a**) What is unusual about the electronic structure of O_2 that may be relevant in this context? (**Hint:** Refer to Problem 25 of Chapter 1.) (**b**) Suggest a process by which O_2 interferes with a key step in alkane halogenation.

31. Typical hydrocarbon fuels (e.g., 2,2,4-trimethylpentane, a common component of gasoline) have very similar heats of combustion when calculated in kilocalories *per gram*. (**a**) Calculate heats of combustion per gram for several representative hydrocarbons in Table 3-7. (**b**) Make the same calculation for ethanol (Table 3-7). (**c**) In evaluating the feasibility of "gasohol" (90% gasoline and 10% ethanol) as a motor fuel, it has been estimated that an automobile running on pure ethanol would get approximately 40% fewer miles per gallon than would an identical automobile running on standard gasoline. Is this estimate consistent with the results in (a) and (b)? What can you say in general about the fuel capabilities of oxygen-containing molecules relative to hydrocarbons?

32. Two simple organic molecules that are in use as fuel additives are methanol (CH_3OH) and 2-methoxy-2-methylpropane [*tert*-butyl methyl ether, $(CH_3)_3COCH_3$]. The $\Delta H°_{comb}$ values for these compounds in the gas phase are

-182.6 kcal mol^{-1} for methanol and -809.7 kcal mol^{-1} for 2-methoxy-2-methylpropane. **(a)** Write balanced equations for the complete combustion of each of these molecules to CO_2 and H_2O. **(b)** Using Table 3-7, compare the ΔH°_{comb} values for these compounds with those for alkanes with similar molecular weights.

33. Ordinary glassblowing torches are fueled by natural gas. However, welders require much hotter temperatures for their work and often use torches fueled by ethyne (acetylene, HC≡CH). **(a)** Write a balanced equation for the combustion of ethyne. **(b)** The heat of combustion of ethyne is -310.7 kcal mol^{-1}. Compare this value with that for propane, an important component of natural gas, both per mole and per gram. Does this explain the hotter flame of ethyne?

34. Figure 3-8 compares the reactions of Cl· with the primary and secondary hydrogens of propane. **(a)** Draw a similar diagram comparing the reactions of Br· with the primary and secondary hydrogens of propane. (**Hint:** First obtain the necessary DH° values from Table 3-1 and calculate ΔH° for both the primary and the secondary hydrogen abstraction reactions. Other data: $E_a = $ 13 kcal mol^{-1} for Br· reacting with a primary C–H bond and $E_a = $ 10 kcal mol^{-1} for Br· reacting with a secondary C–H bond. **(b)** Which among the transition states of these reactions would you call "early," and which "late"? **(c)** Judging from the locations of the transition states of these reactions along the reaction coordinate, should they show greater or lesser radical character than do the corresponding transition states for chlorination (Figure 3-8)? **(d)** Is your answer to (c) consistent with the selectivity differences between Cl· reacting with propane and Br· reacting with propane? Explain.

35. Two of the propagation steps in the Cl·/O_3 system consume ozone and oxygen atoms (which are necessary for the production of ozone), respectively (Section 3-9).

$$Cl + O_3 \longrightarrow ClO + O_2$$
$$ClO + O \longrightarrow Cl + O_2$$

Calculate ΔH° for each of these propagation steps. Use the following data: DH° for ClO = 56 kcal mol^{-1}; DH° for O_2 = 120 kcal mol^{-1}; DH° for an O–O_2 bond in O_3 = 26 kcal mol^{-1}. Write the overall equation described by the combination of these steps and calculate its ΔH°. Comment on the thermodynamic favorability of the process.

Team Problem

36. **(a)** Provide an IUPAC name for each of the isomers that you drew in Exercise 2-1(a). **(b)** For each isomer that you drew and named here, give all the free radical monochlorination and monobromination products that are structurally isomeric. **(c)** Referring to Table 3-6, discuss which starting alkane and which halogen will yield the least number of isomeric products.

Preprofessional Problems

37. The reaction $CH_4 + Cl_2 \rightarrow CH_3Cl + HCl$ is an example of
    (a) neutralization **(b)** an acidic reaction
    (c) an isomerization **(d)** an ionic reaction
    (e) a radical chain reaction

38.

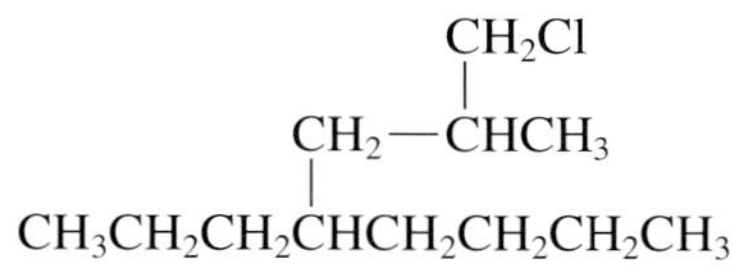

$$CH_2Cl$$
$$|$$
$$CH_2-CHCH_3$$
$$|$$
$$CH_3CH_2CH_2CHCH_2CH_2CH_2CH_3$$

The sum of all the digits that appear in the (IUPAC) name for this compound is which of the following?

(a) five   (b) six   (c) seven   (d) eight   (e) nine

39. In a competition reaction, equimolar amounts of the four alkanes shown were allowed to react with a limited amount of Cl_2 at 300°C. Which one of these alkanes would be depleted most from the mixture?

(a) pentane   (b) 2-methylpropane   (c) butane   (d) propane

40. The reaction of Cl_2 to yield CH_3Cl and HCl is well known. On the basis of the values in the short table below, the enthalpy $\Delta H°$ (kcal mol^{-1}) of this reaction is

(a) $+135$ (b) -135 (c) $+25$ (d) -25

Bond Dissociation Energies $DH°$ (kcal mol^{-1})

H–Cl	103	Cl–Cl	58
H_3C–Cl	85	H_3C–H	105

Cyclic Alkanes

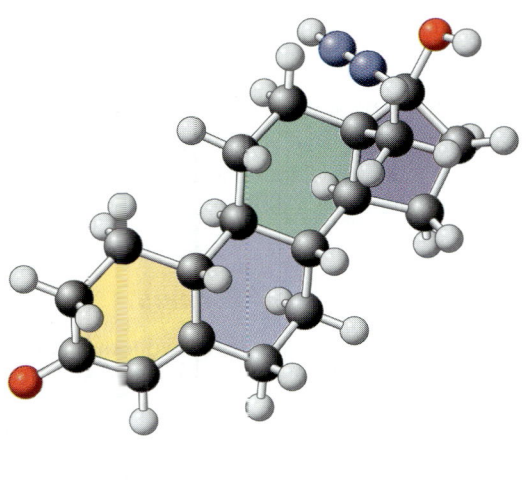

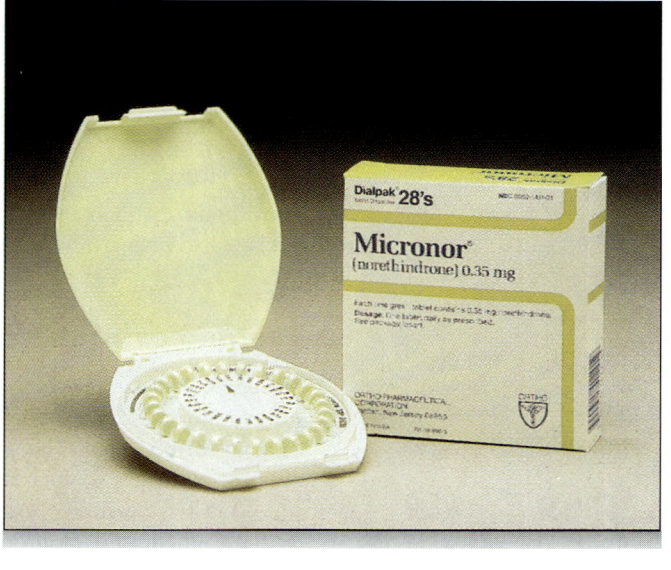

The polycyclic alkane framework of the steroids is exemplified by norethindrone, one of the active ingredients in the birth control pill.

W hen you hear or read the word *steroids,* two things probably come to mind immediately: athletes who illegally "take steroids" to develop their muscles and "the pill" used for birth control. But what do you know about steroids aside from this general association? What is their structure? How does one steroid differ from another? Where are they found in nature?

An example of a naturally occurring steroid is diosgenin, obtained from the Mexican yam and used as a starting material for the synthesis of several commercial steroids. Most striking is the number of *rings* in the compound.

Diosgenin

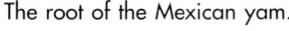

The root of the Mexican yam.

Hydrocarbons containing single-bonded carbon atoms arranged in rings are known as **cyclic alkanes, carbocycles** (in contrast with heterocycles, Chapter 25), or **cycloalkanes.** The majority of organic compounds occurring in nature contain rings. Indeed, so many fundamental biological functions depend on the chemistry of ring-

containing compounds that life as we know it could not exist in their absence. This chapter deals with the names, physical properties, structural features, and conformational characteristics of the cycloalkanes. Members of this class of compounds serve to review and amplify some of the principles presented in Chapter 2 regarding the linear and branched alkanes. We end with the biochemical significance of selected carbocycles and their derivatives, including some common flavoring agents, cholesterol, and other biological regulators.

4-1 Names and Physical Properties of Cycloalkanes

How do the cycloalkanes differ in their names and physical properties from their non-cyclic (also called *acyclic*) analogs containing the same number of carbons?

The names of the cycloalkanes follow IUPAC rules

We can construct a molecular model of a cycloalkane by removing two terminal hydrogen atoms from a model of a straight-chain alkane and allowing the terminal carbons to form a bond. The general formula of a cycloalkane is C_nH_{2n}. The system for naming members of this class of compounds is straightforward: Alkane names are preceded by the prefix **cyclo-.** Three members in the homologous series—starting with the smallest, cyclopropane—are shown in the margin, written both in condensed form and in bond-line notation.

Cyclopropane

Cyclobutane

Cyclohexane

EXERCISE 4-1

Make molecular models of cyclopropane through cyclododecane. Compare the relative conformational flexibility of each ring with that of others within the series and with that of the corresponding straight-chain alkanes.

Naming a substituted cyclic alkane requires the numbering of the individual ring carbons only if more than one substituent is attached to the ring. In monosubstituted systems, the carbon of attachment is defined as carbon 1 of the ring. For polysubstituted compounds, take care to provide the lowest possible numbering sequence. When two such sequences are possible, the alphabetical order of the substituent names takes precedence. Radicals derived from cycloalkanes by abstraction of a hydrogen atom are **cycloalkyl radicals.** Substituted cycloalkanes are therefore sometimes named as cycloalkyl derivatives. In general, the smaller unit is treated as a substituent to the larger one—for example, propylcyclopentane (not cyclopentylpropane) and cyclohexyloctane (not octylcyclohexane).

Methylcyclopropane   **1-Ethyl-1-methylcyclobutane**   **1-Chloro-2-methyl-4-propylcyclopentane**   **Cyclobutylcyclohexane**

(Not 2-chloro-1-methyl-4-propylcyclopentane)

Disubstituted cycloalkanes possess isomers

Inspection of molecular models of disubstituted cycloalkanes in which the two substituents are located on different carbons shows that *two isomers are possible* in each case. In one, the two substituents are positioned on the *same* face, or side, of the ring; in the other, on *opposite* faces. Substituents on the same face are called **cis** (*cis,* Latin, on the same side); those on opposite faces, **trans** (*trans,* Latin, across).

cis-**1,2-Dimethylcyclopropane**

trans-**1,2-Dimethylcyclopropane**

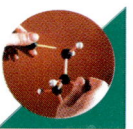

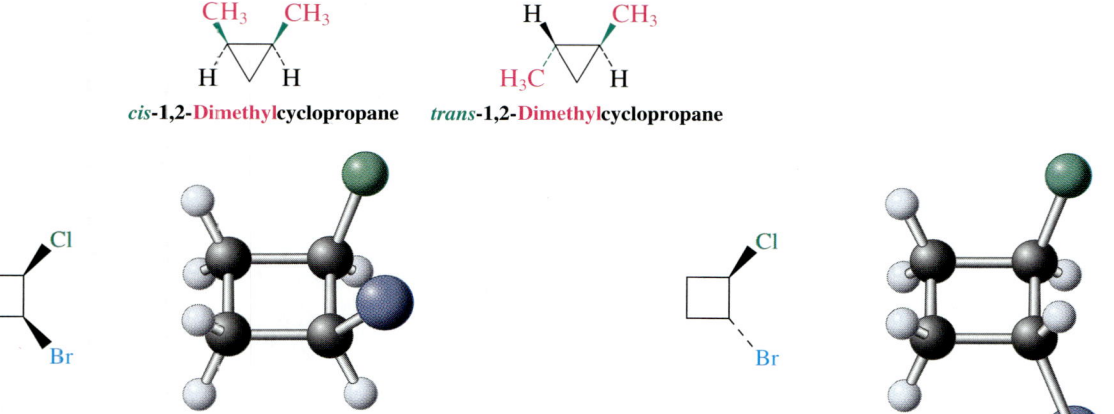

cis-**1-Bromo-2-chlorocyclobutane**

trans-**1-Bromo-2-chlorocyclobutane**

Cis and trans isomers are **stereoisomers**—compounds that have identical connectivities (i.e., their atoms are attached in the same sequence) but differ in the arrangement of their atoms in space. They are distinct from constitutional or structural isomers (Sections 1-9 and 2-2), which are compounds with differing sequences of atoms. Conformations (Sections 2-5 through 2-7) also are stereoisomers by the preceding definition. However, unlike cis and trans isomers, which can be interconverted only by *breaking* bonds (try it on your models), conformers are readily equilibrated by *rotation* about bonds. The subject of stereoisomerism will be discussed in more detail in Chapter 5.

Dashed-wedged line structures can be used to depict the three-dimensional arrangement of substituted cycloalkanes. The positions of any remaining hydrogens are not always shown. Structural and cis-trans isomerisms give rise to a variety of structural possibilities in substituted cycloalkanes. For example, there are eight isomeric bromomethylcyclohexanes (three of which are shown below), all with different and distinct physical and chemical properties.

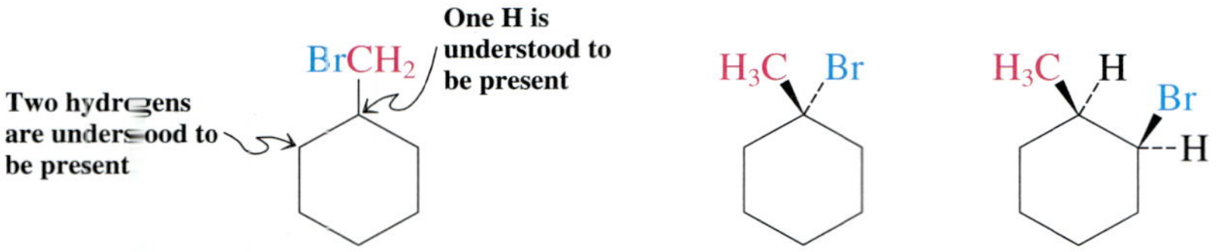

(Bromomethyl)-cyclohexane **1-Bromo-1-methyl-cyclohexane** *cis*-**1-Bromo-2-methylcyclohexane**

EXERCISE 4-2

Give the structures and names of the other five isomeric bromomethylcyclohexanes.

TABLE 4-1	Physical Properties of Various Cycloalkanes		
Cycloalkane	Boiling point (°C)	Melting point (°C)	Density at 20°C (g mL^{-1})
Cyclopropane	−32.7	−127.6	0.617^b
Cyclobutane	−12.5	−50.0	0.720
Cyclopentane	49.3	−93.9	0.7457
Cyclohexane	80.7	6.6	0.7785
Cycloheptane	118.5	−12.0	0.8098
Cyclooctane	148.5	14.3	0.8349
Cyclododecane	160 (100 torr)	64	0.861
Cyclopentadecane	110 (0.1 torr^a)	66	0.860

^{a}Sublimation point.
^{b}At 25°C.

The properties of the cycloalkanes differ from those of their straight-chain analogs

The physical properties of a few cycloalkanes are recorded in Table 4-1. Note that, compared with the corresponding straight-chain alkanes (Table 2-3), the cycloalkanes have higher boiling and melting points, as well as higher densities. These differences are due in large part to increased London interactions of the relatively more rigid and more symmetric cyclic systems. In comparing lower cycloalkanes possessing an odd number of carbons with those having an even number, we find a pronounced alternation in their melting points. This phenomenon has been ascribed to differences in crystal packing forces between the two series.

In summary, names of the cycloalkanes are derived in a straightforward manner from those of the straight-chain alkanes. In addition, the position of a single substituent is defined to be C1. Disubstituted cycloalkanes can give rise to cis and trans isomers, depending on the location of the substituents. Physical properties parallel those of the straight-chain alkanes, except that the individual values for boiling and melting points and for densities are higher for the cyclic compounds of equal carbon number.

4-2 Ring Strain and the Structure of Cycloalkanes

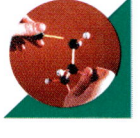

The molecular models made for Exercise 4-1 reveal obvious differences between cyclopropane, cyclobutane, cyclopentane, and so forth, and the corresponding straight-chain alkanes. One notable feature of the first two members in the series is how difficult it is to close the ring without breaking the plastic tubes used to represent bonds. This problem is called **ring strain.** The reason for it lies in the tetrahedral carbon model. The C–C–C bond angles in, for example, cyclopropane (60°) and cyclobutane (90°) differ considerably from the tetrahedral value. As the ring size increases, strain diminishes. Thus, cyclohexane can be assembled without distortion or strain.

Does this observation tell us anything about the relative stability of the cyclo-alkanes—for example, as measured by their heats of combustion, $\Delta H°_{comb}$? How does strain affect structure? This section and Section 4-3 address these questions.

The heats of combustion of the cycloalkanes reveal the presence of ring strain

Section 3-10 introduced one measure of the stability of a molecule: its heat content. We also learned that the heat content of an alkane could be estimated by measuring its heat of combustion, $\Delta H°_{comb}$ (Table 3-7). To determine the stability of each cy-cloalkane, we can compare its heat of combustion with the value measured for the analogous straight-chain molecule. The (negative) $\Delta H°_{comb}$ values for the straight-chain alkanes increase by about the same amount with each successive member of the series.

$\Delta H°_{comb}$ **Values for the Series of Straight-Chain Alkanes**

$$\left. \begin{array}{ll} CH_3CH_2CH_3 \text{ (gas)} & -530.6 \\ CH_3CH_2CH_2CH_3 \text{ (gas)} & -687.4 \\ CH_3(CH_2)_3CH_3 \text{ (gas)} & -845.2 \end{array} \right\} \begin{array}{l} -156.8 \\ \\ -157.8 \end{array} \quad kcal\ mol^{-1}$$

There appears to be a regular increment of about 157 kcal mol^{-1} for each additional CH_2 group. When averaged over a large number of alkanes, this value approaches 157.4 kcal mol^{-1}.

What does this tell us about cycloalkanes? Because these molecules have the general formula $(CH_2)_n$, we might expect their approximate $\Delta H°_{comb}$ to be $-(n \times 157.4)$ kcal mol^{-1} (Table 4-2, column 2). However, the measured heats of combustion turn

TABLE 4-2	Calculated and Experimental Heats of Combustion (kcal mol^{-1}) of Various Cycloalkanes			
Ring size (C_n)	$\Delta H°_{comb}$ (calculated)	$\Delta H°_{comb}$ (experimental)	Total strain	Strain per CH_2 group
3	−472.2	−499.8	27.6	9.2
4	−629.6	−655.9	26.3	6.6
5	−787.0	−793.5	6.5	1.3
6	−944.4	−944.5	0.1	0.0
7	−1101.8	−1108.2	6.4	0.9
8	−1259.2	−1269.2	10.0	1.3
9	−1416.6	−1429.5	12.9	1.4
10	−1574.0	−1586.0	14.0	1.4
11	−1731.4	−1742.4	11.0	1.1
12	−1888.8	−1891.2	2.4	0.2
14	−2203.6	−2203.6	0.0	0.0

Note: The calculated numbers are based on the value of −157.4 kcal mol^{-1} for a CH_2 group.

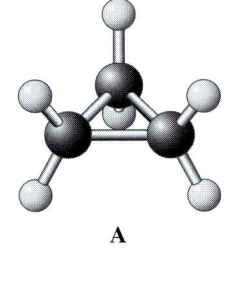

A

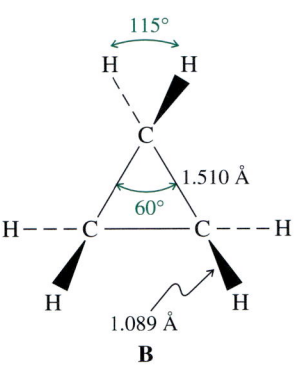

B

FIGURE 4-1 ———

Cyclopropane: (A) molecular model; (B) bond lengths and angles.

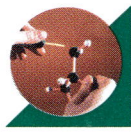

out to be *larger in magnitude* (Table 4-2, column 3). For example, cyclopropane should have a $\Delta H°_{comb}$ of -472.2 kcal mol^{-1}, but the experimental value is -499.8 kcal mol^{-1}. The difference between expected and observed values is 27.6 kcal mol^{-1}. It is attributed to a property of cyclopropane of which we are already aware because of the model we built: *ring strain.* The strain per CH_2 group in this molecule is 9.2 kcal mol^{-1}. A similar calculation for cyclobutane (Table 4-2) reveals a ring strain of 26.3 kcal mol^{-1}, or about 6.6 kcal mol^{-1} per CH_2 group. In cyclopentane, this effect is much smaller, the total strain amounting to only 6.5 kcal mol^{-1}, and cyclohexane is virtually strain free. However, succeeding members of the series again show considerable strain until we reach very large rings. Because of these trends, organic chemists have loosely defined four groups of cycloalkanes.

1. *Small rings* (cyclopropane, cyclobutane)
2. *Common rings* (cyclopentane, cyclohexane, cycloheptane)
3. *Medium rings* (from eight- to twelve-membered)
4. *Large rings* (thirteen-membered and larger)

What kinds of effects contribute to the ring strain in cycloalkanes? We shall answer this question by exploring the detailed structures of several of these compounds.

Strain affects the structures and conformations of the smaller cycloalkanes

As we have just seen, the smallest cycloalkane, *cyclopropane,* is much less stable than expected for three methylene groups. Why should this be? The reason is twofold: torsional strain and bond-angle strain.

The structure of the cyclopropane molecule is represented in Figure 4-1. We notice first that all methylene hydrogens are eclipsed, much like the hydrogens in the eclipsed conformation of ethane (Section 2-5). We know that the energy of the eclipsed form of ethane is raised above that of the more stable staggered conformation because of **eclipsing (torsional) strain.** This effect is also present in cyclopropane. Moreover, the carbon skeleton in cyclopropane is by necessity flat and quite rigid, and bond rotation that might relieve eclipsing strain is very difficult.

Second, we notice that cyclopropane has C–C–C bond angles of 60°, a significant deviation from the "natural" tetrahedral bond angle of 109.5°. How is it possible for three supposedly tetrahedral carbon atoms to maintain a bonding relation at such highly distorted angles? The problem is perhaps best illustrated in Figure 4-2, in which the bonding in the strain-free "open cyclopropane," the trimethylene diradical ·$CH_2CH_2CH_2$·, is compared with that in the closed form. We can see that the two ends of the trimethylene diradical cannot "reach" far enough to close the ring without "bending" the two C–C bonds already present. However, if all three C–C bonds in cyclopropane adopt a bent configuration (interorbital angle 104°, see Figure 4-2B), overlap is sufficient for bond formation. The energy needed to distort the tetrahedral carbons enough to close the ring is called **bond-angle strain.** The ring strain in cyclopropane is derived from a combination of eclipsing and bond-angle contributions.

As a consequence of its structure, cyclopropane has relatively weak C–C bonds ($DH° = 65$ kcal mol^{-1}). This value is low (recall that the C–C strength in ethane is 90 kcal mol^{-1}) because breaking the bond opens the ring and relieves ring strain. Therefore, cyclopropane undergoes several unusual reactions. For example, reaction with hydrogen in the presence of a palladium catalyst opens the ring to give propane.

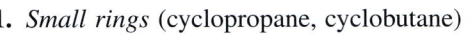

$$\triangle \ + \ H_2 \ \xrightarrow{\text{Pd catalyst}} \ CH_3CH_2CH_3 \qquad \Delta H° = -37.6 \text{ kcal mol}^{-1}$$

Propane

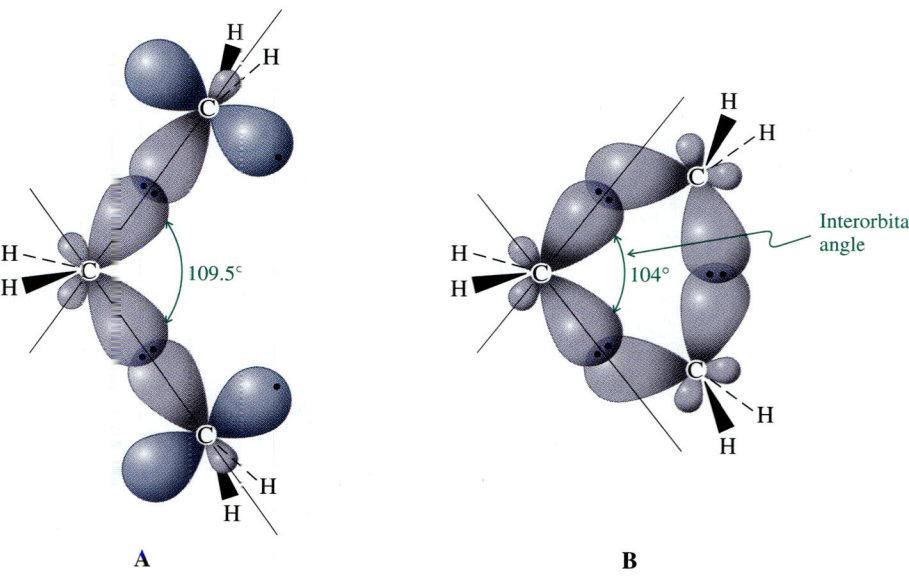

FIGURE 4-2

Molecular-orbital picture of (A) the trimethylene diradical and (B) the bent bonds in cyclopropane. Only the hybrid orbitals forming C–C bonds are shown. Note the interorbital angle of 104° in cyclopropane.

EXERCISE 4-3

Trans-1,2-dimethylcyclopropane is more stable than *cis*-1,2-dimethylcyclopropane. Why? Draw a picture to illustrate your answer. Which isomer liberates more heat on combustion?

What about higher cycloalkanes? The structure of *cyclobutane* (Figure 4-3) reveals that this molecule is not planar but puckered, with an approximate bending angle of 26°. The nonplanar structure of the ring, however, is not very rigid. The molecule "flips" rapidly from one puckered conformation to the other. Construction of a molecular model shows why distorting the four-membered ring from planarity is favorable: It partly relieves the strain introduced by the eight eclipsing hydrogens. Moreover, bond-angle strain is considerably reduced relative to that in cyclopropane,

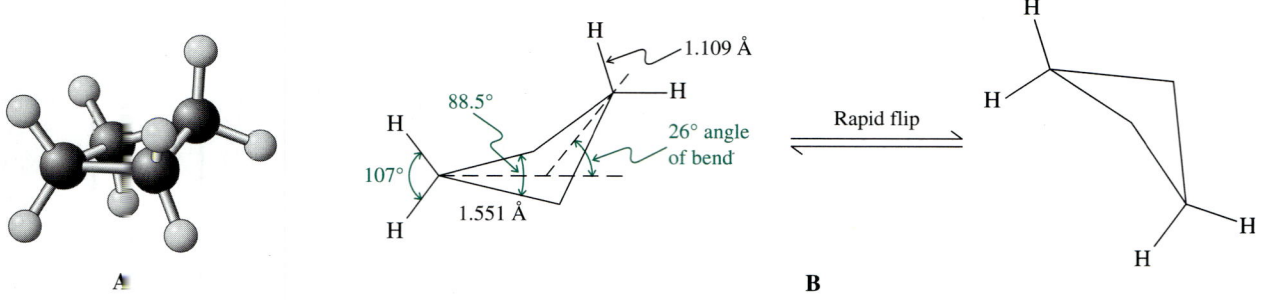

FIGURE 4-3

Cyclobutane (A) molecular model; (B) bond lengths and angles. The nonplanar molecule "flips" rapidly from one conformation to another.

FIGURE 4-4

Cyclopentane: (A) molecular model of the half-chair conformation; (B) bond lengths and angles. The molecule is flexible, with little strain.

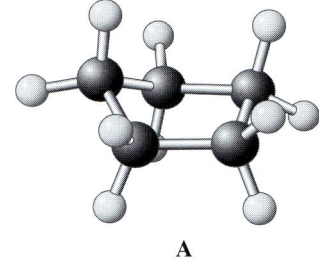

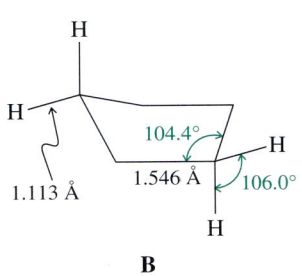

A

B

although maximum overlap is, again, only possible with the use of bent bonds. The C–C bond strength in cyclobutane also is low (about 63 kcal mol^{-1}) because of the release of ring strain on ring opening and the consequences of relatively poor overlap in bent bonds. Cyclobutane is less reactive than cyclopropane but undergoes similar ring-opening processes.

$$\square \quad + \quad H_2 \quad \xrightarrow{\text{Pd catalyst}} \quad CH_3CH_2CH_2CH_3$$
$$\textbf{Butane}$$

Cyclopentane might be expected to be planar because the angles in a regular pentagon are 108°, close to tetrahedral. However, such a planar arrangement would have *ten* H–H eclipsing interactions. The puckering of the ring reduces this effect, as indicated in the structure of the molecule (Figure 4-4). Although puckering relieves eclipsing, it also increases bond-angle strain. The conformation of lowest energy is a compromise in which the energy of the system is minimized. There are two puckered conformations possible for cyclopentane: the **envelope** and the **half chair.**

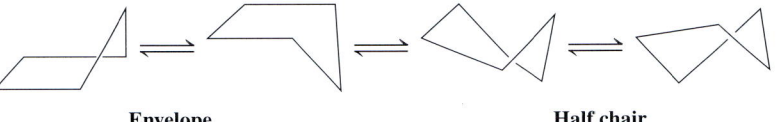

Envelope   **Half chair**

There is little difference in energy between them, and the activation barriers for rapid interconversion are extremely low. Overall, cyclopentane has relatively little ring strain and hence does not show the unusual reactivity of three- or four-membered rings.

4-3 Cyclohexane: A Strain-Free Cycloalkane

The cyclohexane ring is one of the most abundant and important structural units in organic chemistry. Its substituted derivatives exist in many natural products (see Section 4-7), and an understanding of its conformational mobility is an important aspect of organic chemistry. Table 4-2 reveals that, within experimental error, cyclohexane is unusual in that it is free of bond-angle or eclipsing strain. Why?

The chair conformation of cyclohexane is strain free

A hypothetical planar cyclohexane would suffer from twelve H–H eclipsing interactions and sixfold bond-angle strain (a regular hexagon requires 120° bond angles). However, one conformation of cyclohexane, obtained by moving carbons 1 and 4 out

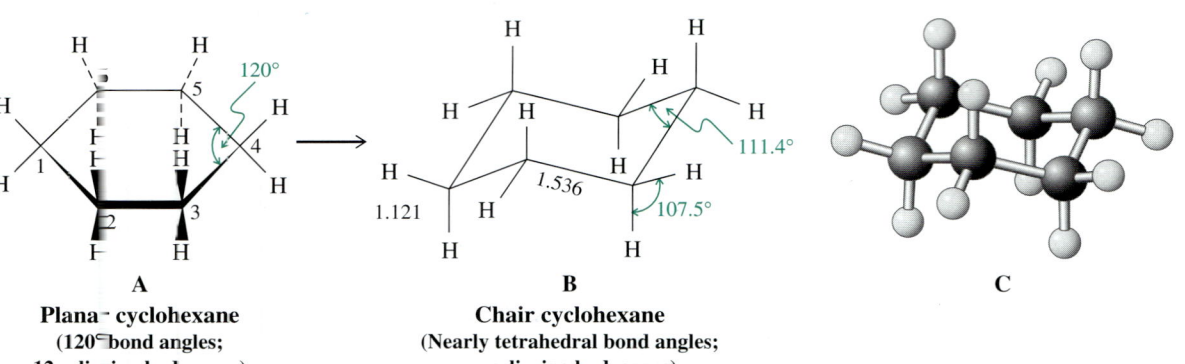

A

Planar cyclohexane
(120° bond angles;
12 eclipsing hydrogens)

B

Chair cyclohexane
(Nearly tetrahedral bond angles;
no eclipsing hydrogens)

C

FIGURE 4-5

Conversion of the (A) hypothetical planar cyclohexane into the (B) chair conformation, showing bond lengths and angles; (C) molecular model. The chair conformation is strain free.

of planarity in opposite directions, is in fact strain free (Figure 4-5). This structure is called the **chair conformation** of cyclohexane (because it resembles a chair), in which eclipsing is completely prevented and the bond angles are very nearly tetrahedral. As seen in Table 4-2, the calculated ΔH°_{comb} of cyclohexane (-944.4 kcal mol^{-1}) based on a strain-free $(CH_2)_5$ model is very close to the experimentally determined value (-944.5 kcal mol^{-1}).

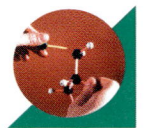

Looking at the molecular model of cyclohexane enables us to recognize the conformational stability of the molecule. If we view it along (any) one C–C bond, we can see the staggered arrangement of all substituent groups along it. We can visualize this arrangement by drawing a Newman projection of that view (Figure 4-6). Because of its lack of strain, cyclohexane is as inert as a straight-chain alkane.

EXERCISE 4-4

Draw Newman projections of the carbon–carbon bonds in cyclopropane, cyclobutane, and cyclopentane in their most stable conformations. Use the models that you prepared for Exercise 4-1 to assist you and refer to Figure 4-6. What are the approximate torsional angles between the C–H bonds in each?

Cyclohexane also has several less stable conformations

Other, less stable conformations of cyclohexane are nevertheless readily accessible to the molecule. One is the **boat form,** in which carbons 1 and 4 are out of the plane

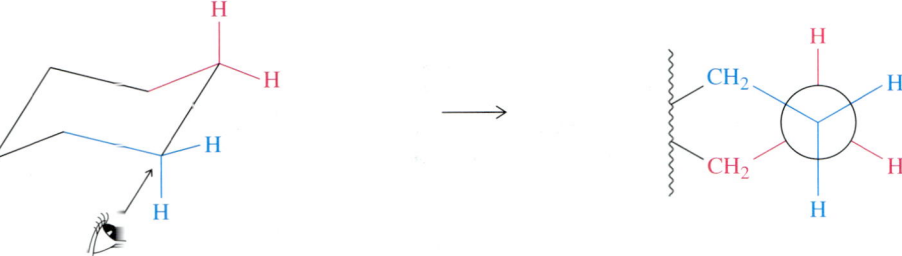

FIGURE 4-6

View along one of the C–C bonds in the chair conformation of cyclohexane. Note the staggered arrangement of all substituents.

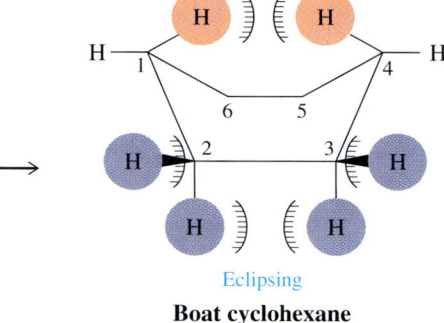

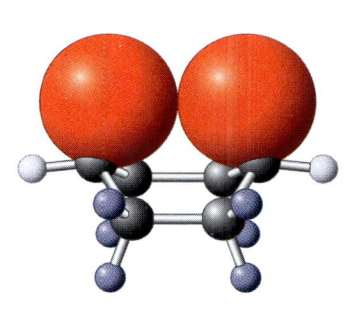

Planar cyclohexane

Boat cyclohexane

FIGURE 4-7

Conversion of the hypothetical planar cyclohexane into the boat form. In the latter form, the hydrogens on carbons 2, 3, 5, and 6 are eclipsed, thereby giving rise to torsional strain. The "inside" hydrogens on carbons 1 and 4 interfere with each other sterically in a transannular interaction. The space-filling size of these two hydrogens is depicted in the ball-and-stick model on the right.

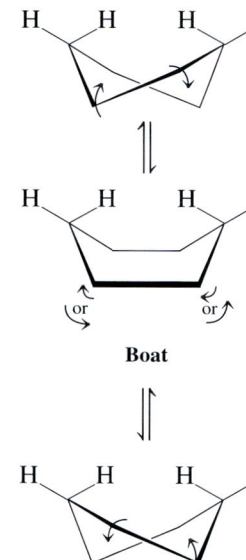

Boat

Twist (skew) boat

FIGURE 4-8

Twist-boat to twist-boat flipping of cyclohexane proceeds through the boat conformation.

in the *same* direction (Figure 4-7). The boat is less stable than the chair form by 6.9 kcal mol^{-1}. One reason for this difference is the eclipsing of eight hydrogen atoms at the base of the boat. Another is steric hindrance (Section 2-7) due to the close proximity of the two inside hydrogens in the boat framework. The distance between these two hydrogens is only 1.83 Å, small enough to create an energy of repulsion of about 3 kcal mol^{-1}. This effect is an example of **transannular strain,** that is, strain resulting from steric crowding of two groups across a ring (*trans,* Latin, across; *anulus,* Latin, ring).

Boat cyclohexane is fairly flexible. If one of the C–C bonds is twisted relative to another, this form can be somewhat stabilized by partial removal of the transannular interaction. The new conformation obtained is called the **twist-boat** (or **skew-boat**) **conformation** of cyclohexane (Figure 4-8). The stabilization relative to the boat form amounts to about 1.4 kcal mol^{-1}. As shown in Figure 4-8, two twist-boat forms are possible. They interconvert rapidly, with the boat conformer acting as a *transition state* (verify this with your model). Thus, the boat cyclohexane is not a normally isolable species, the twist-boat form is present in very small amounts, and the chair form is the major conformer (Figure 4-9). The activation barrier separating the most stable chair from the boat manifold is 10.8 kcal mol^{-1}. We shall see that the equilibration depicted in Figure 4-9 has important structural consequences with respect to the positions of substituents on the cyclohexane ring.

Cyclohexane has axial and equatorial hydrogen atoms

The chair-conformation model of cyclohexane reveals that the molecule has two types of hydrogens. Six carbon–hydrogen bonds are nearly parallel to the principal molecular axis (Figure 4-10) and hence are referred to as **axial;** the other six are nearly perpendicular to the axis and close to the equatorial plane and are therefore called **equatorial.***

---

*An equatorial plane is defined as being perpendicular to the axis of rotation of a rotating body and equidistant from its poles, such as the equator of the planet Earth.

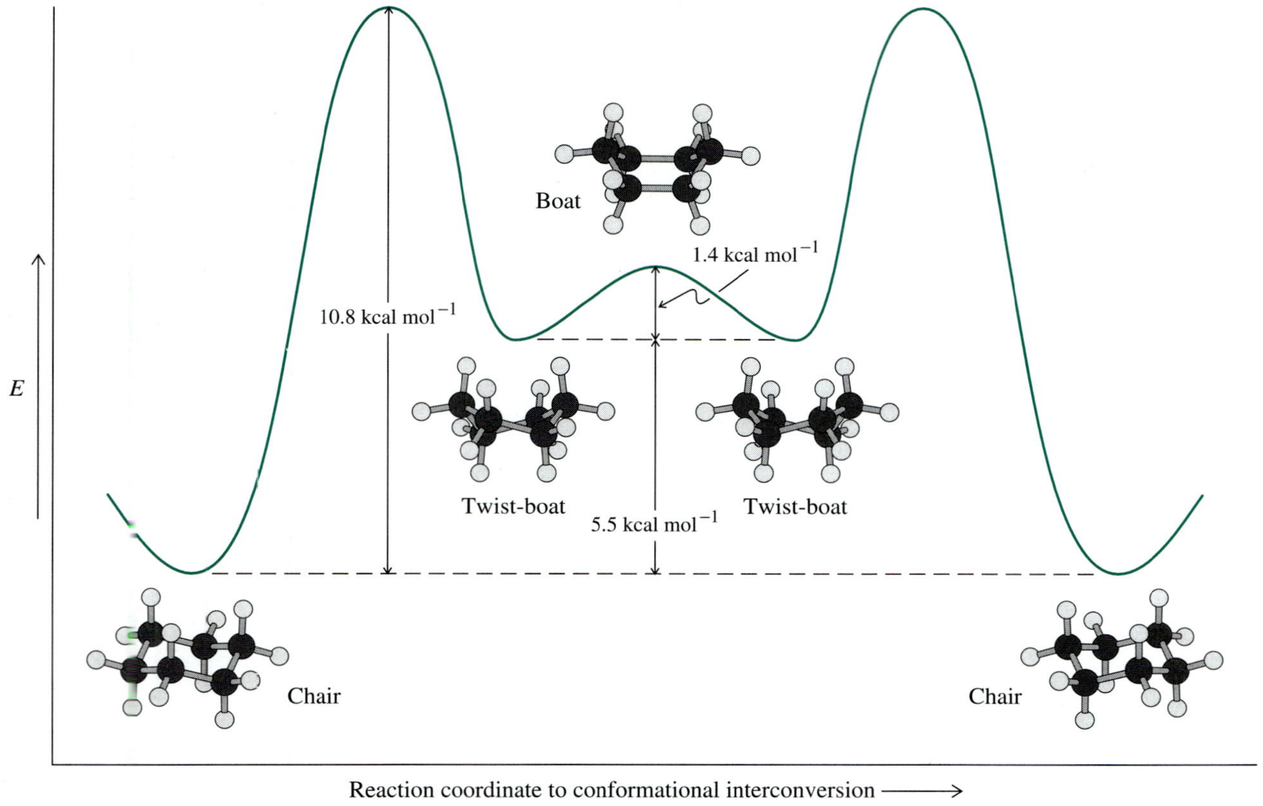

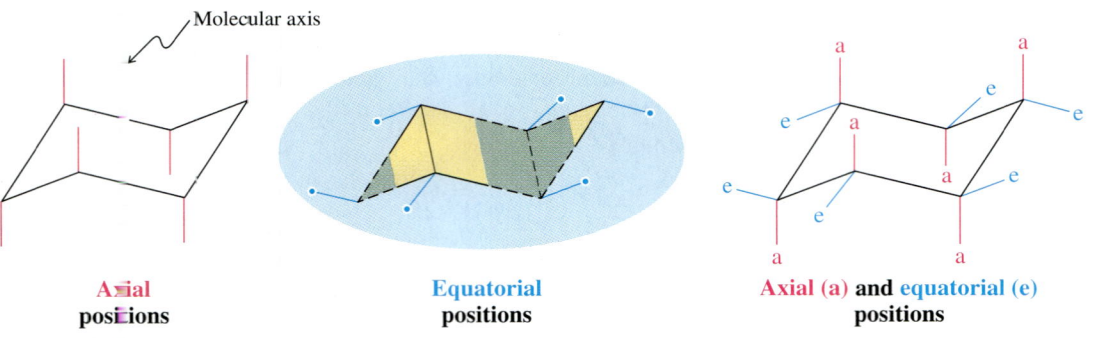

E

10.8 kcal mol^{-1}

Boat

1.4 kcal mol^{-1}

Twist-boat   Twist-boat

5.5 kcal mol^{-1}

Chair   Chair

Reaction coordinate to conformational interconversion ⟶

FIGURE 4-9

Potential energy diagram for the chair–chair interconversion of cyclohexane through the twist-boat and boat forms. In the procession from left to right, the chair is converted into a twist boat (by the twisting of one of the C–C bonds) with an activation barrier of 10.8 kcal mol^{-1}. The twist-boat form flips (as shown in Figure 4-8) through the boat conformer as a transition state (1.4 kcal mol^{-1} higher in energy) into another twist-boat structure, which relaxes back into the (ring-flipped) chair cyclohexane. Use your molecular models to visualize these changes.

Molecular axis

Axial
positions

Equatorial
positions

Axial (a) and equatorial (e)
positions

a   a
e   a   e
e   e
e   a   e
a   a

FIGURE 4-10

The axial and equatorial positions of hydrogens in the chair form of cyclohexane.

Being able to draw cyclohexane chair conformations will help you learn the chemistry of six-membered rings. Several rules are useful.

How to Draw Chair Cyclohexanes

1. Draw the chair so as to place the C2 and C3 atoms below and slightly to the right of C5 and C6, with apex 1 pointing downward on the left and apex 4 pointing upward on the right.

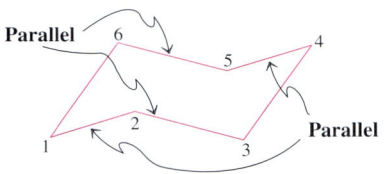

> **The bond between C1 and C6 is also parallel to that between C3 and C4.**

2. Add all the axial bonds as vertical lines, pointing downward at C1, C3, and C5 and upward at C2, C4, and C6.

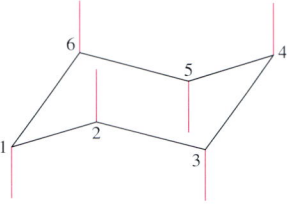

3. Draw the two equatorial bonds at C1 and C4 at a slight angle from horizontal, pointing upward at C1 and downward at C4, parallel to the bond between C2 and C3 (or between C5 and C6).

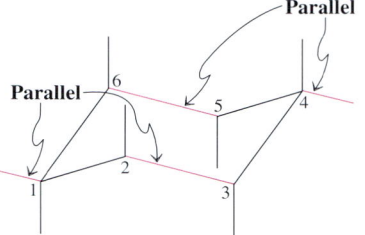

4. This rule is the most difficult to follow: Add the remaining equatorial bonds at C2, C3, C5, and C6 by aligning them *parallel* to the C–C bond "once removed," as shown below.

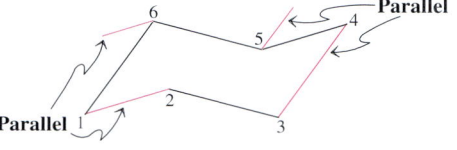
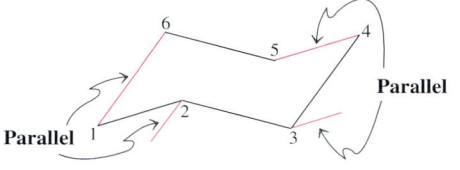

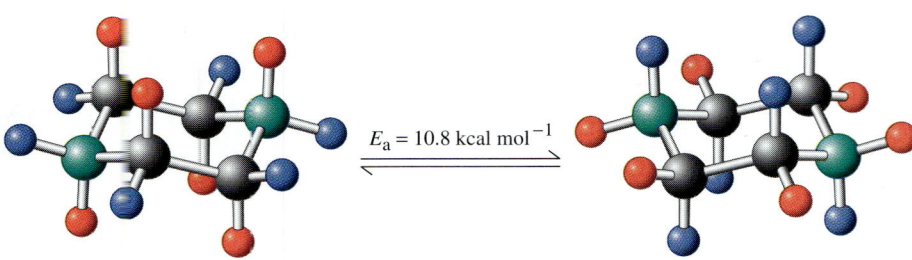

$$E_a = 10.8 \text{ kcal mol}^{-1}$$

FIGURE 4-11

Chair–chair interconversion ("ring flipping") in cyclohexane. In the process, which is rapid at room temperature, a (green) carbon at one end of the molecule moves up, while its counterpart (also green) at the other end moves down. All groups originally in axial positions (red in the structure at the left) become equatorial, and those that start in equatorial positions (blue) become axial.

Conformational flipping interconverts axial and equatorial hydrogens

What happens to the identity of the equatorial and axial hydrogens when we let chair cyclohexane equilibrate with its boat forms? You can follow the progess of conformational interconversion in Figure 4-9 with the help of molecular models. Starting with the chair structure on the left, you can simply "flip" the CH_2 group farthest to the left (C1 in the preceding section) upward through the equatorial plane to generate the boat conformers. If you now return the molecule to the chair form not by a reversal of the movement but by the equally probable alternative—namely, the flipping downward of the opposite CH_2 group (C4)—you will recognize that the original sets of axial and equatorial positions have traded places. In other words, cyclohexane undergoes chair–chair interconversions ("flipping") in which all axial hydrogens in one chair become equatorial in the other and vice versa (Figure 4-11). The activation energy for this process is 10.8 kcal mol^{-1} (Figure 4-9). As suggested in Sections 2-5 through 2-7, this value is so low that, at room temperature, two equivalent chair forms interconvert rapidly (approximately 100,000 times per second).

To summarize, the discrepancy between calculated and measured heats of combustion in the cycloalkanes can be largely attributed to three forms of strain: bond angle (deformation of tetrahedral carbon), eclipsing (torsional), and transannular (across the ring). Because of strain, the small cycloalkanes are chemically reactive, undergoing ring-opening reactions. Cyclohexane is strain free. It has a lowest-energy chair, as well as additional higher-energy conformations, particularly the boat and twist-boat structures. Chair–chair interconversion is rapid at room temperature; it is a process in which equatorial and axial hydrogen atoms interchange their positions.

4-4 Substituted Cyclohexanes

We can now apply our knowledge of conformational analysis to substituted cyclohexanes. Let us look at the simplest alkylcyclohexane, methylcyclohexane.

Axial and equatorial methylcyclohexanes are not equivalent in energy

In methylcyclohexane, the methyl group occupies either an equatorial or an axial position.

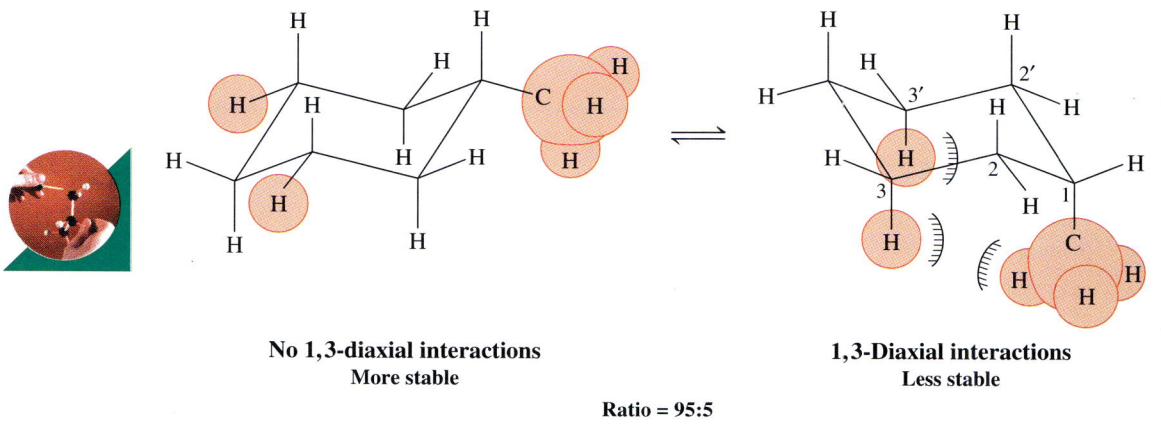

No 1,3-diaxial interactions
More stable

1,3-Diaxial interactions
Less stable

Ratio = 95:5

Are the two forms equivalent? Clearly not. In the equatorial conformer, the methyl group extends into space away from the remainder of the molecule. In contrast, in the axial conformer, the methyl substituent is close to the other two axial hydrogens on the same side of the molecule. The distance to these hydrogens is small enough (about 2.7 Å) to result in steric repulsion, another example of transannular strain. Because this effect is due to axial substituents on carbon atoms that have a 1,3-relation (in the drawing, 1,3 and 1,3'), it is called a **1,3-diaxial interaction.** This interaction is the same as that resulting in the *gauche* conformation of butane (Section 2-7). Thus, the axial methyl group is *gauche* to two of the ring carbons (C3 and C3'); when it is in the equatorial position, it is *anti* to the same nuclei.

The two forms of chair methylcyclohexane are in equilibrium. *The equatorial conformer is more stable* by 1.7 kcal mol^{-1} and is favored by a ratio of 95:5 at 25°C (Section 2-8). The activation energy for chair–chair interconversion is similar to that in cyclohexane itself (about 11 kcal mol^{-1}), and equilibrium between the two conformers is established rapidly at room temperature.

The unfavorable 1,3-diaxial interactions to which an axial substituent is exposed are readily seen in Newman projections of the ring C–C bond bearing that substituent. In contrast with that in the axial form (*gauche* to two ring bonds), the substituent in the equatorial conformer (*anti* to the two ring bonds) is away from the axial hydrogens (Figure 4-12).

EXERCISE 4-5

Calculate K for equatorial versus axial methylcyclohexane from the $\Delta G°$ value of 1.7 kcal mol^{-1}. Use the expression $\Delta G°$ (in kcal mol^{-1}) $= -1.36 \log K$. (**Hint:** If $\log K = x$, then $K = 10x$.) How well does your result agree with the 95:5 conformer ratio stated in the text?

The energy difference between the axial and the equatorial forms of many mono-substituted cyclohexanes has been measured; several are given in Table 4-3. In many cases (but not all), particularly for alkyl substituents, the energy difference between the two forms increases with the size of the substituent, a direct consequence of in-

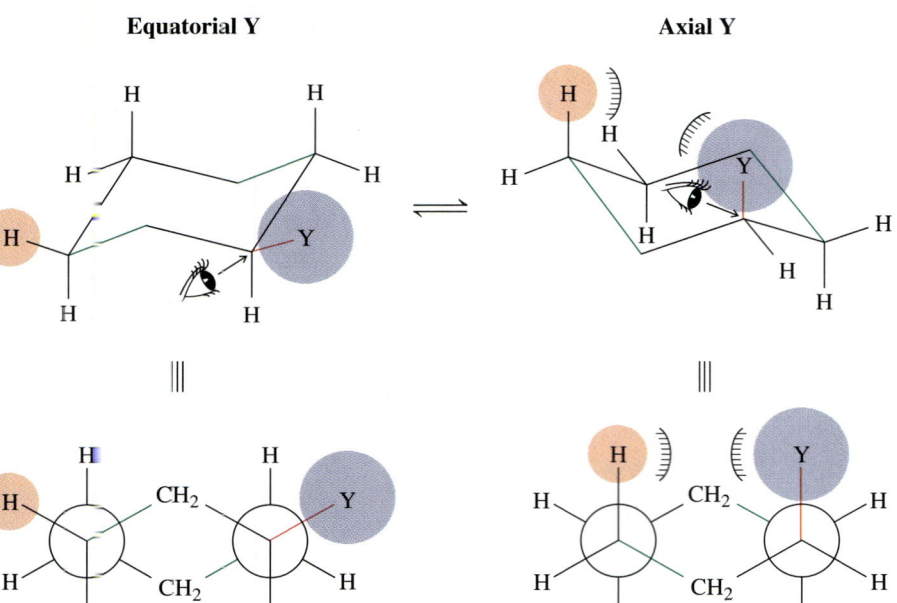

Equatorial Y   **Axial Y**

FIGURE 4-12
Newman projections of a substituted cyclohexane. The conformation with axial Y is less stable because of 1,3-diaxial interactions. Axial Y is *gauche* to the ring bonds shown in green; equatorial Y is *anti*.

creasing unfavorable 1,3-diaxial interactions. This effect is particularly pronounced in (1,1-dimethylethyl)cyclohexane (*tert*-butylcyclohexane). The energy difference here is so large (about 5 kcal mol^{-1}) that very little (about 0.01%) of the axial conformer is present at equilibrium.

TABLE 4-3	Change in Free Energy on Flipping from the Cyclohexane Conformer with the Indicated Substituent Equatorial to the Conformer with the Substituent Axial		
Substituent	$\Delta G°$ **(kcal mol^{-1})**	**Substituent**	$\Delta G°$ **(kcal mol^{-1})**
H	0	F	0.25
CH_3	1.70	Cl	0.52
CH_3CH_2	1.75	Br	0.55
$(CH_3)_2CH$	2.20	I	0.46
$(CH_3)_3C$	≈ 5	HO	0.94
$HO-\overset{\overset{O}{\|\|}}{C}$	1.41	CH_3O	0.75
$CH_3O-\overset{\overset{O}{\|\|}}{C}$	1.29	H_2N	1.4

Note: In all examples, the equatorial form is more stable.

Substituents compete for equatorial positions

To predict the most stable conformer of a more highly substituted cyclohexane, the cumulative effect of placing substituents either axially or equatorially must be considered, in addition to their potential mutual 1,3-diaxial or 1,2-*gauche* (Section 2-7) interactions. For many cases, we can ignore the last two and simply apply the values of Table 4-3 for a prediction.

Let us look at some isomers of dimethylcyclohexane to illustrate this point. In 1,1-dimethylcyclohexane, one methyl group is always equatorial and the other axial. The two chair forms are identical, and hence their energies are equal.

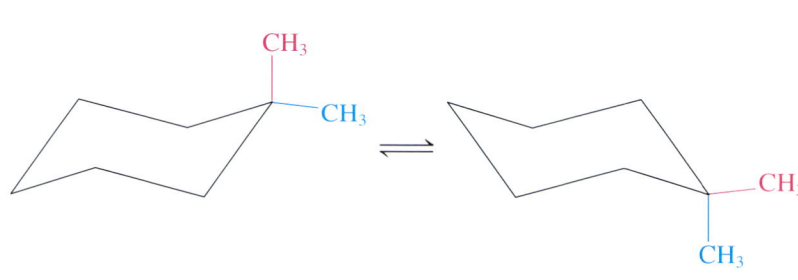

One CH₃ axial
One CH₃ equatorial

One CH₃ axial
One CH₃ equatorial

1,1-Dimethylcyclohexane
(Conformations equal in energy, equally stable)

Similarly, in *cis*-1,4-dimethylcyclohexane, both chairs have one axial and one equatorial substituent and are of equal energy.

> The bonds to both methyl groups point downward; they are cis (i.e., on the same side of the ring) *regardless* of conformation.

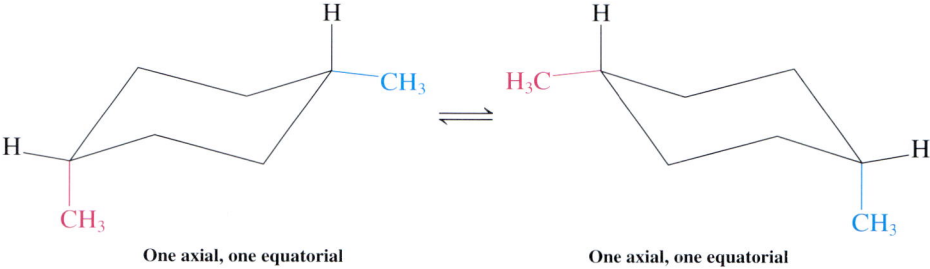

One axial, one equatorial

One axial, one equatorial

cis-1,4-Dimethylcyclohexane

On the other hand, the trans isomer can exist in two different chair conformations: one having two axial methyl groups (diaxial) and the other having two equatorial groups (diequatorial).

> The bond to one methyl group points downward, the other upward. They are trans (i.e., on opposite sides of the ring) *regardless* of conformation.

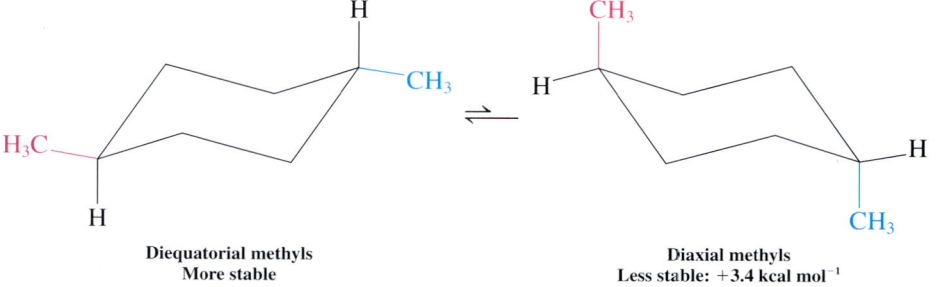

Diequatorial methyls
More stable

Diaxial methyls
Less stable: +3.4 kcal mol⁻¹

trans-1,4-Dimethylcyclohexane

Experimentally, the latter is preferred over the former by 3.4 kcal mol^{-1}, exactly twice the $\Delta G°$ value for monomethylcyclohexane. Indeed, this additive behavior of the data given in Table 4-3 applies to many other substituted cyclohexanes. For example, the $\Delta G°$ (diaxial $\rightleftharpoons$ diequatorial) for *trans*-1-fluoro-4-methylcyclohexane is -1.95 kcal mol^{-1} [$-(1.70$ kcal mol^{-1} for CH_3 plus 0.25 kcal mol^{-1} for F)]. Conversely, in *cis*-1-fluoro-4-methylcyclohexane, the two groups compete for the equatorial positions and the corresponding $\Delta G° = -1.45$ kcal mol^{-1} [$-(1.70$ kcal mol^{-1} minus 0.25 kcal mol^{-1})], with the larger methyl winning out over the smaller fluorine.

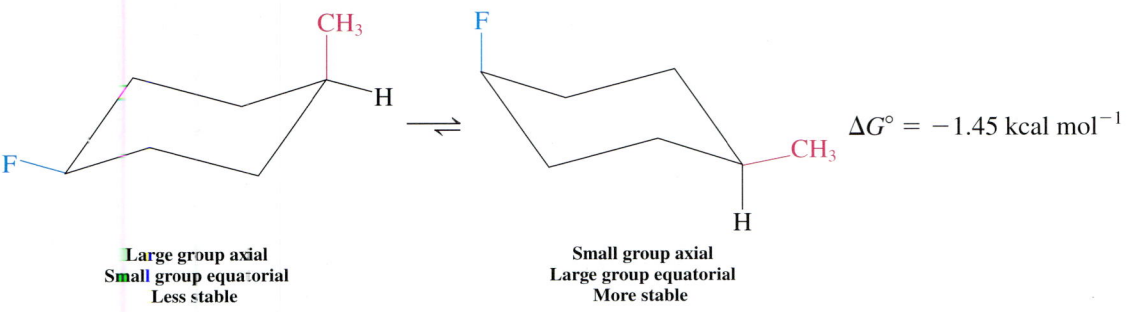

$$\Delta G° = -1.45 \text{ kcal mol}^{-1}$$

Large group axial
Small group equatorial
Less stable

Small group axial
Large group equatorial
More stable

cis-**1-Fluoro-4-methylcyclohexane**

EXERCISE 4-6

Calculate $\Delta G°$ for the equilibrium between the two chair conformers of (a) 1-ethyl-1-methylcyclohexane; (b) *cis*-1-ethyl-4-methylcyclohexane; (c) *trans*-1-ethyl-4-methyl-cyclohexane.

EXERCISE 4-7

Draw both chair conformations for each of the following isomers: (a) *cis*-1,2-dimethyl-cyclohexane; (b) *trans*-1,2-dimethylcyclohexane; (c) *cis*-1,3-dimethylcyclohexane; (d) *trans*-1,3-dimethylcyclohexane. Which of these isomers always have equal numbers of axial and equatorial substituents? Which exist as equilibrium mixtures of diaxial and diequatorial forms?

EXERCISE 4-8

Although the substituent values in Table 4-3 are additive and may be used to indicate the position of the equilibrium between two substituted cyclohexane conformers, the observed $\Delta G°$ values can be perturbed by additional 1,3-diaxial or 1,2-*gauche* interactions between groups. For example, like *trans*-1,4-dimethylcyclohexane, its isomers *cis*-1,3- and *trans*-1,2-dimethylcyclohexane exist in a diequatorial–diaxial equilibrium and hence should exhibit the same $\Delta G°$ value of 3.4 kcal mol^{-1}. However, the measured values are larger (3.7 kcal mol^{-1}) for the former and smaller (2.5 kcal mol^{-1}) for the latter. Explain. (**Hints:** For *cis*-1,3-dimethylcyclohexane, look closely at all 1,3-diaxial interactions and compare them with those of diaxial *trans*-1,4-dimethylcyclohexane. For the *trans*-1,2-isomer, take into consideration the proximity of the two methyl groups [see *gauche–anti* butane, Section 2-7].)

In summary, the conformational analysis of cyclohexane enables us to predict the relative stability of its various conformers and even to approximate the energy differences between two chair conformations. Bulky substituents, particularly a 1,1-dimethylethyl group, tend to shift the chair–chair equilibrium toward the side in which the large substituent is equatorial.

4-5 Larger Cycloalkanes

Do similar relations hold for the larger cycloalkanes? Table 4-2 shows that cycloalkanes with rings larger than that of cyclohexane also have more strain. This strain is due to a combination of bond-angle distortion, partial eclipsing of hydrogens, and transannular steric repulsions. It is not possible for medium-sized rings to relieve all of these strain-producing interactions in a single conformation. Instead, a compromise solution is found in which the molecule equilibrates among several geometries that are very close in energy.

Essentially strain-free conformations are attainable only for large-sized cycloalkanes, such as cyclotetradecane (Table 4-2). In such rings, the carbon chain adopts a structure very similar to that of the straight-chain alkanes (Section 2-4), having staggered hydrogens and an all-*anti* configuration. However, even in these systems, the attachment of substituents usually introduces various amounts of strain. Most cyclic molecules described in this book are not strain free.

4-6 Polycyclic Alkanes

The cycloalkanes discussed so far contain only one ring and therefore may be referred to as monocyclic alkanes. In more complex structures—the bi-, tri-, tetra-, and higher polycyclic hydrocarbons—two or more rings share carbon atoms. We shall see the structural variety possible in these compounds, many of which, when bearing alkyl and functional groups, exist in nature.

Polycyclic alkanes may contain fused or bridged rings

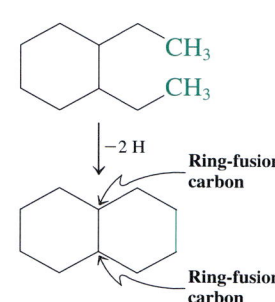

Decalin

Molecular models of polycyclic alkanes can be readily constructed by linking the carbon atoms of two alkyl substituents in a monocyclic alkane. For example, if two hydrogen atoms are removed from the methyl groups in 1,2-diethylcyclohexane, thereby allowing a new C–C bond to form, the result is a new molecule with the common name decalin. In decalin, two cyclohexanes share two adjacent carbon atoms, and the two rings are said to be **fused.** Compounds constructed in this way are called **fused bicyclic** ring systems, and the shared carbon atoms are called the **ring-fusion carbons.** Groups attached to ring-fusion carbons are called **ring-fusion substituents.**

If we treat a molecular model of *cis*-1,3-dimethylcyclopentane in the same way, we obtain another carbon skeleton, that of norbornane. Norbornane is an example of a **bridged bicyclic** ring system. In bridged bicyclic systems, two nonadjacent carbon atoms, the **bridgehead** carbons, belong to both rings.

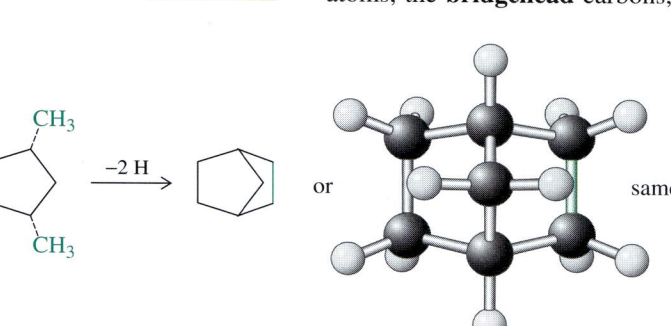

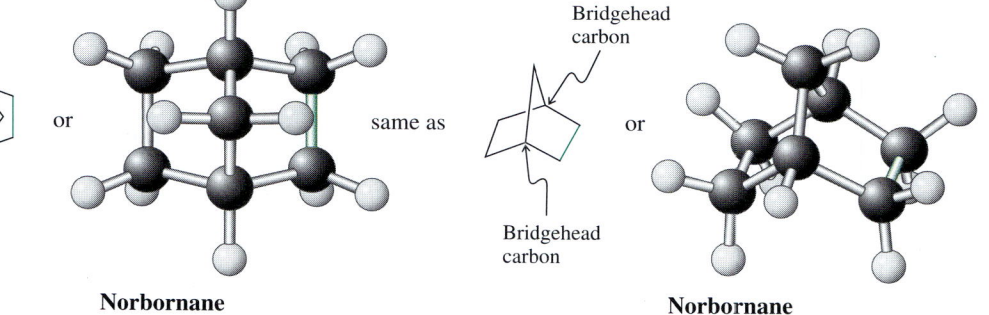

Norbornane

Norbornane

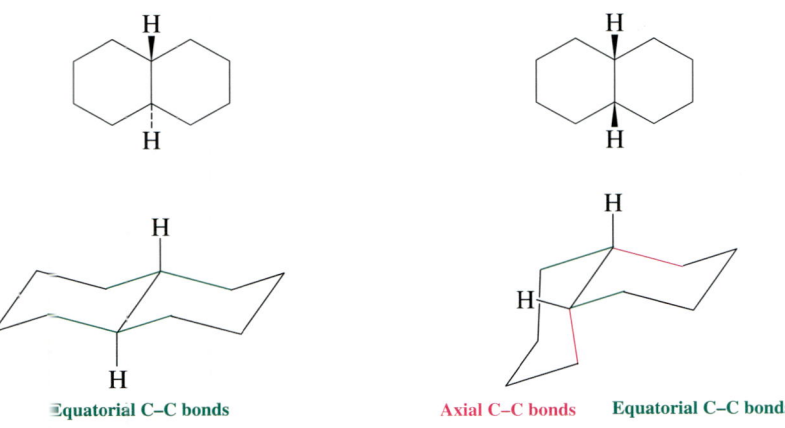

Equatorial C–C bonds

trans-Decalin

Axial C–C bonds   **Equatorial C–C bonds**

cis-Decalin

FIGURE 4-13

Conventional drawings and chair conformations of *trans*- and *cis*-decalin. The trans isomer contains only equatorial carbon–carbon bonds at the ring fusion, whereas the cis isomer possesses two equatorial C–C bonds (green) and two axial C–C linkages (red), one with respect to each ring.

If we think of one of the rings as a substituent on the other, we can identify stereochemical relations at ring fusions. In particular, bicyclic ring systems can be cis or trans fused. The stereochemistry of the ring fusion is most easily determined by inspecting the ring-fusion substituents. For example, the ring-fusion hydrogens of *trans*-decalin are trans with respect to each other, whereas those of *cis*-decalin have a cis relation (Figure 4-13).

EXERCISE 4-9

Construct molecular models of both *cis*- and *trans*-decalin. What can you say about their conformational mobility?

Do hydrocarbons have strain limits?

Seeking the limits of strain in hydrocarbon bonds is a fascinating area of research that has resulted in the synthesis of many exotic molecules. What is surprising is how much bond-angle distortion a carbon atom is able to tolerate. A case in point in the bicyclic series is bicyclobutane, whose strain energy is estimated to be 66.5 kcal mol^{-1}, making it remarkable that the molecule exists at all.

A series of strained compounds attracting the attention of synthetic chemists possess a carbon framework geometrically equivalent to the Platonic solids: the *tetrahedron* (tetrahedrane), the *hexahedron* (cubane), and the pentagonal *dodecahedron* (dodecahedrane). (See their structures in the margin on page 148.) In these polyhedra, all faces are composed of equally sized rings—namely, cyclopropane, cyclobutane, and cyclopentane, respectively. The hexahedron was synthesized first in 1964, a C_8H_8 hydrocarbon shaped like a cube and accordingly named cubane. The experimental strain energy (157 kcal mol^{-1}) is approximately equal to the total strain of six cyclobutanes. Although tetrahedrane itself is unknown, a tetra(1,1-dimethylethyl) derivative was synthesized in 1978. Despite the measured strain (from $DH°_{comb}$) of 129 kcal mol^{-1}, the compound is stable and has a melting point of 135°C. The synthesis

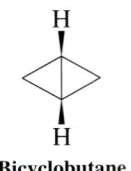

Bicyclobutane

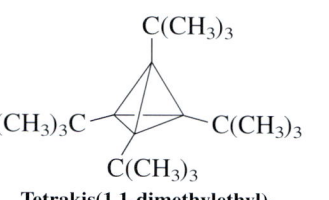

Tetrakis(1,1-dimethylethyl)-tetrahedrane

Tetrahedrane (C₄H₄)

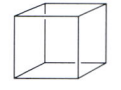

Cubane (C₈H₈)

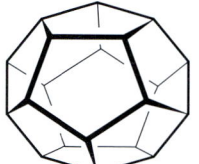

Dodecahedrane (C₂₀H₂₀)

of dodecahedrane was achieved in 1982. It required 23 synthetic operations, starting from a simple cyclopentane derivative. The last step gave 1.5 mg of pure compound. Although small, this amount was sufficient to permit complete characterization of the molecule. Its melting point at 430°C is extraordinarily high for a C_{20} hydrocarbon and is indicative of the symmetry of the compound. For comparison, icosane, also with 20 carbons, melts at 36.8°C (Table 2-3).

In summary, carbon atoms in bicyclic compounds are shared by rings in either fused or bridged arrangements. A great deal of strain may be tolerated by carbon in its bonds, particularly to other carbon atoms. This capability has allowed the preparation of molecules in which carbon is severely deformed from its tetrahedral shape.

4-7 Carbocyclic Products in Nature

Let us now take a brief look at the variety of cyclic molecules created in nature. **Natural products** are organic compounds produced by living organisms. Some of these compounds, such as methane, are extremely simple; others have great structural complexity. Scientists have attempted to classify the multitude of natural products in various ways. Generally, four schemes are followed, in which these products are classified according to (1) chemical structure, (2) physiological activity, (3) organism or plant specificity (taxonomy), and (4) biochemical origin.

There are many reasons why organic chemists are interested in natural products. Many of these compounds are powerful drugs, others function as coloring or flavoring agents, and yet others are important raw materials. A study of animal secretions furnishes information concerning the ways in which animals use chemicals to mark trails, harm their predators, and attract the opposite sex. Investigations of the biochemical pathways by which an organism metabolizes and otherwise transforms a compound are sources of insight into the detailed workings of the organism's bodily functions. Two classes of natural products, terpenes and steroids, have received particularly close attention from organic chemists.

Terpenes are constructed in plants from isoprene units

Most of you have smelled the strong odor emanating from freshly crushed plant leaves or orange peels. This odor is due to the liberation of a mixture of volatile compounds called **terpenes,** usually containing 10, 15, or 20 carbon atoms. Terpenes are used as food flavorings (the extracts from cloves and peppermint), as perfumes (roses, lavender, sandalwood), and as solvents (turpentine).

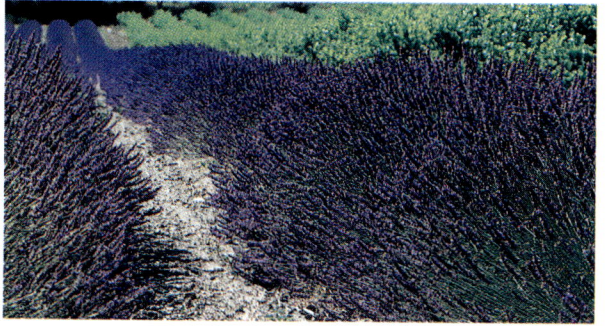

Lavender field in the Côte du Rhone region, France.

Cubane Derivatives Have Potential as Explosives

The synthesis of cubane and its derivatives has been scaled up to the kilogram level because of potential applications of such compounds as explosive materials and highly energetic fuels. A key functionalization step of cubane is the radical chlorocarbonylation of commercial cubanecarboxylic acid to a "tetrahedral" tetrasubstituted system, which can be converted into the powerful explosive tetranitrocubane.

produce a (usually destructive) shock wave in the surrounding medium. The reaction can be initiated by the mechanical impact of friction, by heat (such as sparks), or by detonating shock (blasting caps). The ring strain in cubane clearly contributes to the thermal lability of tetranitrocubane, and the molecular formula, $C_8H_4N_4O_8$, indicates a composition conducive to gaseous product formation (e.g., 8 CO + 2 N_2 + 2 H_2, to mention just one hypothetical

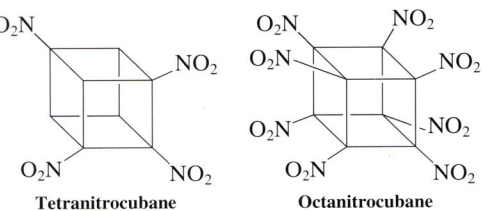

Cubanecarboxylic acid

$\xrightarrow[- HCl, - CO_2]{(COCl)_2, h\nu}$

"Tetrahedral" cubane derivative

Tetranitrocubane

Octanitrocubane

Explosives are generally compounds capable of extremely rapid decomposition with generation of high heat and a large quantity of gaseous products to

outcome). Octanitrocubane is predicted to be the most explosive nitrocubane derivative but is as yet unknown.

Terpenes are synthesized in the plant by the linkage of at least two molecular units containing five carbon atoms. The structure of these units is like that of 2-methyl-1,3-butadiene (isoprene), and so they are referred to as **isoprene units.** Depending on how many isoprene units are incorporated into the structure, terpenes are classified as mono- (C_{10}), sesqui- (C_{15}), and diterpenes (C_{20}). (The isoprene building units are shown in color in the examples given here.)

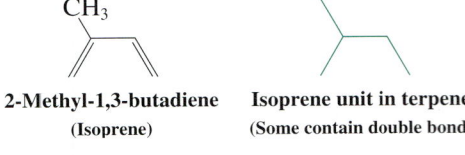

2-Methyl-1,3-butadiene
(Isoprene)

Isoprene unit in terpenes
(Some contain double bonds)

Chrysanthemic acid is a monocyclic terpene containing a three-membered ring. Its esters are found in the flower heads of pyrethrum (*Chrysanthemum cinerariae-folium*) and are naturally occurring insecticides. A cyclobutane is present in grandisol, the sex-attracting chemical used by the male boll weevil (*Anthonomus grandis*). Menthol (peppermint oil) is an example of a substituted cyclohexane natural product, whereas camphor (from the camphor tree) and β-cadinene (from juniper and cedar) are simple bicyclic terpenes, the first a norbornane system, the second a decalin derivative. **Taxol** (paclitaxel) is a complex, functionalized diterpene isolated from the bark of the Pacific yew tree, *Taxus brevifolia,* in 1962 as part of a National Cancer Institute program in search of natural products exhibiting anticancer activity. Taxol proved to be perhaps the most interesting of more than 100,000 compounds extracted from more

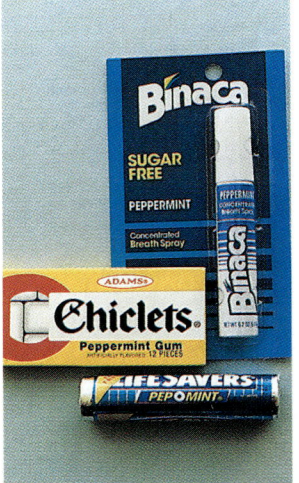

The flavor of peppermint in the products shown is due mainly to menthol in the extracts of the peppermint plant (*Mentha piperita*).

than 35,000 plant species and constitutes a clinically approved, powerful new weapon in the arsenal against human cancerous tumor growth. Because roughly six trees must be sacrificed to treat one patient, many efforts are underway to improve efficacy and availability and to increase yields. Many of these efforts have been undertaken by synthetic organic chemists, leading to the first two total syntheses of taxol in 1994.

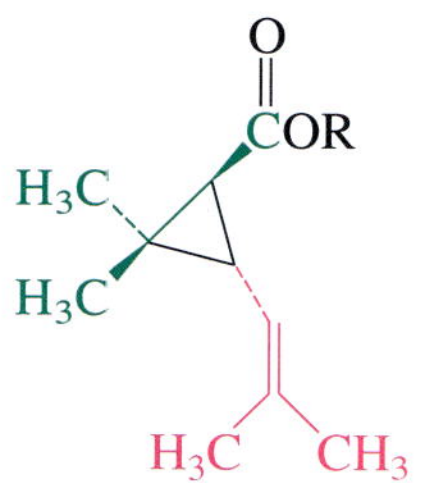

trans-**Chrysanthemic acid (R = H)**
trans-**Chrysanthemic esters (R ≠ H)**

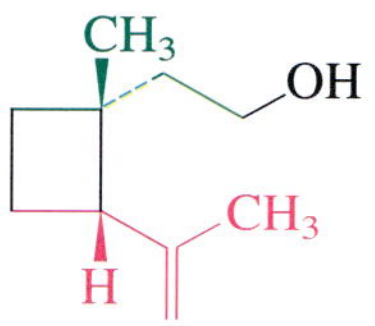

Grandisol

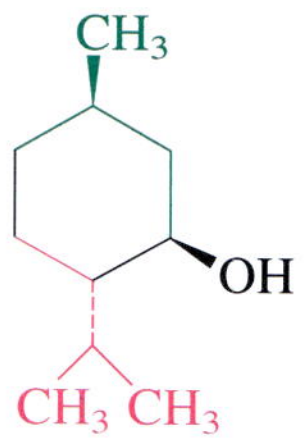

Menthol

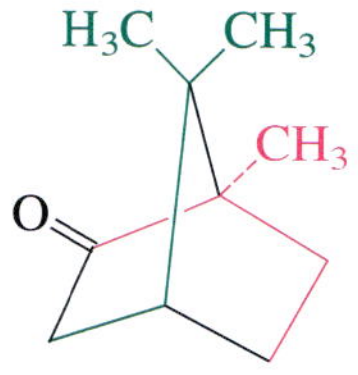

Camphor

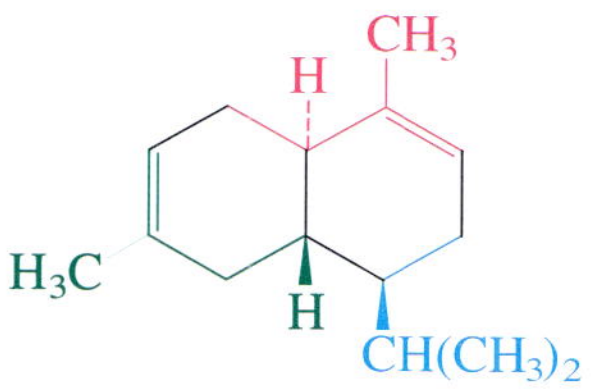

β-Cadinene

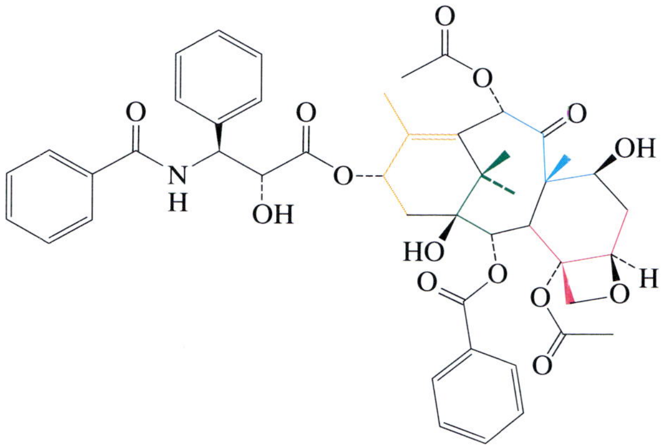

Taxol

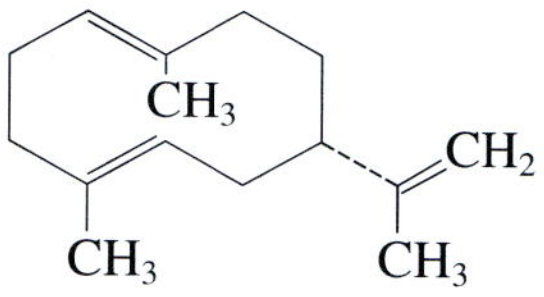

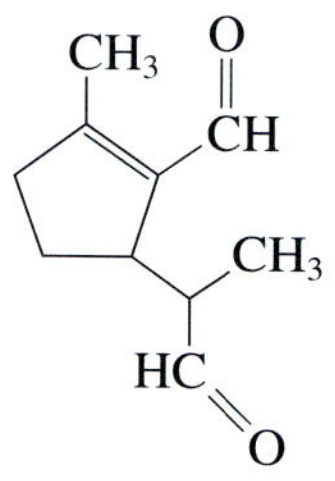

EXERCISE 4-10

Draw the preferred chair conformation of menthol.

EXERCISE 4-11

The structures of two terpenes utilized by insects in defense are shown in the margin. Classify them as mono-, sesqui-, or diterpenes. Identify the isoprene units in each.

EXERCISE 4-12

After reviewing Section 2-1, specify the functional groups present in the terpenes shown in Section 4-7.

Steroids are tetracyclic natural products with powerful physiological activities

Steroids are abundant in nature, and many derivatives have physiological activity. Steroids frequently function as **hormones,** which are regulators of biochemical activity. In the human body, for example, they control sexual development and fertility, in addition to other functions. Because of this feature, many steroids, often the products of laboratory synthesis, are used in medicines in, for example, the treatment of cancer, arthritis, or allergies, and in birth control.

In the steroids, three cyclohexane rings are fused in such a way as to form an angle. The ring junctions are usually trans, as in *trans*-decalin. The fourth ring is a cyclopentane; its addition gives the typical tetracyclic structure. The four rings are labeled A, B, C, D, and the carbons are numbered according to a scheme specific to steroids. Many steroids have methyl groups attached to C10 and C13 and oxygen at C3 and C17. In addition, longer side chains may be found at C17. The trans fusion of the rings allows for a least-strained all-chair configuration in which the methyl groups and hydrogen atoms at the ring junctions occupy axial positions.

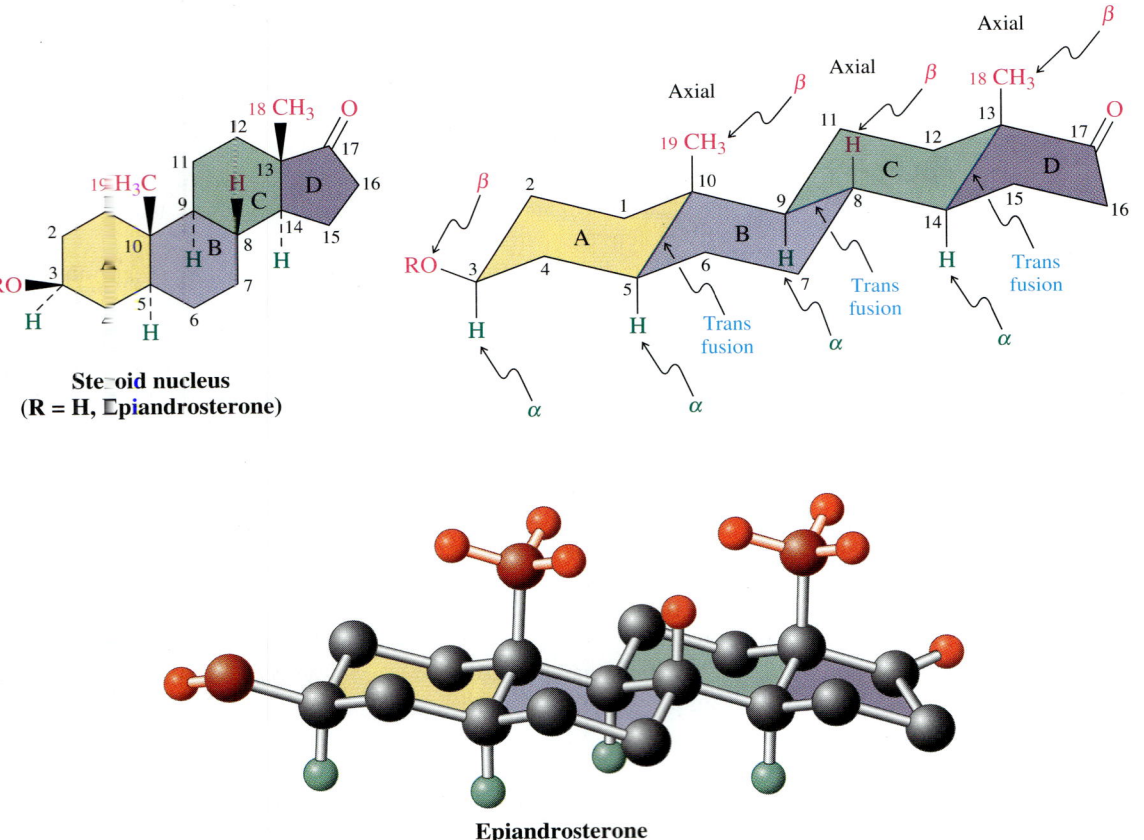

Steroid nucleus
(R = H, Epiandrosterone)

Epiandrosterone

Groups attached above the plane of the steroid molecule as written are β substituents, whereas those below are referred to as α. Thus, the structure of the steroid nucleus shown here has a 3β-OR, 5α-H, 10β-CH₃, and so forth. The axial methyl groups are also called **angular** methyls because they protrude sharply from the general framework (*angulus,* Latin, at an angle; being at a sharp corner).

Among the most abundant steroids, cholesterol is present in almost all human and animal tissue, particularly in the brain and the spinal cord.

Cholesterol

Cholic acid

Cortisone

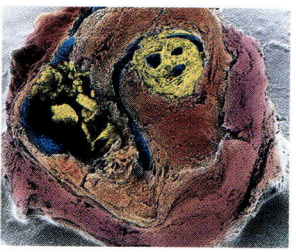

Cross section of a human coronary artery of the heart blocked by cholesterol plaques. The twisted artery wall is shown in brown, the inner open space in blue, and the deposits in yellow.

In fact, it is so concentrated in the spinal cord of cattle that it is isolated from that tissue by simple extraction. The adult human body contains from 200 to 300 g of cholesterol; gallstones may consist entirely of it. This steroid has been implicated in several circulatory diseases because it precipitates in the arteries, thereby causing arteriosclerosis and heart disease. Although its biological function in the body is not completely understood, it is a precursor of steroid hormones and bile acids. Bile acids are produced in the liver as part of a fluid delivered to the duodenum to aid in the emulsification, digestion, and absorption of fats. An example is cholic acid.

Cortisone, used extensively in the treatment of rheumatoid inflammations, is one of the adrenocortical hormones produced by the outer part (cortex) of the adrenal glands. These hormones participate in regulating the electrolyte and water balance in the body, as well as protein and carbohydrate metabolism.

The sex hormones are divided into three groups: (1) the male sex hormones, or *androgens;* (2) the female sex hormones, or *estrogens;* and (3) the pregnancy hormones, or *progestins.* Testosterone is the principal male sex hormone. Produced by the testes, it is responsible for male (masculine) characteristics (deep voice, facial hair, general physical constitution). Synthetic testosterone analogs are used in medicine to promote muscle and tissue growth (anabolic steroids; *ana-,* Greek, up—i.e., "anabolic," the opposite of "metabolic"), for example, in patients with muscular atrophy. Unfortunately, such steroids are also abused and consumed illegally, most commonly by "body builders" and athletes, even though the health risks are numerous, including liver cancer, coronary heart disease, and sterility. Estradiol is the principal female sex hormone. It was first isolated by extraction of four tons of sow ovaries, yielding only a few milligrams of pure steroid. Estradiol is responsible for the development of the secondary female characteristics and participates in the control of the menstrual cycle. An example of a progestin is progesterone, responsible for preparing the uterus for implantation of the fertilized egg.

Testosterone

Estradiol

Progesterone

CHEMICAL HIGHLIGHT 4-2

Controlling Fertility: From "the Pill" to RU-486

The menstrual cycle is controlled by three protein hormones from the pituitary gland. The follicle-stimulating hormone (FSH) induces the growth of the egg, and the luteinizing hormone (LH) induces its release from the ovaries and the formation of an ovarian tissue called the *corpus luteum*. The third pituitary hormone (luteotropic hormone, also called luteotropin or prolactin), stimulates the corpus luteum and maintains its function.

As the cycle begins and egg growth is initiated, the tissue around the egg secretes increasing quantities of estrogens. When a certain concentration of estrogen in the bloodstream has been reached, the production of FSH is turned off. The egg is released at this stage in response to LH. At the time of ovulation, LH also triggers the formation of the corpus luteum, which in turn begins to secrete increasing amounts of progesterone. This last hormone suppresses any further ovulation by turning off the production of LH. If the egg is not fertilized, the corpus luteum regresses and the ovum and the *endometrium* (uterine lining) are expelled (menstruation). Pregnancy, on the other hand, leads to increased production of estrogens and progesterone to prevent pituitary hormone secretion and thus renewed ovulation.

The birth control pill consists of a mixture of synthetic potent estrogen and progesterone derivatives (more potent than the natural hormones), which, when taken throughout most of the menstrual cycle, prevent both development of the ovum and ovulation by turning off production of both FSH and LH. The female body is essentially being tricked into believing that it is pregnant. Some of the commercial pills contain a combination of norethindrone and

Norethindrone

Ethynylestradiol

ethynylestradiol. Other preparations consist of similar analogs with minor structural variations.

RU-486 (mifepristone) is a synthetic steroid that blocks the effects of progesterone. The fertilized egg is not implanted, because the necessary preparation of the endometrium has been prevented. RU-486 has been used in France since 1988 as a "morning after" pill. After much discussion and testing, the Food and Drug Administration approved the drug for the U.S. market in 1997.

RU-486 (mifepristone)

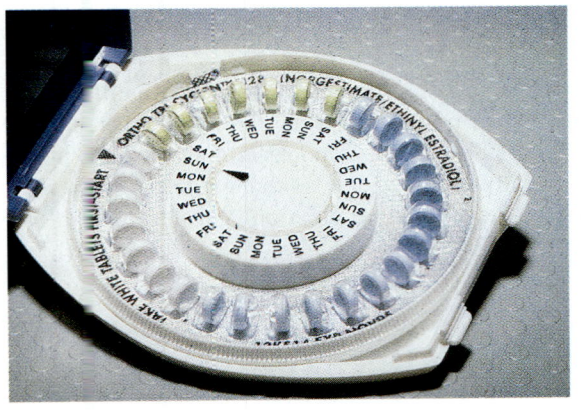

"Pill" dispenser.

The structural similarity of the steroid hormones is remarkable, considering their widely divergent activity. Steroids are the active ingredients of "the pill," functioning as an antifertility agent for the control of the female menstrual cycle and ovulation. It is estimated that 50 to 60 million women throughout the world take "the pill" as the primary form of contraception.

In summary, there is great variety in the structure and function of naturally occurring organic products, as manifested in the terpenes and the steroids. Natural products will be frequently introduced in subsequent chapters to illustrate the presence and chemistry of a functional group, to demonstrate synthetic strategy or the use of reagents, to picture three-dimensional relations, and to exemplify medicinal applications. Several classes of natural products will be discussed more extensively: fats (Sections 19-12 and 20-4), carbohydrates (Chapter 24), alkaloids (Section 25-7), and amino and nucleic acids (Chapter 26).

CHAPTER INTEGRATION PROBLEMS

a. 1, 2, 3, 4, 5, 6-Hexachlorocyclohexane exists as a number of cis-trans isomers. Using the flat cyclohexane stencil and dashed-wedged lines, draw all of them.

SOLUTION

Before starting to solve this part by random trial and error, consider a more systematic approach. In this approach, we start with the simplest case of all chlorines positioned cis to one another (using all wedged lines) and then look at the various permutations obtained by placing a progressively increasing number of the substituents trans (dashed lines). We can stop at the stage of three chlorines "up" and three chlorines "down" because "two up and four down" is the same as "four up and two down," and so on. The first two cases, "six up" (A) and "five up and one down" (B), are unique because only one structural possibility exists for each:

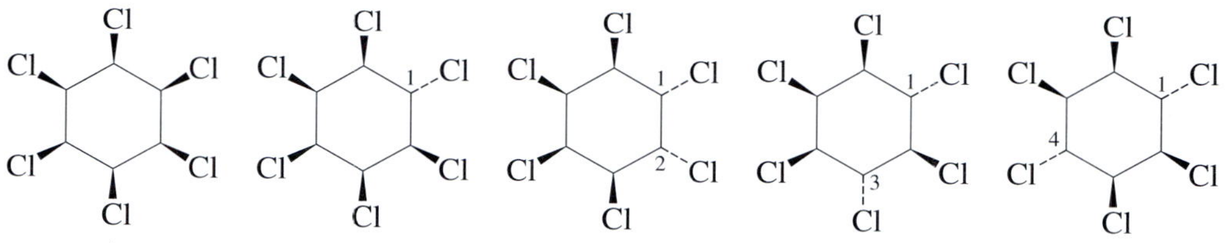

Now we look at the "four up and two down" isomers. There are three differing ways of placing two chlorines "down": their positioning along the six-membered ring can be only 1, 2 (C), 1, 3 (D), or 1, 4 (E), because the options 1, 5 and 1, 6 are the same as 1, 3 and 1, 2, respectively. Finally, a similar line of thought leads to the realization that placing three chlorines "down" can be done only in three unique ways: 1, 2, 3 (F), 1, 2, 4 (G), and 1, 3, 5 (H):

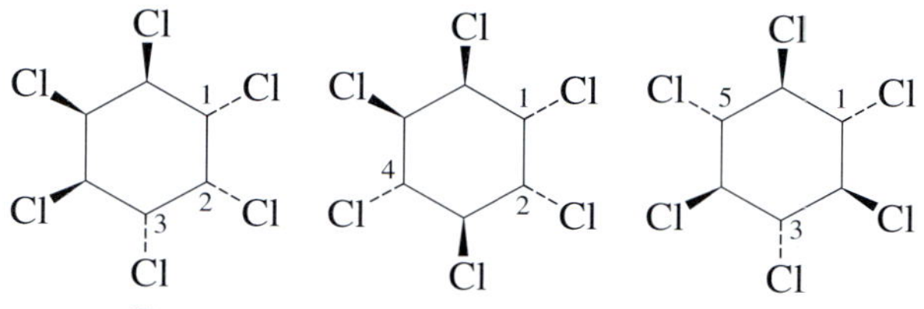

b. The so-called γ-isomer is an insecticide (lindane, gammexane, kwell) with the following structure:

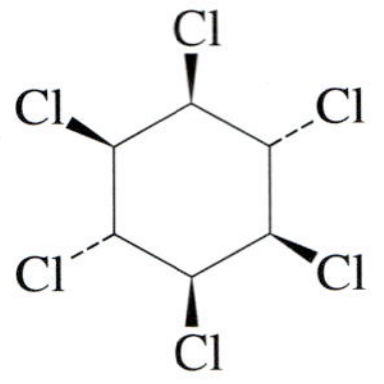

γ-Hexachlorocyclohexane

Draw the two chair conformations of this compound. Which one is more stable?

SOLUTION

Note that the γ-isomer corresponds to our structure E in (a). To help in converting the flat cyclohexane stencil structures into their chair renditions, look at Figure 4-11 and note the alternating relation of the members of the two sets of hydrogens. As we proceed around the ring, we can see that neighboring members (i.e., those in a "1,2"-relation) of either set (i.e., axial or equatorial neighbors) always have a trans disposition. On the other hand, the 1,3-relation is always cis, whereas that describable as 1,4 is again trans. Conversely, in a dashed-wedged line structure, when two neighboring substituents (i.e., 1,2-relation) are cis, one of the substituents will be axial and the other equatorial in either of the two chair conformations of the molecule. When they are trans, one chair picture will show them diequatorial, the other diaxial. This assignment alternates in 1,3-related substituents: cis results in a diaxial-diequatorial pair of chair conformers, trans in an axial-equatorial or equatorial-axial pair, and so on. These relations are summarized in Table 4-4.

The two chair forms of γ-hexachlorocyclohexane thus look as shown here:

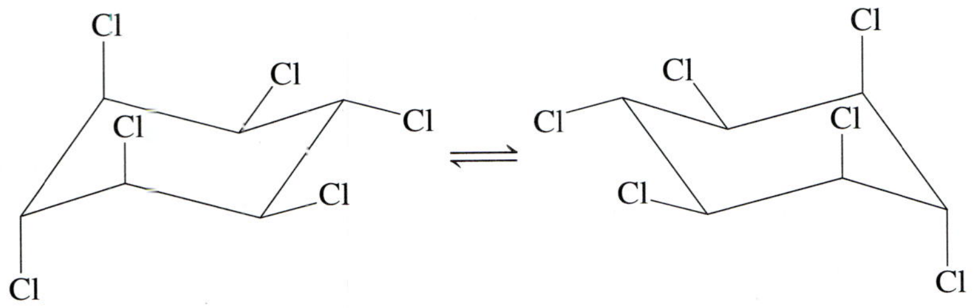

$\Delta G° = 0 \text{ kcal mol}^{-1}$

Either structure has three equatorial and three axial chlorine substituents; hence they are equal in energy.

TABLE 4-4	Relation of *Cis-Trans* Stereochemistry in Substituted Cyclohexanes to Equatorial-Axial Positions in the Two-Chair Forms	
cis-1,2	Axial-equatorial	Equatorial-axial
trans-1,2	Axial-axial	Equatorial-equatorial
cis-1,3	Axial-axial	Equatorial-equatorial
trans-1,3	Axial-equatorial	Equatorial-axial
cis-1,4	Axial-equatorial	Equatorial-axial
trans-1,4	Axial-axial	Equatorial-equatorial

c. For which isomer do you expect the energy difference between the two cyclohexane chair forms to be largest? Estimate the $\Delta G°$ value.

SOLUTION

The biggest $\Delta G°$ for chair–chair flip is that between an all-equatorial and all-axial hexachlorocyclohexane. Application of Table 4-4 and inspection of Figure 4-11 reveal that this relation holds only for the all-trans isomer. Table 4-3 gives the $\Delta G°$ (equatorial-axial) for Cl as 0.52 kcal mol^{-1}; hence, for our example, $\Delta G° = 6 \times 0.52 = 3.12$ kcal mol^{-1}.

$$\Delta G° = +3.12 \text{ kcal mol}^{-1}$$

all-*trans*-Hexachlorohexane

Note that this value is only an estimate. For example, it ignores the six Cl–Cl *gauche* interactions in the all-equatorial form, which would reduce the energy difference between the two conformers. However, it also disregards the six 1,3-diaxial interactions in the all-axial chair, which would counteract this effect.

IMPORTANT CONCEPTS

1. **Cycloalkane** nomenclature is derived from that of the straight-chain alkanes.

2. All but the 1,1-disubstituted cycloalkanes exist as two isomers: If both substituents are on the same face of the molecule, they are **cis;** if they are on opposite faces, they are **trans.** Cis and trans isomers are **stereoisomers**—compounds that have identical connectivities but differ in the arrangement of their atoms in space.

3. Some cycloalkanes are **strained.** Distortion of the bonds about tetrahedral carbon introduces **bond-angle strain. Eclipsing (torsional) strain** results from the inability of a structure to adopt staggered conformations about C–C bonds. Steric repulsion between atoms across a ring leads to **transannular strain.**

4. Bond-angle strain in the small cycloalkanes is largely accommodated by the formation of **bent bonds.**

5. Bond-angle, eclipsing, and other strain in the cycloalkanes larger than cyclopropane (which is by necessity flat) can be accommodated by deviations from planarity.

6. Ring strain in the small cycloalkanes gives rise to reactions that result in opening of the ring.

7. Deviations from planarity lead to conformationally mobile structures, such as **chair, boat,** and **twist-boat** cyclohexane. Chair cyclohexane is almost strain free.

8. Chair cyclohexane contains two types of hydrogens: **axial** and **equatorial.** These interconvert rapidly at room temperature by a conformational **chair–chair ("flip") interconversion,** with an activation energy of 10.8 kcal mol^{-1}.

9. In monosubstituted cyclohexanes, the $\Delta G°$ of equilibration between the two chair conformations is substituent dependent. Axial substituents are exposed to **1,3-diaxial interactions.**

10. In more highly substituted cyclohexanes, substituent effects are often **additive,** the bulkiest substituents being the most likely to be equatorial.

11. Completely strain-free cycloalkanes are those that can readily adopt an all-*anti* conformation and lack transannular interactions.

12. **Bicyclic** ring systems may be **fused** or **bridged.** Fusion can be cis or trans.

13. Natural products are generally classified according to structure, physiological activity, taxonomy, and biochemical origin. Examples of the last class are the **terpenes,** of the first the **steroids.**

14. Terpenes are made up of **isoprene** units of five carbons.

15. Steroids contain three angularly fused cyclohexanes (A, B, C rings) attached to the cyclopentane D ring. Beta substituents are above the molecular plane, alpha substituents below.

16. An important class of steroids are the **sex hormones,** which have a number of physiological functions, including the control of fertility.

PROBLEMS

13. Write as many structures as you can that have the formula C_5H_{10} and contain one ring. Name them.

14. Name the following molecules according to the IUPAC nomenclature system.

(a) I — cyclopropane structure

(b) H_3C — cyclopentane — $CH(CH_3)_2$

(c) cyclobutane with Cl, Cl

(d) cyclohexyl-substituted cyclohexane with CH_3

(e) cyclohexane with Br, Br

(f) cyclohexane with Br, Br

15. Draw structural representations of each of the following molecules: (a) *trans*-2-chloro-1-ethylcyclopropane; (b) *cis*-1-bromo-2-chlorocyclopentane; (c) 2-chloro-1,1-diethylcyclopropane; (d) *trans*-2-bromo-3-chloro-1,1-diethylcyclopropane; (e) *cis*-1,3-dichloro-2,2-dimethylcyclobutane; (f) *cis*-2-chloro-1,1-difluoro-3-methylcyclopentane.

16. The kinetic data for the radical chain chlorination of several cycloalkanes (see the adjoining table) illustrate that the C–H bonds of cyclopropane and, to a lesser extent, cyclobutane are somewhat abnormal. (a) What do these data tell you about the strength of the cyclopropane C–H bond and the stability of the cyclopropyl radical? (b) Suggest a reason for the stability characteristics of the cyclopropyl radical. (**Hint:** Consider bond-angle strain in the radical relative to cyclopropane itself.)

17. Use the data in Tables 3-2 and 4-2 to estimate the $DH°$ value for a C–C bond in (a) cyclopropane; (b) cyclobutane; (c) cyclopentane; and (d) cyclohexane.

18. Draw each of the following substituted cyclobutanes in its two interconverting "puckered" conformations (Figure 4-3). When the two conformations differ in energy, identify the more stable shape and indicate the form(s) of strain that raise the relative energy of the less stable one. (**Hint:** Puckered cyclobutane has axial and equatorial positions similar to those in chair cyclohexane.)
(a) Methylcyclobutane
(b) *cis*-1,2-Dimethylcyclobutane
(c) *trans*-1,2-Dimethylcyclobutane
(d) *cis*-1,3-Dimethylcyclobutane
(e) *trans*-1,3-Dimethylcyclobutane

Reactivity per Hydrogen Toward Cl·

Cycloalkane	Reactivity
Cyclopentane	0.9
Cyclobutane	0.7
Cyclopropane	0.1

Note: Relative to cyclohexane = 1.0; at 68°C, hv, CCl_4 solvent.

Which is more stable: *cis-* or *trans-*1,2-dimethylcyclobutane; *cis-* or *trans-*1,3-dimethylcyclobutane?

19. For each of the following cyclohexane derivatives, indicate (i) whether the molecule is a cis or trans isomer and (ii) whether it is in its most stable conformation. If your answer to (ii) is no, flip the ring and draw its most stable conformation.

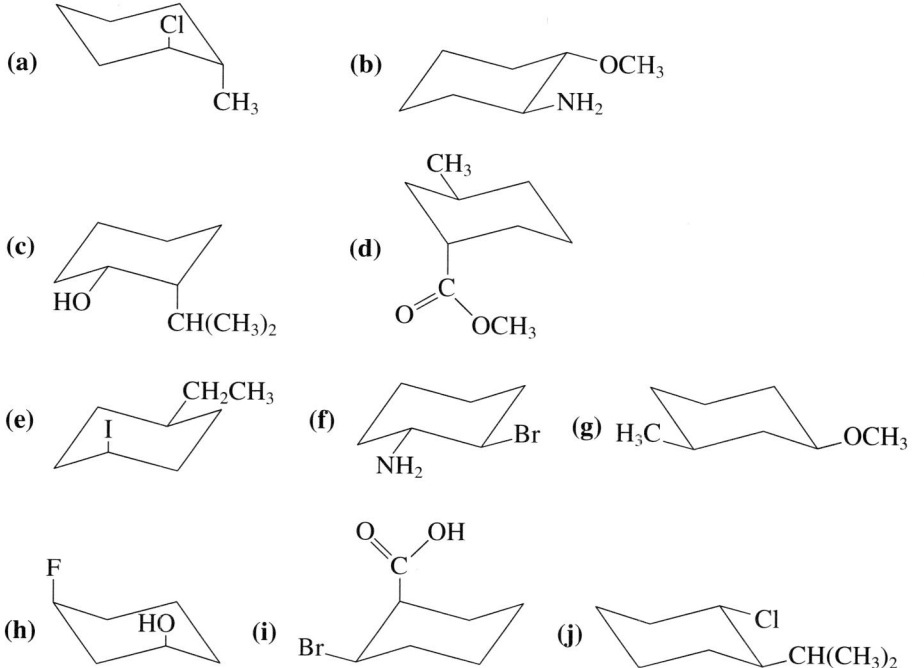

20. Using the data in Table 4-3, calculate the $\Delta G°$ for ring flip to the other conformation of the molecules depicted in Problem 19. Make sure that the sign (i.e., positive or negative) of your values is correct.

21. Draw the most stable conformation for each of the following substituted cyclohexanes; then, in each case, flip the ring and redraw the molecule in the higher energy chair conformation: **(a)** cyclohexanol; **(b)** *trans-*3-methylcyclohexanol (see structures in the margin); **(c)** *cis-*1-(1-methylethyl)-3-methylcyclohexane; **(d)** *trans-*1-ethyl-3-methoxycyclohexane; **(e)** *trans-*1-chloro-4-(1,1-dimethylethyl)cyclohexane.

22. For each molecule in Problem 21, estimate the energy difference between the most stable and next best conformation. Calculate the approximate ratio of the two at 300 K.

23. Sketch a potential-energy diagram (similar to that in Figure 4-9) for methylcyclohexane showing the two possible chair conformations at the left and right ends of the reaction coordinate for conformational interconversion.

24. Draw all the possible all-chair conformers of cyclohexylcyclohexane.

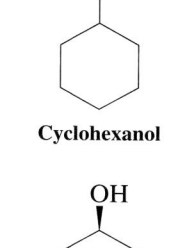

OH

Cyclohexanol

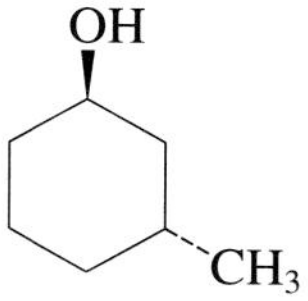

OH

CH₃

***trans-*3-Methylcyclohexanol**

25. What is the most stable of the four *boat* conformations of methylcyclohexane, and why?

26. The most stable conformation of *trans*-1,3-bis(1,1-dimethylethyl)cyclohexane is not a chair. What conformation would you predict for this molecule? Explain.

27. The bicyclic hydrocarbon formed by the fusion of a cyclohexane ring with a cyclopentane ring is known as hexahydroindane. Using the drawings of *trans-* and *cis-*decalin for reference (Figure 4-13), draw the structures of *trans-* and *cis*-hexahydroindane, showing each ring in its most stable conformation.

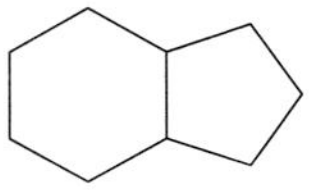

Hexahydroindane

28. On viewing the drawings of *cis-* and *trans*-decalin in Figure 4-13, which do you think is the more stable isomer? Estimate the energy difference between the two isomers. Answer the same question for cis and trans dimethyl-substituted decalins.

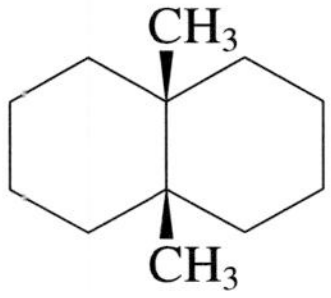

cis-9,10-Dimethyldecalin 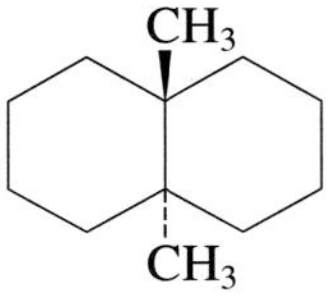   *trans*-9,10-Dimethyldecalin

29. Identify each of the following molecules as a monoterpene, a sesquiterpene, or a diterpene (all names are common).

(a) Geraniol

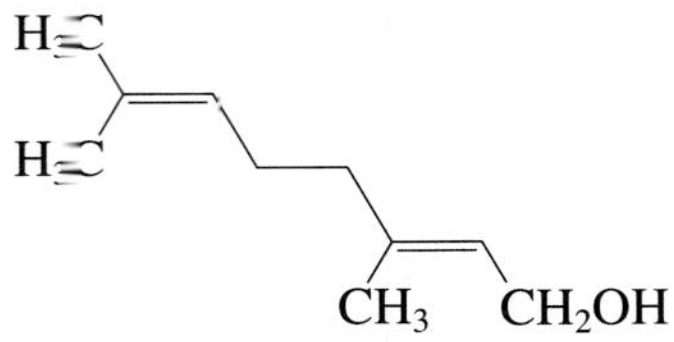

(b) Eremanthin

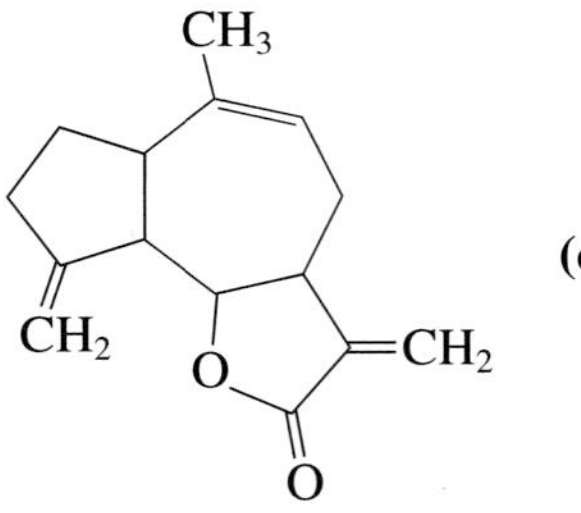

(c) Eudesmol

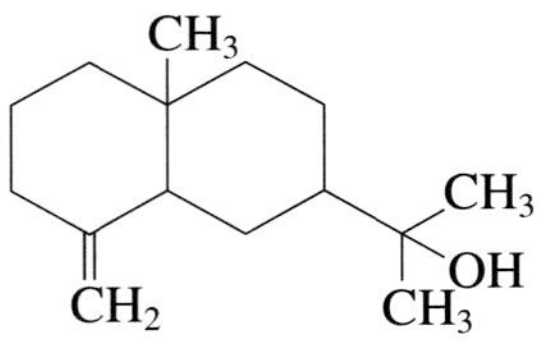

(d) Ipomeamarone

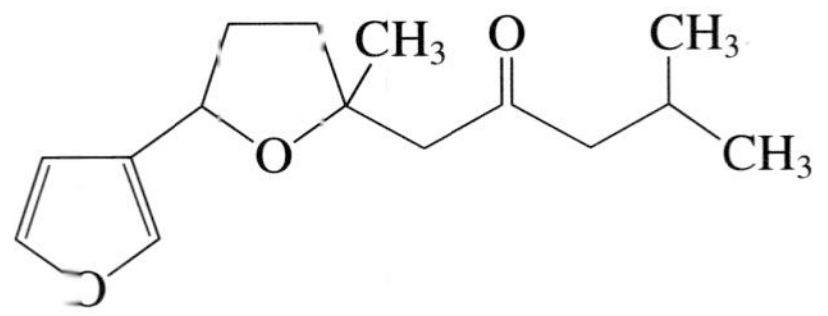

(e) Genipin

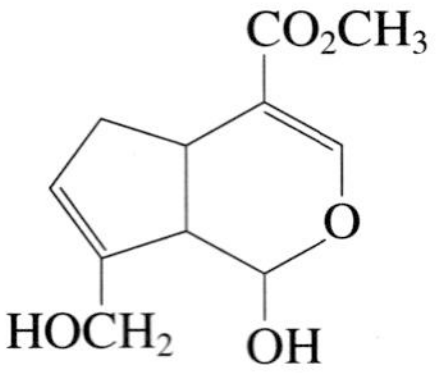

(f) Castoramine

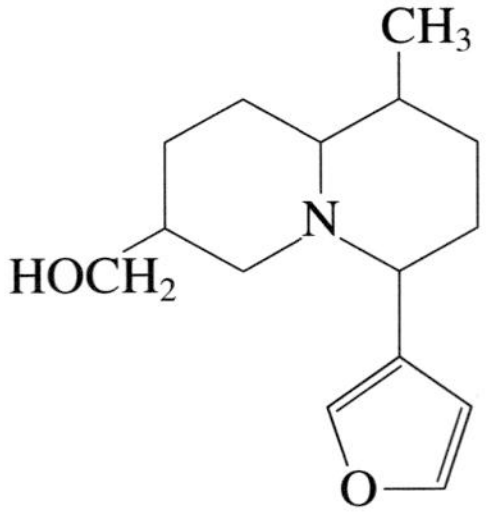

(g) Cantharidin

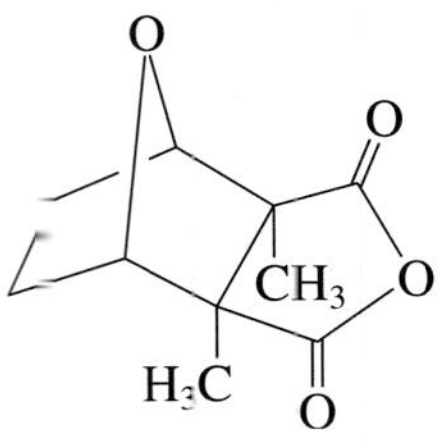

(h) Vitamin A

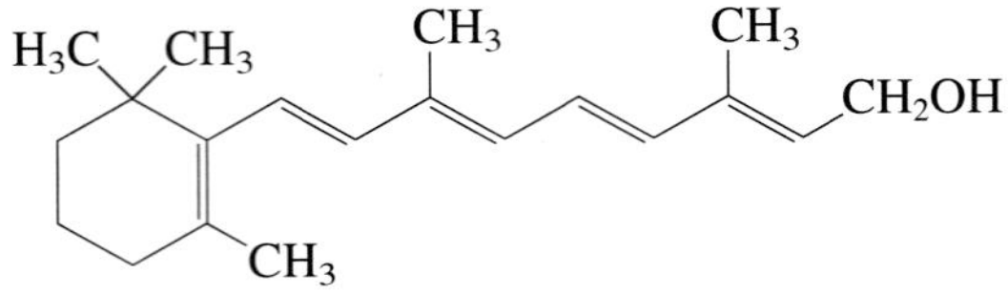

30. Find the 2-methyl-1,3-butadiene (isoprene) units in each of the naturally occurring organic molecules pictured in Problem 29.

31. Circle and identify by name all the functional groups in any three of the steroids illustrated in Section 4-7. Label any polarized bonds with partial positive and negative charges (δ^+ and δ^-).

32. Several additional examples of naturally occurring molecules with strained ring structures are shown here.

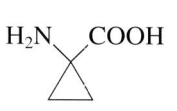

1-Aminocyclopropane-carboxylic acid

(Present in plants, this molecule plays a role in the ripening of fruits and the dropping of autumn leaves)

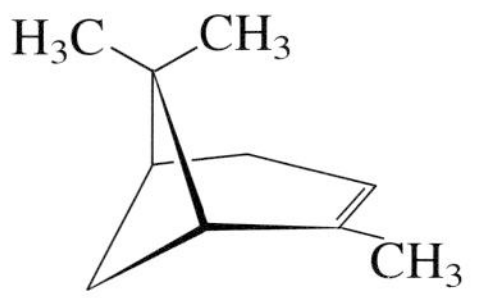

α-Pinene

(Present in cedar-wood oil)

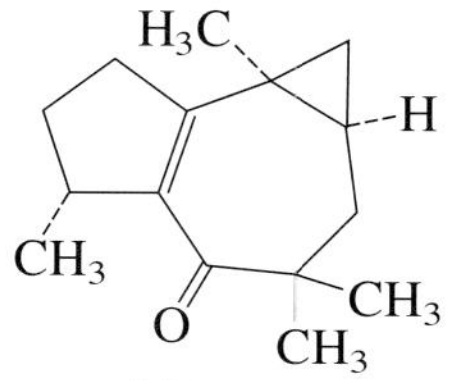

Africanone

(Also a plant-leaf oil)

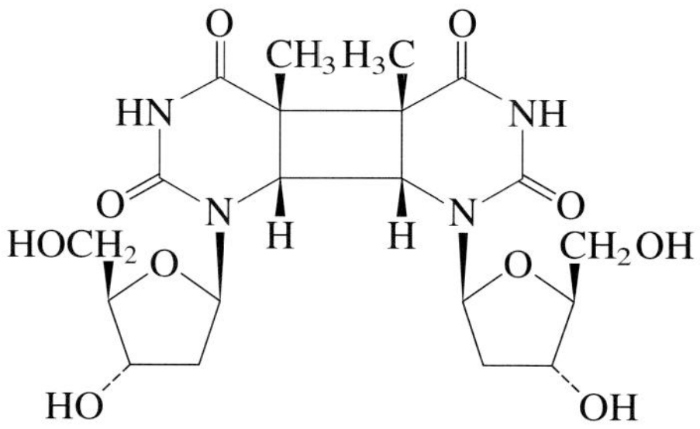

Thymidine dimer

(A component of DNA that has been exposed to ultraviolet light)

Identify the terpenes (if any) in the preceding group of structures. Find the 2-methyl-1,3-butadiene units in each structure and classify the latter as a mono-, sesqui-, or diterpene.

33. If cyclobutane were flat, it would have exactly 90° C–C–C bond angles and could conceivably use pure *p* orbitals in its C–C bonds. What would be a possible hybridization for the carbon atoms of the molecule that would allow all the C–H bonds to be equivalent? Exactly where would the hydrogens on each carbon be located? What are the real structural features of the cyclobutane molecule that contradict this hypothesis?

34. Compare the structure of cyclodecane in an all-chair conformation with that of *trans*-decalin. Explain why all-chair cyclodecane is highly strained, and yet *trans*-decalin is nearly strain free. Make models.

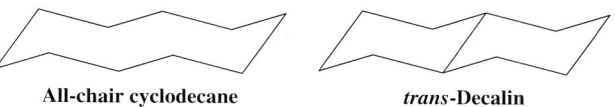

All-chair cyclodecane   ***trans*-Decalin**

35. Fusidic acid is a steroidlike microbial product that is an extremely potent antibiotic with a broad spectrum of biological activity. Its molecular shape is

most unusual and has supplied important clues to researchers investigating the methods by which steroids are synthesized in nature.

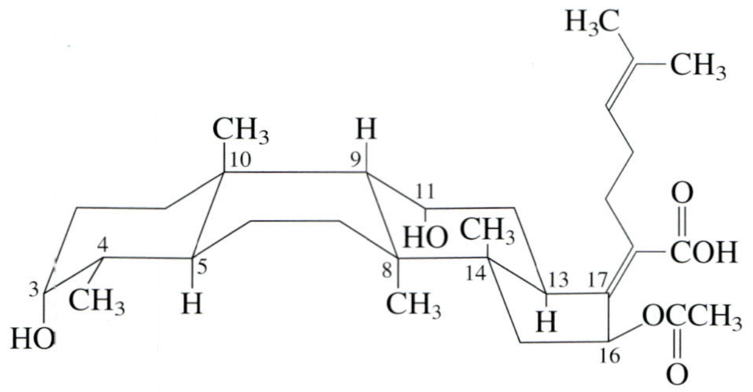

Fusidic acid

(a) Locate all the rings in fusidic acid and describe their conformations. (b) Identify all ring fusions in the molecule as having either cis or trans geometry. (c) Identify all groups attached to the rings as being either α- or β-substituents. (d) Describe in detail how this molecule differs from the typical steroid in structure and stereochemistry. (As an aid to answering these questions, the carbon atoms of the framework of the molecule have been numbered.)

36. The enzymatic oxidation of alkanes to produce alcohols (see Chemical Highlight 3-4) is a simplified version of the reactions that produce the adrenocortical steroid hormones. In the biosynthesis of corticosterone from progesterone (Section 4-7), two such oxidations take place successively. It is thought that the monooxygenase enzymes act as complex oxygen-atom donors in these reactions. A suggested mechanism, as applied to cyclohexane, consists of the two steps shown below the biosynthesis.

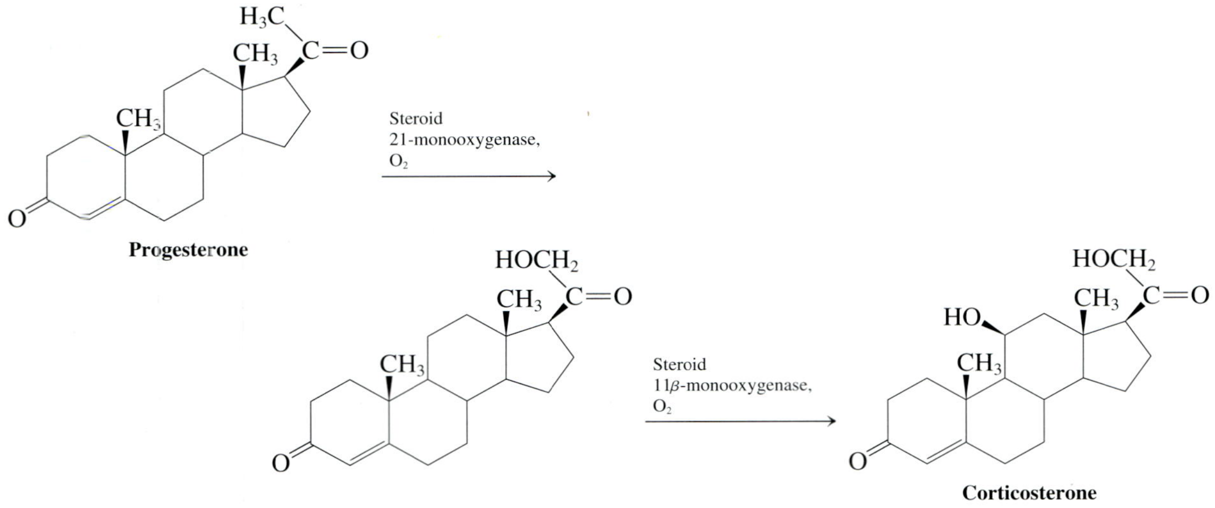

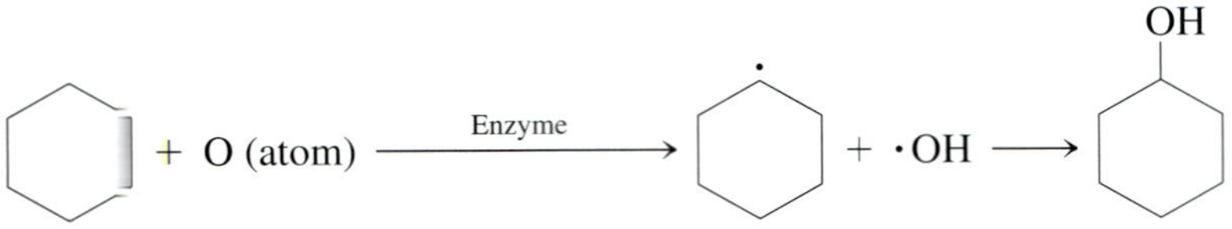

Calculate $\Delta H°$ for each step and for the overall oxidation reaction of cyclo-hexane. Use the following $DH°$ values: cyclohexane C–H bond, 96.5 kcal mol^{-1}; bond in O–H radical, 102.5 kcal mol^{-1}; cyclohexanol C–O bond, 93 kcal mol^{-1}.

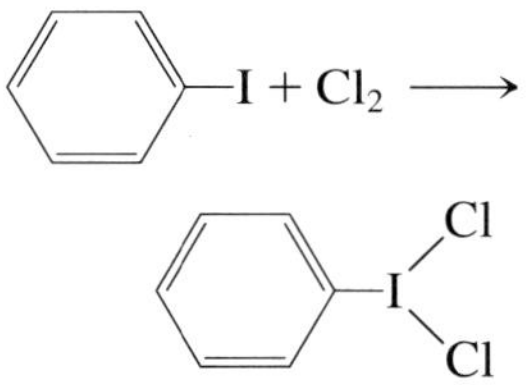

37. Like sulfuryl chloride and NCS (Section 3-8), iodobenzene dichloride, formed by the reaction of iodobenzene and chlorine, is a reagent for the chlorination of alkane C–H bonds. Chlorinations in which iodobenzene dichloride is used are initiated by light.

(a) Propose a radical chain mechanism for the chlorination of a typical alkane RH by iodobenzene dichloride. To get you started, the overall equation for the reaction is given below, as is the initiation step.

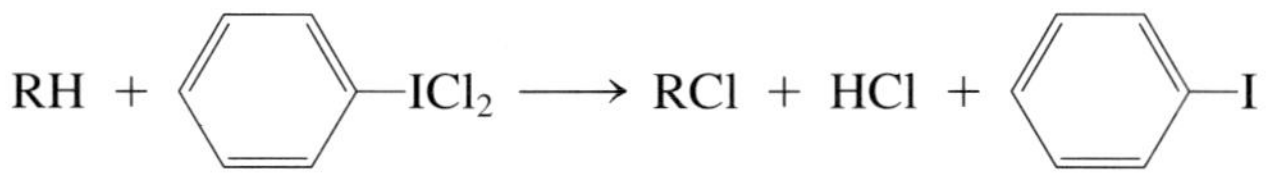

Iodobenzene dichloride

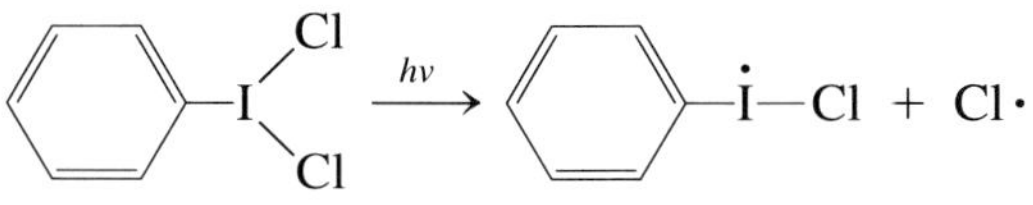

(b) Radical chlorination of typical steroids by iodobenzene dichloride gives, predominantly, three isomeric monochlorination products. On the basis of both reactivity (tertiary, secondary, primary) considerations and steric effects (which might hinder the approach of a reagent toward a C–H bond that might other-wise be reactive), predict the three major sites of chlorination in the steroid molecule. Either make a model or carefully analyze the drawings of the steroid nucleus in Section 4-7.

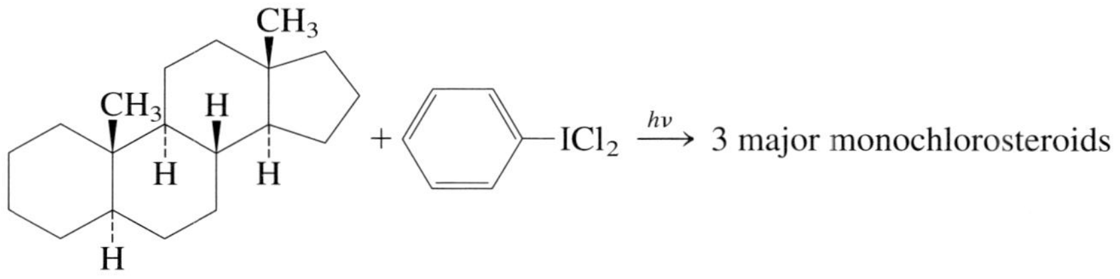

38. As Problem 36 indicates, the enzymatic reactions that introduce functional groups into the steroid nucleus in nature are highly selective, unlike the labora-tory chlorination described in Problem 37. However, by means of a clever adaptation of this reaction, it is possible to partly mimic nature's selectivity in the laboratory. Two such examples are illustrated below and on the next page.

(a)

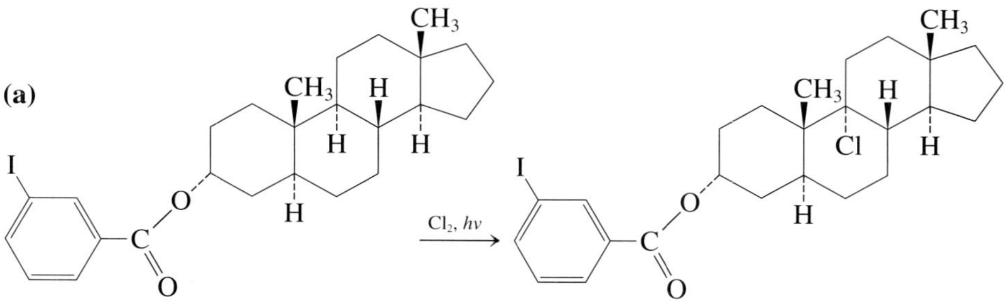

(b)

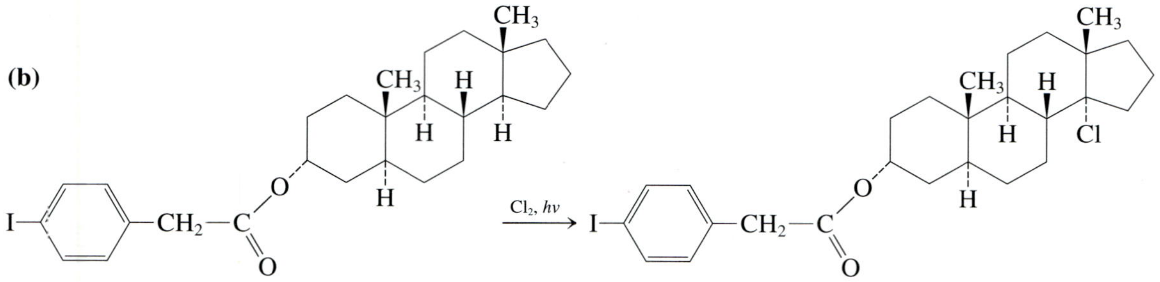

Propose reasonable explanations for the results of these two reactions. Make a model of the product of the addition of Cl_2 to each iodocompound (compare Problem 37) to help in analyzing each system.

Team Problem

39. Consider the following compounds:

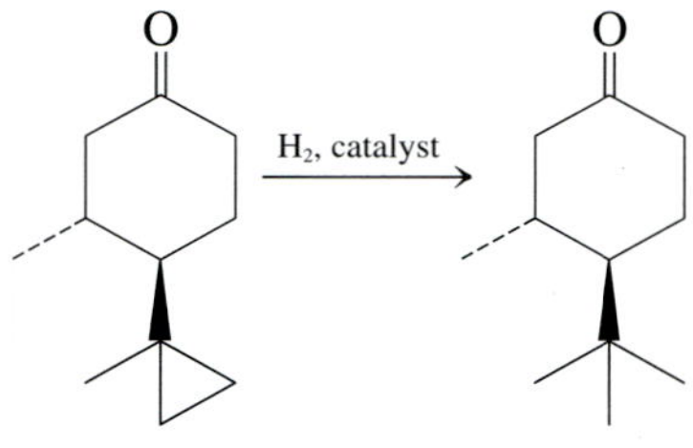

Conformational analysis reveals that, though compound A exists in a chair conformation, compound B does not.
(a) Make a model of A. Draw chair conformations and label the substituents as equatorial or axial. Circle the most stable conformation. (Note that the carbonyl carbon is sp^2 hybridized and therefore the attached oxygen is neither equatorial nor axial. Do not let that lead you astray.)
(b) Make a model of B. Consider both transannular and *gauche* interactions in your analysis of its two chair forms. Discuss the steric problems of these conformations in comparison with those of A. Illustrate the key points of your discussion with Newman projections. Suggest a less sterically encumbered conformation for B.

Preprofessional Problems

40. Which of the following cycloalkanes has the greatest ring strain?

(a) Cyclopropane **(b)** Cyclobutane **(c)** Cyclohexane **(d)** Cycloheptane

41. The following molecule has

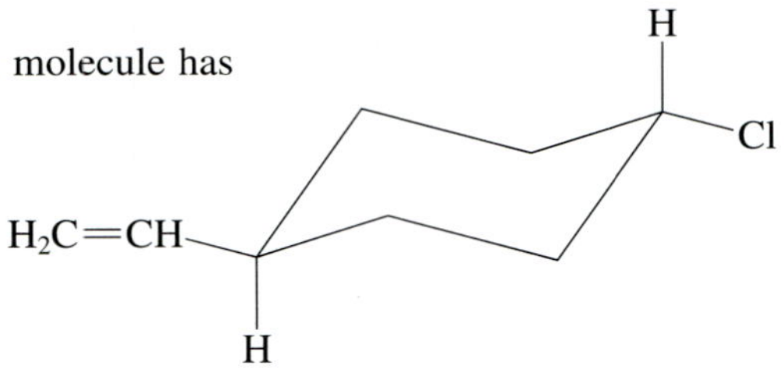

(a) one axial chlorine and one sp^2 carbon (b) one axial chlorine and two sp^2 carbons (c) one equatorial chlorine and one sp^2 carbon (d) one equatorial chlorine and two sp^2 carbons.

42. In this compound

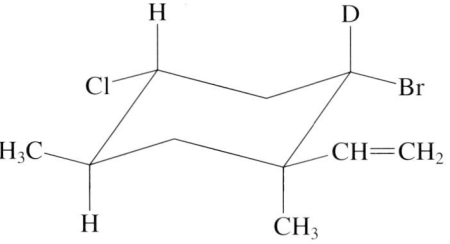

(a) the D is equatorial   (b) the methyls are both equatorial
(c) the Cl is axial   (d) the deuterium is axial

43. Which of the following isomers has the smallest heat of combustion?

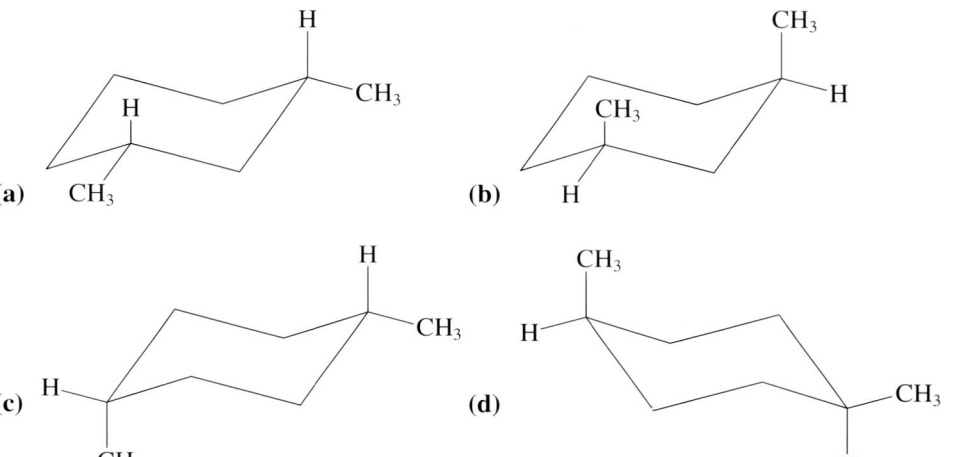

Stereoisomers

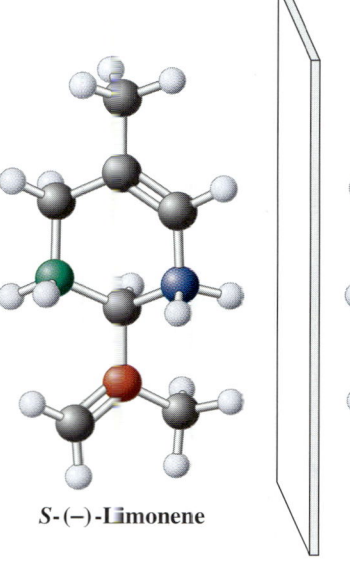
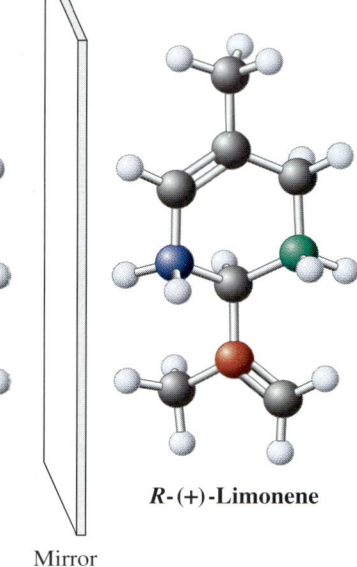

S-(−)-Limonene   R-(+)-Limonene

Mirror

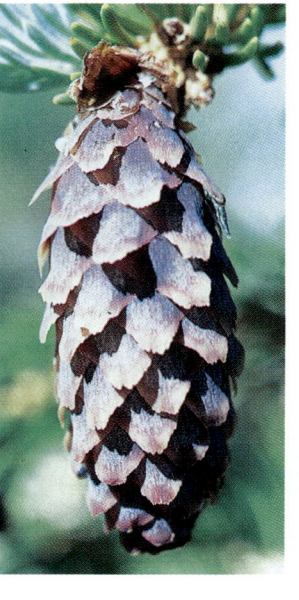

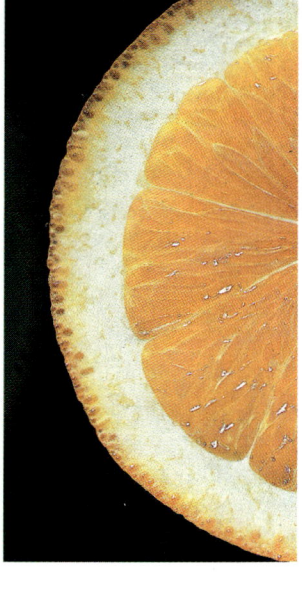

Have you ever looked at yourself in the mirror in the morning and exclaimed: "That can't be me!" Well, you were right. What you see, your mirror image, is not identical with your image: Your image and mirror image are *nonsuperimposable*. You can demonstrate this fact by trying to shake hands with your mirror counterpart: As you reach out with your right hand, your mirror image will offer you its left hand! We shall see that many molecules have this property—namely, that image and mirror image are nonsuperimposable and therefore not identical. How do we classify such structures?

Because they have the same molecular formula, these molecules are isomers but of a different kind from those encountered so far. The preceding chapters dealt with two kinds of isomerism: constitutional (also called structural) and stereo (Figure 5-1). **Constitutional isomerism** describes compounds that have identical molecular formulas but differ in the order in which the individual atoms are connected (Section 1-9).

Image and mirror image of the hydrocarbon limonene smell quite different. The S-image is present in the cones of fir trees and has a turpentine-like odor; the R-mirror image gives oranges their characteristic fragrance.

Constitutional Isomers

CH_4H_{10} $CH_3CH_2CH_2CH_3$ $H_3C-\overset{\displaystyle CH_3}{\underset{\displaystyle CH_3}{CH}}$

Butane   2-Methylpropane

C_2H_6O CH_3CH_2OH CH_3OCH_3

Ethanol   Methoxymethane
(Dimethyl ether)

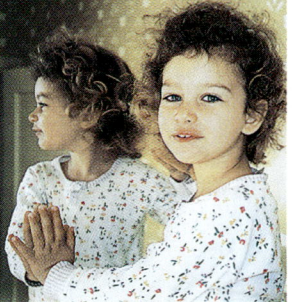

Paloma Sat-Vollhardt and her mirror image.

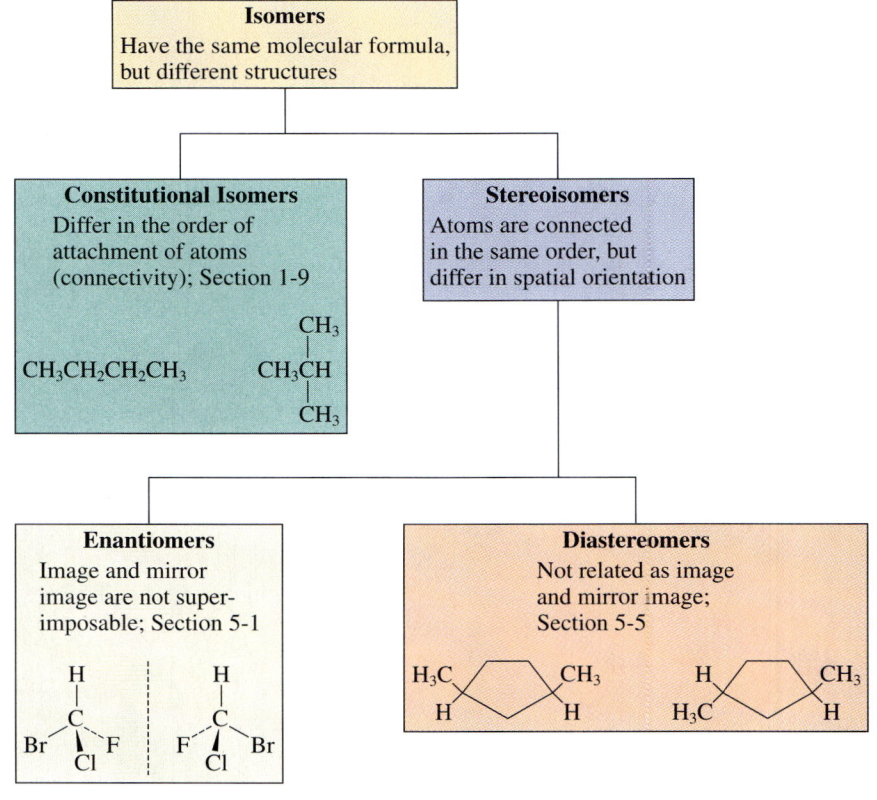

FIGURE 5-1
Relations among isomers of
various types.

EXERCISE 5-1

Are cyclopropylcyclopentane and cyclobutylcyclobutane isomers?

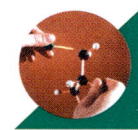

Stereoisomerism describes isomers whose atoms are connected in the same order
but differ in their spatial arrangement. Examples of stereoisomers include the relatively stable cis-trans isomers and the rapidly equilibrating conformational ones
(Section 4-1).

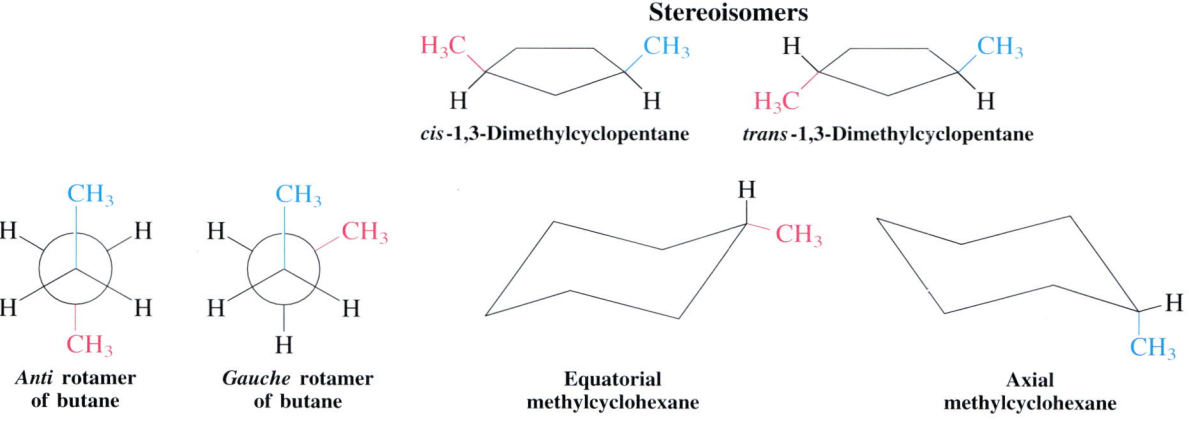

EXERCISE 5-2

Draw additional stereoisomers of methylcyclohexane. (**Hint:** Use molecular models in
conjunction with Figure 4-8.)

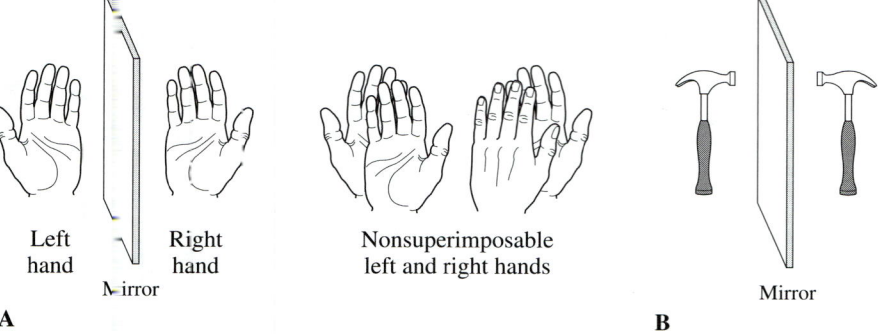

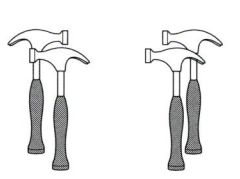

FIGURE 5-2

(A) Left and right hands as models for enantiomeric relations. Like these mirror images, chiral molecules cannot be superimposed on their corresponding enantiomers. (B) Image and mirror image of an achiral hammer are superimposable.

This chapter introduces another type of stereoisomerism, **mirror-image stereoiso-merism.** Molecules in this class are said to possess "handedness," referring to the fact that your left hand is not superimposable on your right hand, yet one hand can be viewed as the mirror image of the other (Figure 5-2A). The property of handedness in molecules is very important in nature, because most biologically relevant com-pounds are either "left-" or "right-handed." As such, they will react differently with each other, much as shaking your friend's right hand is very different from shaking his or her left hand. A summary of isomeric relations is depicted in Figure 5-1.

5-1 Chiral Molecules

How can a molecule exist as two nonsuperimposable mirror images? Consider the radical bromination of butane. This reaction proceeds mainly at one of the secondary carbons to furnish 2-bromobutane. A molecular model of the starting material *seems* to show that either of the two hydrogens on that carbon may be replaced to give only one form of 2-bromobutane (Figure 5-3). Is this really true, however?

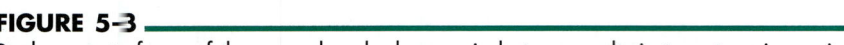

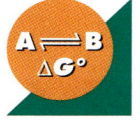

FIGURE 5-3

Replacement of one of the secondary hydrogens in butane results in two stereoisomeric forms of 2-bromobutane.

Chiral molecules cannot be superimposed on their mirror images

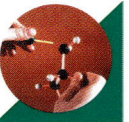

Look more closely at the 2-bromobutanes obtained by replacing either of the methylene hydrogens with bromine. In fact, the two structures are nonsuperimposable and there-fore *not identical* (see page 168). The two molecules are related as object and mirror im-age, and to convert one into the other would require the breaking of bonds. A molecule

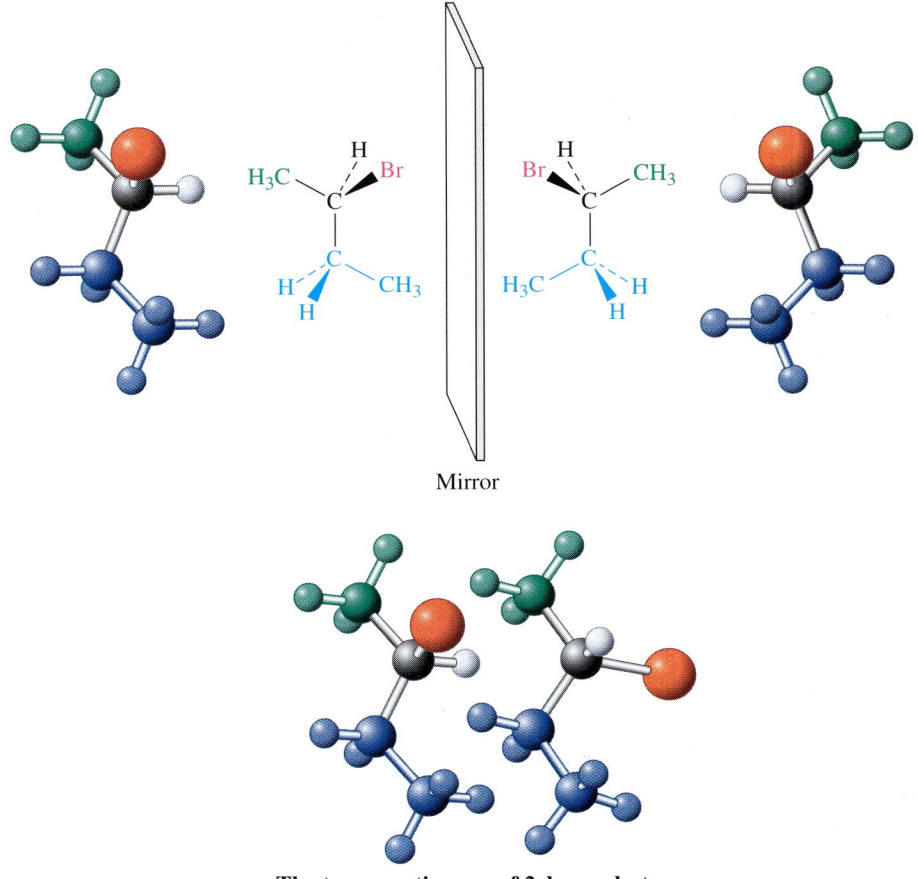

The two enantiomers of 2-bromobutane
are nonsuperimposable

that is not superimposable on its mirror image is said to be **chiral.** Each isomer of the image–mirror image pair is called an **enantiomer** (*enantios,* Greek, opposite). In our example of the bromination of butane, a 1:1 mixture of enantiomers is formed.

In contrast with chiral molecules, such as 2-bromobutane, compounds having structures that *are* superimposable on their mirror images are **achiral.** Examples of chiral

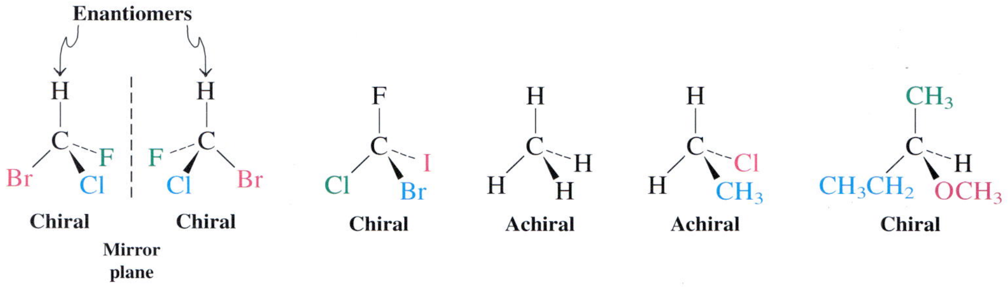

Enantiomers

Chiral Chiral Chiral Achiral Achiral Chiral

Mirror
plane

and achiral molecules are shown above. The first two chiral structures depicted are enantiomers of each other.

All the chiral examples contain an atom that is connected to four *different* substituent groups. Such a nucleus is called an **asymmetric atom** (e.g., asymmetric carbon) or a **stereocenter.** Centers of this type are sometimes denoted by an asterisk. *Molecules with one stereocenter are always chiral.* (We shall see in Section 5-5 that structures incorporating more than one such center need *not* be chiral.)

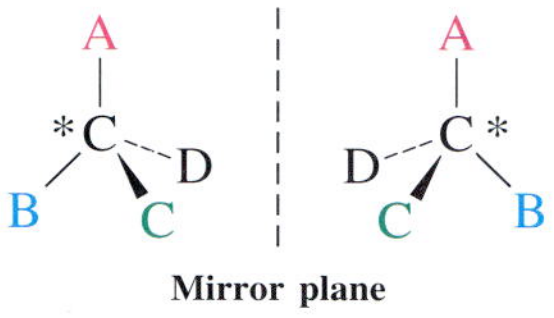

Mirror plane

(C* = a stereocenter based on
asymmetric carbon)

EXERCISE 5-3

Among the natural products shown in Section 4-7, which are chiral and which are achiral? Give the number of stereocenters in each case.

The symmetry in molecules helps to distinguish chiral structures from achiral ones

The word *chiral* is derived from the Greek *cheir,* meaning "hand" or "handedness." Human hands have the mirror-image relation that is typical of enantiomers (see Figure 5-2A). Among the many other objects that are chiral are shoes, ears, screws, and spiral staircases. On the other hand, there are many achiral objects, such as balls, ordinary water glasses, hammers (Figure 5-2B), and nails.

Many chiral objects, such as spiral staircases, do not have stereocenters. This statement is true for many chiral molecules. *Remember that the only criterion for chirality is the nonsuperimposable nature of object and mirror image.* In this chapter, we shall confine our discussion to molecules that are chiral as a result of the presence of stereocenters. But how do we determine whether a molecule is chiral or not? As you have undoubtedly already noticed, it is not always easy to tell. A foolproof way is to construct molecular models of the molecule and its mirror image and look for superimposability. However, this procedure is very time consuming. A simpler method is to look for symmetry in the molecule under investigation.

For most organic molecules, we have to consider only one test for chirality: the presence of a plane of symmetry. A **plane of symmetry (mirror plane)** is one that

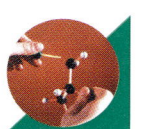

CHEMICAL HIGHLIGHT 5-1   **Chiral Substances in Nature**

Many organic compounds exist in nature as only one enantiomer, some as both. For example, natural *alanine* (systematic name: 2-aminopropanoic acid) is an abundant amino acid that is found in only one form. *Lactic acid* (2-hydroxypropanoic acid), however, is present in blood and muscle fluid as one enantiomer but in sour milk and some fruits and plants as a mixture of the two.

may be thought of as bearing four different groups, if we consider the ring itself to be two separate and different substituents. They are different because, starting from the stereocenter, the clockwise sequence of atoms differs from the counterclockwise sequence. Carvone is found in nature in both enantiomeric forms. The characteristic odor of caraway and dill seed is due to the enantiomer shown, whereas the flavor of spearmint is due to the other enantiomer.

H	H
$\overset{\displaystyle \mid}{\underset{\displaystyle CH_3}{H_2N-\overset{*}{C}-COOH}}$	$\overset{\displaystyle \mid}{\underset{\displaystyle CH_3}{HO-\overset{*}{C}-COOH}}$
2-Aminopropanoic acid (**Alanine**)	**2-Hydroxypropanoic acid** (**Lactic acid**)

Another case is *carvone* [2-methyl-5-(1-methylethenyl)-2-cyclohexenone], which contains a stereocenter in a six-membered ring. This carbon atom

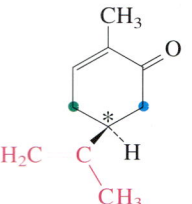

2-Methyl-5-(1-methylethenyl)-2-cyclohexenone (**Carvone**)

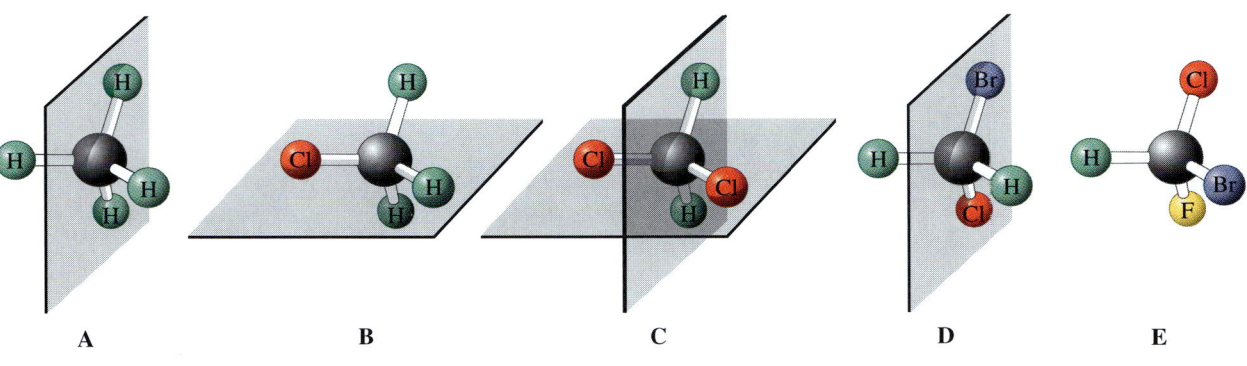

A   B   C   D   E

FIGURE 5-4

Examples of planes of symmetry: (A) methane has four planes of symmetry, only one of which is shown; (B) chloromethane has three such planes, only one of which is shown; (C) dichloromethane has only two; (D) bromochloromethane only one; and (E) bromochlorofluoromethane has none. Chiral molecules cannot have a plane of symmetry.

bisects the molecule so that the part of the structure lying on one side of the plane mirrors the part on the other side. For example, methane has six planes of symmetry, chloromethane has three, dichloromethane two, bromochloromethane one, and bromochlorofluoromethane none (Figure 5-4).

How do we use this idea to distinguish a chiral molecule from an achiral one? *Chiral molecules cannot have a plane of symmetry.* For example, the first four methanes in Figure 5-4 are clearly achiral because of the presence of a mirror plane. You will be able to classify most molecules in this book as chiral or achiral simply by identifying the presence or absence of a plane of symmetry.

Implicit in our definition of achirality and the associated mirror plane is a practical understanding of the "structure" of a molecule. Typically, for the practicing organic chemist, this understanding translates into what is distinctly observable or, better, isolable at room temperature as a stable entity. As we learned in Chapters 2 and 4, such an entity may exist as a mixture of rapidly equilibrating rotamers or conformational isomers. These processes may render a compound achiral if they include rapid equilibration between enantiomeric forms (e.g., through the intermediacy of an achiral isomer or the intervention of an achiral transition state). For example, although we recognize that *gauche* butane is chiral, butane is defined as achiral because rotation about the C2–C3 bond leads to enantiomerization. Similarly, *cis*-1,2-dimethylcyclohexane (Section 4-4) is defined as achiral because of rapid chair–chair interconversion between the two enantiomers through an achiral form (see also Section 5-6). Such molecules are said to have "average symmetry." Most chiral molecules we will encounter in this book owe their asymmetry to a rigid stereocenter, such as a carbon atom bearing four different substituents.

EXERCISE 5-4

Draw pictures of the following common achiral objects, indicating the plane of symmetry in each: a ball, an ordinary water glass, a hammer, a chair, a suitcase, a toothbrush.

EXERCISE 5-5

Write the structures of all dimethylcyclobutanes. Specify those that are chiral. Show the mirror planes in those that are not.

To summarize, a chiral molecule exists in either of two stereoisomeric forms called enantiomers. These enantiomers are related as object and nonsuperimposable mirror image. Most chiral organic molecules contain stereocenters, although chiral structures that lack such centers do exist. A molecule that contains a plane of symmetry is achiral.

5-2 Optical Activity

Considering their close similarity, we may wonder how it is possible to distinguish one enantiomer from another. This task is indeed a very difficult one, because most *physical* properties of enantiomers are identical. A notable exception is the interaction with a special type of light.

Our first example of a chiral molecule was the two enantiomers of 2-bromobutane. If we were to isolate each enantiomer in pure form, we would find that we cannot distinguish between them on the basis of their physical properties, such as boiling points, melting points, and densities. This result should not surprise us: Their bonds are identical and so are their energy contents. However, when plane-polarized light (which will be defined shortly) is passed through a sample of one of the enantiomers, the plane of polarization of the incoming light is *rotated* in one direction (either clockwise or counterclockwise). When the same experiment is repeated with the other enantiomer, the plane of the polarized light is rotated by exactly the same amount *but in the opposite direction.*

An enantiomer that rotates the plane of light in a clockwise sense as the viewer faces the light source is **dextrorotatory** (*dexter,* Latin, right), and the compound is (arbitrarily) referred to the (+) enantiomer. Consequently, the other enantiomer, which will effect counterclockwise rotation, is **levorotatory** (*laevus,* Latin, left) and called the (−) enantiomer. This special interaction with light is called **optical activity,** and enantiomers are frequently called **optical isomers.**

Optical rotation is measured with a polarimeter

What is plane-polarized light, and how is its rotation measured? Ordinary light can be thought of as bundles of electromagnetic-field waves that oscillate simultaneously in all planes perpendicular to the direction of the light beam. When such light is passed through a material called a polarizer, all but one of these light waves are "filtered" away, and the resulting beam oscillates in only one plane: **plane-polarized light** (Figure 5-5).

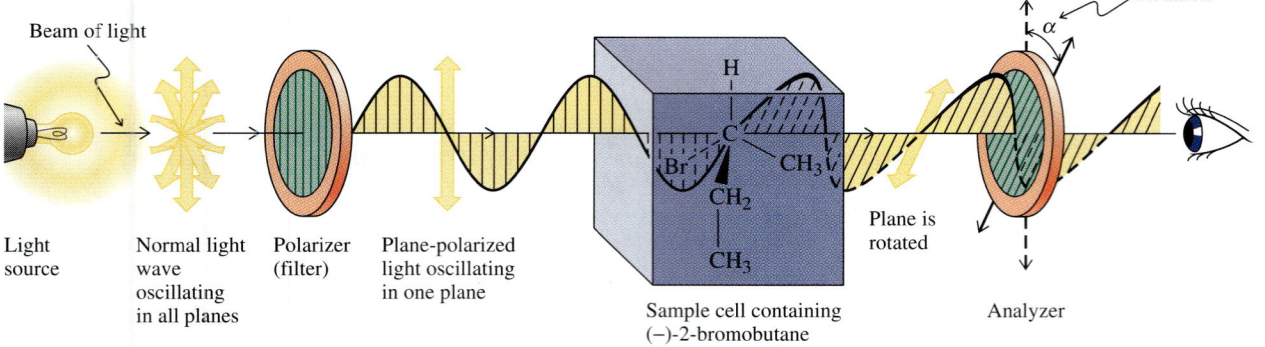

FIGURE 5-5

Measuring the optical rotation of the (−) enantiomer of 2-bromobutane with a polarimeter.

When light travels through a molecule, the electrons around the nuclei and in the various bonds interact with the electric field of the light beam. If a beam of plane-polarized light is passed through a chiral substance, the electric field interacts differently with, say, the "left" and "right" halves of the molecule. This interaction results in a rotation of the plane of polarization, called **optical rotation;** the sample giving rise to it is referred to as **optically active.**

Optical rotations are measured by using a **polarimeter** (Figure 5-5). In this instrument, light is first plane polarized and subsequently traverses a cell containing the sample. Rotation of the plane is detected by another polarizer—the analyzer—to maximize transmittance of the light beam to the eye of the observer. The measured rotation (in degrees) is the **observed optical rotation,** α, of the sample. Its value depends on the concentration and structure of the optically active molecule, the length of the sample cell, the wavelength of the light, the solvent, and the temperature. To avoid ambiguities, chemists have agreed on a standard value of α, the **specific rotation,** $[\alpha]$, for each compound. This quantity (which is solvent dependent) is defined as

> **Specific Rotation***
>
> $$[\alpha]_\lambda^{t^\circ} = \frac{\alpha}{l \cdot c}$$

where $[\alpha]$ = specific rotation
   t = temperature in degrees Celsius
   λ = wavelength of incident light; for the sodium D lamp, indicated simply by "D"; it is 589 mm, the yellow emission line of hot sodium vapor
   α = observed optical rotation in degrees
   l = length of sample container in decimeters; its value is frequently 1 (i.e., 10 cm)
   c = concentration (grams per milliliter of solution)

EXERCISE 5-6

A solution of 0.1 g mL^{-1} of common table sugar (the naturally occurring form of sucrose) in water in a 10-cm cell exhibits a clockwise optical rotation of 6.65°. Calculate $[\alpha]$. Does this information tell you $[\alpha]$ for the enantiomer of natural sucrose?

The specific rotation of an optically active molecule is a physical constant characteristic of that molecule, just like its melting point, boiling point, and density. Four specific rotations are recorded in Table 5-1.

Optical rotation indicates enantiomeric composition

As mentioned, enantiomers rotate plane-polarized light by equal amounts but in opposite directions. Thus, in 2-bromobutane the ($-$) enantiomer rotates this plane coun-

---

*The dimensions of $[\alpha]$ are deg cm^2 g^{-1}, the units (for l = 1) 10^{-1} deg cm^2 g^{-1}. Because of their awkward appearance, it is common practice to give $[\alpha]$ without units, in contrast with the observed rotation α (degrees).

TABLE 5-1	Specific Rotations of Various Chiral Compounds $[\alpha]_D^{25°C}$	

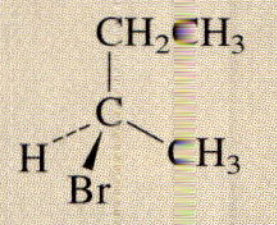

CH₂CH₃

(structure) —23.1

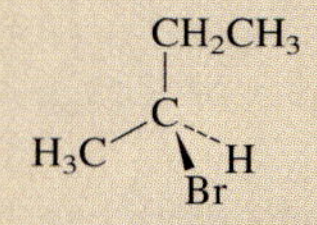

(−)-2-Bromobutane

CH₂CH₃

(structure) +23.1

(+)-2-Bromobutane

(structure) +8.5

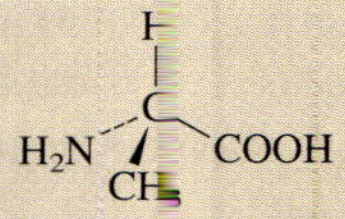

(+)-2-Aminopropanoic acid
[(+)-Alanine]

(structure) −3.8

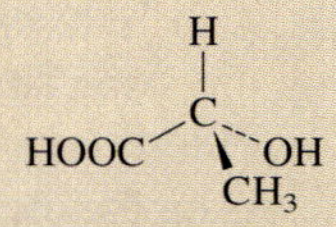

(−)-2-Hydroxypropanoic acid
[(−)-Lactic acid]

Note: Pure liquid for the haloalkane; in aqueous solution for the acids.

terclockwise by 23.1°, its mirror image (+)-2-bromobutane clockwise by 23.1°. It follows that a 1:1 mixture of (+) and (−) enantiomers shows no rotation and is therefore optically inactive. Such a mixture is called a **racemic mixture.** If one enantiomer equilibrates with its mirror image it is said to undergo **racemization.** For example, optically active acids such as (+)-alanine (Table 5-1) have been found to undergo very slow racemization in fossil deposits by tertiary C–H bond-breaking, a process resulting in reduced optical activity.

Optical activity can be measured in a mixture of enantiomers, but only if the enantiomers are present in unequal amounts. Using the value of the measured rotation, we can calculate the composition of such a mixture. For example, if a solution of (+)-alanine from a fossil exhibits an $[\alpha]$ of only +4.25 (i.e., one-half the value for the pure enantiomer), we can deduce that 50% of the sample is pure (+) isomer and the other 50% is racemic. It is said to have 50% **enantiomer excess.** Because the racemic portion consists of equal amounts of (+) and (−), the actual composition of the sample is 75% (+) and 25% (−), as shown here.

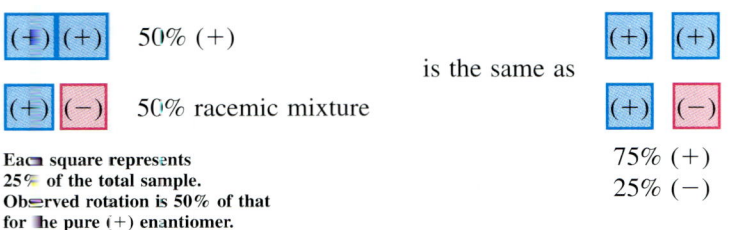

The 25% (−) enantiomer cancels the rotation of a corresponding amount of (+). This mixture is called 50% (i.e., 75% − 25%) *optically pure:* The observed optical rotation is one-half that of the pure dextrorotatory enantiomer.

Optical Purity

$$\% \text{ optical purity} = \left(\frac{[\alpha]_{observed}}{[\alpha]} \cdot 100 \right) = \text{enantiomer excess}$$

EXERCISE 5-7

What is the optical rotation of a sample of (+)-2-bromobutane that is 75% optically pure? What percentages of (+) and (−) enantiomers are present in this sample? Answer the same questions for samples of 50% and 25% optical purity.

In summary, two enantiomers can be distinguished by their optical activity; that is, their interaction with plane-polarized light as measured in a polarimeter. One enantiomer always rotates such light clockwise (dextrorotatory), the other counterclockwise (levorotatory) by the same amount. The specific rotation, $[\alpha]$, is a physical constant possible only for chiral molecules. The interconversion of enantiomers leads to racemization and the disappearance of optical activity.

5-3 Absolute Configuration: *R–S* Sequence Rules

How do we establish the structure of one pure enantiomer of a chiral compound? And, once we know the answer, is there a way to name it unambiguously and distinguish it from its mirror image?

X-ray diffraction can establish the absolute configuration

Virtually all the physical characteristics of one enantiomer are identical with those of its mirror image, except for the sign of optical rotation. Is there a correlation between the sign of optical rotation and the actual spatial arrangement of the substituent groups, the **absolute configuration?** Is it possible to determine the structure of an enantiomer by measuring its $[\alpha]$ value? The answer to both questions is, unfortunately, no. *There is no straightforward correlation between the sign of rotation and the structure of the particular enantiomer.* For example, conversion of lactic acid (Table 5-1) into its sodium salt changes the sign (and degree) of rotation, even though the absolute configuration at the stereocenter is unchanged.

If the sign of rotation does not tell us anything about structure, how do we know which enantiomer of a chiral molecule is which? Or, to put it differently, how do we know that the levorotatory enantiomer of 2-bromobutane has the structure indicated in Table 5-1 (and therefore the dextrorotatory enantiomer the mirror-image configuration)? The answer is that such information can be obtained only through single-crystal X-ray diffraction analysis (Section 1-9). This does not mean that every chiral compound must be submitted to X-ray analysis to ascertain its structure. Absolute configuration can also be established by chemical correlation with a molecule whose own structure has been proved by this method. For example, knowing the stereocenter in (−)-lactic acid by X-ray also provides the absolute configuration of the (+)-sodium salt (i.e., the same).

$[\alpha]_D^{25°C} = -3.8$

(−)-Lactic acid

NaOH, H$_2$O

$[\alpha]_D^{25°C} = +13.5$

(+)-Sodium lactate

Stereocenters are labeled *R* or *S*

To name enantiomers unambiguously, we need a system that allows us to indicate the handedness in the molecule, a sort of "left hand" versus "right hand" nomenclature. Such a system was developed by three chemists, R. S. Cahn, C. Ingold, and V. Prelog.*

---

*Dr. Robert S. Cahn (1899–1981), Fellow of the Royal Institute of Chemistry, London; Professor Christopher Ingold (1893–1970), University College, London; Professor Vladimir Prelog (b. 1906), Swiss Federal Institute of Technology (ETH), Zürich, Nobel Prize 1975 (chemistry).

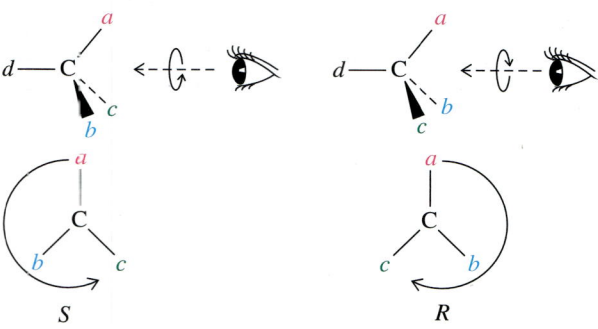

FIGURE 5-6
Assignment of R or S configuration at a tetrahedral stereocenter. The group of lowest priority is placed as far away from the observer as possible. In many of the structural drawings in this chapter, the color scheme shown here is used to indicate the priority of substituents—in decreasing order, red > blue > green > black.

Let us see how the handedness around an asymmetric carbon atom is labeled. The first step is to rank all four substituents in the order of decreasing priority, the rules of which will be described shortly. Substituent *a* has the highest priority, *b* the second highest, *c* the third, and *d* the lowest. Next, we position the molecule (mentally, on paper, or by using a molecular model set) so that the lowest priority substituent is placed as far away from us as possible (Figure 5-6). This process results in two (and

---

CHEMICAL HIGHLIGHT 5-2

Absolute Configuration: A Historical Note

$[\alpha]_D^{25°C} = +8.7$

D-(+)-2,3-Dihydroxypropanal
[D-(+)-Glyceraldehyde]

$[\alpha]_D^{25°C} = -8.7$

L-(−)-2,3-Dihydroxypropanal
[L-(−)-Glyceraldehyde]

Before the X-ray diffraction technique was developed, the absolute configurations of chiral molecules were unknown. Amusingly, the first assignment of a three-dimensional structure to a chiral molecule was a

guess made more than a century ago. The naturally occurring dextrorotatory enantiomer of 2,3-dihydroxypropanal (glyceraldehyde) was arbitrarily assigned the structure shown above and labeled "D-glyceraldehyde." The label "D" was not used to refer to the sign of rotation of plane-polarized light but to the relative arrangement of the substituent groups.

The other isomer was called L-glyceraldehyde. All chiral compounds that could be converted into D-(+)-glyceraldehyde by reactions that did not affect the configuration at the stereocenter were assigned the D configuration, and their mirror images the L. In 1951, the absolute configurations of these compounds became known and the original guess was found to be correct. D,L nomenclature is still used for sugars (Chapter 24) and amino acids (Chapter 26).

D-Configurations

L-Configurations

only two) possible arrangements of the remaining substituents. If the progression from *a* to *b* to *c* is counterclockwise, the configuration at the stereocenter is named *S* (*sinister,* Latin, left). Conversely, if the progression is clockwise, the center is *R* (*rectus,* Latin, right). The symbol *R* or *S* is added as a prefix in parentheses to the name of the chiral compound, as in (*R*)-2-bromobutane and (*S*)-2,3-dihydroxypropanal. A racemic mixture is designated *R,S*, as in (*R,S*)-bromochlorofluoromethane. The sign of the rotation of plane-polarized light may be added if it is known, as in (*S*)-(+)-2-bromobutane and (*R*)-(+)-2,3-dihydroxypropanal. It is important to remember, however, that the symbols *R* and *S* are *not* necessarily correlated with either sign of α.

Sequence rules assign priorities to substituents

Before applying the *R,S* nomenclature to a stereocenter, we must first assign priorities by using sequence rules.

RULE 1. We look first at the atoms attached directly to the stereocenter. A substituent atom of higher atomic number takes precedence over one of lower atomic number. Consequently, the substituent of lowest priority is hydrogen. In regard to isotopes, the atom of higher atomic mass receives higher priority.

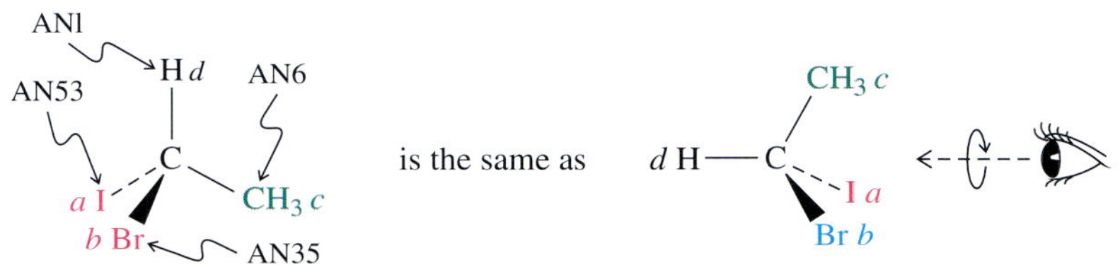

(R)-1-Bromo-1-iodoethane

RULE 2. What if two substituents have the same rank when we consider the atoms directly attached to the stereocenter? In such a case, we proceed along the two respective substituent chains until we reach a point of difference.

For example, an ethyl substituent takes priority over methyl. Why? At the point of attachment to the stereocenter, each substituent has a carbon nucleus, equal in priority. Farther from that center, however, methyl has only hydrogen atoms, but ethyl has a carbon atom (higher in priority).

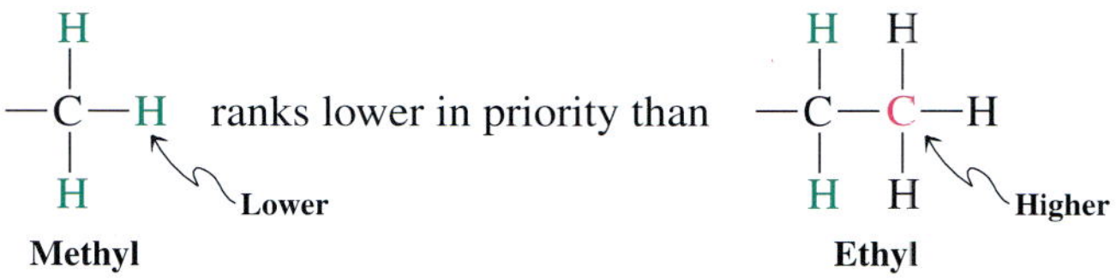

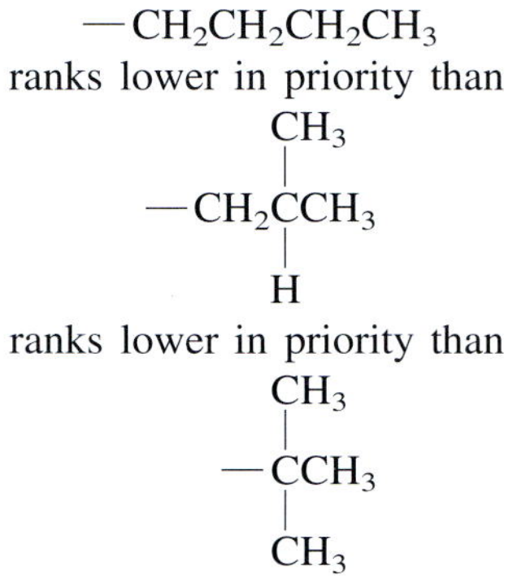

However, 1-methylethyl takes precedence over ethyl because, at the first carbon, ethyl bears only one other carbon substituent, but 1-methylethyl bears two. Similarly, 2-methylpropyl takes priority over butyl but ranks lower than 1,1-dimethylethyl.

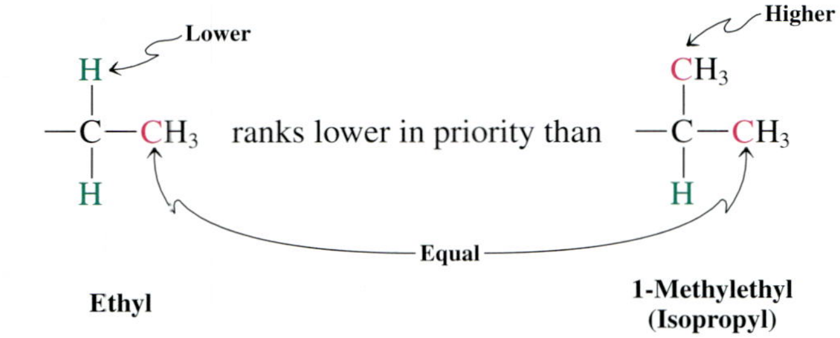

Ethyl   **1-Methylethyl (Isopropyl)**

We must remember that the decision on priority is made at the *first* point of difference along otherwise similar substituent chains. When that point has been reached, the constitution of the remainder of the chain is irrelevant.

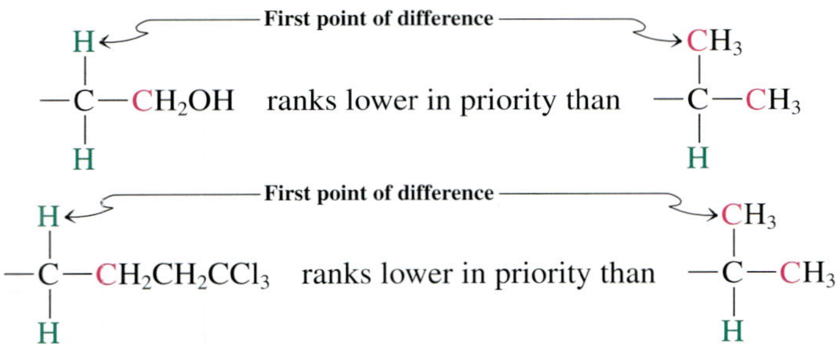

When we reach a point along a substituent chain at which it branches, we choose the branch that is higher in priority. When two substituents have similar branches, we rank the elements in those branches until we reach a point of difference.

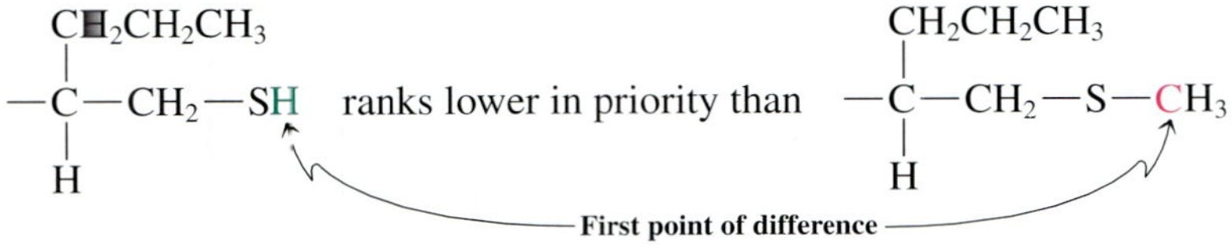

Some examples are shown below.

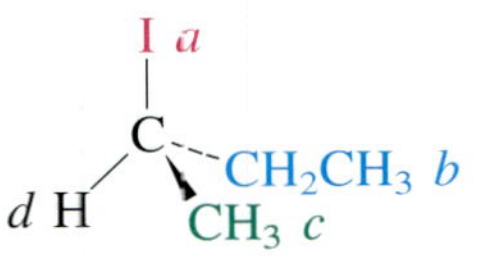

(R)-2-Iodobutane 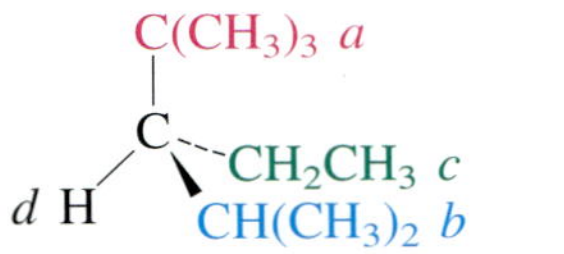   **(S)-3-Ethyl-2,2,4-trimethylpentane**

RULE 3. Double and triple bonds are treated as if they were single, and the atoms in them are duplicated or triplicated at each end by the respective atoms at the other end of the multiple bond.

The red atoms shown in the groups on the right side of the display are not really there. They are added only for the purpose of assigning a relative priority to each of the corresponding groups to their left.

$$H_2C=CH_2 \text{ is treated as } -\overset{|}{\underset{|}{C}}-\overset{|}{\underset{|}{C}}-R$$

(illustration: $\underset{R}{\overset{H}{\diagdown}}C=C\underset{R}{\overset{H}{\diagup}}$ is treated as $-\overset{H}{\underset{C}{C}}-\overset{H}{\underset{C}{C}}-R$)

$-C\equiv C-R$ is treated as $-\overset{C}{\underset{C}{C}}-\overset{C}{\underset{C}{C}}-R$

$-\overset{O}{\overset{\|}{C}}-H$ is treated as $-\overset{O}{\underset{H}{C}}-\overset{C}{O}$

$-\overset{O}{\overset{\|}{C}}-OH$ is treated as $-\overset{O}{\underset{OH}{C}}-\overset{C}{O}$

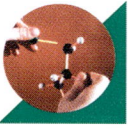

Examples are shown in the margin.

EXERCISE 5-8

Draw the structures of the following substituents and within each group rank them in order of decreasing priority. **(a)** Methyl, bromomethyl, trichloromethyl, ethyl; **(b)** 2-methylpropyl (isobutyl), 1-methylethyl (isopropyl), cyclohexyl; **(c)** butyl, 1-methylpropyl (*sec*-butyl), 2-methylpropyl (isobutyl), 1,1-dimethylethyl (*tert*-butyl); **(d)** ethyl, 1-chloroethyl, 1-bromoethyl, 2-bromoethyl.

EXERCISE 5-9

Assign the absolute configuration of the molecules depicted in Table 5-1.

EXERCISE 5-10

Draw one enantiomer of your choice (specify which, *R* or *S*) of 2-chlorobutane, 2-chloro-2-fluorobutane, and (HC≡C)(CH₂=CH)C(Br)(CH₃).

To correctly assign the stereostructure of stereoisomers, we must develop a fair amount of three-dimensional "vision," or "stereoperception." In the structures that have been used to illustrate the priority rules, the lowest priority substituent is located at the left of the carbon center and in the plane of the page and the remainder of the substituents at the right, the upper-right group also being positioned in this plane. However, this is not the only way of drawing dashed-wedged line structures; others are equally correct. Consider some of the structural drawings of (*S*)-2-bromobutane. These are simply different views of the same molecule.

Six Ways of Depicting (S)-2-Bromobutane

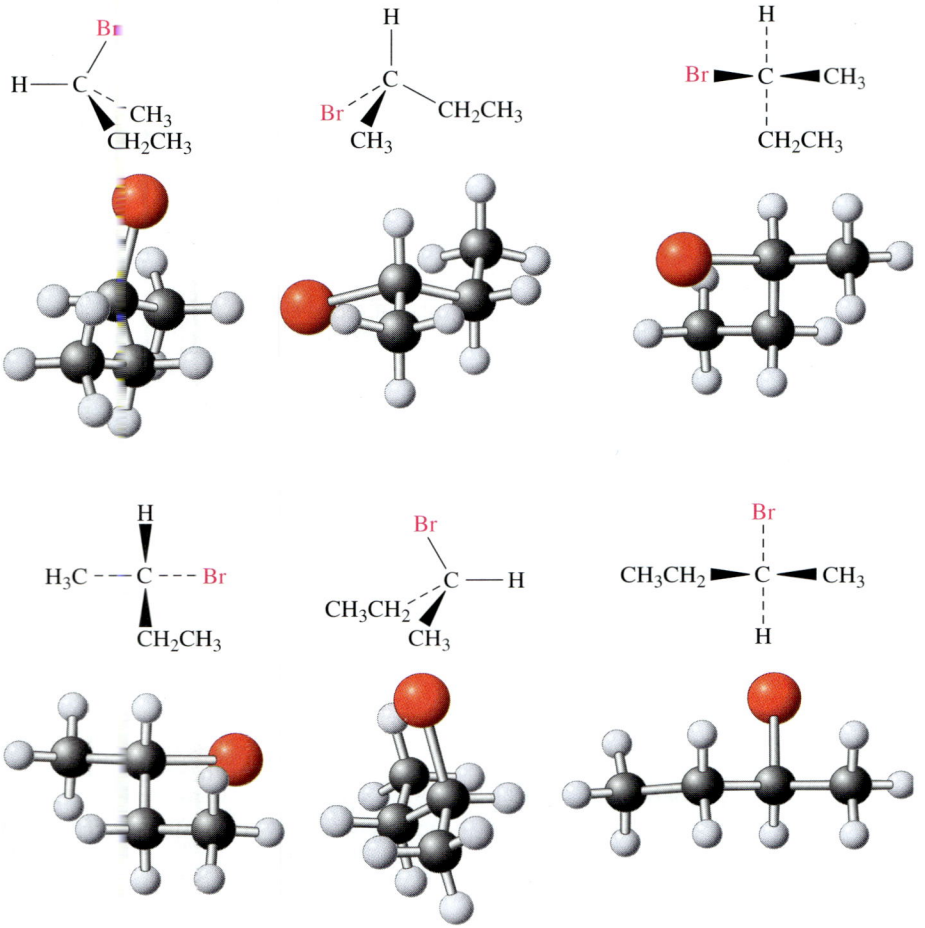

To summarize, the sign of optical rotation cannot be used to establish the absolute configuration of a stereoisomer. Instead, X-ray diffraction (or chemical correlations) must be used. We can express the absolute configuration of the chiral molecule as *R* or *S* by applying the sequence rules, which allow us to rank all substituents in order of decreasing priority. Turning the structures so as to place the lowest priority group at the back causes the remaining substituents to be arranged in clockwise (*R*) or counterclockwise (*S*) fashion.

5-4 Fischer Projections

A **Fischer* projection** is a simplified way of depicting tetrahedral carbon atoms and their substituents in two dimensions. With this method, the molecule is drawn in the form of a cross, the central carbon being at the point of intersection. The horizontal

---

*Professor Emil Fischer (1852–1919), University of Berlin, Nobel Prize 1902 (chemistry).

lines signify bonds directed *toward* the viewer; the vertical lines are pointing *away*. Dashed-wedged line structures have to be arranged in this way to facilitate their conversion into Fischer projections.

Conversion of the Dashed-Wedged Line Structures of 2-Bromobutane into Fischer Projections (of the Stereocenter)

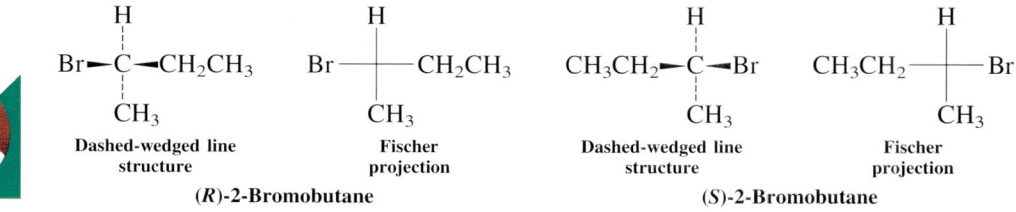

Dashed-wedged line structure   **Fischer projection**
     (R)-2-Bromobutane

Dashed-wedged line structure   **Fischer projection**
     (S)-2-Bromobutane

You will notice that just as there are several ways of depicting a molecule in the dashed-wedged line notation, there are several correct Fischer projections of the same stereocenter.

Two Additional Projections of (R)-2-Bromobutane

A simple mental procedure that will allow you to safely convert any dashed-wedged line structure into a Fischer projection is to picture yourself at the molecular level and grasp any two substituents with your hands while facing the central carbon. If you then imagine descending on the page while holding the molecule, the two remaining substituents positioned behind that center will dock on the surface and then submerge below it, and your left and right hands will position the two horizontal, wedged-line groups in the proper orientation. This procedure places the two remaining dashed-line groups (juxtaposing your head and feet, respectively) vertically.

A Simple Mental Exercise: Conversion of Dashed-Wedge Line Structures into Fischer Projections

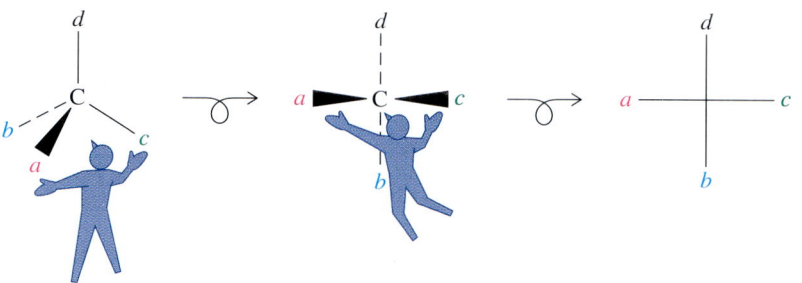

Having achieved this conversion, you can change one Fischer projection into another of the same molecule by using certain manipulations: rotations and substituent switches. However, we shall see next that care has to be taken not to inadvertently convert *R* into *S* configurations.

Rotating a Fischer projection may or may not change the absolute configuration

What happens when we rotate a Fischer projection by 90°? Does the result depict the spatial arrangement of the original molecule? The definition of a Fischer projection—horizontal bonds are pointed above, vertical ones below the plane of the page—tells us that the answer is clearly no, because this rotation has *switched* the relative spatial disposition of the two sets: the result is a picture of the enantiomer. On the other hand, rotation by 180° is fine, because horizontal and vertical lines have not been interchanged; the resulting drawing represents the same enantiomer.

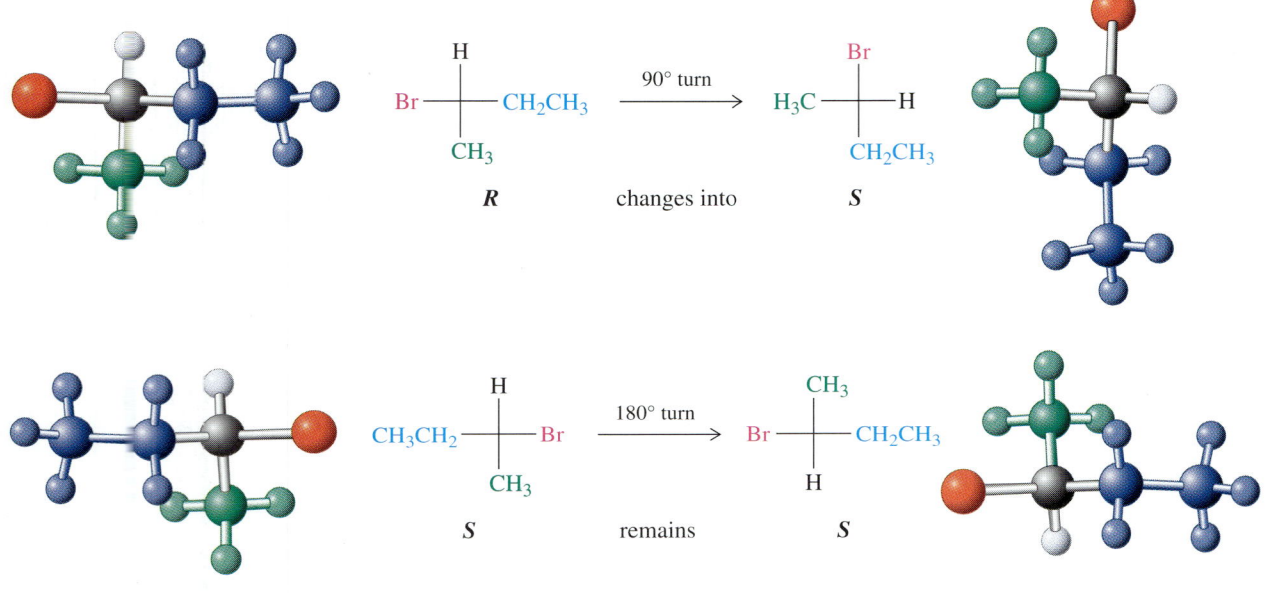

EXERCISE 5-11

Draw Fischer projections for all the molecules in Exercises 5-9 and 5-10.

Exchanging substituents in a Fischer projection also changes the absolute configuration

As is the case for dashed-wedged line structures, there are several Fischer projections of the same enantiomer, a situation that may lead to confusion. How can we quickly ascertain whether two Fischer projections are depicting the same enantiomer or two mirror images? We have to find a sure way to convert one Fischer projection into another in a manner that either leaves the absolute configuration unchanged or converts it into its opposite. It turns out that this task can be achieved by simply making substituent groups trade places. As we can readily verify by using molecular models, any *single* such exchange turns one enantiomer into its mirror image. Two such exchanges (we may select different substituents every time) produce the original absolute configuration. As shown by the dashed-wedged line structures below, this operation merely results in a different view of the same molecule, rotated 120° about the C–Cl bond.

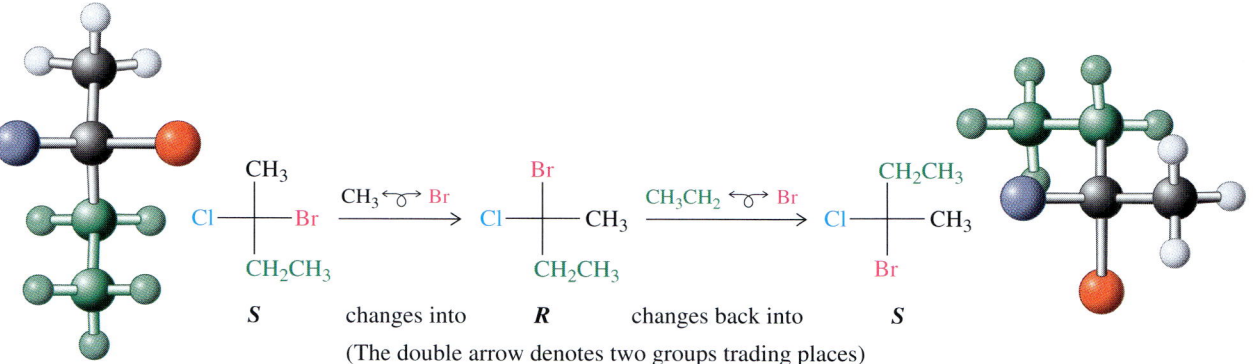

S   changes into   **R**   changes back into   **S**

(The double arrow denotes two groups trading places)

We now have a simple way of establishing whether two different Fischer projections depict the same or opposite configurations. If the conversion of one structure into another takes an even number of exchanges, the structures are identical. If it requires an odd number of such moves, the structures are mirror images of each other.

Consider, for example, the two Fischer projections A and B. Do they represent molecules having the same configuration? The answer is found quickly. We convert A into B by two exchanges; so A equals B.

Draw the dashed-wedged line structures corresponding to Fischer projections A and B, above. Is it possible to transform A into B by means of a rotation about a single bond? If so, identify the bond and the degree of rotation required. Use models if necessary.

Fischer projections tell us the absolute configuration

When we deal with stereochemical problems, an accurate perception of space is very useful. However, Fischer projections allow us to assign absolute configurations without having to visualize the three-dimensional arrangement of the atoms. For this purpose, we first draw the molecule as a (any) Fischer projection. Next, we rank all the substituents in accord with the sequence rules. Finally, we exchange two groups so that the lowest priority substituent is at the top, and then we exchange any other pair (to make sure that the absolute configuration stays unchanged from the original). On completion of these manipulations, we find that the three groups of priority, *a*, *b*, and *c*, are arranged in either clockwise or counterclockwise fashion, in turn corresponding to either the *R* or the *S* configuration.

Although this procedure provides you with a fail-safe mechanism with which to assign the absolute configuration of a stereocenter, you must not forget its three-dimensional origin. It is important that you continue to work with molecular models or other three-dimensional representations of molecules to improve your ability to think of them in space.

EXERCISE 5-13

What is the absolute configuration of the following molecules?

$$H{-}{\overset{\displaystyle Br}{\underset{\displaystyle CH_3}{|}}}{-}D \qquad F{-}{\overset{\displaystyle Cl}{\underset{\displaystyle I}{|}}}{-}Br \qquad H_2N{-}{\overset{\displaystyle CH_3}{\underset{\displaystyle H}{|}}}{-}\overset{}{\underset{O}{C}}OH$$

EXERCISE 5-14

Convert the Fischer projections in Exercise 5-13 into dashed-wedged line formulas and determine their absolute configurations by using the procedure described in Section 5-3. When the lowest priority group is at the top in a Fischer projection, is it in front of the plane of the page or behind it? Does this explain why the procedure outlined above for determination of configuration from Fischer projections succeeds?

In summary, a Fischer projection is a convenient way of drawing chiral molecules. We can rotate such projections in the plane by 180° but not by 90°. Switching substituents reverses absolute configuration, if done an odd number of times, but leaves it intact when the number of such exchanges is even. By placing the substituent of lowest priority on top, we can readily assign the absolute configuration.

5-5 Molecules Incorporating Several Stereocenters: Diastereomers

Many molecules contain several stereocenters. Because the configuration about each center can be R or S, several possible structures emerge, all of which are isomeric.

Two stereocenters can give four stereoisomers: chlorination of 2-bromobutane at C3

Section 5-1 described how a carbon-based stereocenter can be created by the radical halogenation of butane. Let us now consider the chlorination of racemic 2-bromobutane to give (among other products) 2-bromo-3-chlorobutane. The introduction of a chlorine atom at C3 produces a new stereocenter in the molecule. This center may have either the R or the S configuration, assignable by using the sequence rules that apply to molecules with only one such center.

How many stereoisomers are possible for 2-bromo-3-chlorobutane? There are four, as can be seen by completing a simple exercise in permutation. Each stereocenter can be either R or S, and, hence, the possible combinations are *RR*, *RS*, *SR*, and *SS* (Figure 5-7).

Because all horizontal lines in Fischer projections signify bonds directed toward the viewer, the result is a representation of a molecule in an *eclipsed* conformation. Therefore, the first step in converting a staggered Newman or dashed-wedged line representation of a molecule into a Fischer projection is to rotate the molecule to form an eclipsed rotamer. To make stereochemical assignments, one treats each stereocenter

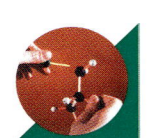

$$CH_3\overset{*}{C}CH_2CH_3$$
$$\underset{Br}{\overset{H}{|}}$$

One stereocenter

$$Cl_2,\ hv\ \Big\downarrow\ {-}HCl$$

$$CH_3\overset{**}{C}{-}\overset{}{C}CH_3$$

$$\underset{Br\ \ H}{\overset{H\ \ Cl}{|\ \ \ |}}$$

Two stereocenters

2-Bromo-3-chlorobutane

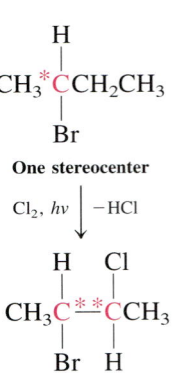

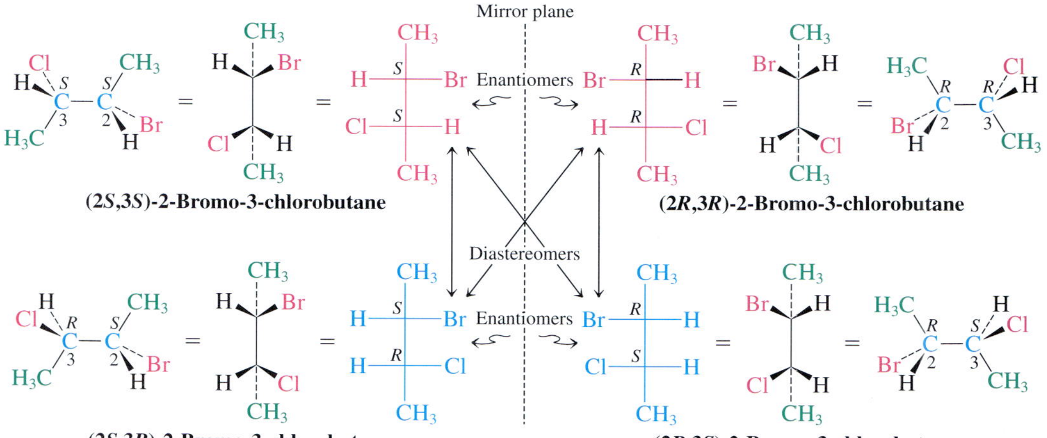

FIGURE 5-7

The four stereoisomers of 2-bromo-3-chlorobutane. Each molecule is the enantiomer of one of the other three (its mirror image) and is at the same time a diastereomer of each of the remaining two. For example, the 2R,3R isomer is the enantiomer of the 2S,3S compound and is diastereomerically related to both the 2S,3R and the 2R,3S structures. Notice that two structures are enantiomers only when they possess the opposite configuration at *every* stereocenter.

separately, and the group containing the other stereocenter is regarded as a simple substituent (Figure 5-8).

By looking closely at the structures of the four stereoisomers (Figure 5-7), we see that there are two related pairs of compounds: an *R,R/S,S* pair and an *R,S/S,R* pair. The members of each individual pair are mirror images of each other and therefore enantiomers. Conversely, each member of one pair is not a mirror image of either member of the other pair; hence, they are not enantiomeric with respect to each other. Stereoisomers that are not related as object and mirror image are called **diastereomers** (*dia,* Greek, across).

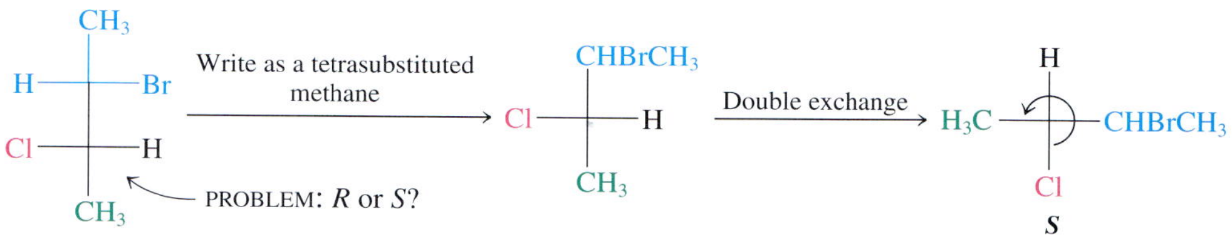

SOLUTION: The center under scrutiny is *S*.

FIGURE 5-8

Assigning the absolute configuration at C3 in 2-bromo-3-chlorobutane. We consider the group containing the stereocenter C2 merely as one of the four substituents. Priorities (also noted in color) are assigned in the usual way ($Cl > CHBrCH_3 > CH_3 > H$), giving rise to the representation shown in the center. Two exchanges place the substituent of lowest priority (hydrogen) at the top of the Fischer projection to facilitate assignment.

EXERCISE 5-15

The two amino acids isoleucine and alloisoleucine are depicted below in staggered conformations. Convert both into Fischer projections. (Keep in mind that Fischer projections are views of molecules *in eclipsed conformations.*) Are these two compounds enantiomers or diastereomers?

Isoleucine   Alloisoleucine

In contrast with enantiomers, diastereomers, because they are *not* mirror images of each other, are distinct molecules with *different physical and chemical properties* (see, e.g., Chemical Highlight 5-3). Their steric interactions and energies differ. They can be separated by fractional distillation or crystallization or by chromatography. They have different melting and boiling points and different densities, just as constitutional isomers do. In addition, they have different specific rotations.

EXERCISE 5-16

What are the stereochemical relations (identical, enantiomers, diastereomers) of the following four molecules? Assign absolute configurations at each stereocenter.

Cis and trans isomers are cyclic diastereomers

It is instructive to compare the stereoisomers of 2-bromo-3-chlorobutane with those of a cyclic analog, 1-bromo-2-chlorocyclobutane (Figure 5-9). In both cases, there are four stereoisomers: *R,R*, *S,S*, *R,S*, and *S,R*. In the cyclic compound, however, the stereoisomeric relation of the first pair to the second is easily recognized: One pair has cis stereochemistry, the other trans. Cis and trans isomers (Section 4-1) in cycloalkanes are in fact diastereomers.

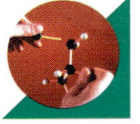
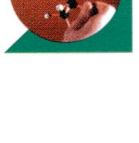

FIGURE 5-9
(A) The diastereomeric relation of *cis*- and *trans*-bromo-2-chlorobutane. (B) Stereochemical assignment of the *R,R* stereoisomer. Recall that the color scheme indicates the priority order of the groups around each stereocenter: red > blue > green > black.

More than two stereocenters means still more stereoisomers

What structural variety do we expect for a compound having three stereocenters? We may again approach this problem by permuting the various possibilities. If we label the three centers consecutively as either *R* or *S*, the following sequence emerges:

$$RRR \quad\quad RRS \quad\quad RSR \quad\quad SRR \quad\quad RSS \quad\quad SRS \quad\quad SSR \quad\quad SSS$$

a total of eight stereoisomers. They can be arranged to reveal a division into four enantiomer pairs of diastereomers.

Image	*RRR*	*RRS*	*RSS*	*SRS*
Mirror image	*SSS*	*SSR*	*SRR*	*RSR*

Generally, *a compound with n stereocenters can have a maximum of 2^n stereoisomers.* Therefore, a compound having three such centers gives rise to a maximum of eight stereoisomers; one having four produces sixteen; one having five, thirty-two; and so forth. The structural possibilities are quite staggering for larger systems.

EXERCISE 5-17

Draw all the stereoisomers of 2-bromo-3-chloro-4-fluoropentane.

In summary, the presence of more than one stereocenter in a molecule gives rise to diastereomers. These are stereoisomers that are not related to each other as object and mirror image. Whereas enantiomers have opposite configurations at every respective stereocenter, two diastereomers do not. A molecule with *n* stereocenters may

---

CHEMICAL HIGHLIGHT 5-3

Stereoisomers of Tartaric Acid

(+)-Tartaric acid

$[\alpha]_D^{20°C} = +12.0$
m.p. 168–170°C
Density (g mL^{-1}) $d = 1.7598$

(−)-Tartaric acid

$[\alpha]_D^{20°C} = -12.0$
m.p. 168–170°C
$d = 1.7598$

meso-Tartaric acid

$[\alpha]_D^{20°C} = 0$
m.p. 146–148°C
$d = 1.666$

Tartaric acid (systematic name: 2,3-dihydroxy-butanedioic acid) is a naturally occurring dicarboxylic acid containing two stereocenters with identical substitution patterns. Therefore it exists as a pair of enantiomers (which have identical physical properties but which rotate plane-polarized light in opposite directions) and an achiral meso compound (with different physical and chemical properties from those of the chiral diastereomers).

The dextrorotatory enantiomer of tartaric acid is widely distributed in nature. It is present in many fruits (fruit acid), and its monopotassium salt is found as a deposit during the fermentation of grape juice. Pure levorotatory tartaric acid is rare, as is the meso isomer.

Tartaric acid is of historical significance, because it was the first chiral molecule whose racemate was separated into the two enantiomers. This happened in

exist in as many as 2^n stereoisomers. In cyclic compounds, cis and trans isomers are diastereomers.

5-6 Meso Compounds

We saw that the molecule 2-bromo-3-chlorobutane contains two distinct stereocenters, each with a *different* halogen substituent. How many stereoisomers are to be expected if both centers are identically substituted?

Two identically substituted stereocenters give rise to only three stereoisomers

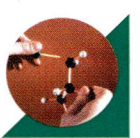

Consider, for example, 2,3-dibromobutane, which can be obtained by the radical bromination of 2-bromobutane. As we did for 2-bromo-3-chlorobutane, we have to consider four structures, resulting from the various permutations in R and S configurations (Figure 5-10; see page 188).

$$CH_3\overset{*}{C}CH_2CH_3 \quad \underset{-HBr}{\overset{Br_2,\ h\nu}{\longrightarrow}} \quad H_3C-\overset{*}{C}-\overset{*}{C}-CH_3$$

One stereocenter

Two stereocenters
2,3-Dibromobutane

1848, long before it was recognized that carbon could be tetrahedral in organic molecules. By 1848, natural tartaric acid had been shown to be dextrorotatory, and the racemate had been isolated from grapes. The words "racemate" and "racemic" are in fact derived from an old common name for this form of tartaric acid, *racemic acid* (*racemus,* Latin, cluster of grapes). The French chemist Louis Pasteur* obtained a sample of the sodium ammonium salt of this acid and noticed that there were two types of crystals: One set was the mirror image of the second. In other words, the crystals were chiral.

By manually separating the two sets, dissolving them in water, and measuring their optical rotation,

Pasteur found one of the crystalline forms to be the pure salt of (+)-tartaric acid and the other to be the levorotatory form. Remarkably, the chirality of the individual molecules in this rare case had given rise to the macroscopic property of chirality in the crystal. He concluded from his observation that the molecules themselves must be chiral. These findings and others led in 1874 to the first proposal, by van't Hoff and Le Bel† independently, that saturated carbon has a tetrahedral—and not, for example, a square planar—bonding arrangement. (Why is the idea of a planar carbon incompatible with that of a stereocenter?)

†Professor Jacobus H. van't Hoff (1852–1911), University of Amsterdam, Nobel Prize 1901 (chemistry); Dr. J. A. Le Bel (1847–1930), Ph.D., Sorbonne, Paris.

*Professor Louis Pasteur (1822–1895), Sorbonne, Paris.

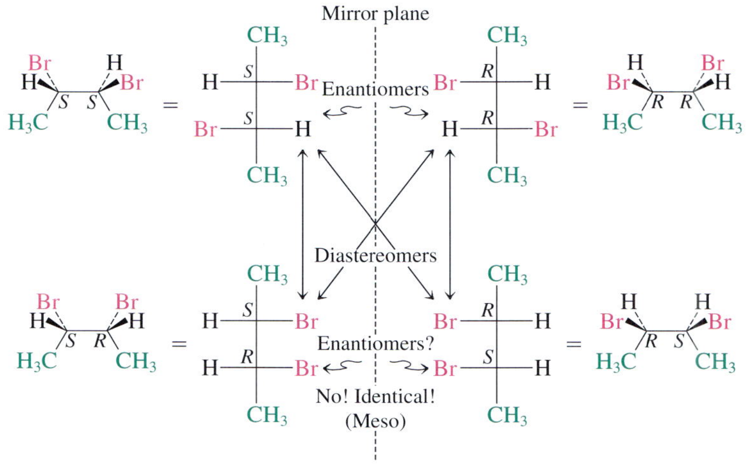

FIGURE 5-10

The stereochemical relations of the stereoisomers of 2,3-dibromobutane. The lower pair consists of identical structures. (Make a model.)

The first pair of stereoisomers, with *R,R* and *S,S* configurations, is clearly recognizable as a pair of enantiomers. However, a close look at the second pair reveals that (*S,R*) and mirror image (*R,S*) are superimposable and therefore identical. Thus, the *S,R* diastereomer of 2,3-dibromobutane is achiral and not optically active, even though it contains two stereocenters. The identity of the two structures can be readily confirmed by using molecular models.

A compound that contains two (or, as we shall see, even more than two) stereocenters but is superimposable with its mirror image is a **meso compound** (*mesos*, Greek, middle). A characteristic feature of a meso compound is the *presence of a mirror plane,* which divides the molecule such that one half is the mirror image of the other half. For example, in 2,3-dibromobutane, the 2*R* center is the reflection of the 3*S* center. This arrangement is best seen in an eclipsed dashed-wedged line structure (Figure 5-11). The presence of a mirror plane in *any* energetically accessible conformation of a molecule (Sections 2-5 and 2-7) is sufficient to make it achiral (Section 5-1). As a consequence, 2,3-dibromobutane exists in the form of three stereoisomers only: a pair of (necessarily chiral) enantiomers and an achiral meso diastereomer.

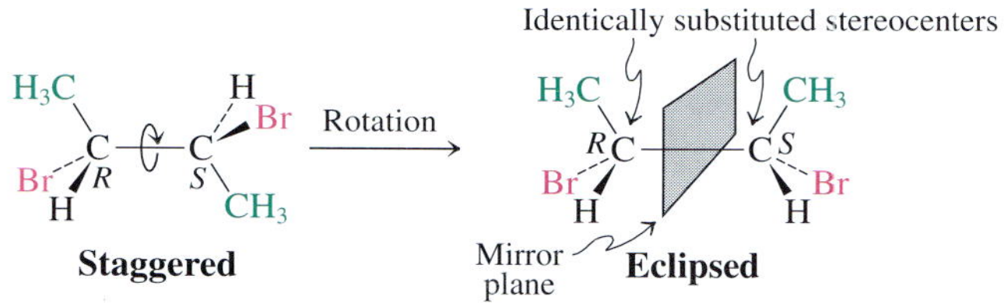

FIGURE 5-11

meso-2,3-Dibromobutane contains a mirror plane when rotated into the eclipsed conformation shown. A molecule with more than one stereocenter is meso and achiral as long as it contains a mirror plane in any readily accessible conformation. Meso compounds possess identically substituted stereocenters.

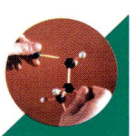

Meso diastereomers can exist in molecules with more than two stereocenters. Examples are 2,3,4-tribromopentane and 2,3,4,5-tetrabromohexane.

Meso Compounds with Multiple Stereocenters

Draw all the stereoisomers of 2,4-dibromo-3-chloropentane.

Cyclic compounds may also be meso

It is again instructive to compare the stereochemical situation in 2,3-dibromobutane with that in an analogous cyclic molecule: 1,2-dibromocyclobutane. We can see that *trans*-1,2-dibromocyclobutane exists as two enantiomers (*R,R* and *S,S*) and may therefore be optically active. The cis isomer, however, has a mirror plane and is meso, achiral, and optically inactive (Figure 5-12).

Notice that we have drawn the ring in a planar shape in order to illustrate the mirror symmetry, although we know from Chapter 4 that cycloalkanes with four or more carbons in the ring are not flat. Is this justifiable? Generally yes, because such compounds, like their acyclic analogs, possess a variety of conformations that are readily accessible at room temperature (Sections 4-2 through 4-4 and Section 5-1). At least one of these conformations will contain the necessary mirror plane to render achiral any cis-disubstituted cycloalkane with identically constituted stereocenters. For simplicity, cyclic compounds may usually be treated *as if they were planar* for the purpose of identifying a mirror plane.

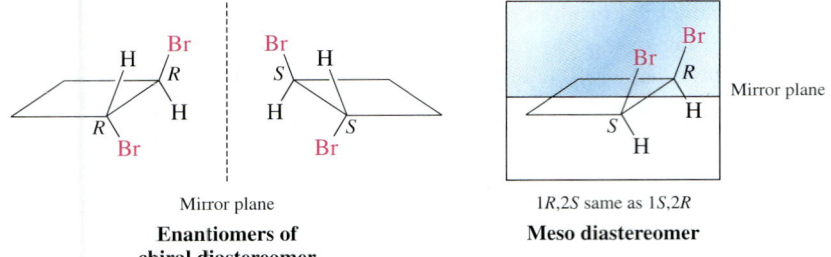

Enantiomers of chiral diastereomer
trans-**1,2-Dibromcyclobutane**

1R,2S same as 1S,2R
Meso diastereomer
cis-**1,2-Dibromocyclobutane**

FIGURE 5-12
The trans isomer of 1,2-dibromocyclobutane is chiral; the cis isomer is a meso compound and optically inactive.

Draw each of the following compounds, representing the ring as planar. Which ones are chiral? Which are meso? Indicate the location of the mirror plane in each meso compound. **(a)** *cis*-1,2-Dichlorocyclopentane; **(b)** its trans isomer; **(c)** *cis*-1,3-dichlorocyclopentane; **(d)** its trans isomer; **(e)** *cis*-1,2-dichlorocyclohexane; **(f)** its trans isomer; **(g)** *cis*-1,3-dichlorocyclohexane; **(h)** its trans isomer.

For each meso compound in Exercise 5-19, draw the conformation that contains the mirror plane. Refer to Sections 4-2 and 4-3 to identify energetically accessible conformations of these ring systems.

In summary, molecules with two or more identically substituted stereocenters may exist as meso stereoisomers. Meso compounds are superimposable on their mirror images and therefore achiral.

5-7 Stereochemistry in Chemical Reactions

We have seen that a chemical reaction can introduce chirality into a molecule. Let us examine more closely the conversion of achiral butane into chiral-2-bromobutane, which gives racemic material. We shall also see that the chiral environment of a stereocenter already present in a molecule exerts control on the stereochemistry of a reaction that introduces a second stereocenter. We begin with another look at the radical bromination of butane.

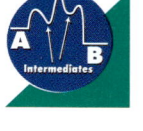

The radical mechanism explains why the bromination of butane results in a racemate

The radical bromination of butane at C2 creates a chiral molecule (Figure 5-3). This happens because one of the methylene hydrogens is replaced by a new group, furnishing a stereocenter—a carbon atom with four different substituents.

In the first step of the mechanism for radical halogenation (Sections 3-6 and 3-7), one of these two hydrogens is abstracted by the attacking bromine atom. It does not matter which of the two is removed: This step does not generate a stereocenter. It furnishes a planar, sp^2-hybridized, and therefore achiral radical. The radical center has two equivalent reaction sites—the two lobes of the p orbital (Figure 5-13)—equally susceptible to attack by bromine in the second step. We can see that the two transition states resulting in the respective enantiomers of 2-bromobutane are mirror images of each other. They are enantiomeric and therefore energetically equivalent. The rates of formation of R and S products are hence equal, and a racemate is formed. In general, *the formation of chiral compounds* (e.g., 2-bromobutane) *from achiral reactants* (e.g., butane and bromine) *yields racemates.* Or, *optically inactive starting materials furnish optically inactive products.**

---

*We shall see later that it *is* possible to generate optically active products from optically inactive starting materials if we use an optically active reagent.

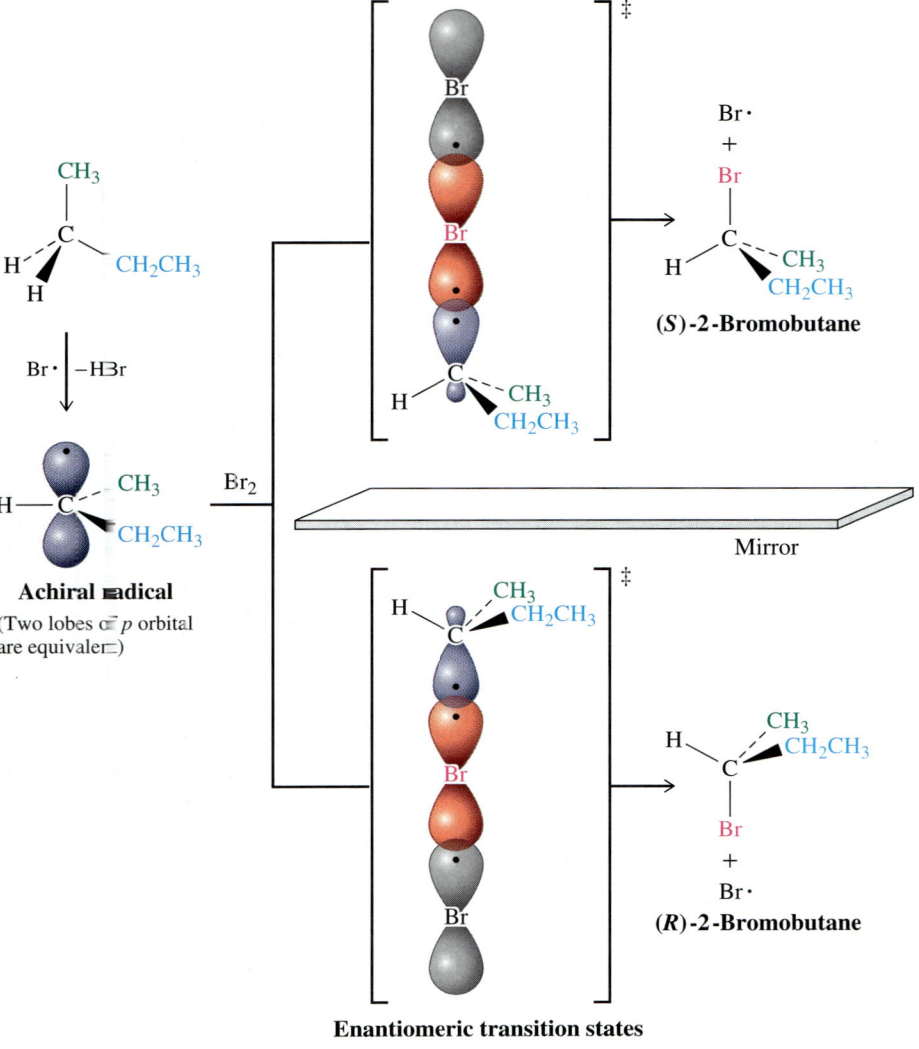

FIGURE 5 13

The creation of racemic 2-bromobutane from butane by radical bromination at C2. Abstraction of either methylene hydrogen by bromine gives an achiral radical. Reaction of Br_2 with this radical is equally likely at either the top or the bottom face, a condition leading to a racemic mixture of products.

The presence of a stereocenter affects the outcome of the reaction: chlorination of (S)-2-bromobutane

Now we understand why the halogenation of an achiral molecule gives a racemic halide. What products can we expect from the halogenation of a chiral, enantiomerically pure molecule?

For example, consider the radical chlorination of the S enantiomer of 2-bromobutane. In this case, the chlorine atom has several options for attack: the two terminal

methyl groups, the single hydrogen at C2, and the two hydrogens on C3. Let us examine each of these reaction paths.

Chlorination of (S)-2-Bromobutane at Either C1 or C4

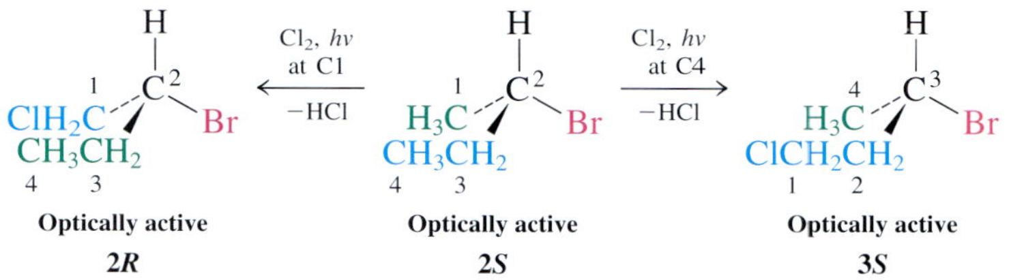

Optically active 2R   Optically active 2S   Optically active 3S

Chlorination of either terminal methyl group is straightforward, proceeding at C1 to give 2-bromo-1-chlorobutane or at C4 to give 3-bromo-1-chlorobutane. In the latter, the original C4 has now become C1, to maintain the lowest possible substituent numbering. *Both of these chlorination products are optically active because the original stereocenter is left intact.* Note, however, that conversion of the C1 methyl into a chloromethyl unit changes the sequence of priorities around C2. Thus, although the stereocenter itself does not participate in the reaction, its designated configuration changes from *S* to *R*.

What about halogenation at C2, the stereocenter? The product from chlorination at C2 of (S)-2-bromobutane is 2-bromo-2-chlorobutane. Even though the substitution pattern at the stereocenter has changed, the molecule remains chiral. However, an attempt to measure the [α] value for the product would reveal the absence of optical activity: *Halogenation at the stereocenter leads to a racemic mixture.* How can this be explained? For the answer, we must look again at the structure of the radical formed in the course of the reaction mechanism.

A racemate forms in this case because hydrogen abstraction from C2 furnishes a planar, sp^2-hybridized, achiral radical.

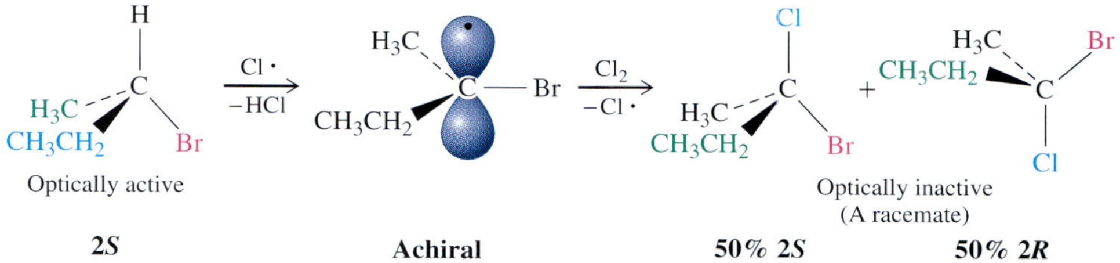

Optically active 2S   Achiral   50% 2S   50% 2R
                                                      Optically inactive
                                                   (A racemate)

Chlorination can occur from either side through enantiomeric transition states of equal energy, as in the bromination of butane (Figure 5-13), producing (S)- and (R)-2-bromo-2-chlorobutane at equal rates and in equal amounts. The reaction is an example of a transformation in which an optically active compound leads to an optically inactive product (a racemate).

What other halogenations of (S)-2-bromobutane would furnish optically inactive products?

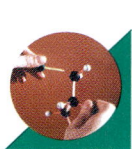

The chlorination of (S)-2-bromobutane at C3 does not affect the existing chiral center. However, *the formation of a second stereocenter gives rise to diastereomers.* Specifically, attachment of chlorine to the left side of C3 in the drawing on the next page gives (2S,3S)-2-bromo-3-chlorobutane, whereas attachment to the right side gives its 2S,3R diastereomer.

Chlorination of (S)-2-Bromobutane at C3

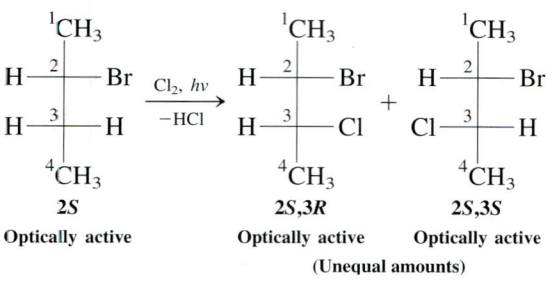

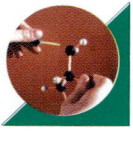

The chlorination at C2 gives a 1:1 mixture of enantiomers. Does the reaction at C3 also give an equimolar mixture of diastereomers? The answer is no. This finding is readily explained on inspection of the two transition states leading to the product (Figure 5-14). Abstraction of either one of the hydrogens results in a radical center at C3. In contrast with the radical formed in the chlorination at C2, however, the two

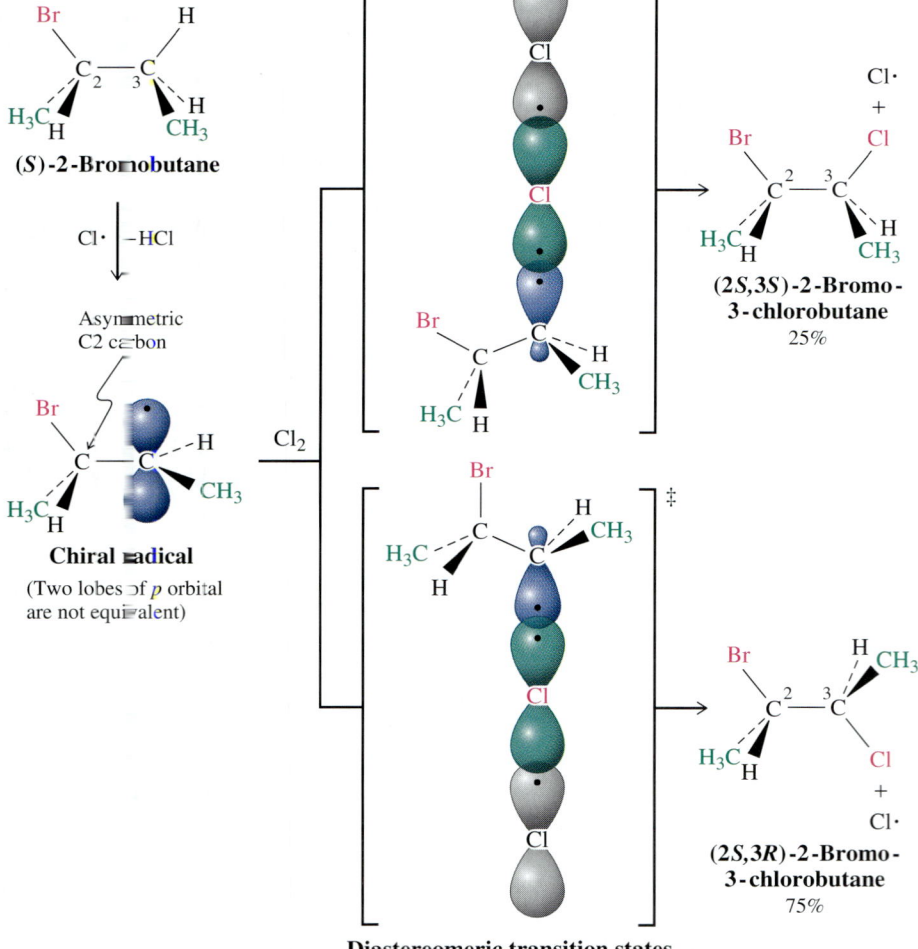

Diastereomeric transition states
(Different in energy)

FIGURE 5-14
The chlorination of (S)-2-bromobutane at C3 produces the two diastereomers of 2-bromo-3-chlorobutane in unequal amounts as a result of the chirality at C2.

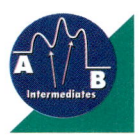

faces of this radical are *not* mirror images of each other, because the radical retains the asymmetry of the original molecule as a result of the presence of the stereocenter at C2. Thus, the two sides of the *p* orbital are not equivalent.

What are the consequences of this nonequivalency? If the rate of attack at the two faces of the radical differ, as one would predict on steric grounds, then the rates of formation of the two diastereomers should be different, as is indeed found: (2*S*,3*R*)-2-Bromo-3-chlorobutane is preferred over the 2*S*,3*S* isomer by a factor of 3 (see Figure 5-14). The two transition states leading to products are not mirror images of each other and are not superimposable: They are diastereomeric. They therefore have different energies and represent different pathways.

EXERCISE 5-22

Write the structures of the products of monobromination of (*S*)-2-bromopentane at each carbon atom. Name the products and specify whether they are chiral or achiral, whether they will be formed in equal or unequal amounts, and which will be in optically active form.

Stereoselectivity is the preference for one stereoisomer

A reaction that leads to the predominant (or exclusive) formation of one of several possible stereoisomeric products is **stereoselective.** For example, the chlorination of (*S*)-2-bromobutane at C3 is stereoselective, as a result of the chirality of the radical intermediate. The corresponding chlorination at C2, however, is not stereoselective: The intermediate is achiral and a racemate is formed.

How much stereoselectivity is possible? The answer depends very much on substrate, reagents, the particular reaction in question, and conditions. Enzymes in nature manage to convert achiral compounds into chiral molecules with very high stereoselectivity. They are capable of this task because enzymes themselves have handedness and therefore convert achiral materials into those that are compatible with their own chirality. An example is the enzyme-catalyzed oxidation of dopamine to (−)-norepinephrine discussed in detail in Problem 52 at the end of the chapter. The chiral reaction environment created by the enzyme gives rise to 100% stereoselectivity in favor of the enantiomer shown. The situation is very similar to shaping flexible achiral objects with your hands. For example, clasping a piece of modeling clay with your left hand furnishes a shape that is the mirror image of that made with your right hand.

$$\text{Dopamine} \xrightarrow[\beta\text{-monooxygenase, } O_2]{} (-)\text{-Norepinephrine}$$

Dopamine   **(−)-Norepinephrine**

In summary, chemical reactions, as exemplified by radical halogenation, can be stereoselective or not. Starting from achiral materials, such as butane, a racemic (non-stereoselective) product is formed by halogenation at C2. The two hydrogens at the methylene carbons of butane are equally susceptible to substitution, the halogenation step in the mechanism of radical bromination proceeding through an achiral intermediate and two enantiomeric transition states of equal energy. Similarly, starting from chiral and enantiomerically pure 2-bromobutane, chlorination of the stereocen-

CHEMICAL HIGHLIGHT 5-4

Chiral Drugs: Racemic or Enantiomerically Pure?

Until recently, most synthetic chiral medicines were prepared as racemic mixtures and sold as such. The reasons were mainly of a practical nature. Reactions that convert an achiral into a chiral molecule ordinarily produce racemates (Section 5-7). In addition, often both enantiomers have comparable physiological activity or one of them (the "wrong" one) is inactive; therefore resolution was deemed unnecessary. Finally, resolution of racemates on a large scale is expensive and substantially adds to the cost of drug development.

However, in several cases, one of the enantiomers of a drug has been found to act as a blocker of the biological receptor site, thus diminishing the activity of the other enantiomer. Worse, one of the enantiomers may have a completely different, and sometimes toxic, spectrum of activity. A tragic example of the latter is the sedative *thalidomide,* which was marketed in 1960

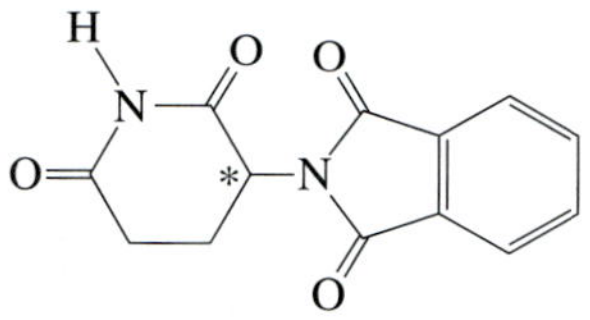

Thalidomide

in Europe as a racemic mixture. Consumption of this drug by pregnant women led to serious birth defects in hundreds of babies. Subsequent studies showed that the *S* enantiomer is teratogenic in some laboratory animals, whereas the *R* form is not. The problem is further complicated by the finding that each enantiomer can racemize at physiological pH.

Because of these (and other) developments, the U.S. Food and Drug Administration revised its guidelines for the commercialization of chiral drugs, making it more advantageous for companies to produce single enantiomers of medicinal products. The result has been a flurry of research activities designed to improve resolution of racemates or, even better, to develop methods of *enantioselective synthesis.* The essence of this approach is that used by nature in enzyme-catalyzed reactions (see the oxidation of dopamine in Section 5-7): An achiral starting material is converted into the chiral product in the presence of an enantiopure environment, often a chiral catalyst. Because in such an environment enantiomeric transition states (Figure 5-13) become diastereomeric (Figure 5-14; note that, in this case, the chiral "environment" of the reacting carbon is provided by the neighboring stereocenter), high stereoselectivity can be achieved. As shown below, such methods have been applied to the syntheses of drugs such as the antiarthritic *naproxen* and the antihypertensive *propanolol* in high enantiomeric purity.

To give you an idea of the importance of this emerging technology, the worldwide market for chiral drugs is estimated to be in excess of $40 billion in 1997.

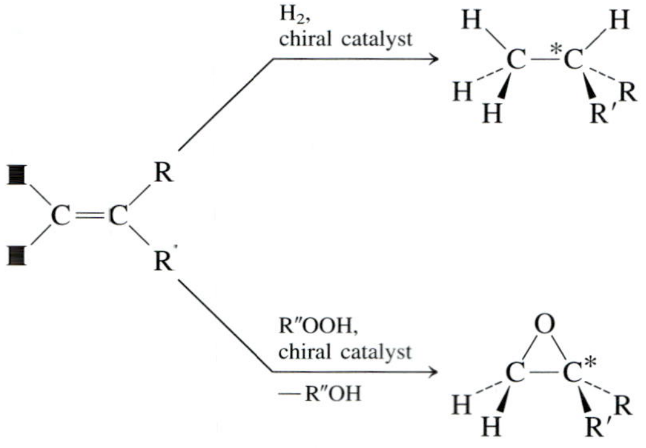

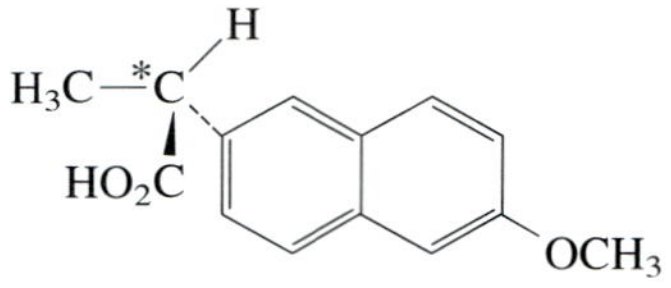

(R)-Naproxen

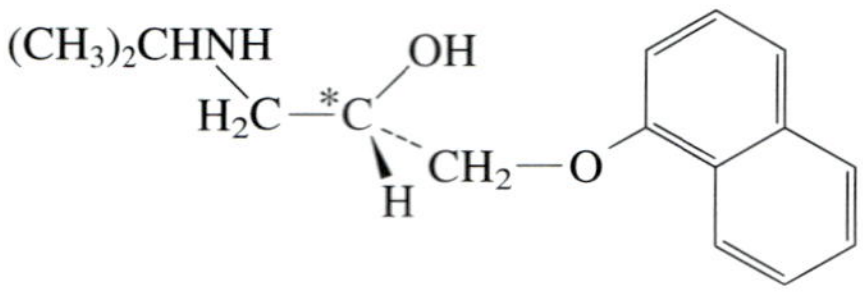

(S)-Propanolol

(C* = a new stereocenter)

ter also gives a racemic product. However, stereoselectivity is possible in the formation of a new stereocenter, because the chiral environment retained by the molecule results in two unequal modes of attack on the intermediate radical. The two transition states have a diastereomeric relation, a condition that leads to the formation of products at unequal rates.

5-8 Resolution: Separation of Enantiomers

As we know, the generation of a chiral structure from an achiral starting material furnishes a racemic mixture. How, then, can pure enantiomers of a chiral compound be obtained?

One possible approach is to start with the racemate and separate one enantiomer from the other. This process is called the **resolution** of enantiomers. Some enan-

---

CHEMICAL HIGHLIGHT 5-5　　**Why Is Nature "Handed"?**

In this chapter, we have seen that many of the organic molecules in nature are chiral. More importantly, most natural compounds in living organisms not only are chiral, but also are present in only one enantiomeric form. An example of an entire class of such compounds consists of the *amino acids,* which are the component units of *polypeptides.* The large polypeptides in nature are called *proteins* or, when they catalyze biotransformations, *enzymes.*

Absolute Configuration of Natural Amino Acids and Polypeptides

Amino acid
(R variable)

Amino acid 1　Amino acid 2　Amino acid 3

Polypeptide

Being made up of smaller chiral pieces, enzymes arrange themselves into bigger conglomerates that also are chiral and show handedness. Thus, much as a right hand will readily distinguish another right hand from a left hand, enzymes (and other biomolecules) have "pockets" that, by virtue of their stereochemically defined features, are capable of recognizing and processing only one of the enantiomers in a racemate. The differences in physiological activity of the two enantiomers of a

Schematized Enantiomer Recognition in the Receptor Site of an Enzyme

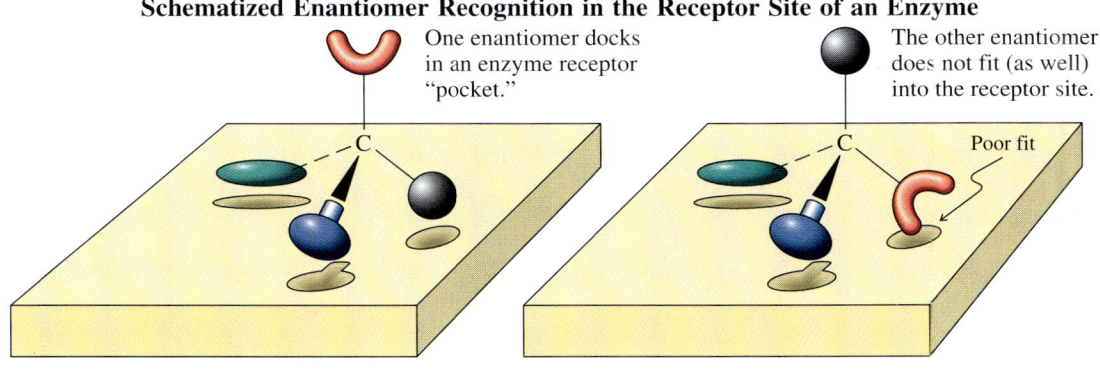

One enantiomer docks in an enzyme receptor "pocket."

The other enantiomer does not fit (as well) into the receptor site.

Poor fit

tiomers, such as those of tartaric acid, crystallize into mirror-image shapes, which can be manually separated (as done by Pasteur; see Chemical Highlight 5-3). However, this process is time consuming, uneconomic for anything but minute-scale separations, and applicable only in rare cases.

A better strategy for resolution is based on the difference in the physical properties of diastereomers. Suppose we can find a reaction that converts a racemate into a mixture of diastereomers. All the R forms of the original enantiomer mixture should then be separable from the corresponding S forms by fractional crystallization, distillation, or chromatography of the diastereomers. How can such a process be developed? The trick is to add an enantiomerically pure reagent that will attach itself to the components of the racemic mixture. For example, we can imagine reaction of a racemate, $X_{R,S}$ (in which X_R and X_S are the two enantiomers), with an optically pure compound Y_S (the choice of the S configuration is arbitrary; the pure R mirror image would work just as well). The reaction produces two optically active diastereomers,

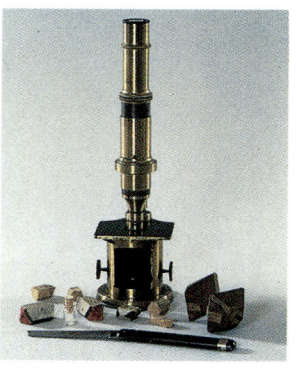

Pasteur's polarimeter and crystals of (+)- and (−)-tartaric acid.

chiral drug are based on this recognition (Chemical Highlight 5-4). A good analogy is that of a chiral key fitting only its image (not mirror image) lock. The

chiral environment provided by these structures is also able to effect highly enantioselective conversions of achiral starting materials into enantiopure, chiral

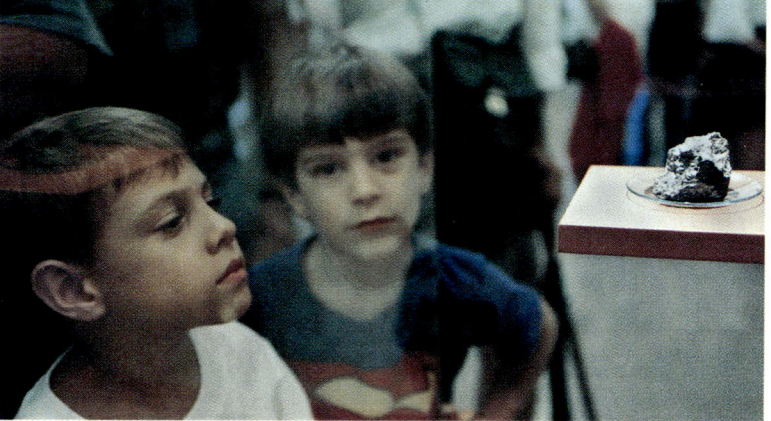

The 4.5-billion-year-old meteorite from Mars which has led scientists to believe that Mars may have once harbored life, on exhibit at the Smithsonian's National Museum of Natural History, Washington, D.C.

products see Chemical Highlights 5-4 and 23-2). In this way, how nature preserves and proliferates its own built-in chirality can be readily understood (at least in principle).

What is more difficult to understand is how the enantiomeric homogeneity of nature arose in the first place; in other words, why was only one stereochemical configuration of the amino acids chosen but not the other? Trying to understand this mystery has fascinated many scientists, because it is very likely linked to the evolution of life as we know

it. Speculation ranges from the invocation of a chance separation of enantiomers ("spontaneous resolution") to the postulate of the operation of a chiral physical force, such as handed radiation (as observed during the decay of radioactive elements). Another hypothesis suggests that enantiomeric excess (and perhaps life itself) was simply imported from another planet, with meteorites as carriers (thus really begging the question). A lot of effort has been expended in trying to detect nonracemic amino acids in meteor (and other planetary) samples, so far without success.

$X_R Y_S$ and $X_S Y_S$, separable by standard techniques (Figure 5-15). Now the bond between X and Y in each of the separated and purified diastereomers is broken, liberating X_R and X_S in their enantiomerically pure states. In addition, the optically active agent Y_S may be recovered and reused in further resolutions.

What we need, then, is a readily available, enantiomerically pure compound, Y, that can be attached to the molecule to be resolved in an easily reversible chemical reaction. In fact, nature has provided us with a large number of pure optically active molecules that can be used. An example is (+)-2,3-dihydroxybutanedioic acid [(+)-(R,R)-tartaric acid]. A popular reaction employed in the resolution of enantiomers is salt formation between acids and bases. For example, (+)-tartaric acid functions as an effective resolving agent of racemic amines. Figure 5-16 shows how this works for 3-butyn-2-amine. The racemate is first treated with (+)-tartaric acid to form two diastereomeric tartrate salts. The salt incorporating the R-amine crystallizes on standing and can be filtered away from the solution, which contains the more soluble salt of the S-amine. Treatment of the (+)-salt with aqueous base liberates the free amine, (+)-(R)-3-butyn-2-amine. Similar treatment of the solution gives the (−)-S enantiomer (evidently slightly less pure: Note the slightly lower optical rotation). This process is just one of many ways in which the formation of diastereomers can be used in the resolution of racemates.

A very convenient way of separating enantiomers without the necessity of isolating diastereomers is by so-called **chiral chromatography.** The principle is the same

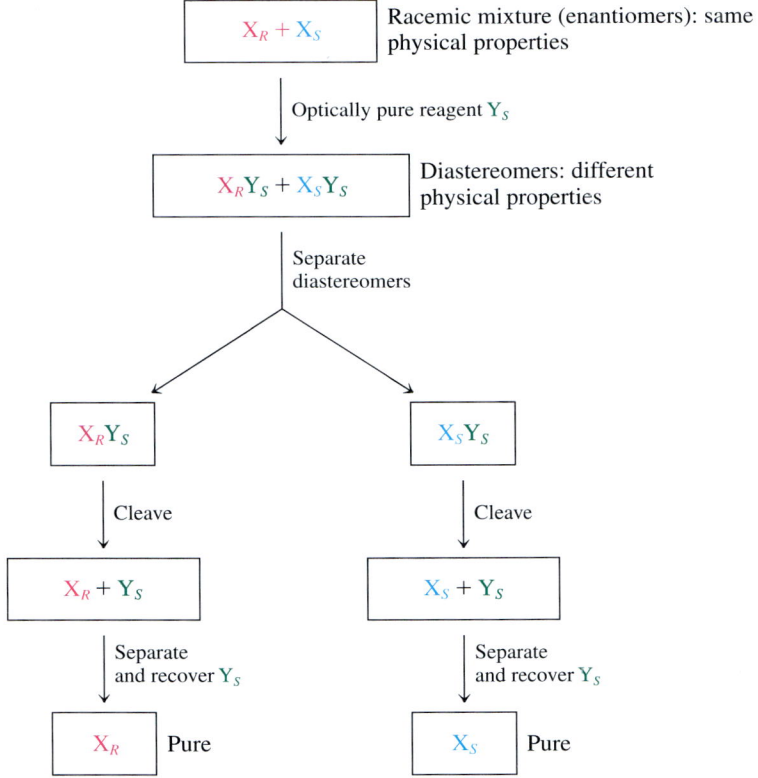

FIGURE 5-15

Flowchart for the separation (resolution) of two enantiomers. The procedure is based on conversion into separable diastereomers by means of reaction with an optically pure reagent.

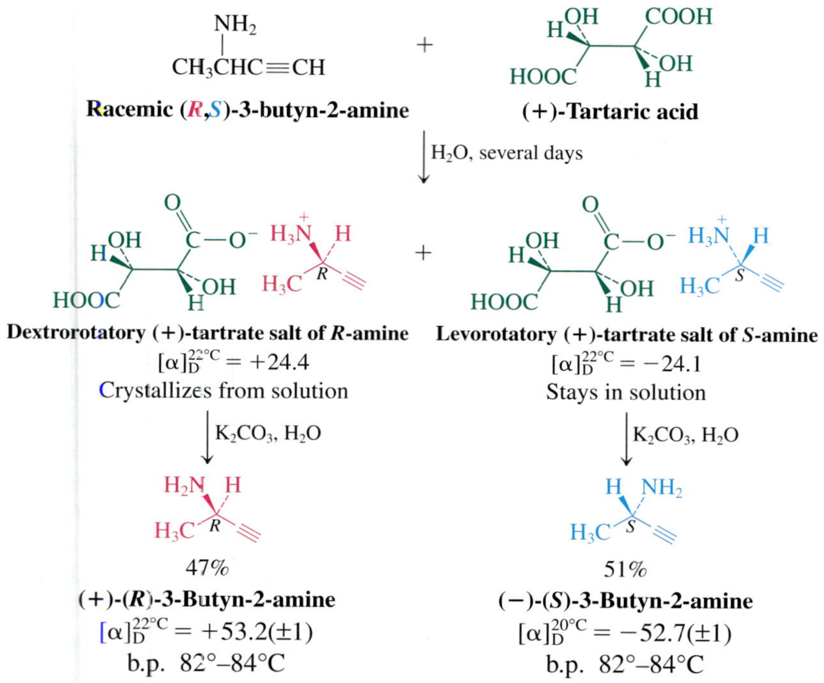

FIGURE 5-16

Resolution of 3-butyn-2-amine with (+)-2,3-dihydroxybutanedioic [(+)-tartaric] acid. It is purely accidental that the [α] values for the two diastereomeric tartrate salts are similar in magnitude and of opposite sign.

as that illustrated in Figure 5-16, except that the optically active auxiliary [such as (+)-tartaric acid or any other suitable cheap optically active compound] is immobilized on a solid support (such as silica gel, SiO_2, or aluminum oxide, Al_2O_3). This material is then used to fill a tube ("column") of varying length, and a solution of the racemate is allowed to pass through it. The individual enantiomers will reversibly bind to the chiral support to different extents (because this interaction is diastereomeric) and therefore be held on the column for different lengths of time (retention time). Therefore, one enantiomer will elute from the column before the other, enabling separation.

CHAPTER INTEGRATION PROBLEM

Selectivity in chemical reactions is a primary goal of the synthetic chemist. We have learned how such selectivity may be achieved, at least to some extent, in radical halogenations: in Sections 3-6 and 3-7, with respect to the type of hydrogen to be replaced (e.g., primary versus secondary versus tertiary) and, in Section 5-7, with respect to stereochemistry. You will have recognized that, because of the reactivity of radicals and the planarity of the carbon-centered radical intermediates, radical halogenations often lack selectivity and any synthetic plan considering them must take into account all possible outcomes of a proposed conversion. For example, looking again at our picture of the generalized steroid nucleus (Section 4-7), you can see that

there are many types of hydrogens, all of which would, in principle, be susceptible to abstraction by a halogen atom.

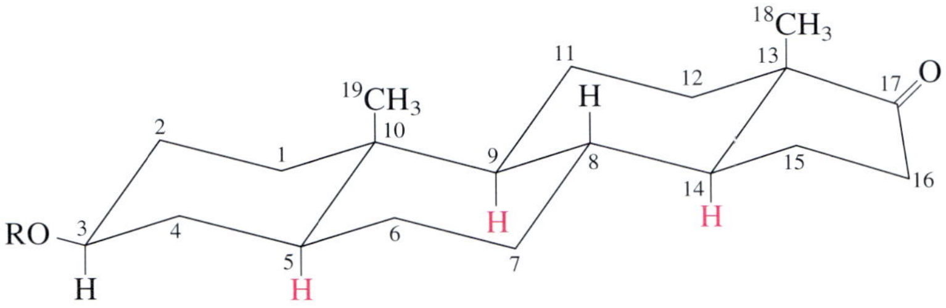

Because the steroids are important biological molecules, their selective functionalization has been the focus of the attention of many researchers and, by developing carefully controlled conditions with special halogenating agents, they have been able to restrict attack not only to the tertiary centers, but also selectively to either C5, C9, or C14 (see also Problems 36–38 in Chapter 4). The following problem illustrates the kind of analyses that they undertook with a less complex cyclohexane fragment of the steroid nucleus.

How many products are there of the radical monobromination of (S)-1-bromo-2,2-dimethylcyclohexane at C1 and C3? Draw the structure of the starting material, name the resulting dibromodimethylcyclohexanes, label them as chiral or achiral, specify whether they will be formed in equal or unequal amounts, and state whether they will be optically active or not.

SOLUTION

We begin by drawing the structure of our starting material, first ignoring stereochemistry (A).

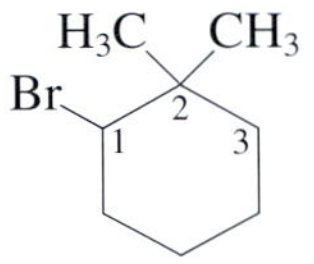

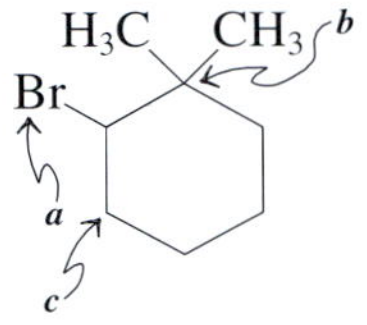

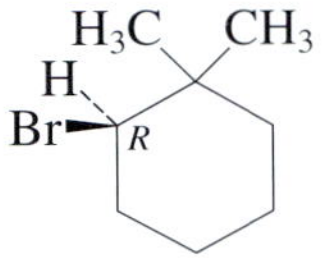

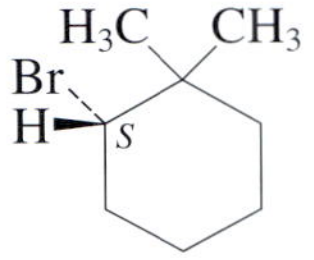

We then designate the priority sequence (B) according to the rules in Section 5-3. We now have a choice of two enantiomeric arrangements (C and D), and the task is to orient the molecule in our minds in such a way as to place the substituent of lowest priority (H atom) as far away as possible. To assist in this mental exercise, picture yourself at the molecular scale (shrunk by a factor of 10^{10}) and stand on the stereocenter in question with the C–H bond pointing away from you. The three remaining substituents will now surround you either in clockwise (R) or counterclockwise (S) fashion: D is the correct structure of the S enantiomer.

Now we are ready to introduce bromine at either C1 or C3. It is important here to remember the mechanism of the free-radical halogenation: The crucial intermediate is a radical center—in our case, at either C1 (E) or C3 (F)—that can be attacked by halogen from either side of the p orbital (Section 3-4).

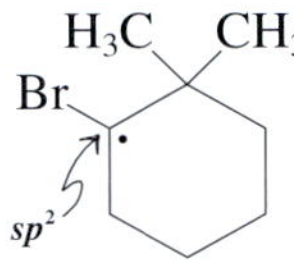

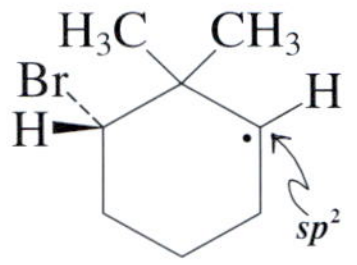

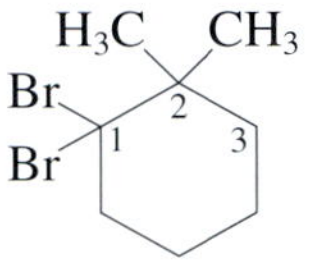

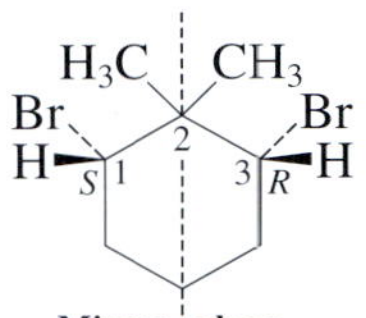

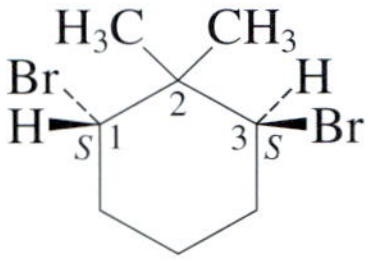

In E, the molecule is symmetrical, and the rate of attack from the top will be equal to that from the bottom. If the halogenation were to be executed by using F_2 or Cl_2, C1 would remain a stereocenter, the R and S enantiomers having been formed in equal amounts (racemate; Section 5-7, Figure 5-13). However, in our case, bromination at C1 removes the asymmetry of this carbon: Compound G, 1,1-dibromo-2-2-dimethylcyclohexane, is achiral and hence not optically active.

Turning to F, the situation is different. Here, the presence of the unchanged original stereocenter (C1 in D) makes the two faces of the intermediate radical center unequal: Two diastereomers (H and I) are formed at unequal rates and therefore in unequal amounts (Section 5-7, Figure 5-14). In H, *cis*-1,3-dibromo-2,2-dimethylcyclohexane, the second bromine is attached in such a way as to introduce a mirror plane into the molecule. H is a meso compound, achiral, and hence not optically active (Section 5-6). Another way of describing what has happened is that the chirality of C1 in D—namely, S—is canceled out by the introduction of its "mirror image" at C3—namely, R. The two stereoisomers are indistinguishable because (1S,3R)-H is the same as (1R,3S)-H. (You can verify that statement by simply rotating compound H about the dashed line representing the mirror plane.)

On the other hand, compound I, (1S,3S)-1,3-dibromo-2,2-dimethylcyclohexane, contains no mirror plane: The molecule is chiral, enantiomerically pure, and, hence, optically active. In other words, the reaction leaves the stereochemical integrity and identity of C1 intact, generating only one enantiomer of the product, which is non-superimposable with its mirror image, the (1R,3R) diastereomer (Section 5-5).

IMPORTANT CONCEPTS

1. Isomers have the same molecular formula but are different compounds. Constitutional (structural) isomers differ in the order in which the individual atoms are connected. Stereoisomers have the same connectivity but differ in the three-dimensional arrangement of the atoms. **Mirror-image stereoisomers** are related to each other as image and mirror image.

2. An object that is not superimposable on its mirror image is **chiral.**

3. A carbon atom bearing four different substituents (**asymmetric carbon**) is an example of a **stereocenter.**

4. Two stereoisomers that are related to each other as image–nonsuperimposable mirror image are called **enantiomers.**

5. A compound containing one stereocenter is chiral and exists as a pair of enantiomers. A 1:1 mixture of enantiomers is a **racemate (racemic mixture).**

6. Chiral molecules cannot have a plane of symmetry (mirror plane). If a molecule has a **mirror plane,** then it is **achiral.**

7. **Diastereomers** are stereoisomers that are not related to each other as object to mirror image. Cis and trans isomers of cyclic compounds are examples of diastereomers.

8. Two stereocenters in a molecule result in as many as four stereoisomers—two diastereomerically related pairs of enantiomers. The maximum number of stereoisomers that a compound with n stereocenters can have is 2^n. This number is reduced when equivalently substituted stereocenters give rise to a plane of symmetry. A molecule containing stereocenters *and* a mirror plane is identical with its mirror image (achiral) and is called a **meso compound.** The presence of a mirror plane in any energetically accessible conformation of a molecule is sufficient to make it achiral.

9. Most of the physical properties of enantiomers are the same. A major exception is their interaction with **plane-polarized light:** One enantiomer will rotate the polarization plane clockwise (**dextrorotatory**), the other counterclockwise (**levorotatory**). This phenomenon is called **optical activity.** The extent of the rotation is measured in degrees and is expressed by the **specific rotation, $[\alpha]$.** Racemates and meso compounds show zero rotation. The **optical purity** of an unequal mixture of enantiomers is given by

$$\% \text{ optical purity} = \left(\frac{[\alpha]_{observed}}{[\alpha]} \right) \cdot 100$$

10. The "handedness" of a stereocenter (its absolute configuration) is revealed by X-ray diffraction and can be assigned as **R** or **S** by using the **sequence rules** of Cahn, Ingold, and Prelog.

11. **Fischer projections** provide stencils for the quick drawing of molecules with stereocenters.

12. Chirality can be introduced into an achiral compound by radical halogenation. When the transition states are enantiomeric (related as object and mirror image), the result is a racemate because the faces of the planar radical react at equal rates.

13. Radical halogenation of a chiral molecule containing one stereocenter will give a racemate if the reaction takes place at the stereocenter. When reaction elsewhere leads to two diastereomers, they will be formed in unequal amounts.

14. The preference for the formation of one stereoisomer, when several are possible, is called **stereoselectivity.**

15. The separation of enantiomers is called **resolution.** It is achieved by the reaction of the racemate with the pure enantiomer of a chiral compound to yield separable diastereomers. Chemical removal of the chiral reagent frees both enantiomers of the original racemate. Another way of separating enantiomers is by **chiral chromatography** on an optically active support

PROBLEMS

23. Classify each of the following common objects as being either chiral or achiral. Assume in each case that the object is in its simplest form, without decoration or printed labels. **(a)** A ladder; **(b)** a door; **(c)** an electric fan; **(d)** a refrigerator; **(e)** Earth; **(f)** a baseball; **(g)** a baseball bat; **(h)** a baseball glove; **(i)** a flat sheet of paper; **(j)** a fork; **(k)** a spoon; **(l)** a knife.

24. Each part of this problem lists two objects or sets of objects. As precisely as you can, describe the relation between the two sets, using the terminology of this chapter; that is, specify whether they are identical, enantiomeric, or diastereomeric. **(a)** An American toy car compared with a British toy car (same color and design but steering wheels on opposite sides); **(b)** two left shoes compared with two right shoes (same color, size, and style); **(c)** a pair of skates compared with two left skates (same color, size, and style); **(d)** a right glove on top of a left glove (palm to palm) compared with a left glove on top of a right glove (palm to palm; same color, size, and style).

25. For each pair of the following molecules, indicate whether its members are identical, structural isomers, conformers, or stereoisomers. How would you describe the relation between conformations when they are maintained at a temperature too low to permit them to interconvert?

(a) $CH_3CH_2CH_2\overset{\overset{\displaystyle CH_3}{|}}{\underset{\underset{\displaystyle CH_3}{|}}{CH}}$ and $CH_3CH_2\overset{\overset{\displaystyle CH_3}{|}}{CH}CH_2CH_3$

(b) and

(c) and

(d) and

(e) $CH_3\overset{\overset{\displaystyle Cl}{|}}{\underset{\underset{\displaystyle Br}{|}}{C}}CH_2CH_2CH_3$ and $CH_3\overset{\overset{\displaystyle Br}{|}}{CH}CH_2\overset{\overset{\displaystyle Cl}{|}}{CH}CH_3$

(f) and

(g)

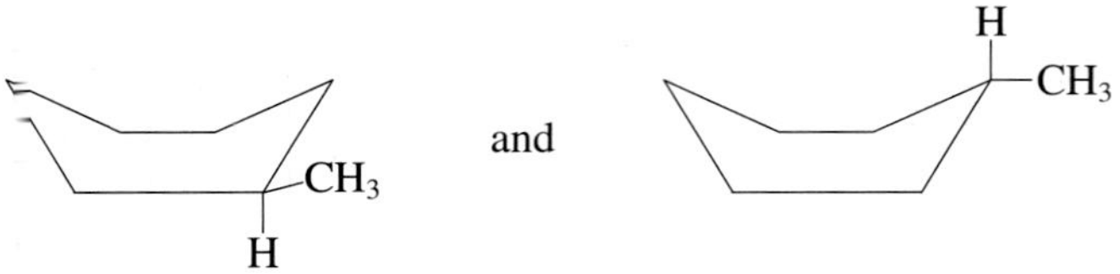

and

(h) 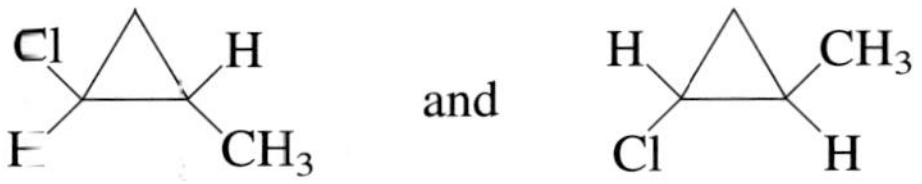 and

26. Which of the following compounds are chiral? (**Hint:** Look for stereocenters.)
    (**a**) 2-Methylheptane   (**b**) 3-Methylheptane   (**c**) 4-Methylheptane
    (**d**) 1,1-Dibromopropane   (**e**) 1,2-Dibromopropane   (**f**) 1,3-Dibromopropane
    (**g**) Ethene, $H_2C{=}CH_2$ (**h**) Ethyne, $HC{\equiv}CH$

(**i**) Benzene, (Note: Like ethene, benzene contains all sp^2-hybridized carbons and is therefore planar.)

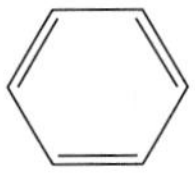

(**j**) Epinephrine,

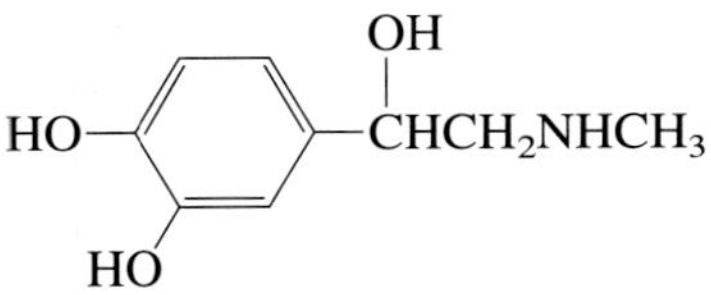

(**k**) Vanillin,

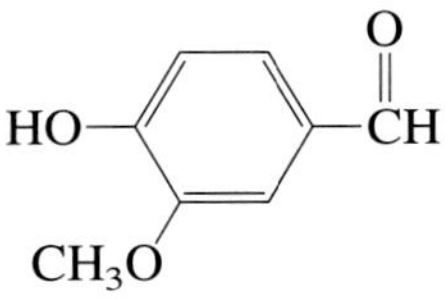

(**l**) Citric acid,

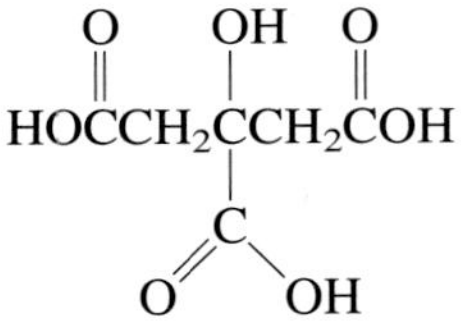

(**m**) Ascorbic acid,

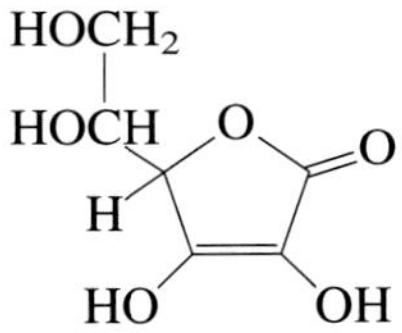

(**n**) *p*-Menthane-1,8-diol (terpin hydrate),

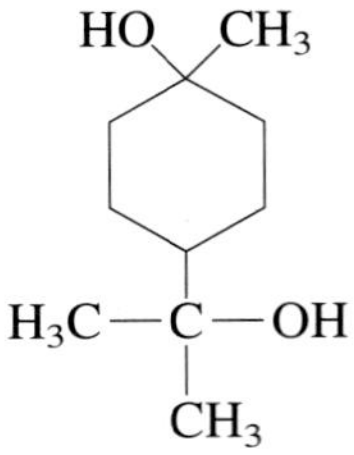

(**o**) Meperidine (demerol),

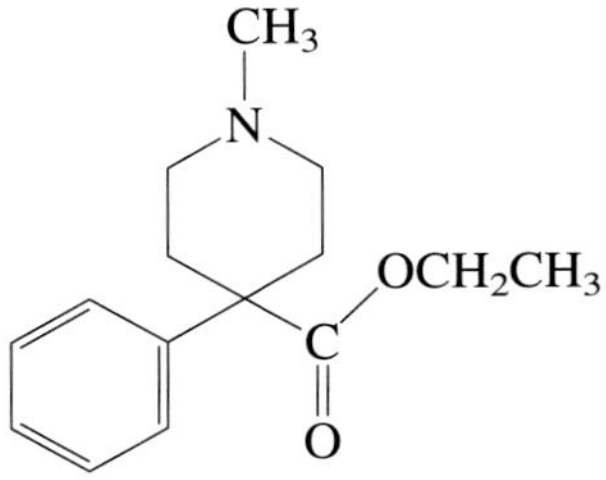

27. Each of the following molecules has the molecular formula $C_5H_{12}O$ (check for yourself). Which ones are chiral?

(**a**) 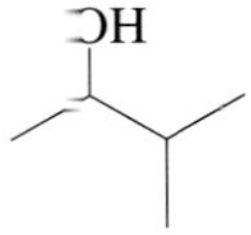   (**b**) 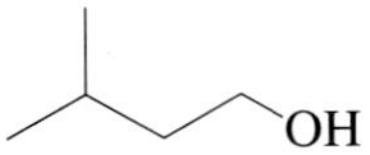   (**c**)

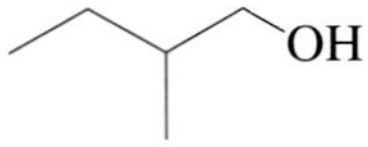

(**d**) 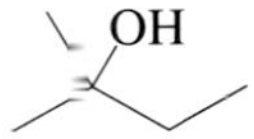   (**e**) 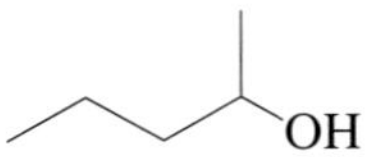   (**f**)

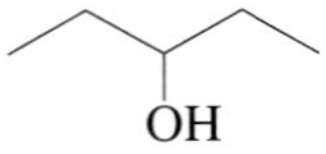

28. Which of the following cyclohexane derivatives are chiral? For the purpose of determining the chirality of a cyclic compound, the ring may generally be treated as if it were planar.

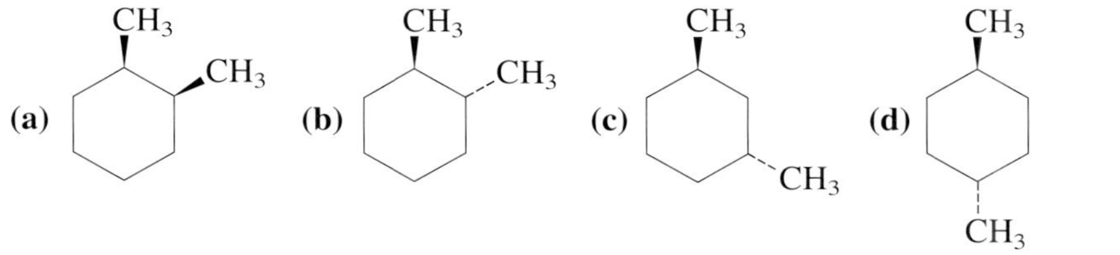

29. For each pair of structures shown below, indicate whether the two species are constitutional isomers, enantiomers, diastereomers of one another, or identical molecules.

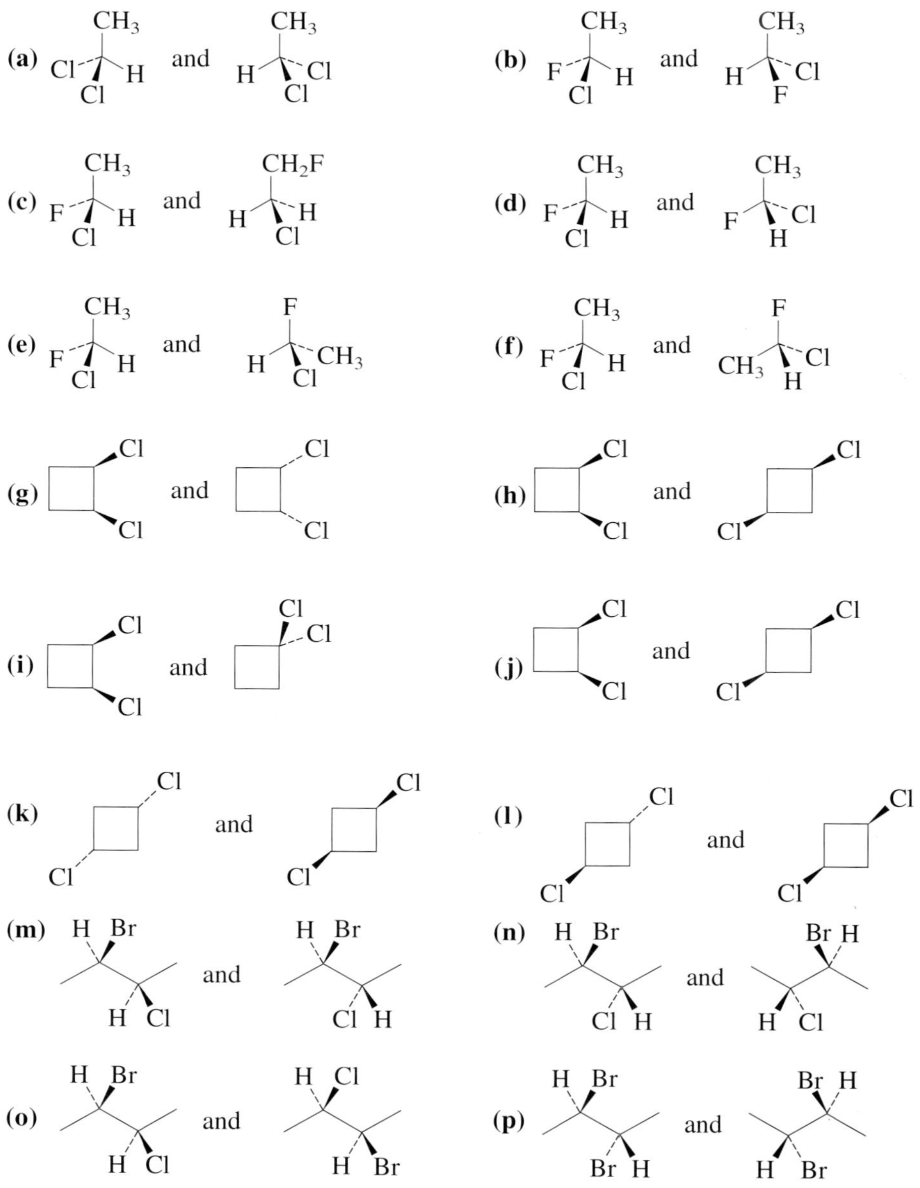

30. For each of the following formulas, identify every structural isomer containing one or more stereocenters, give the number of stereoisomers for each, and draw and fully name at least one of the stereoisomers in each case.
(a) C_7H_{16} (b) C_8H_{18} (c) C_5H_{10}, with one ring

31. Assign the appropriate designation of configuration (R or S) to the stereocenter in each of the following molecules. (**Hint:** Regarding cyclic structures containing stereocenters, treat the ring as if it were two separate substituents that happen to be attached to each other at the far end of the molecule—look for the first point of difference, just as you would for acyclic structures.)

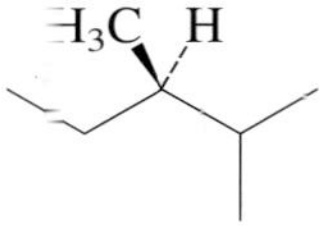

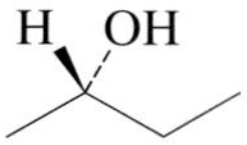

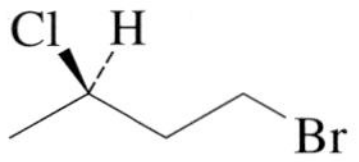

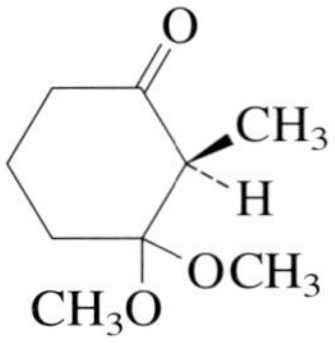

32. Mark the stereocenters in each of the chiral molecules in Problem 27. Draw any single stereoisomer of each of these molecules, and assign the appropriate designation (R or S) to each stereocenter.

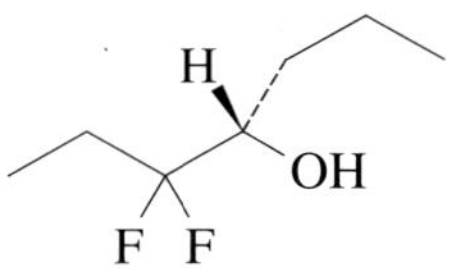

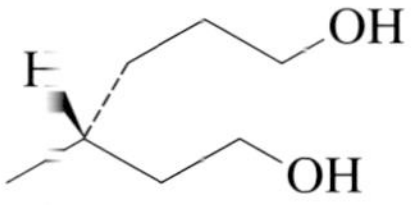

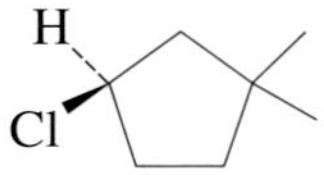

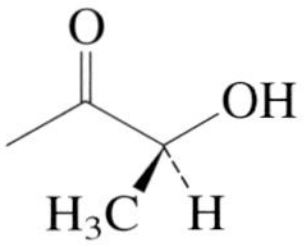

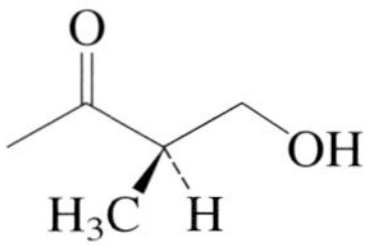

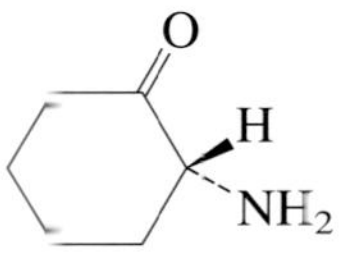

33. The two isomers of carvone [systematic name: 2-methyl-5-(1-methylethenyl)-2-cyclohexenone] are drawn here. Which is R and which is S?

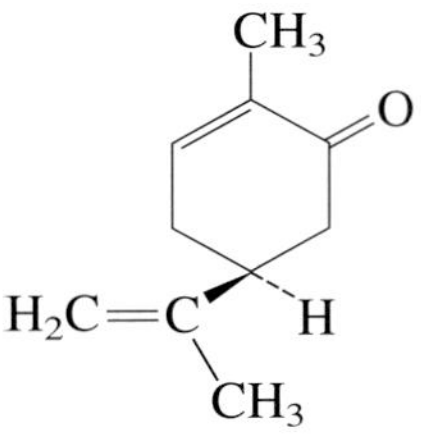

(+)-**Carvone**
(In caraway seeds)

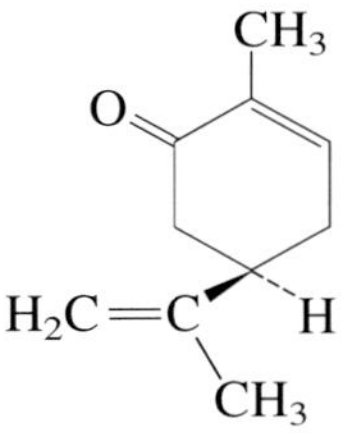

(−)-**Carvone**
(In spearmint)

34. Draw structural representations of each of the following molecules. Be sure that your structure clearly shows the configuration at the stereocenter. (**Hint:** You may find it useful to first draw the enantiomer whose configuration is easiest for you to determine and then, if necessary, modify your structure to fit the one requested in the problem.) (a) (R)-2-chloropentane; (b) (S)-2-methyl-3-bromohexane; (c) (S)-1,3-dichlorobutane; (d) (R)-2-chloro-1,1,1-trifluoro-3-methylbutane.

35. Draw structural representations of each of the following molecules. Be sure that your structure clearly shows the configuration at each stereocenter.
(a) (R)-3-bromo-3-methylhexane; (b) ($3R,5S$)-3,5-dimethylheptane;
(c) ($2R,3S$)-2-bromo-3-methylpentane; (d) (S)-1,1,2-trimethylcyclopropane;
(e) ($1S,2S$)-1-chloro-1-trifluoromethyl-2-methylcyclobutane; (f) ($1R,2R,3S$)-1,2-dichloro-3-ethylcyclohexane.

36. For each of the following questions, assume that all measurements are made in 10-cm polarimeter sample containers. **(a)** A 10-mL solution of 0.4 g of optically active 2-butanol in water displays an optical rotation of −0.56°. What is its specific rotation? **(b)** The specific rotation of sucrose (common sugar) is +66.4. What would be the observed optical rotation of such a solution containing 3 g of sucrose? **(c)** A solution of pure (S)-2-bromobutane in ethanol is found to have an observed $\alpha = 57.3°$. If $[\alpha]$ for (S)-2-bromobutane is 23.1, what is the concentration of the solution?

37. Natural epinephrine, $[\alpha]_D^{25°C} = -50$, is used medicinally. Its enantiomer is medically worthless and is, in fact, toxic. You, a pharmacist, are given a solution said to contain 1 g of epinephrine in 20 mL of liquid, but the optical purity is not specified. You place it in a polarimeter (10-cm tube) and get a reading of −2.5°. What is the optical purity of the sample? Is it safe to use medicinally?

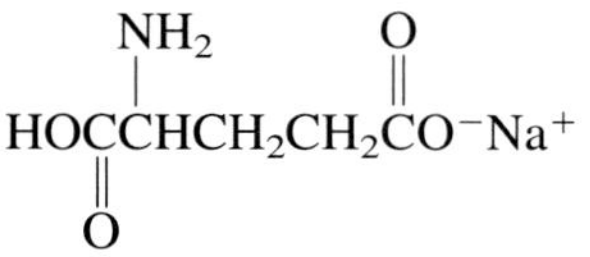

$$NH_2 \quad\quad O$$
$$HOCCHCH_2CH_2CO^-Na^+$$
$$\underset{\|}{O}$$

38. Sodium hydrogen (S)-glutamate [(S)-monosodium glutamate], $[\alpha]_D^{25°C} = +24$, is the active flavor enhancer known as MSG. The condensed formula of MSG is shown in the margin. **(a)** Draw the structure of the S enantiomer of MSG. **(b)** If a commercial sample of MSG were found to have a $[\alpha]_D^{25°C} = +8$, what would be its optical purity? What would be the percentages of the S and R enantiomers in the mixture? **(c)** Answer the same questions for a sample with $[\alpha]_D^{25°C} = +16$.

39. For each of the following compounds, mark each stereocenter, assign an R or S designation, and draw a clear picture of the molecule's enantiomer.

(a) 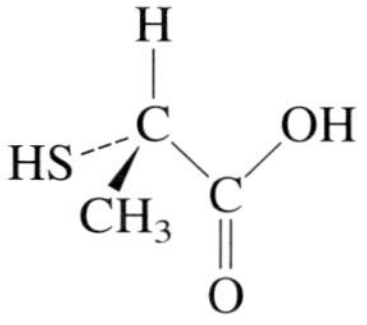   (b) 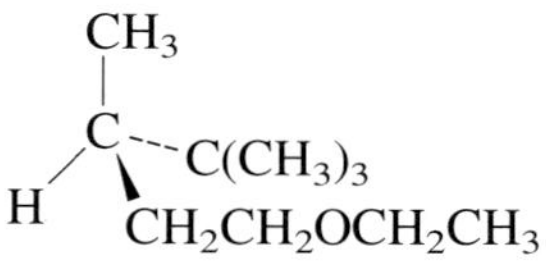   (c)

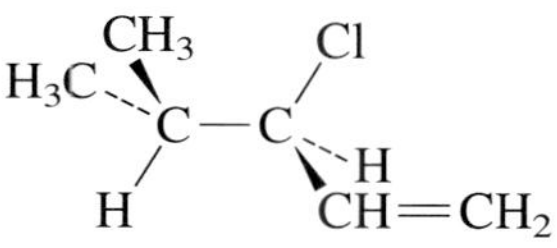

(d) 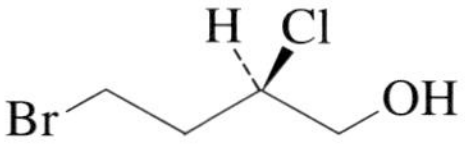   (e) 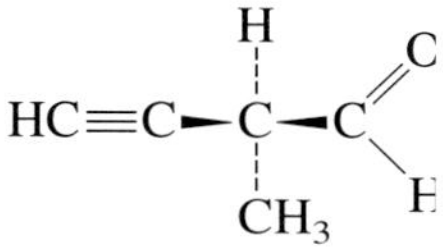   (f)

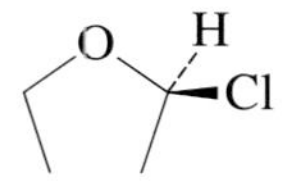

(g)

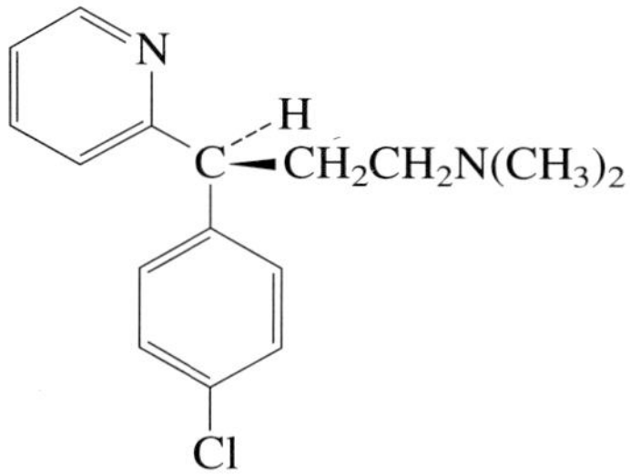

(Note: The carbons in benzene or benzenelike rings are treated in the same way as those in alkenes. Use sequence rule 3 from Section 5-3.)

Chlorpheniramine
(As in Coricidin decongestant)

(h)

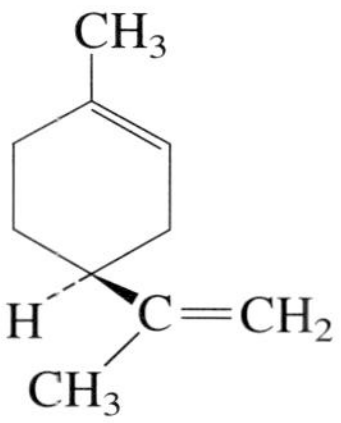

Limonene
(From trees, fruits, etc.)

40. For each of the following pairs of structures, indicate whether the two compounds are identical or enantiomers of each other.

(a) 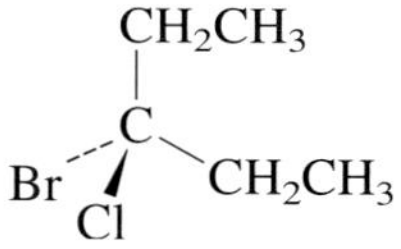   and

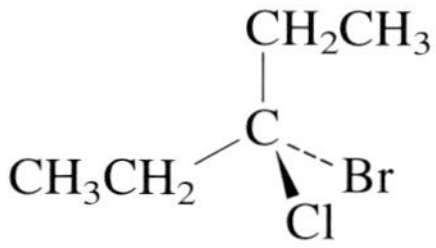

(b) 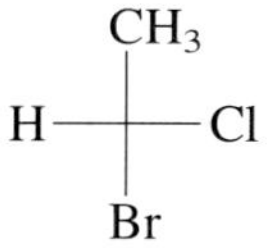   and

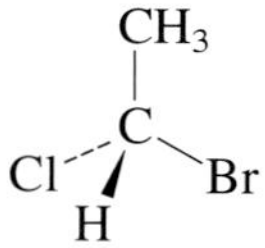

(c) Cl—⊢—CF$_3$ and F$_3$C—⊢—CH$_3$
  CH$_3$ / OCH$_3$ (top), OCH$_3$ / Cl (bottom)

(d) H$_2$N—C—CO$_2$H and H—⊢—CH(CH$_3$)$_2$
  H / CH(CH$_3$)$_2$; NH$_2$ / CO$_2$H

41. Determine the *R* or *S* designation for each stereocenter in the structures in Problem 40.

42. Redraw each of the following molecules as a Fischer projection; then assign *R* or *S* designations to each stereocenter.

(a) H$_3$C, Cl, C—C, H, Cl, H, CH$_3$

(b) OHC, HO, CO$_2$H, CH$_3$, CH$_3$, OH

(c) H$_2$N, OH, C—C, H, H, CH$_3$, COOH

(d) H$_3$C, Br, C—C, H, H, Cl, CH$_3$

43. The compound pictured in the margin is a sugar called (−)-arabinose. Its specific rotation is −105. **(a)** Draw the enantiomer of (−)-arabinose. **(b)** Does (−)-arabinose have any other enantiomers? **(c)** Draw a diastereomer of (−)-arabinose. **(d)** Does (−)-arabinose have any other diastereomers? **(e)** If possible, predict the specific rotation of the structure that you drew for (a). **(f)** If possible, predict the specific rotation of the structure that you drew for (c). **(g)** Does (−)-arabinose have any optically inactive diastereomers? If it does, draw one.

44. Write the complete IUPAC name of the following compound (do not forget stereochemical designations).

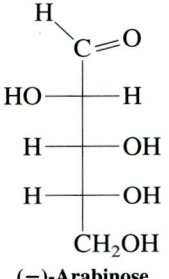

(−)-Arabinose

CH$_2$CH$_3$ — C — H / CH$_2$CH$_2$Cl ; Cl   C$_5$H$_{10}$Cl$_2$

Reaction of this compound with 1 mol of Cl$_2$ in the presence of light produces several isomers of the formula C$_5$H$_9$Cl$_3$. For each part of this problem, give the following information: How many stereoisomers are formed? If more than one is formed, are they generated in equal or unequal amounts? Designate every stereocenter in each stereoisomer as *R* or *S*.
(a) Chlorination at C3   **(b)** Chlorination at C4   **(c)** Chlorination at C5

45. Monochlorination of methylcyclopentane can result in several products. Give the same information as that requested in Problem 44 for the monochlorination of methylcyclopentane at C1, C2, and C3.

46. Draw all possible products of the chlorination of (*S*)-2-bromo-1,1-dimethyl-cyclobutane. Specify whether they are chiral or achiral, whether they are formed in equal or unequal amounts, and which are optically active when formed.

47. Illustrate how to resolve racemic 1-phenylethanamine (shown in the margin), using the method of reversible conversion into diastereomers.

48. Draw a flowchart that diagrams a method for the resolution of racemic 2-hydroxypropanoic acid (lactic acid, Table 5-1), using (*S*)-1-phenylethanamine.

NH$_2$
C$_6$H$_5$CHCH$_3$
1-Phenylethanamine

49. How many different stereoisomeric products are formed in the monobromina-tion of (**a**) racemic *trans*-1,2-dimethylcyclohexane and (**b**) pure (*R,R*)-1,2-dimethylcyclohexane? (**c**) For your answers to (a) and (b), indicate whether you expect equal or unequal amounts of the various products to be formed. In-dicate to what extent products can be separated on the basis of having different physical properties (e.g., solubility, boiling point).

50. Make a model of *cis*-1,2-dimethylcyclohexane in its most stable conformation. If the molecule were rigidly locked into this conformation, would it be chiral? (Test your answer by making a model of the mirror image and checking for superimposability.)

Flip the ring of the model. What is the stereoisomeric relation between the original conformation and the conformation after flipping the ring? How do the results that you have obtained in this problem relate to your answer to Problem 28(a).

51. Morphinane is the parent substance of the broad class of chiral molecules known as the morphine alkaloids. Interestingly, the (+) and (−) enantiomers of the compounds in this family have rather different physiological properties. The (−) compounds, such as morphine, are "narcotic analgesics" (painkillers), whereas the (+) compounds are "antitussives" (ingredients in cough syrup). Dextromethorphan is one of the simplest and most common of the latter.

Morphinane   **Dextromethorphan**

(**a**) Locate and identify all the stereocenters in dextromethorphan. (**b**) Draw the enantiomer of dextromethorphan. (**c**) As best you can (it is not easy), assign *R* and *S* configurations to all the stereocenters in dextromethorphan.

52. The enzymatic introduction of a functional group into a biologically important molecule is not only specific with regard to the location at which the reaction occurs in the molecule (see Chapter 4, Problem 36), but also usually specific in the stereochemistry obtained. The biosynthesis of epinephrine first requires that a hydroxy group be introduced specifically to produce (−)-norepinephrine from the achiral substrate dopamine. (The completion of the synthesis of epi-nephrine will be presented in Problem 59 of Chapter 9.) Only the (−) enan-tiomer is functional in the appropriate physiological manner, so the synthesis must be highly stereoselective.

Dopamine   **(−)-Norepinephrine**

(**a**) Is the configuration of (−)-norepinephrine *R* or *S*? (**b**) In the absence of an enzyme, would the transition states of a radical oxidation leading to (−)- and

(+)-norepinephrine be of equal or unequal energy? What term describes the relation between these transition states? **(c)** In your own words, describe how the enzyme must affect the energy of these transition states to favor production of the (−) enantiomer. Does the enzyme have to be chiral or can it be achiral?

Team Problem

53. Studies have shown that one isomeric form of compound A is an effective agent against certain types of neurodegenerative disorders. Recognize that structure A contains a decalin-type system, as illustrated in structure B, and that the nitrogen can be treated just like a carbon.

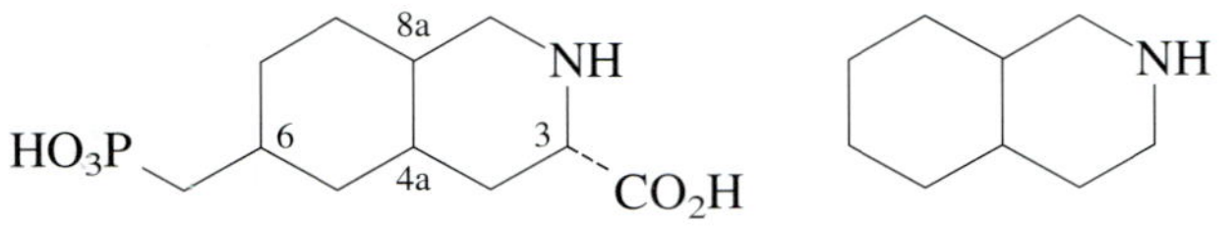

(a) Use your model kits to analyze the ring juncture. Make models of the cis as well as the trans ring juncture of structure B. You should have four different models. Identify the stereochemical relation between them as diastereomeric or enantiomeric. Draw the isomers and assign the R or S configuration to the stereocenters at the ring fusion.
(b) Although the trans ring juncture is the energetically more favorable one, the compound with cis ring juncture is the stereoisomer of structure A that shows biological activity. Make models of structure A that have the cis ring juncture exclusively. Set the stereochemistry of C3 as shown in structure A and vary the center at C6 in relation to that at C3. Again, there are four different models. Draw them and convince yourselves that none of them are enantiomers by assigning the R or S configuration to all four of the stereocenters in each of the compounds.
(c) The stereoisomer of compound A that shows the greatest biological activity has a cis ring fusion with substituents at C3 and C6 that are both equatorial. Which of the stereoisomers that you drew encompasses these constraints? Identify it by recording the absolute configuration at C3, C4a, C6, and C8a.

Preprofessional Problems

54. Which compound will *not* exhibit optical activity? (Note that these are all Fischer projections.)

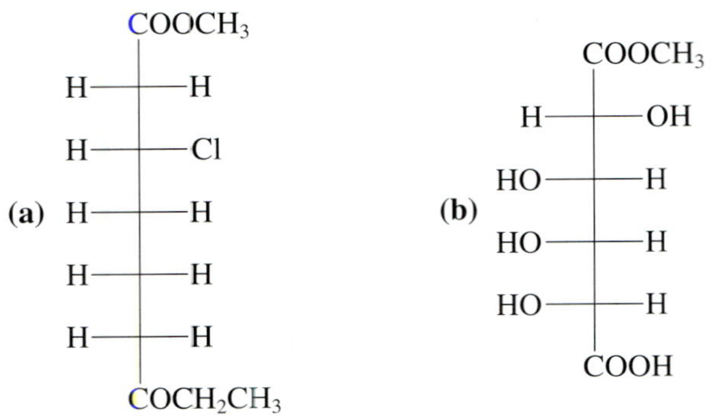

(c)

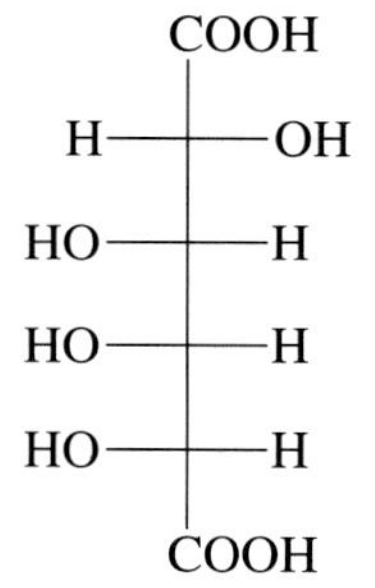

$$\begin{array}{c} \text{COOH} \\ \text{H}\!-\!\!-\!\!-\!\text{OH} \\ \text{HO}\!-\!\!-\!\!-\!\text{H} \\ \text{HO}\!-\!\!-\!\!-\!\text{H} \\ \text{HO}\!-\!\!-\!\!-\!\text{H} \\ \text{COOH} \end{array}$$

(d)

$$\begin{array}{c} \text{COOH} \\ \text{H}\!-\!\!-\!\!-\!\text{OH} \\ \text{Cl}\!-\!\!-\!\!-\!\text{H} \\ \text{Cl}\!-\!\!-\!\!-\!\text{H} \\ \text{H}\!-\!\!-\!\!-\!\text{OH} \\ \text{COOH} \end{array}$$

55. The enantiomer of

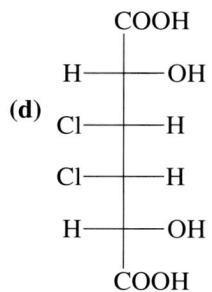

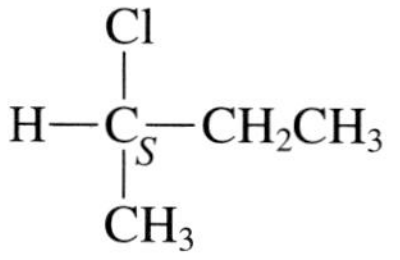

$$\begin{array}{c} \text{Cl} \\ | \\ \text{H}\!-\!\underset{S}{\text{C}}\!-\!\text{CH}_2\text{CH}_3 \\ | \\ \text{CH}_3 \end{array}$$

(a) is $\begin{array}{c} \text{Cl} \\ | \\ \text{CH}_3\text{CH}_2\!-\!\underset{R}{\text{C}}\!-\!\text{H} \\ | \\ \text{CH}_3 \end{array}$

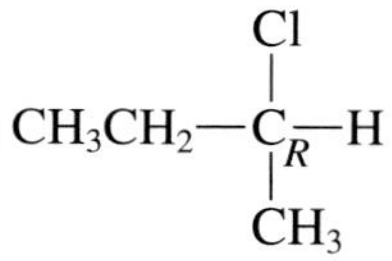

(b) can exist only at low temperatures

(c) is nonisomeric

(d) is incapable of existence

56. The molecule that is of the *R* configuration according to the Cahn-Ingold-Prelog convention is (remember these are Fischer projections):

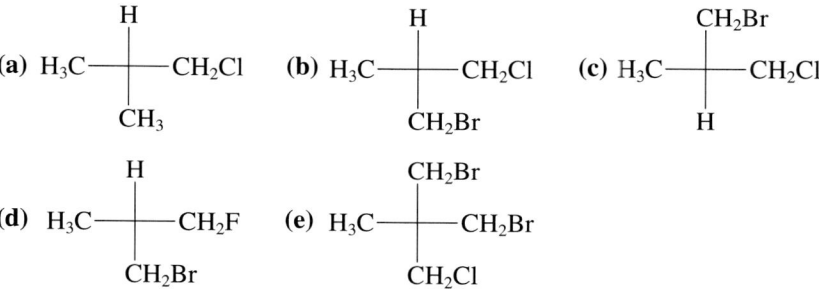

(a) $\text{H}_3\text{C}\!-\!\!\!\!\!\begin{array}{c}\text{H}\\|\\|\\\text{CH}_3\end{array}\!\!\!\!\!-\!\text{CH}_2\text{Cl}$

(b) $\text{H}_3\text{C}\!-\!\!\!\!\!\begin{array}{c}\text{H}\\|\\|\\\text{CH}_2\text{Br}\end{array}\!\!\!\!\!-\!\text{CH}_2\text{Cl}$

(c) $\text{H}_3\text{C}\!-\!\!\!\!\!\begin{array}{c}\text{CH}_2\text{Br}\\|\\|\\\text{H}\end{array}\!\!\!\!\!-\!\text{CH}_2\text{Cl}$

(d) $\text{H}_3\text{C}\!-\!\!\!\!\!\begin{array}{c}\text{H}\\|\\|\\\text{CH}_2\text{Br}\end{array}\!\!\!\!\!-\!\text{CH}_2\text{F}$

(e) $\text{H}_3\text{C}\!-\!\!\!\!\!\begin{array}{c}\text{CH}_2\text{Br}\\|\\|\\\text{CH}_2\text{Cl}\end{array}\!\!\!\!\!-\!\text{CH}_2\text{Br}$

57. Which compound below is *not* a meso compound?

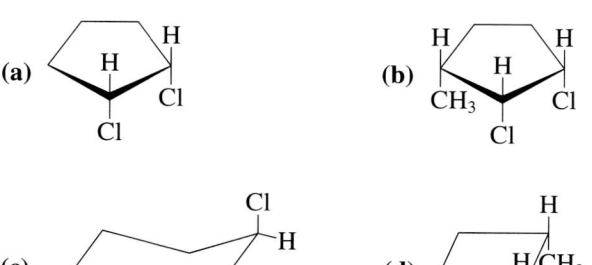

(a)

(b)

(c)

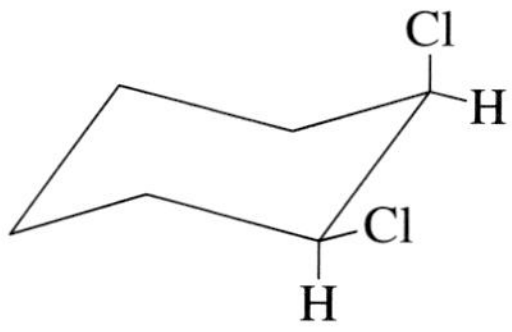

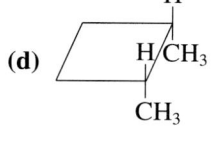

(d)

(e)

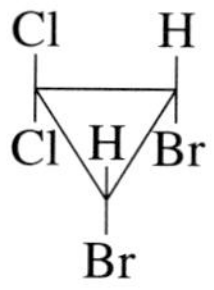

Bimolecular Nucleophilic Substitution

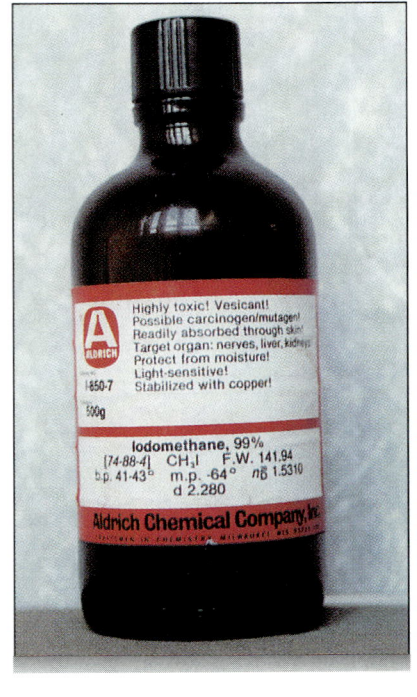

The nucleophilic substitution reaction between iodomethane and a nitrogen atom in the molecule guanine, a constituent of DNA, may cause abnormalities in cell function serious enough to cause cancer.

Organic chemistry provides us with myriad ways to convert one substance into another. The products of these transformations are literally all around us. Recall from Chapter 2, however, that functional groups are the centers of reactivity in organic molecules; before we can make practical use of organic chemistry, we must develop our ability to work with these functional groups. In Chapter 3 we examined halogenation of alkanes, a process by which the carbon–halogen group is introduced into an initially unfunctionalized structure. Where do we go from here?

In this chapter we turn to the chemistry of the products of halogenation, the haloalkanes. We shall see how the polarized carbon–halogen bond governs the reactivity of these substances and how it can be converted into other functional groups. On the basis of the kinetics observed for a common reaction of haloalkanes, we introduce a new mechanism and learn the effects of different solvents on its progress. We shall also learn principles that apply in general to the mechanistic behavior of molecules with polar functional groups. Finally, we shall begin to practice the application of these principles and see the role that they play in many conversions of halogenated organic compounds into other substances, such as amino acids, the building blocks of proteins.

Laboratory Preparation of the Amino Acid Alanine (from Chapter 26)

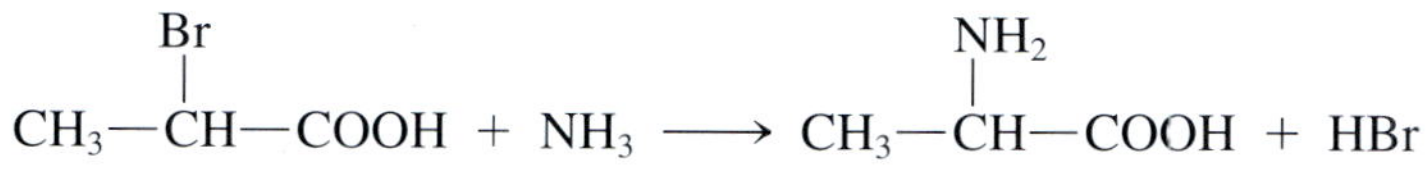

$$\underset{\displaystyle \overset{\displaystyle Br}{|}}{CH_3-CH-COOH} + NH_3 \longrightarrow \underset{\displaystyle \overset{\displaystyle NH_2}{|}}{CH_3-CH-COOH} + HBr$$

Alanine

We start with the rules for naming haloalkanes.

6-1 Naming the Haloalkanes

We learned in Chapter 2 that alkanes are depicted by the general formula R–H, where R denotes an alkyl group. In a similar way, the **haloalkanes** are represented as R–X, in which X corresponds to a halogen atom.

In the systematic (IUPAC) nomenclature, the halogen is treated as a substituent to the alkane framework.

CH_3I

Iodomethane   **Fluoro**cyclohexane   **2-Bromo**-**2-methyl**propane

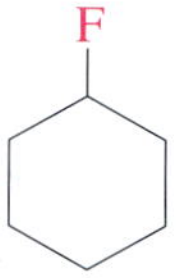

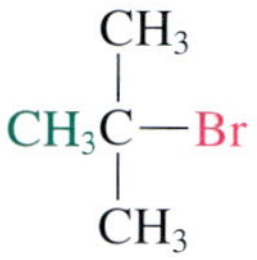

The longest alkane chain is numbered so that the first substituent from either end receives the lowest number. As usual, substituents are ordered according to the alphabet. Complex appendages are named according to the rules used for complex alkyl substituents (Section 2-3).

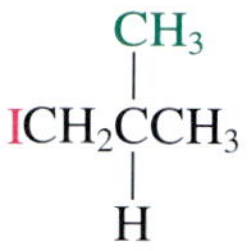

ICH_2CCH_3

1-Iodo-**2-methyl**propane   **(1-Iodoethyl)cyclooctane** 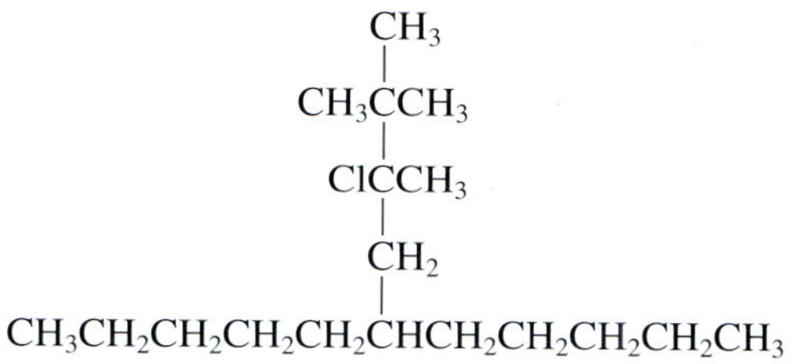   **6-(2-Chloro-2,3,3-trimethylbutyl)undecane**

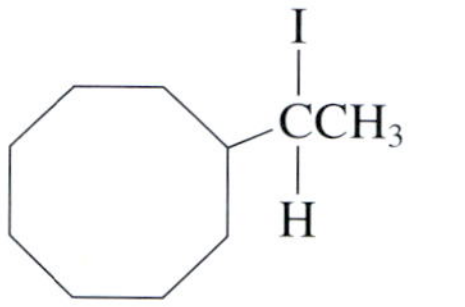

Common names are based on the older term *alkyl halide*. For example, the three structures at the beginning of this section have the common names methyl iodide, cyclohexyl fluoride, and *tert*-butyl bromide, respectively. Some chlorinated solvents have common names: for example, carbon tetrachloride, CCl_4; chloroform, $CHCl_3$; and methylene chloride, CH_2Cl_2.

EXERCISE 6-1

Draw the structures of (2-iodoethyl)cyclooctane and 5-butyl-3-chloro-2,2,3-trimethyl-decane.

In summary, haloalkanes are named in accord with the rules that apply to naming the alkanes (Section 2-3), the halo substituent being treated the same as alkyl groups.

6-2 Physical Properties of Haloalkanes

The physical properties of the haloalkanes are quite distinct from those of the corresponding alkanes. To understand these differences, we must consider the size of the halogen substituent and the polarity of the carbon–halogen bond. Let us see how these factors affect bond strength, bond length, molecular polarity, and boiling point.

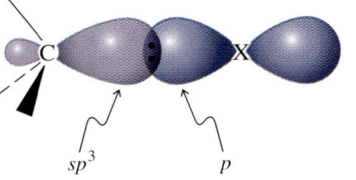

FIGURE 6-1 ——————
Bond between an alkyl carbon and a halogen. The size of the p orbital is substantially larger than that shown for X = Cl, Br, or I.

The bond strength of C–X decreases as the size of X increases

The bond between carbon and a halogen is made up mainly by the overlapping of an sp^3 hybrid orbital on carbon with a p orbital on the halogen (Figure 6-1). In the progression from fluorine to iodine in the periodic table, the size of the halogen p orbital increases, and the electron cloud around the halogen atom becomes more diffuse. Consequently, its overlap with the carbon orbital and hence the C–X bond strength diminish. For example, the C–X bond dissociation energies in the halomethanes, CH_3X, decrease along the series; at the same time, the C–X bond lengths increase (Table 6-1).

The C–X bond is polarized

A characteristic of the haloalkanes is their polar C–X bond. Recall from Section 1-3 that the halogens are more electronegative than is carbon. Thus, the electron density along the C–X bond is displaced in the direction of X, thereby giving the halogen a partial negative charge (δ^-) and the carbon a partial positive charge (δ^+). How does this bond polarization govern the chemical behavior of the haloalkanes? We shall see, for example, that anions and other electron-rich species can attack the positively polarized carbon atom. Cations and other electron-deficient species, however, attack the halogen.

TABLE 6-1	C–X Bond Lengths and Bond Strengths in CH_3X	
Halo-methane	Bond length (Å)	Bond strength (kcal mol^{-1})
CH_3F	1.385	110
CH_3Cl	1.784	85
CH_3Br	1.929	71
CH_3I	2.139	57

The Polar Character of the C–X Bond

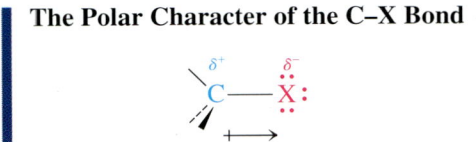

Haloalkanes have higher boiling points than the corresponding alkanes

Does the polarity of the C–X bond affect the physical properties of the haloalkanes? Yes, their boiling points are generally higher than those of the corresponding alkanes (Table 6-2). The most important contributor to this effect is Coulombic attraction between the δ^+ and δ^- ends of C–X bond dipoles in the liquid state (*dipole–dipole interaction*).

Boiling points also rise with increasing size of X, the result of increased molecular weight and greater London interactions (Section 2-4). Recall that London forces arise from mutual correlation of electrons among molecules. This effect is strongest when the outer electrons are not held very tightly around the nucleus, as in the heavier atoms. To measure it, we define the **polarizability** of an atom as the degree to

Dipole–Dipole Attraction

TABLE 6-2	Boiling Points of Haloalkanes (R–X)				
		Boiling point (°C)			
R	X =	H	F	Cl	Br	I
CH_3		−161.7	−78.4	−24.2	3.6	42.4
CH_3CH_2		−88.6	−37.7	12.3	38.4	72.3
$CH_3(CH_2)_2$		−42.1	−2.5	46.6	71.0	102.5
$CH_3(CH_2)_3$		−0.5	32.5	78.4	101.6	130.5
$CH_3(CH_2)_4$		36.1	62.8	107.8	129.6	157.0
$CH_3(CH_2)_7$		125.7	142.0	182.0	200.3	225.5

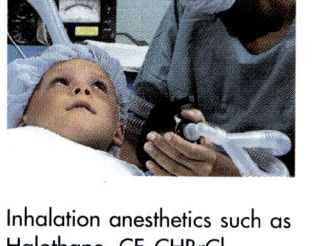

Inhalation anesthetics such as Halothane, $CF_3CHBrCl$, derive their biological activity from the polar nature of their C–X bonds.

which its electron cloud is deformed under the influence of an external electric field. The more polarizable an atom, the more effectively it will enter into London interactions, and the higher will be the boiling point.

To summarize, the halogen orbitals become increasingly diffuse along the series F, Cl, Br, I. Hence, (1) the C–X bond strength decreases; (2) the C–X bond becomes longer; (3) for the same R, the boiling points increase; (4) the polarizability of X becomes greater; and (5) London interactions improve. We shall see next that these interrelated effects also play an important role in the reactions of haloalkanes.

6-3 Nucleophilic Substitution

Haloalkanes often react with substances containing an unshared electron pair. This reagent can be an anion, such as iodide ($:\overset{..}{\underset{..}{I}}:^-$), or a neutral species, such as ammonia ($:NH_3$). Each reagent can attack the haloalkane and replace the halide, a process called **nucleophilic substitution.** A great many species are transformed in this way, particularly in solution. The process occurs widely in nature and can be controlled effectively even on an industrial scale. Let us see how it works.

Nucleophiles attack electrophilic centers

We noted that in the polarization of the carbon–halogen bond, the carbon atom acquires a partial positive charge. As a result, this center exhibits a tendency to react with species possessing unshared pairs of electrons: The carbon is said to be **electrophilic** (literally, "electron loving"; *philos,* Greek, loving). In turn, atoms bearing lone pairs are described as **nucleophilic** ("nucleus loving"). By definition, the terms *nucleophile* and *Lewis base* (Section 2-9) are synonymous: *All nucleophiles are Lewis bases.* In common usage, we employ the term *nucleophile* to describe a Lewis basic species that is attacking a *nonhydrogen* electrophilic atom. Nucleophiles, often denoted by the abbreviation Nu, may be negatively charged or neutral, but every nucleophile contains at least one unshared pair of electrons.

Halogenated Organic Compounds and the Environment

Today, more than 15,000 halogenated organic compounds are manufactured for commercial use. The largest single industrial use of chlorine is in the preparation of polychloroethene-containing plastics [polyvinyl chloride (PVC), Section 12-14]. More than 6 million tons of PVC-based materials are produced annually in the United States. Other significant applications of chlorinated organics include solvents, industrial lubricants and insulators, herbicides, and insecticides. Some of these substances persist in the environment after disposal and are suspected of causing a variety of adverse health effects in both humans and wildlife. Insecticides such as DDT (Chemical Highlight 3-3), chlordane, and the hexachlorocyclohexanes (e.g., lindane) are known endocrine disruptors, causing reproductive abnormalities in animals. These environmentally very persistent substances are still present in significant amounts, despite their use in the United States having been restricted or banned since the early 1970s. In the same category are the PCBs (polychlorinated biphenyls), widely used as insulating fluids in electrical transmission equipment since the 1920s

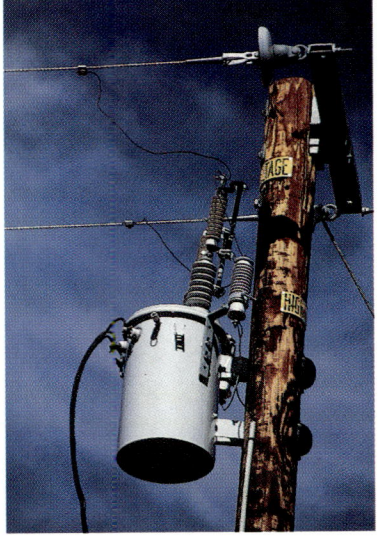

PCBs, formerly used as insulating fluids in power transformers, have been banned since the 1970s.

until their restriction in 1976, and a major contaminant in the waters of the Great Lakes basin.

Several solutions to the problems of use and disposal of chlorinated organics have been developed. For example, carbon dioxide at moderately elevated pressure and temperature converts into a fluid that can extract the caffeine from coffee beans, replacing dichloromethane for this purpose. Properly controlled incineration can destroy halocarbon wastes with minimal environmental effect. The difficult problem of decontamination of polluted sites is being addressed by several innovative technologies. One of them, bioremediation, employs microorganisms that feed on chlorinated organics. Indeed, by the early 1990s, exotic anaerobic organisms naturally present in the upper Hudson River had removed most of the chlorine from the PCBs in the river sediments, converting these molecules into substances more readily biodegradable by conventional aerobes. Whether human efforts to develop bioremediation into a practical technology will ultimately succeed on a large scale is very much an open question at the present time. For discussions on related topics, see Chemical Highlights 6-2 and 22-1.

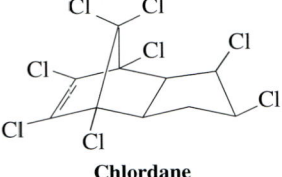

Chlordane

Lindane

Decachlorobiphenyl (a PCB)

The nucleophilic substitution of a haloalkane is described by either of two general equations.

Color code
Nucleophiles: red
Electrophiles: blue
Leaving groups: green

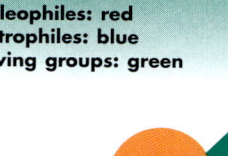

Nucleophilic Substitutions

$$Nu:^- \quad + \quad R-\overset{\delta^+}{\underset{}{\ddot{X}}}:^{\delta^-} \quad \longrightarrow \quad R-Nu \quad + \quad :\ddot{X}:^-$$

Nucleophile   **Electrophile**   **Leaving group**

$$Nu: \quad + \quad R-\overset{\delta^+}{\underset{}{\ddot{X}}}:^{\delta^-} \quad \longrightarrow \quad [R-Nu]^+ \quad + \quad :\ddot{X}:^-$$

Nucleophile   **Electrophile**   **Leaving group**

In the first example, a negatively charged nucleophile reacts with a haloalkane to yield a neutral substitution product. In the second example, an uncharged Nu produces a positively charged product. In both cases, the group displaced is the halide ion, $:\ddot{X}:^-$, which is called the **leaving group.** Specific examples of these two types of nucleophilic substitution are shown in Table 6.3. As will be the case in many equations and mechanisms that follow, nucleophiles, electrophiles, and leaving groups are shown here in red, blue, and green, respectively. The general term **substrate** (*substratus,* Latin, to have been subjected) is applied to the organic starting material—in this case, the haloalkane—which is the target of attack by a nucleophile.

Nucleophilic substitution exhibits considerable diversity

Note that Table 6-3 depicts only primary and secondary halides. Tertiary substrates behave differently toward these nucleophiles, and even secondary halides may give other products in addition to those of substitution. These reactions will be addressed in Chapter 7. The "cleanest" nucleophilic substitutions are obtained with methyl and primary haloalkanes.

Let us inspect these transformations in greater detail. In reaction 1, a hydroxide ion, typically derived from sodium or potassium hydroxide, displaces chloride from chloromethane to give methanol. This substitution is a general synthetic method for converting a primary haloalkane into an alcohol.

A variation of this transformation is reaction 2. Methoxide ion reacts with iodoethane to give methoxyethane, an example of the synthesis of an ether (Section 9-6).

In reactions 1 and 2, the species attacking the haloalkane is an anionic oxygen nucleophile. Reaction 3 shows that a halide ion may function not only as a leaving group, but also as a nucleophile.

Reaction 4 depicts a carbon nucleophile, cyanide (often supplied as sodium cyanide, $Na^+{}^-CN$), and leads to the formation of a new carbon–carbon bond, an important means of lengthening the carbon chain.

Reaction 5 shows the sulfur analog of reaction 2, demonstrating that nucleophiles in the same column of the periodic table react similarly to give analogous products. This conclusion is also borne out by reactions 6 and 7. However, the nucleophiles in these two reactions are *neutral,* and the expulsion of the negatively charged leaving group results in a cationic species, an ammonium or phosphonium salt.

All of the nucleophiles shown in Table 6-3 are quite reactive, but not all for the same reasons. Some are reactive because they are strongly basic (HO^-, CH_3O^-). Oth-

| TABLE 6-3 | The Diversity of Nucleophilic Substitution | | | |

Reaction number	Substrate	Nucleophile	Product	Leaving group
1.	CH_3Cl **Chloromethane**	$+$ HO^-	$\longrightarrow$ CH_3OH **Methanol**	$+$ Cl^-
2.	CH_3CH_2I **Iodoethane**	$+$ CH_3O^-	$\longrightarrow$ $CH_3CH_2OCH_3$ **Methoxyethane**	$+$ I^-
3.	$\overset{\overset{\textstyle H}{\vert}}{CH_3CCH_2CH_3}$ $\underset{\textstyle Br}{\vert}$ **2-Bromobutane**	$+$ I^-	$\longrightarrow$ $\overset{\overset{\textstyle H}{\vert}}{CH_3CCH_2CH_3}$ $\underset{\textstyle I}{\vert}$ **2-Iodobutane**	$+$ Br^-
4.	$\overset{\overset{\textstyle H}{\vert}}{CH_3CCH_2I}$ $\underset{\textstyle CH_3}{\vert}$ **1-Iodo-2-methyl-propane**	$+$ $N{\equiv}C^-$	$\longrightarrow$ $\overset{\overset{\textstyle H}{\vert}}{CH_3CCH_2C{\equiv}N}$ $\underset{\textstyle CH_3}{\vert}$ **3-Methylbutane-nitrile**	$+$ I^-
5.	**Bromocyclohexane** (ring with Br)	$+$ CH_3S^-	$\longrightarrow$ **Methylthiocyclohexane** (ring with SCH_3)	$+$ Br^-
6.	CH_3CH_2I **Iodoethane**	$+$ $:NH_3$	$\longrightarrow$ $\overset{\overset{\textstyle H}{\vert+}}{CH_3CH_2N\!H}$ $\underset{\textstyle H}{\vert}$ **Ethylammonium iodide**	$+$ I^-
7.	CH_3Br **Bromomethane**	$+$ $:P(CH_3)_3$	$\longrightarrow$ $\overset{\overset{\textstyle CH_3}{\vert+}}{CH_3P\!CH_3}$ $\underset{\textstyle CH_3}{\vert}$ **Tetramethylphosphonium bromide**	$+$ Br^-

Note: Remember that nucleophiles are red, electrophiles are blue, and leaving groups are green.

ers are weak bases (I^-) whose nucleophilicity derives from other characteristics. Notice that, in each example, the leaving group is a halide ion. Halides are unusual in that they may serve as leaving groups as well as nucleophiles (therefore making reaction 3 reversible). However, the same is *not* true of some of the other nucleophiles in Table 6-3 (in particular, the strong bases); the equilibria of their reactions lie strongly in the direction shown. These topics are addressed in Sections 6-8 and 6-9, as are factors that affect the reversibility of displacement reactions. First, however, we shall examine the mechanism of nucleophilic substitution.

EXERCISE 6-2

What are the substitution products of the reaction of 1-bromobutane with (a) $:\!\ddot{I}\!:^-$; (b) $CH_3CH_2\!:\!\ddot{O}\!:^-$; (c) N_3^-; (d) $:As(CH_3)_3$; (e) $(CH_3)_2\ddot{S}e$?

EXERCISE 6-3

Suggest starting materials for the preparation of **(a)** $(CH_3)_4N^+I^-$; **(b)** $CH_3SCH_2CH_3$.

In summary, nucleophilic substitution is a fairly general reaction for primary and secondary haloalkanes. The halide functions as the leaving group, and several types of nucleophilic atoms enter into the process.

6-4 Reaction Mechanisms Involving Polar Functional Groups: Using "Electron-Pushing" Arrows

In our consideration of radical halogenation in Chapter 3, we found that a knowledge of its mechanism was helpful in explaining the experimental characteristics of the process. The same will be the case for nucleophilic substitution, and, indeed, virtually every chemical process that we encounter. Nucleophilic substitution is an example of a polar reaction: it includes charged species and polarized bonds. Recall (Chapter 2) that an understanding of electrostatics is essential if we are to comprehend how such processes take place. Opposite charges attract—nucleophiles are attracted to electrophiles—and this principle provides us with a basis for understanding the mechanisms of polar organic reactions. In this section, we shall expand the concept of *electron flow* first introduced in the context of acid-base reactions (Section 2-9) and learn the conventional methods for illustrating polar reaction mechanisms by *moving electrons* from electron-rich to electron-poor sites. In subsequent sections, we shall apply these ideas specifically to nucleophilic substitution.

Curved arrows depict the movement of electrons

As we learned in Section 2-9, acid-base processes require electron movement. Let us briefly examine the Brønsted-Lowry process in which the acid HCl donates a proton to a molecule of water in aqueous solution:

$$HCl + H_2O \rightleftharpoons H_3O^+ + Cl^-$$

In this process, a lone pair originally on the oxygen atom of water has become part of a new bond in the hydronium ion. Likewise, the pair of electrons that constituted the H–Cl bond has shifted to the chlorine atom, converting it into a negatively charged chloride ion. We employ two curved arrows to denote the movement of these two pairs of electrons:

Depiction of a Brønsted-Lowry Acid-Base Reaction by Using Curved Arrows

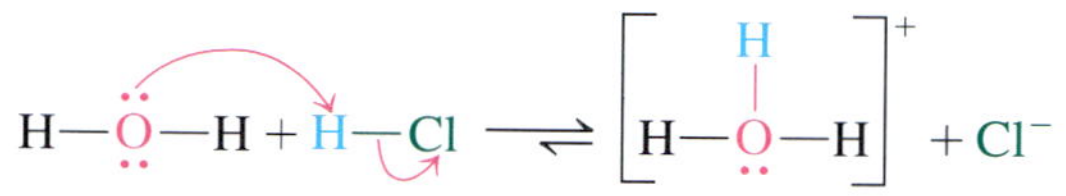

Notice that the arrow starting at the lone pair on oxygen and ending at the hydrogen of HCl does *not* imply that the lone electron pair departs from oxygen completely; it just becomes a *shared* pair between that oxygen atom and the atom to which the arrow points. In contrast, however, the arrow beginning at the H–Cl bond and pointing toward the chlorine atom *does* signify cleavage of the bond; that electron pair becomes separated from hydrogen and ends up entirely on the chloride ion.

EXERCISE 6-4

Use curved arrows to depict the flow of electrons in each of the following acid-base reactions. **(a)** Hydrogen ion + hydroxide ion; **(b)** fluoride ion + boron trifluoride, BF_3; **(c)** ammonia + hydrogen chloride; **(d)** hydrogen sulfide, H_2S, + sodium methoxide, $NaOCH_3$; **(e)** dimethyloxonium ion, $(CH_3)_2OH^+$, + water; **(f)** the self-ionization of water to give hydronium ion and hydroxide ion.

What about mechanisms in organic chemistry? We shall find that they share many of the features of acid-base reactions. As stated in Section 6-3, a nucleophile is nothing more than a Lewis base that attacks an electrophilic atom other than hydrogen. Electrophiles and Lewis acids are related but not in quite the same way: Lewis acids constitute the *subset* of electrophiles that are at least two electrons short of a closed shell. As we have seen, the carbon atom in a haloalkane possesses a filled outer shell; it is electrophilic by virtue of its partial positive charge, which renders it susceptible to attack by nucleophiles. Nucleophilic substitution is just one of many kinds of processes in which electrophiles and nucleophiles interact. Several examples are shown here, with curved arrows representing electron-pair movement.

Curved-Arrow Representations of Several Common Types of Mechanisms

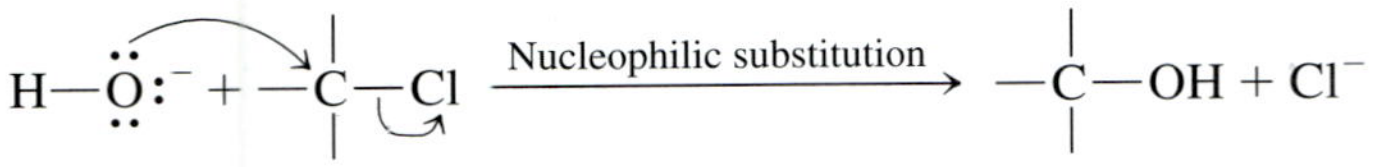

Compare with Brønsted acid-base reaction

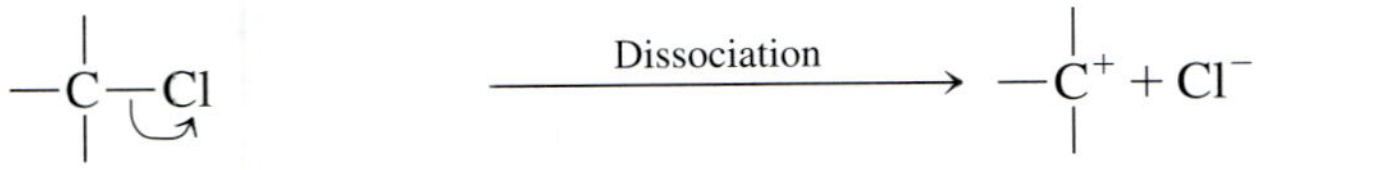

Reverse of Lewis acid–Lewis base reaction

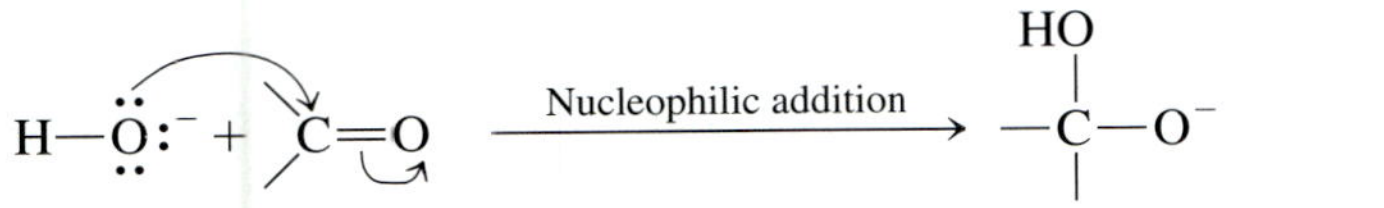

Only one of the two bonds between C and O is cleaved

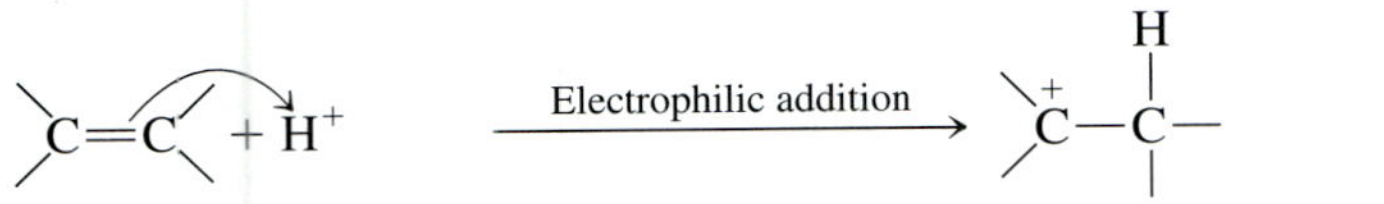

Carbon–carbon double bond acting as a Lewis base

The first and third examples illustrate a characteristic property of electron movement: If an electron pair moves toward an atom, that atom must have a "place to put that electron pair," so to speak. In nucleophilic substitution, the carbon atom in a haloalkane begins with a filled outer shell; another electron pair cannot be added without displacement of the electron pair bonding carbon to halogen. The two electron pairs can be viewed as "flowing" in a synchronous manner: As one pair arrives at the closed-shell atom, the other departs, thereby preventing violation of the octet rule. When you use the curved-arrow method to depict electron movement, *it is absolutely essential to keep in mind the rules for drawing Lewis structures.* Proper use of electron-pushing arrows, however, helps in drawing correct structures, because all electrons are moved to their proper destinations.

There are other types of processes, but, surprisingly, *not that many.* One of the most powerful consequences of studying organic chemistry from a mechanistic point of view is the way in which this approach highlights similarities between types of polar reactions even if the specific atoms and bonds are not the same.

EXERCISE 6-5

Identify the electrophilic and nucleophilic sites in the four mechanisms shown earlier as curved-arrow representations.

EXERCISE 6-6

Propose a curved-arrow depiction of the flow of electrons in the following processes, which will be considered in detail in this chapter and in Chapter 7.

(a) $-\overset{/}{\underset{\backslash}{\text{C}}}{}^{+}$ + Cl$^-$ ⟶ $-\overset{|}{\underset{|}{\text{C}}}-$Cl   **(b)** HO$^-$ + $\overset{\text{H}}{\underset{/}{\overset{|}{\text{C}}}}-\overset{|}{\underset{|}{\text{C}}}-$ ⟶ H$_2$O + $\overset{\backslash}{/}$C=C$\overset{/}{\backslash}$

In summary, curved arrows depict movement of electron pairs in reaction mechanisms. Electrons move from nucleophilic, or Lewis basic, atoms toward electrophilic, or Lewis acidic, sites. If a pair of electrons approaches an atom already containing a closed shell, a pair of electrons must depart from that atom so as not to exceed the maximum capacity of its valence orbitals.

6-5 A First Look at the Nucleophilic Substitution Mechanism: Kinetics

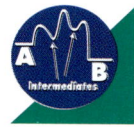

Many questions can be raised at this stage. What are the kinetics of nucleophilic substitution, and how does this information help us determine the underlying mechanism? What happens with optically active haloalkanes? Can we predict relative rates of substitution? These questions will be addressed in the remainder of this chapter.

When a mixture of chloromethane and sodium hydroxide in water is heated (denoted by the uppercase Greek letter *delta*, Δ, at the right of the arrow in the equation in the margin), a high yield of two compounds—methanol and sodium chloride—is the result. This outcome, however, does not tell us anything about *how* starting materials are converted into products. What experimental methods are available for answering this question?

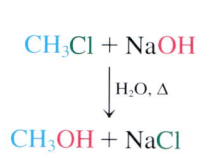

One of the most powerful techniques employed by chemists is the measurement of the *kinetics* of the reaction (Section 2-8). By comparing the rate of product formation beginning with several different concentrations of the starting materials, we can establish the **rate law** for a chemical process. Let us see what this experiment tells us about the reaction of chloromethane with sodium hydroxide.

The reaction of chloromethane with sodium hydroxide is bimolecular

We can monitor rates by measuring either the disappearance of one of the reactants or the appearance of one of the products. When we apply this method to the reaction of chloromethane with sodium hydroxide, we find that the rate depends on the initial concentrations of *both* of the reagents. For example, doubling the concentration of hydroxide doubles the rate at which the reaction proceeds. Likewise, at a fixed hydroxide concentration, doubling the concentration of chloromethane has the same effect. Doubling the concentrations of both increases the rate by a factor of 4. These results are consistent with a *second-order* process (Section 2-8), which is governed by the following rate equation.

$$\text{Rate} = k[\text{CH}_3\text{Cl}][\text{HO}^-] \text{ mol L}^{-1}\text{ s}^{-1}$$

All the examples given in Table 6-3 exhibit such second-order kinetics: Their rates are directly proportional to the concentration of both substrate and nucleophile.

EXERCISE 6-7

When a solution containing 0.01 M sodium azide ($Na^+N_3^-$) and 0.01 M iodomethane in methanol at 0°C is monitored kinetically, the results reveal that iodide ion is produced at a rate of 4.0×10^{-10} mol L^{-1} s^{-1}. Write the formula of the organic product of this reaction and calculate its rate constant k. What would be the rate of appearance of I^- for each of the following initial concentrations of reactants? **(a)** $[NaN_3] = 0.02$ M; $[CH_3I] = 0.01$ M. **(b)** $[NaN_3] = 0.02$ M; $[CH_3I] = 0.02$ M. **(c)** $[NaN_3] = 0.03$ M; $[CH_3I] = 0.03$ M.

What kind of mechanism is consistent with a second-order rate law? The simplest is one in which the two reactants interact in a single step. We call such a process **bimolecular,** and the general term applied to substitution reactions of this type is **bimolecular nucleophilic substitution,** abbreviated as **S_N2** (S stands for substitution, N for nucleophilic, and 2 for bimolecular).

Bimolecular nucleophilic substitution is a concerted, one-step process

Bimolecular nucleophilic substitution is a one-step process: The nucleophile attacks the haloalkane, with simultaneous expulsion of the leaving group. Bond-making takes place *at the same time* as bond-breaking. Because the two events occur "in concert," we call this process a **concerted** reaction.

We can envisage two stereochemically distinct alternatives for such concerted displacements. The nucleophile could approach the substrate from the same side as the leaving group, one group exchanging for the other. This pathway is called **frontside displacement** (Figure 6-2). The second possibility is a **backside displacement,** in which the nucleophile approaches carbon from the side opposite the leaving group

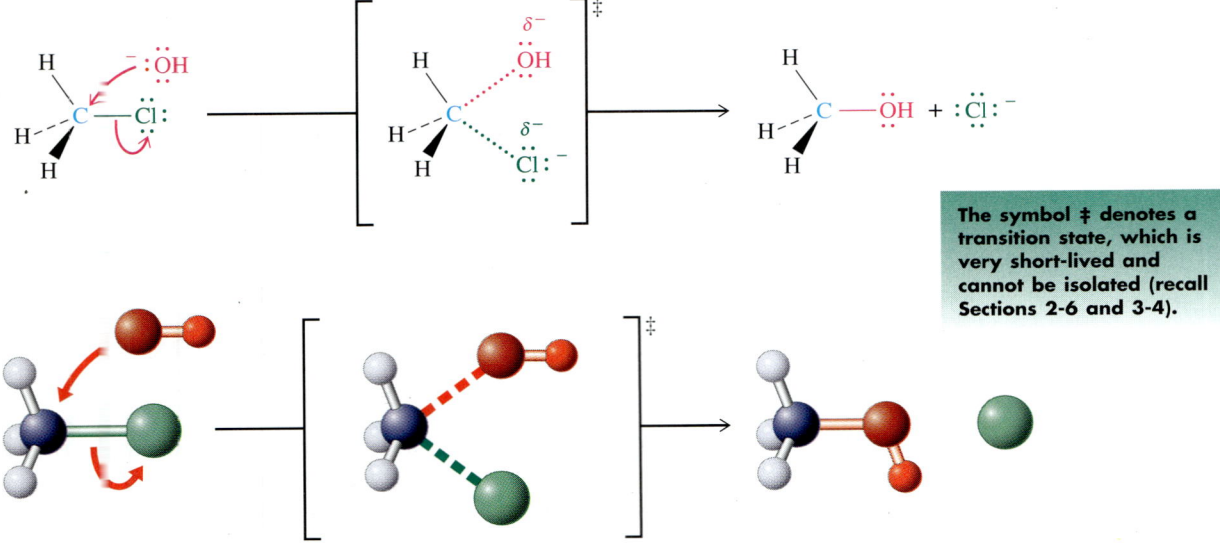

The symbol ‡ denotes a transition state, which is very short-lived and cannot be isolated (recall Sections 2-6 and 3-4).

FIGURE 6-2
Frontside nucleophilic substitution. The concerted nature of bond-making (to OH) and bond-breaking (from Cl) is indicated by the dotted lines.

FIGURE 6-3

Backside nucleophilic substitution. Attack is from the side *opposite* the leaving group.

(Figure 6-3). In both equations, we use curved arrows to denote the *movement of electron pairs.* An electron pair from the negatively charged hydroxide oxygen moves toward carbon, creating the C–O bond, while that of the C–Cl linkage shifts onto chlorine, thereby expelling the latter as $:\ddot{C}l:^-$. In either of the two respective transition states, the negative charge is distributed over both the oxygen and the chlorine atoms.

EXERCISE 6-8

Draw representations of the hypothetical frontside and backside displacement mechanisms for the S_N2 reaction of sodium iodide with 2-bromobutane (Table 6-3). Use arrows like those shown in Figures 6-2 and 6-3 to represent electron-pair movement.

In summary, the reaction of chloromethane with hydroxide to give methanol and chloride, as well as the related transformations of a variety of nucleophiles with haloalkanes, are examples of the bimolecular process known as the S_N2 reaction. Two single-step mechanisms—frontside attack and backside attack—may be envisioned for the reaction. Both are concerted processes, consistent with the second-order kinetics obtained experimentally. Can we distinguish between the two? To answer this question, we return to a topic that we have considered in detail: stereochemistry.

6-6 Frontside or Backside Attack? Stereochemistry of the S_N2 Reaction

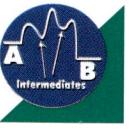

When we compare the structural drawings in Figures 6-2 and 6-3 with respect to the arrangement of their component atoms in space, we note immediately that, in the first conversion, the three hydrogens stay put and to the left of the carbon, whereas, in the second, they have "moved" to the right. In fact, the two methanol pictures are related as object and mirror image. In this example, the two are superimposable and therefore indistinguishable—properties of an achiral molecule. The situation is entirely different for a chiral haloalkane in which the electrophilic carbon is a stereocenter.

The S$_N$2 reaction is stereospecific

Consider the reaction of (S)-2-bromobutane with iodide ion. Frontside displacement should give rise to 2-iodobutane with the *same* configuration as that of the starting material; backside displacement should furnish a product with the *opposite* configuration.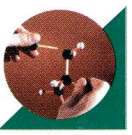

What is actually observed? It is found that (S)-2-bromobutane gives (R)-2-iodobutane on treatment with iodide: *This and all other S$_N$2 reactions proceed with* **inversion of configuration.** A process whose mechanism requires that each stereoisomer of the starting material transform into a specific stereoisomer of product is described as **stereospecific.** The S$_N$2 reaction is therefore stereospecific, proceeding by a backside displacement mechanism to give inversion of configuration at the site of the reaction.

Stereochemistry of the Backside Displacement Mechanism for S$_N$2 Reactions

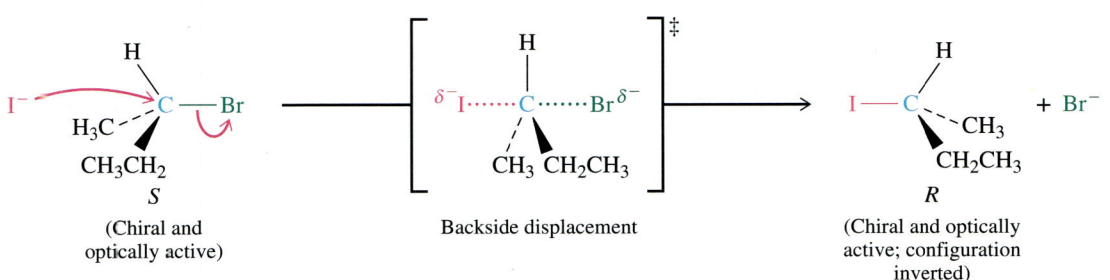

(Chiral and optically active)

Backside displacement

(Chiral and optically active; configuration inverted)

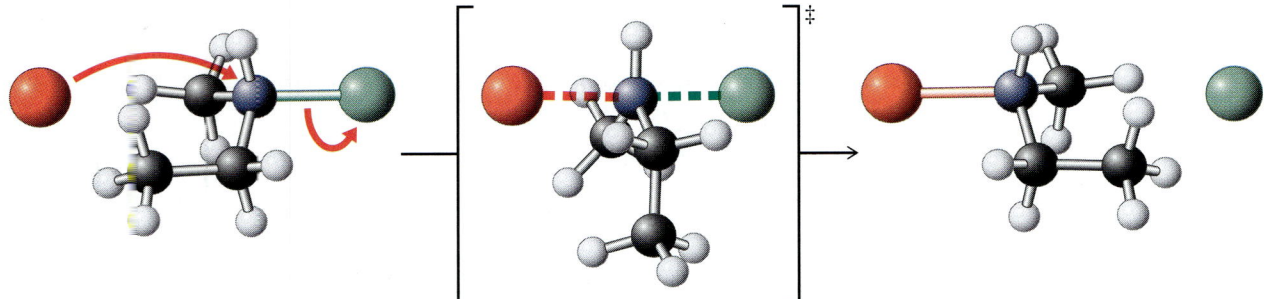

EXERCISE 6-9

Write the products of the following S$_N$2 reactions. **(a)** (R)-3-chloroheptane + Na$^+$ $^-$SH; **(b)** (S)-2-bromooctane + N(CH$_3$)$_3$; **(c)** (3R,4R)-4-iodo-3-methyloctane + K$^+$ $^-$SeCH$_3$.

EXERCISE 6-10

Write the structures of the products of the S$_N$2 reactions of cyanide ion with **(a)** *meso*-2,4-dibromopentane (double S$_N$2 reaction); **(b)** *trans*-1-iodo-4-methylcyclohexane.

The transition state of the S$_N$2 reaction can be described in an orbital picture

The transition state for the S$_N$2 reaction can be described in orbital terms, as shown in Figure 6-4. As the nucleophile approaches the back lobe of the sp^3 hybrid orbital used by carbon to bind the halogen atom, the rest of the molecule becomes planar at the transition state by changing the hybridization at carbon to sp^2. The negative charge is no longer located entirely on the nucleophile; it is also partly on the leaving group.

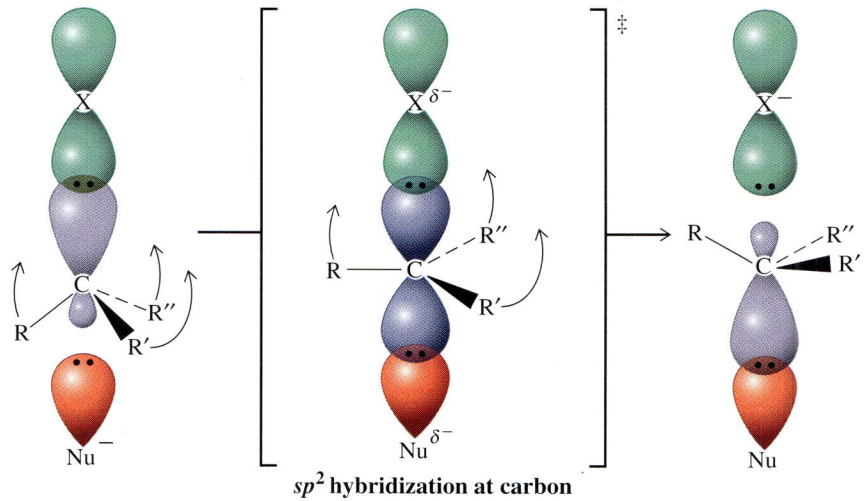

sp^2 **hybridization at carbon**

FIGURE 6-4
Molecular-orbital description of the S_N2 reaction. The process is reminiscent of the inversion of an umbrella exposed to gusty winds.

As the reaction proceeds to products, the inversion motion is completed, the carbon returns to the tetrahedral sp^3 configuration, and the leaving group becomes a fully charged anion. A depiction of the course of the reaction using a potential energy–reaction coordinate diagram is shown in Figure 6-5.

6-7 Consequences of Inversion in S_N2 Reactions

What are the consequences of the inversion of stereochemistry in the S_N2 reaction? Because the reaction is stereospecific, we can design ways to use displacement reactions to synthesize a desired stereoisomer.

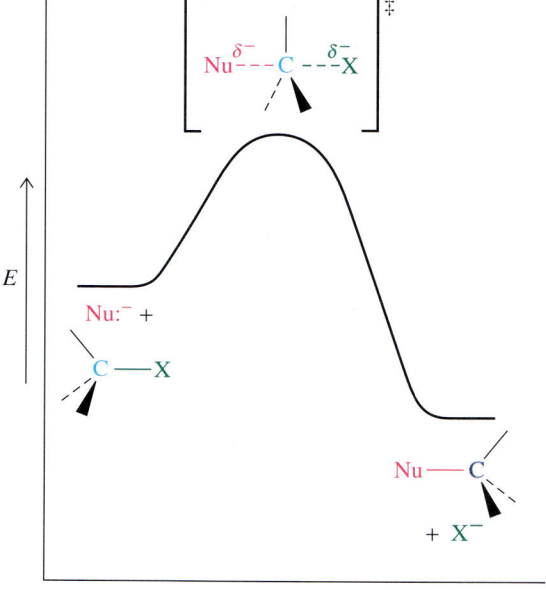

FIGURE 6-5
Potential energy diagram for an S_N2 reaction. The process takes place in a single step, with a single transition state.

We can synthesize a specific enantiomer by using S$_N$2 reactions

Consider the conversion of 2-bromooctane into 2-octanethiol in its reaction with hydrogen sulfide ion, HS⁻. If we were to start with optically pure *R* bromide, we would obtain only *S* thiol and none of its *R* enantiomer.

Inversion of Configuration of an Optically Pure Compound by S$_N$2 Reaction

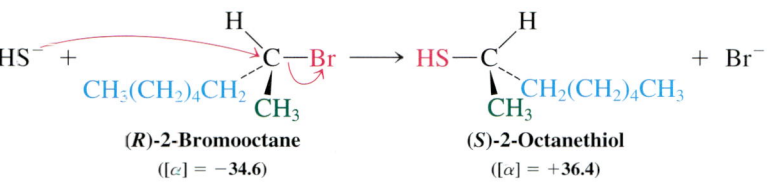

(*R*)-2-Bromooctane   (*S*)-2-Octanethiol

([α] = −34.6)   ([α] = +36.4)

> **Color code for priorities (see Section 5-3)**
> **Highest:** red
> **Second highest:** blue
> **Third highest:** green
> **Lowest:** black

But what if we wanted to convert (*R*)-2-bromooctane into the *R* thiol? One technique uses a sequence of *two* S$_N$2 reactions, each resulting in inversion of configuration at the stereocenter. For example, an S$_N$2 reaction with iodide would first generate (*S*)-2-iodooctane. We would then use this halide with an inverted configuration as the substrate in a second displacement, now with HS⁻ ion, to furnish the *R* thiol. This double inversion sequence of two S$_N$2 processes gives us the result we desire, a net **retention of configuration.**

Using Double Inversion to Give Net Retention of Configuration

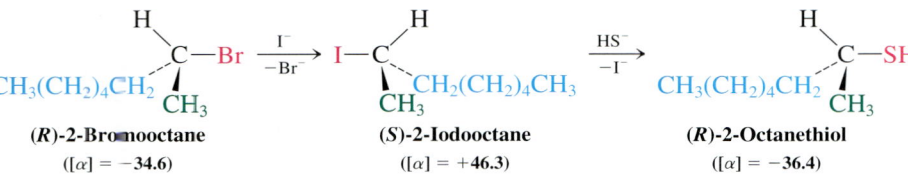

(*R*)-2-Bromooctane   (*S*)-2-Iodooctane   (*R*)-2-Octanethiol

([α] = −34.6)   ([α] = +46.3)   ([α] = −36.4)

EXERCISE 6-11

As we saw for carvone (Chapter 5, Problem 33), enantiomers can sometimes be distinguished by odor and flavor. 3-Octanol and some of its derivatives are examples: The dextrorotatory compounds are found in natural peppermint oil, whereas their (−) counterparts contribute to the essence of lavender. Show how you would synthesize optically pure samples of each enantiomer of 3-octyl acetate, starting with (*S*)-3-iodooctane.

$$O$$
$$\parallel$$
$$OCCH_3$$
$$|$$
$$CH_3CH_2CHCH_2CH_2CH_2CH_2CH_3$$
3-Octyl acetate

EXERCISE 6-12

Treatment of (*S*)-2-iodooctane with NaI causes the optical activity of the starting material to disappear. Explain.

In substrates bearing more than one stereocenter, inversion will take place *only* at the carbons that undergo reaction with the incoming nucleophile. Note that the reaction of (2*S*,4*R*)-2-bromo-4-chloropentane with excess cyanide ion results in a meso product.

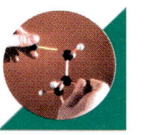

S$_N$2 Reactions of Molecules with Two Stereocenters

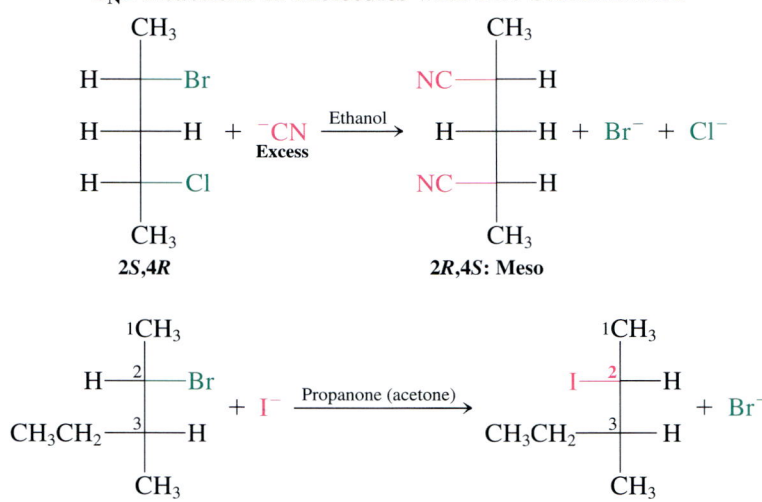

First structure: Fischer projection
CH3 top
H—Br
H—H
H—Cl
CH3 bottom
2S,4R

+ ⁻CN (Excess), Ethanol →

NC—H
H—H
NC—H
CH3
2R,4S: Meso
+ Br⁻ + Cl⁻

Second:
¹CH3
H—²—Br
CH3CH2—³—H
CH3
2S,3R
+ I⁻ Propanone (acetone) →
¹CH3
I—²—H
CH3CH2—³—H
CH3
2R,3R
+ Br⁻

EXERCISE 6-13

As an aid in the prediction of stereochemistry, organic chemists often use the guideline "diastereomers produce diastereomers." Replace the starting compound in each of the two preceding examples with one of its diastereomers, and write the product of S$_N$2 displacement with the nucleophile shown. Are the resulting structures in accord with this "rule"?

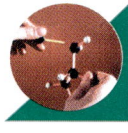

Similarly, nucleophilic substitution of a substituted halocycloalkane may change the stereochemical relation between the substituents.

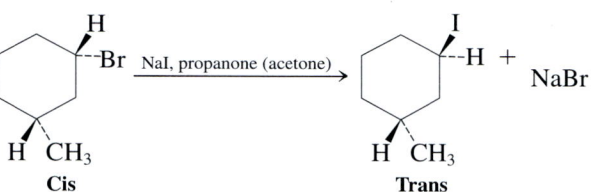

Cis   **Trans**

In summary, inversion of configuration in the S$_N$2 reaction has distinct stereochemical consequences. Optically active substrates give optically active products, unless the nucleophile and the leaving group are the same or meso compounds are formed. In cyclic systems, cis and trans stereochemical relations may be interconverted.

6-8 S$_N$2 Reactivity and Leaving-Group Ability

The relative facility of S$_N$2 displacements depends on a variety of factors, including the nature of the leaving group, the relative reactivity of the nucleophile, and the structure of the alkyl group in the substrate. We can enhance our understanding of the mechanism of the process by measuring relative reaction rates as we systematically vary each component. We shall begin with leaving groups, first considering the halides and then generalizing to other groups that can function in this capacity as well. Subsequent sections will address the nucleophile and the alkyl part of the substrate.

Leaving-group ability is a measure of the ease of its displacement

As a general rule, nucleophilic substitution will occur only when the group being displaced, X, is readily able to depart, taking with it the electron pair of the C–X bond. Are there structural features that might allow us to predict, at least qualitatively, whether a leaving group is "good" or "bad"? Not surprisingly, the relative ease with which it can be displaced, its **leaving-group ability,** can be correlated with its capacity to accommodate a negative charge. Remember that a certain amount of negative charge is transferred to the leaving group in the transition state of the reaction (Figure 6-4).

For the halogens, leaving-group ability increases along the series from fluorine to iodine. Thus, iodide is regarded as a "good" leaving group; fluoride, however, is so "poor" that S_N2 reactions of fluoroalkanes are rarely observed.

Leaving-Group Ability

$$I^- > Br^- > Cl^- > F^-$$
Best   Worst

EXERCISE 6-14

Predict the product of the reaction of 1-chloro-6-iodohexane with one equivalent of sodium methylselenide ($Na^{+-}SeCH_3$).

Halides are not the only groups that can be displaced by nucleophiles in S_N2 reactions. Other examples of good leaving groups are sulfur derivatives of the type $ROSO_3^-$ and RSO_3^- such as methyl sulfate ion, $CH_3OSO_3^-$, and various sulfonate ions. Alkyl sulfate and sulfonate leaving groups are used so often that trivial names, such as mesylate, triflate, and tosylate, have found their way into the chemical literature.

Sulfate and Sulfonate Leaving Groups

| Methyl sulfate ion | Methanesulfonate ion (Mesylate ion) | Trifluoromethanesulfonate ion (Triflate ion) | 4-Methylbenzenesulfonate ion (p-Toluenesulfonate ion, tosylate ion) |

Weak bases are good leaving groups

Is there some characteristic property that distinguishes good leaving groups from poor ones? Yes: *Leaving-group ability is inversely related to base strength.* Weak bases are best able to accommodate negative charge and are the best leaving groups. Among the halides, iodide is the weakest base and therefore the best leaving group in the series. Sulfates and sulfonates are weak bases as well. Table 6-4 (see page 228) lists a variety of species in ascending order of base strength, which is quantified by a **basicity constant,** K_b.

Is there a way to recognize weak bases readily? The weaker X^- is as a base, the stronger is its conjugate acid HX. Therefore, *good leaving groups are the conjugate bases of strong acids.* This rule applies to the four halides: HF is the weakest of the conjugate acids, HCl is stronger, and HBr and HI are stronger still. In our review of acids and bases (Section 2-9), we outlined some guidelines that may be used to evaluate relative strengths of acids. It is perhaps worthwhile to repeat them: The strength of an acid, HA, increases with the following structural features.

1. The increasing *size* of A in the procession down a column in the periodic table. Therefore, the acid strengths of the hydrogen halides follow the order HI >

TABLE 6-4	Base Strengths and Leaving Groups
Leaving group	K_b
Good leaving groups (weaker bases)	
I^-	6.3×10^{-20}
HSO_4^-	1.0×10^{-19}
Br^-	2.0×10^{-19}
Cl^-	6.3×10^{-17}
H_2O	2.0×10^{-16}
$CH_3SO_3^-$	6.3×10^{-16}
Poor leaving groups (stronger bases)	
F^-	1.6×10^{-11}
$CH_3CO_2^-$	5.0×10^{-10}
NC^-	1.6×10^{-5}
CH_3S^-	1.0×10^{-4}
CH_3O^-	32
HO^-	50
H_2N^-	1.0×10^{21}
H_3C^-	$\sim 1.0 \times 10^{36}$

HBr > HCl > HF. Consequently, I^- is the weakest conjugate base and, as noted earlier, the best leaving group, whereas F^-, the strongest base in the series, is difficult to displace in S_N2 processes.

2. The ability of the conjugate base, A^-, to accommodate the negative charge in either or both of two ways:

   (a) The increasing *electronegativity* of A in the procession from left to right across a row in the periodic table. For example, the decreasing order of acidity in the series HF > H_2O > NH_3 > CH_4 parallels the decreasing electronegativity of A. The leaving-group ability of the corresponding conjugate bases follows suit: F^- is a poor leaving group, but HO^- is worse—much too strong a base to depart in ordinary S_N2 processes. The even stronger bases NH_2^- and CH_3^- are similarly incapable of leaving in displacement reactions.

   (b) The resonance in A^- that allows delocalization of charge over several atoms. The sulfate and sulfonate anions illustrated earlier in this section possess significant resonance stabilization, thus explaining the considerable strength of their conjugate acids (Table 2-6) and their own value as excellent leaving groups.

EXERCISE 6-15

Predict the relative acidities within each of the following groups. **(a)** H_2S, H_2Se; **(b)** PH_3, H_2S; **(c)** $HClO_3$, $HClO_2$; **(d)** HBr, H_2Se; **(e)** NH_4^+, H_3O^+. Within each of the groups, identify the conjugate bases and predict their relative leaving-group abilities.

EXERCISE 6-16

Predict the relative basicities within each of the following groups. **(a)** ^-OH, ^-SH; **(b)** $^-PH_2$, ^-SH; **(c)** I^-, ^-SeH; **(d)** $HOSO_2^-$, $HOSO_3^-$. Predict the relative acidities of the conjugate acids within each group.

In summary, the leaving-group ability of a substituent is roughly proportional to the strength of its conjugate acid. Both depend on the ability of the leaving group to accommodate negative charge. In addition to the halides Cl^-, Br^-, and I^-, sulfates and sulfonates (such as methane- and 4-methylbenzenesulfonates) are good leaving groups. Good leaving groups are weak bases, the conjugate bases of strong acids. We shall return to uses of sulfates and sulfonates as leaving groups in synthesis in Section 9-4.

6-9 Effect of Nucleophilicity on the S_N2 Reaction

Now that we have looked at the effect of the leaving group, let us turn to a consideration of nucleophiles. How can we predict their relative nucleophilic strength, their **nucleophilicity?** We shall see that nucleophilicity depends on a variety of factors: charge, basicity, solvent, polarizability, and the nature of substituents. To grasp the relative importance of these effects, let us analyze the outcome of a series of comparative experiments.

Increasing negative charge increases nucleophilicity

If the same nucleophilic atom is used, does charge play a role in the reactivity of a given nucleophile as determined by the rate of its S_N2 reaction? The following experiments answer this question.

EXPERIMENT 1

$$CH_3Cl + HO^- \longrightarrow CH_3OH \quad + Cl^- \quad \text{Fast}$$
$$CH_3Cl + H_2O \longrightarrow CH_3OH_2^+ + Cl^- \quad \text{Very slow}$$

EXPERIMENT 2

$$CH_3Cl + H_2N^- \longrightarrow CH_3NH_2 \quad + Cl^- \quad \text{Very fast}$$
$$CH_3Cl + H_3N \longrightarrow CH_3NH_3^+ + Cl^- \quad \text{Slower}$$

Conclusion Of a pair of nucleophiles containing the same reactive atom, the species with a negative charge is the more powerful nucleophile. Or, of a base and its conjugate acid, the base is always more nucleophilic. This finding is intuitively very reasonable. Because nucleophilic attack is characterized by the formation of a bond with an electrophilic carbon center, the more negative the attacking species, the faster the reaction should be.

EXERCISE 6-17

Predict which member in each of the following pairs is a better nucleophile. **(a)** HS^- or H_2S; **(b)** CH_3SH or CH_3S^-; **(c)** CH_3NH^- or CH_3NH_2; **(d)** HSe^- or H_2Se.

Nucleophilicity decreases to the right in the periodic table

Experiments 1 and 2 compared pairs of nucleophiles containing the same nucleophilic element (e.g., oxygen in H_2O versus HO^- and nitrogen in H_3N versus H_2N^-). What about nucleophiles of similar structure but with different nucleophilic atoms? Let us examine the elements along one row of the periodic table.

EXPERIMENT 3

$$CH_3CH_2Br + H_3N \longrightarrow CH_3CH_2NH_3^+ + Br^- \quad \text{Fast}$$
$$CH_3CH_2Br + H_2O \longrightarrow CH_3CH_2OH_2^+ + Br^- \quad \text{Very slow}$$

EXPERIMENT 4

$$CH_3CH_2Br + H_2N^- \longrightarrow CH_3CH_2NH_2 + Br^- \quad \text{Very fast}$$
$$CH_3CH_2Br + HO^- \longrightarrow CH_3CH_2OH \quad + Br^- \quad \text{Slower}$$

Conclusion Nucleophilicity again appears to correlate with basicity: The more basic species is the more reactive nucleophile. Therefore, in the procession from the left to the right of the periodic table, nucleophilicity decreases. The approximate order of reactivity for nucleophiles in the first row is

$$H_2N^- > HO^- > NH_3 > F^- > H_2O$$

EXERCISE 6-18

In each of the following pairs of molecules, predict which is the more nucleophilic. **(a)** Cl^- or CH_3S^-; **(b)** $P(CH_3)_3$ or $S(CH_3)_2$; **(c)** $CH_3CH_2Se^-$ or Br^-; **(d)** H_2O or HF.

Should basicity and nucleophilicity be correlated?

The parallels between nucleophilicity and basicity first described in Section 6-3 are intuitively reasonable: Strong bases typically make good nucleophiles. However, a fundamental difference between the two properties is based on how they are measured. Basicity is a *thermodynamic* property, measured by an equilibrium constant:

$$A^- + H_2O \underset{}{\overset{K}{\rightleftharpoons}} AH + HO^- \qquad K = \text{equilibrium constant}$$

In contrast, nucleophilicity is a *kinetic* phenomenon, quantified by comparing rates of reactions:

$$Nu^- + R{-}X \xrightarrow{k} Nu{-}R + X^- \qquad k = \text{rate constant}$$

Despite these inherent differences, we have observed good correlation between basicity and nucleophilicity in the cases of charged versus neutral nucleophiles along a row of the periodic table. What happens if we look at nucleophiles in a column of the periodic table?

Solvation impedes nucleophilicity

If it is a general rule that nucleophilicity correlates with basicity, then the elements considered from top to bottom of a column of the periodic table should show decreasing nucleophilic power. Recall (Section 6-8) that basicity decreases in an analogous fashion. To test this prediction, let us consider another series of experiments.

EXPERIMENT 5

$$CH_3CH_2CH_2O\overset{\overset{O}{\|}}{\underset{\underset{O}{\|}}{S}}CH_3 + Cl^- \xrightarrow{CH_3OH} CH_3CH_2CH_2Cl + {}^-O_3SCH_3 \qquad \textbf{Slow}$$

$$CH_3CH_2CH_2O\overset{\overset{O}{\|}}{\underset{\underset{O}{\|}}{S}}CH_3 + Br^- \xrightarrow{CH_3OH} CH_3CH_2CH_2Br + {}^-O_3SCH_3 \qquad \textbf{Faster}$$

$$CH_3CH_2CH_2O\overset{\overset{O}{\|}}{\underset{\underset{O}{\|}}{S}}CH_3 + I^- \xrightarrow{CH_3OH} CH_3CH_2CH_2I + {}^-O_3SCH_3 \qquad \textbf{Fastest}$$

EXPERIMENT 6

$$CH_3CH_2CH_2Br + CH_3O^- \xrightarrow{CH_3OH} CH_3CH_2CH_2OCH_3 + Br^- \qquad \textbf{Not very fast}$$

$$CH_3CH_2CH_2Br + CH_3S^- \xrightarrow{CH_3OH} CH_3CH_2CH_2SCH_3 + Br^- \qquad \textbf{Very fast}$$

Conclusion. Nucleophilicity *increases* in the procession down the periodic table, a trend *directly opposing* that expected from the basicity of the nucleophiles tested. Sulfur nucleophiles are more reactive than the analogous oxygen systems but less so than their selenium counterparts. Similarly, among the halides, iodide is the fastest, although it is the weakest base.

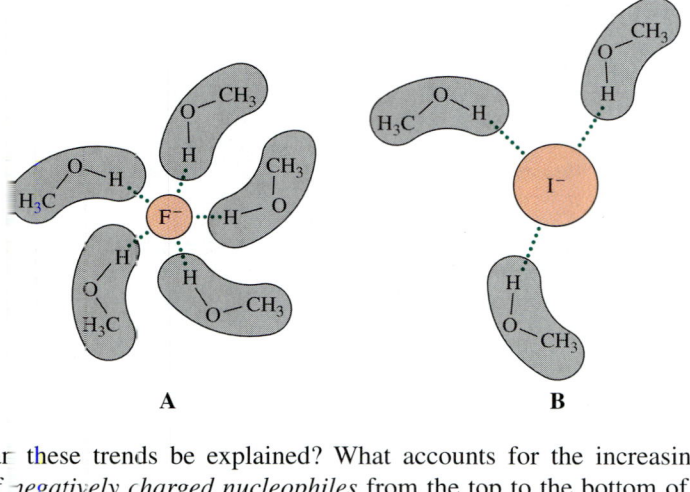

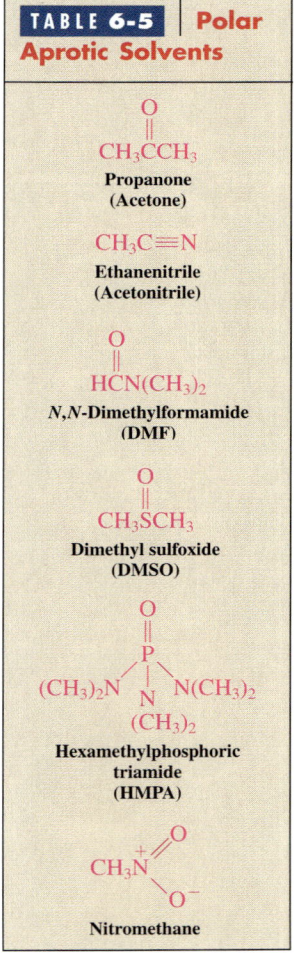

FIGURE 6-6

Approximate representation of the difference between solvation of (A) a small anion (F^-) and (B) a large anion (I^-). The tighter solvent shell around the smaller F^- impedes its ability to participate in nucleophilic substitution reactions.

How can these trends be explained? What accounts for the increasing nucleophilicity of *negatively charged nucleophiles* from the top to the bottom of a column of the periodic table? An important explanation is the interaction of solvent with the different anions.

When a solid dissolves, the intermolecular forces that held it together (Section 2-4) are replaced by interactions between molecules and solvent. In solution, molecules or ions are surrounded by solvent molecules; they are said to be **solvated.** Salts dissolve only in polar solvents, because only in these solvents is there enough solvation to separate the oppositely charged ions.

How does the solvation affect the strength of a nucleophile? Generally, solvation weakens the nucleophile by forming a shell of solvent molecules around the nucleophile and impeding its ability to attack an electrophile. Moreover, *smaller anions are more tightly solvated than are larger ones* because their charge is more concentrated. Figure 6-6 depicts this effect for methanol. The smaller fluoride ion is much more heavily solvated than is the larger iodide. Is this true in other solvents as well?

Protic and aprotic solvents: the effect of hydrogen bonding

In some polar solvents, such as methanol, ethanol, and water, a hydrogen is attached to an electronegative atom, Y. These solvents contain highly polarized $\overset{\delta+}{H}–\overset{\delta-}{Y}$ bonds, in which the hydrogen has protonlike character and can interact particularly strongly with anionic nucleophiles (Figure 6-6). (We shall study these interactions, called **hydrogen bonds,** more closely in Chapter 8.) Such solvents, called **protic,** are commonly used in nucleophilic substitutions.

Other solvents useful in S_N2 reactions are highly polar but **aprotic:** They lack positively polarized hydrogens. Polar, aprotic solvents such as propanone (acetone) are capable of dissolving salts. However, because they do not form hydrogen bonds, these solvent molecules solvate anionic nucleophiles relatively weakly. The result is that the reactivity of the nucleophile is raised, sometimes dramatically. For example, bromomethane reacts with potassium iodide 500 times as fast in propanone (acetone) as in methanol. Several other polar aprotic solvents are shown in Table 6-5; all lack protons capable of hydrogen bonding.

Table 6-6 (see page 232) compares the rates of S_N2 reactions of iodomethane with chloride in three protic solvents—methanol, formamide, and *N*-methylformamide—and one *aprotic* solvent, *N,N*-dimethylformamide (DMF). The rate of reaction in DMF is more than a million times as great as it is in methanol. Recall that small anionic nucleophiles are most heavily solvated by protic solvents, explaining why the nucleophilicity of the halides increases from top to bottom of the periodic table *in protic*

TABLE 6-5 | **Polar Aprotic Solvents**

$$CH_3\overset{\overset{O}{\|}}{C}CH_3$$

Propanone (Acetone)

$$CH_3C{\equiv}N$$

Ethanenitrile (Acetonitrile)

$$H\overset{\overset{O}{\|}}{C}N(CH_3)_2$$

***N,N*-Dimethylformamide (DMF)**

$$CH_3\overset{\overset{O}{\|}}{S}CH_3$$

Dimethyl sulfoxide (DMSO)

$$(CH_3)_2N\overset{\overset{O}{\|}}{\underset{\underset{(CH_3)_2}{|}}{P}}N(CH_3)_2$$

Hexamethylphosphoric triamide (HMPA)

$$CH_3\overset{+}{N}\overset{\nearrow O}{\underset{\searrow O^-}{}}$$

Nitromethane

TABLE 6-6	Relative Rates of S_N2 Reactions of Iodomethane with Chloride Ion in Various Solvents

$$CH_3I + Cl^- \xrightarrow[k_{rel}]{\text{Solvent}} CH_3Cl + I^-$$

Solvent			
Formula	Name	Classification	Relative rate (k_{rel})
CH_3OH	Methanol	Protic	1
$HCONH_2$	Formamide	Protic	12.5
$HCONHCH_3$	N-Methylformamide	Protic	45.3
$HCON(CH_3)_2$	N,N-Dimethylformamide	Aprotic	1,200,000

solvents. Switching to an *aprotic* solvent reduces the solvation and increases the reactivity of all anions, but the effect is greatest for small anionic nucleophiles. Indeed, in polar aprotic solvents, the opposing factors of base strength and polarizability substantially reduce the differences in nucleophilicity among the halides and may even lead to inversion of the reactivity order under some conditions.

Increasing polarizability improves nucleophilic power

FIGURE 6-7

Comparison of I^- and F^- in the S_N2 reaction. (A) The larger iodide is a better nucleophile, because its polarizable $5p$ orbital is distorted toward the electrophilic carbon atom. (B) The tight, less polarizable $2p$ orbital on fluoride does not interact as effectively with the electrophilic carbon at a point along the reaction coordinate comparable to the one for (A).

The solvation effects just described should be very pronounced only for charged nucleophiles. Nevertheless, the degree of nucleophilicity increases down the periodic table, even for *uncharged nucleophiles,* for which solvent effects should be much less strong: For example, $H_2Se > H_2S > H_2O$, and $PH_3 > NH_3$. Therefore, there must be an additional explanation for the observed trend in nucleophilicity.

This explanation lies in the polarizability of the nucleophile. Larger elements have larger, more diffuse, and more polarizable electron clouds. These electron clouds allow for more effective orbital overlap in the S_N2 transition state (Figure 6-7). The result is a lower transition-state energy and faster nucleophilic substitution.

EXERCISE 6-19

Which species is more nucleophilic: **(a)** CH_3SH or CH_3SeH; **(b)** $(CH_3)_2NH$ or $(CH_3)_2PH$?

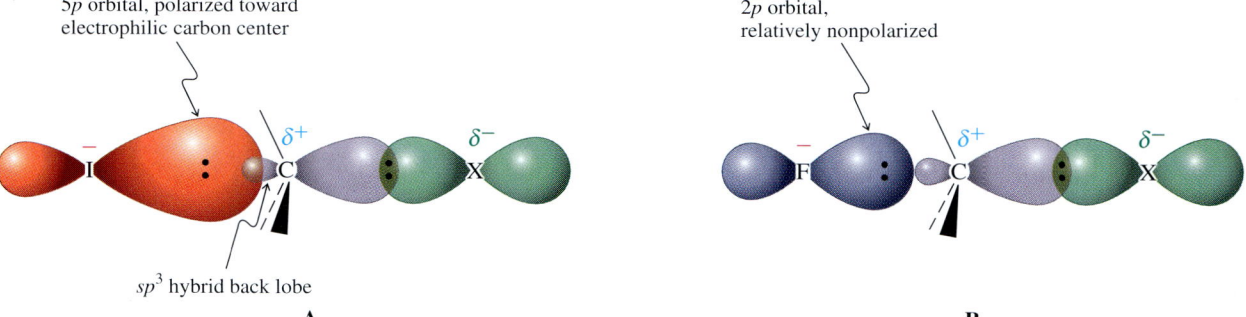

$5p$ orbital, polarized toward electrophilic carbon center

sp^3 hybrid back lobe

A

$2p$ orbital, relatively nonpolarized

B

Sterically hindered nucleophiles are poorer reagents

We have seen that the bulk of the surrounding solvent may adversely affect the power
of a nucleophile, another example of steric hindrance (Section 2-7). Such hindrance
may also be built into the nucleophile itself in the form of bulky substituents. The effect on the rate of reaction can be seen in Experiment 7.

EXPERIMENT 7

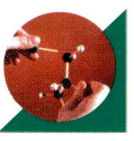

$$CH_3I \ + \ CH_3O^- \ \longrightarrow \ CH_3OCH_3 \ + \ I^- \quad \text{Fast}$$

$$CH_3I \ + \ CH_3\overset{\displaystyle CH_3}{\underset{\displaystyle CH_3}{C}}O^- \ \longrightarrow \ CH_3O\overset{\displaystyle CH_3}{\underset{\displaystyle CH_3}{C}}CH_3 \ + \ I^- \quad \text{Slower}$$

Conclusion. Sterically bulky nucleophiles react more slowly.

EXERCISE 6-20

Which of the two nucleophiles in the following pairs will react more rapidly with bromomethane?

(a) CH_3S^- or $CH_3\overset{\displaystyle CH_3}{\underset{}{C}}HS^-$ **(b)** $(CH_3)_2NH$ or $(CH_3\overset{\displaystyle CH_3}{\underset{}{C}}H)_2NH$

Nucleophilic substitutions may be reversible

The halide ions Cl$^-$, Br$^-$, and I$^-$ are both good nucleophiles and good leaving groups.
Therefore, their S$_N$2 reactions are reversible. In such situations, the directions in which
the equilibria lie must be determined experimentally. Bond-strength calculations alone
cannot be used to determine the thermodynamics, because these calculations do not
take into account the energetics of ion solvation, a complex problem. For example,
in propanone (acetone), the reactions between lithium chloride and primary bromo-
and iodoalkanes are reversible, with the equilibrium on the side of the chloroalkane
products:

$$CH_3CH_2CH_2CH_2I + LiCl \ \underset{\text{(acetone)}}{\overset{\text{Propanone}}{\rightleftharpoons}} \ CH_3CH_2CH_2CH_2Cl + LiI$$

This result correlates with the relative bond strengths in the product and starting material ($DH°_{C-Cl} = 80$ kcal mol^{-1} versus $DH°_{C-I} = 53$ kcal mol^{-1}). However, this
equilibrium may be driven in the reverse direction by a simple "trick": Whereas all
of the lithium halides are soluble in acetone, solubility of the sodium halides decreases
dramatically in the order NaI > NaBr > NaCl, the last being virtually insoluble in
this solvent. Indeed, the reaction between NaI and a primary or secondary chloroalkane
in acetone is *completely* driven to the side of the iodoalkane (the reverse of the reaction just shown) by the precipitation of NaCl:

$$CH_3CH_2CH_2CH_2Cl + NaI \ \underset{\text{(acetone)}}{\overset{\text{Propanone}}{\rightleftharpoons}} \ CH_3CH_2CH_2CH_2I + NaCl\downarrow$$

Insoluble
in acetone

TABLE 6-7 Relative Rates of Reaction of Various Nucleophiles with Iodomethane in Methanol	
Nucleophile	**Relative rate**
CH_3OH	1
NO_3^-	~ 32
F^-	500
$CH_3\overset{\overset{\displaystyle O}{\|}}{C}O^-$	20,000
Cl^-	23,500
$(CH_3CH_2)_2S$	219,000
NH_3	316,000
CH_3SCH_3	347,000
N_3^-	603,000
Br^-	617,000
CH_3O^-	1,950,000
CH_3SeCH_3	2,090,000
CN^-	5,010,000
$(CH_3CH_2)_3As$	7,940,000
I^-	26,300,000
HS^-	100,000,000

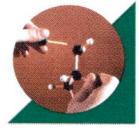

The direction of the equilibrium in reaction 3 of Table 6-3 may be manipulated in exactly the same way. However, when the nucleophile in an S_N2 reaction is a strong base (e.g., HO^- or CH_3O^-; see Table 6-4), it will be incapable of acting as a leaving group. In such cases, K_{eq} will be very large, and displacement will essentially be an irreversible process (Table 6-3, reactions 1 and 2).

To summarize, nucleophilicity is controlled by a number of factors. Increased negative charge and progression from right to left and down the periodic table generally increase nucleophilic power. Table 6-7 compares the reactivity of a range of nucleophiles relative to that of methanol (arbitrarily set at 1). We can confirm the validity of the conclusions of this section by inspecting the various entries. The use of aprotic solvents improves nucleophilicity, especially of smaller anions, by eliminating hydrogen bonding.

6-10 Effect of the Alkyl Group on the S_N2 Reaction

Finally, does the structure of the substrate, particularly in the vicinity of the center bearing the leaving group, affect the rate of nucleophilic attack? Once again, we can get a sense of comparative reactivities by looking at relative rates of reaction. Let us examine the kinetic data that have been obtained.

Branching at the reacting carbon decreases the rate of the S_N2 reaction

What happens if we successively replace each of the hydrogens in a halomethane with a methyl group? Will this affect the rate of its S_N2 reactions? In other words, what are the relative bimolecular nucleophilic reactivities of methyl, primary, secondary, and tertiary halides? Kinetic experiments show that reactivities rapidly decrease in the order shown in Table 6-8.

We can find an explanation by comparing the transition states for these three substitutions. Figure 6-8A shows this structure for the reaction of chloromethane with hydroxide ion. The carbon is surrounded by the incoming nucleophile, the outgoing leaving group, and three substituents (all hydrogen in this case). Although the presence of these five groups increases the crowding about the carbon relative to that in the starting halomethane, the hydrogens do not give rise to serious steric interactions with the nucleophile, because of their small size. However, replacement of one hydrogen by a methyl group, as in a haloethane, creates substantial steric repulsion with the incoming nucleophile, thereby raising the transition-state energy (Figure 6-8B). This effect significantly retards nucleophilic attack. If we continue to replace hydrogen atoms with methyl groups, we find that steric hindrance to nucleophilic attack increases dramatically. The two methyl groups in the secondary substrate severely shield the backside of the carbon attached to the leaving group; the rate of reaction diminishes considerably (Figure 6-8C and Table 6-8). Finally, in the tertiary substrate, in which a third methyl group is present, access to the backside of the halide-bearing carbon is entirely blocked (Figure 6-8D); the transition state for S_N2 substitution is energetically inaccessible, and displacement of a tertiary halide by this mechanism is not observed. To summarize, as we successively replace the hydrogens of a halomethane by methyl groups (or alkyl groups in general), S_N2 reactivity decreases in the following order:

Relative S$_N$2 Displacement Reactivity of Haloalkanes

Methyl	>	primary	>	secondary	>	tertiary
Fast		**Slower**		**Very slow**		**Not at all**

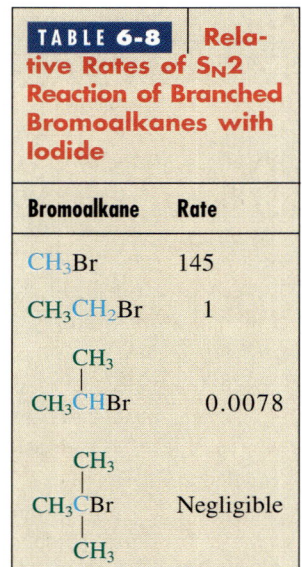

TABLE 6-8 Relative Rates of S$_N$2 Reaction of Branched Bromoalkanes with Iodide

Bromoalkane	Rate
CH_3Br	145
CH_3CH_2Br	1
$CH_3\overset{CH_3}{\underset{}{CHBr}}$	0.0078
$CH_3\overset{CH_3}{\underset{CH_3}{CBr}}$	Negligible

EXERCISE 6-21

Predict the relative rates of the S$_N$2 reaction of cyanide with these pairs of substrates.

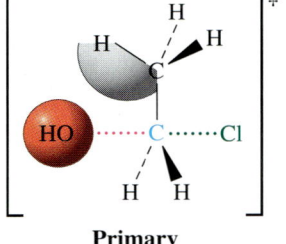

(a) [structure] and [structure]

(b) $CH_3CH_2\overset{CH_3}{\underset{CH_3}{CBr}}$ and $CH_3CH_2CH_2Br$

Now that we have seen the effect of major structural changes on substrate reactivity in the S$_N$2 process, we are in a position to evaluate the effects of more subtle structural modifications. In all cases, we shall find that steric hindrance to attack at the backside of the reacting carbon is the most important consideration.

Lengthening the chain by one or two carbons reduces S$_N$2 reactivity

As we have seen, the replacement of one hydrogen atom in a halomethane by a methyl group (Figure 6-8B) causes significant steric hindrance and reduction of the rate of S$_N$2 reaction. Chloroethane is about two orders of magnitude less reactive than chloromethane in S$_N$2 displacements. Will elongation of the chain of the primary alkyl substrate by the addition of methylene (CH_2) groups further reduce S$_N$2 reactivity? Kinetic experiments reveal that 1-chloropropane reacts about half as fast as chloroethane with nucleophiles such as I^-.

Does this trend continue as the chain gets longer? The answer is *no:* Higher haloalkanes such as 1-chlorobutane and 1-chloropentane, react at about the same rate as does 1-chloropropane.

FIGURE 6-8

Transition states for S$_N$2 reactions of hydroxide ion with (A) chloromethane, (B) chloroethane, (C) 2-chloropropane, and (D) 2-chloro-2-methylpropane.

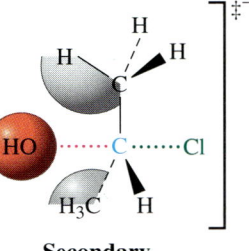

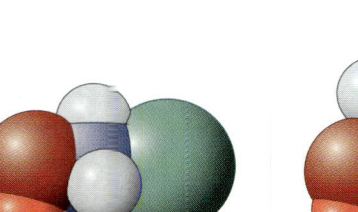

Methyl	Primary	Secondary	Tertiary
		(Slow reaction: hydrogens on two methyl groups interfere)	(No S$_N$2 reaction; too much steric hindrance)

A | B | C | D

The Dilemma of Bromomethane: Highly Useful but Also Highly Toxic

Bromomethane, CH_3Br, is a substance with numerous uses. Easy and inexpensive to prepare, it is employed as an insect fumigant for large storage spaces such as warehouses and railroad boxcars. It is also effective in eradicating insect infestations in soil and around a number of major crops, including potatoes and tomatoes. Not surprisingly, it owes part of its value to its high toxicity, which can be attributed largely to its S_N2 reactivity. The chemistry of life is highly dependent on several classes of molecules containing nucleophilic groups such as amines (–NH_2 and related functions) and thiols (–SH). The biochemical roles of these substituents are many and varied, as well as being critical to the survival of living organisms. Highly reactive electrophiles such as bromomethane wreak havoc on this biochemistry by indiscriminately *alkylating* such nucleophilic atoms—by reacting through the S_N2 mechanism to attach alkyl groups (in this case, a methyl group) to them (see, for example, the reaction below). Some of these processes can generate HBr as a by-product, which amplifies the danger posed by this material to living systems.

The toxicity of bromomethane is not limited to insects. Human exposure is known to cause numerous health problems: Direct contact causes burns to the skin, chronic exposure leads to kidney, liver, and central nervous system damage, and inhalation of high concentrations can lead to the destruction of lung tissue, to pulmonary edema, and to death. The limit set for bromomethane exposure in the workplace is a concentration of 20 parts per million of bromomethane vapor in ambient air. As is the case for so many substances that have been found to be useful in large-scale applications in our society, bromomethane's toxicity poses a dilemma that requires the most responsible control of its use. The resolution between the issues of utility and safety does not always come easily, and the costs—human, environmental, and economic—must be assessed most carefully.

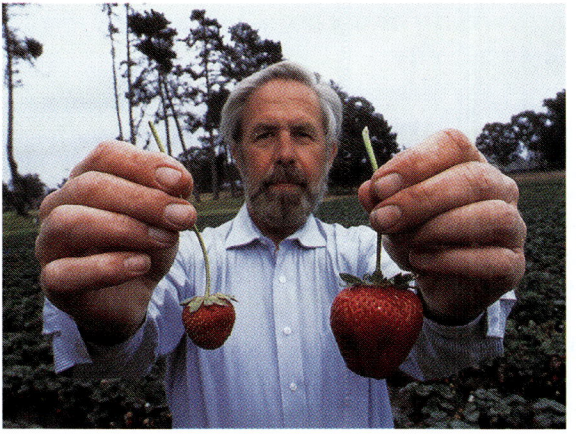

Plant pathologist Frank Westerlund demonstrates the "bromomethane difference" between strawberries grown in a fumigated plot (*right*) and those not (*left*). The latter are withered by verticillium wilt, a fungus.

$$R-\ddot{S}-H \;+\; CH_3-Br \;\longrightarrow\; R-\overset{+}{\underset{\cdots}{S}}-H \;+\; Br^- \;\longrightarrow\; R-\ddot{S}-CH_3 \;+\; HBr$$

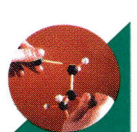

Again, an examination of the transition states to backside displacement provides an explanation for these observations. In Figures 6-9A and 6-9B, one of the hydrogens on the methyl carbon of chloroethane is partially obstructing the path of attack of the incoming nucleophile. The 1-halopropanes have an additional methyl group near the reacting carbon center. If reaction occurs from the most stable *anti* rotamer of the substrate, the incoming nucleophile faces severe steric hindrance (Figure 6-9C). However, rotation to a *gauche* conformation before attack gives an S_N2 transition state similar to that derived from a haloethane (Figure 6-9D). The propyl substrate exhibits only a small decrease in reactivity relative to the ethyl, the decrease

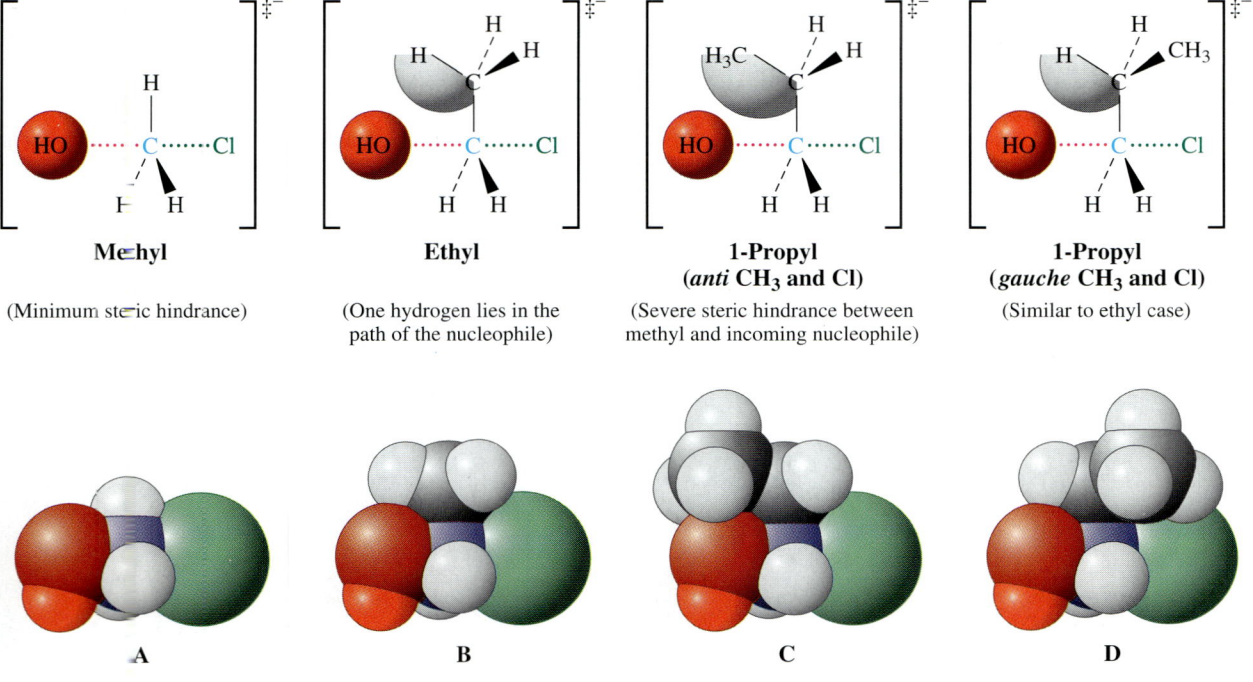

Methyl

(Minimum steric hindrance)

Ethyl

(One hydrogen lies in the path of the nucleophile)

1-Propyl
(*anti* CH₃ and Cl)

(Severe steric hindrance between methyl and incoming nucleophile)

1-Propyl
(*gauche* CH₃ and Cl)

(Similar to ethyl case)

A   **B**   **C**   **D**

FIGURE 6-9

Dashed-wedged line and space-filling drawings of the transition states for S_N2 reactions of hydroxide ion with (A) chloromethane; (B) chloroethane; and (C and D) two rotamers of 1-chloropropane: (C) *anti* and (D) *gauche*. The shaded areas in the dashed-wedged line drawings highlight steric interference with the incoming nucleophile. This interference is illustrated strikingly in the space-filling drawing. Partial charges have been omitted for clarity. (See Figure 6-3.)

resulting from the energy input needed to attain a *gauche* conformation. Further chain elongation has no effect, because the added carbon atoms do not increase steric hindrance around the reacting carbon in the transition state.

Branching next to the reacting carbon also retards substitution

What about multiple substitution at the position *next to* the electrophilic carbon? Let us compare the reactivities of bromoethane and its derivatives (Table 6-9). A dramatic decrease in rate is seen on further substitution: 1-Bromo-2-methylpropane is two orders of magnitude less reactive toward iodide than is 1-bromopropane, and 1-bromo-2,2-dimethylpropane is virtually inert. Branching at positions farther from the site of reaction has a much smaller effect.

Recalling Figure 6-9, we know that rotation into a *gauche* conformation is necessary to permit nucleophilic attack on a 1-halopropane. We can use the same picture to understand the data in Table 6-9. For a 1-halo-2-methylpropane, the only conformation that permits the nucleophile to approach the backside of the reacting carbon experiences *two gauche* methyl–halide interactions, a considerably worse situation (Figure 6-10B). With the addition of a third methyl group, as in a 1-halo-2,2-dimethylpropane, backside attack is blocked almost completely (Figure 6-10C).

TABLE 6-9 **Relative Reactivities of Branched Bromoalkanes with Iodide**

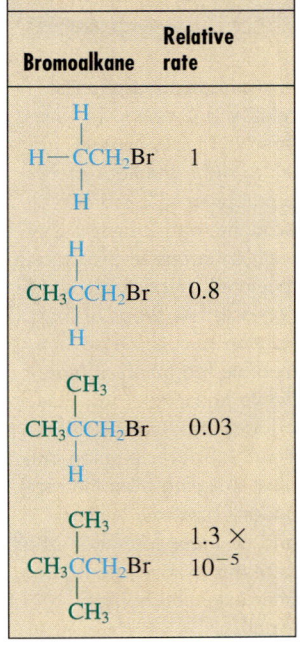

Bromoalkane	Relative rate
H—CCH₂Br (with H above and below)	1
CH₃CCH₂Br (with H above and below)	0.8
CH₃CCH₂Br (with CH₃ above and H below)	0.03
CH₃CCH₂Br (with CH₃ above and CH₃ below)	1.3×10^{-5}

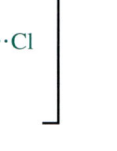

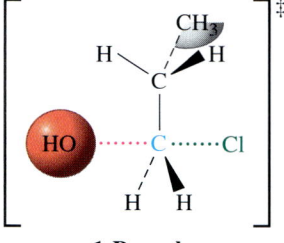
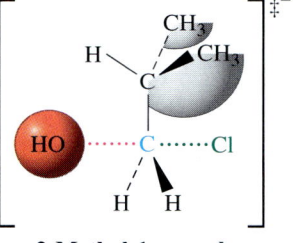
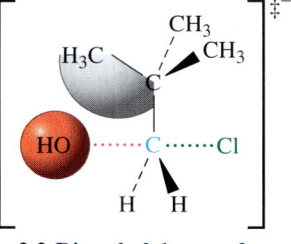

1-Propyl
(*gauche* CH₃ and Cl)

2-Methyl-1-propyl
(**two** *gauche* CH₃ and Cl)

(High energy transition state:
reaction is slower)

2,2-Dimethyl-1-propyl

(All conformations experience
severe steric hindrance)

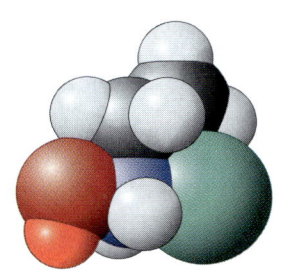

A   B   C

FIGURE 6-10

Dashed-wedged line and space-filling renditions of the transition states for S_N2 reactions of hydroxide ion with (A) 1-chloropropane, (B) 1-chloro-2-methylpropane, and (C) 1-chloro-2,2-dimethylpropane. Increasing steric hindrance from a second *gauche* interaction reduces the rate of reaction in (B). S_N2 reactivity in (C) is eliminated almost entirely because a methyl group prevents backside attack by the nucleophile in all accessible conformations of the substrate. (See also Figures 6-8 and 6-9.)

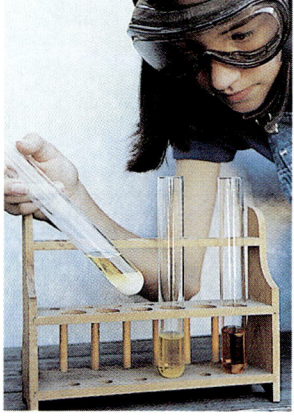

A visual demonstration of relative S_N2 reactivity. The three test tubes contain, from left to right, solutions of 1-bromobutane, 2-bromo-propane, and 2-bromo-2-methylpropane in propanone (acetone), respectively. Addition of a few drops of NaI solution to each causes immediate formation of NaBr (white precipitate) from the primary bromoalkane (left), slow NaBr precipitation only after warming from the secondary substrate (center), and no NaBr formation at all from the tertiary halide even after extended heating (right).

EXERCISE 6-22

Predict the order of reactivity in the S_N2 reaction of

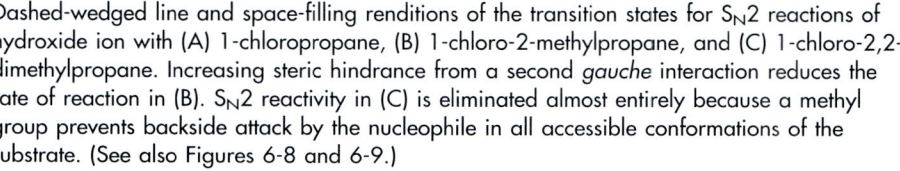

versus

In summary, the structure of the alkyl part of a haloalkane can have a pronounced effect on nucleophilic attack. Simple chain elongation beyond three carbons has little effect on the rate of the S_N2 reaction. However, increased branching leads to strong steric hindrance and rate retardation.

CHAPTER INTEGRATION PROBLEM

a. Write a mechanism and final product for the reaction between sodium ethoxide, NaOCH₂CH₃, and bromoethane, CH₃CH₂Br, in ethanol solvent, CH₃CH₂OH.

SOLUTION

The mechanism is backside attack in which the nucleophilic atom of the reagent attacks the atom of the substrate that contains the leaving group (Section 6-6). We begin by identifying each of these components. The nucleophilic atom is the negatively charged oxygen atom in ethoxide ion, $CH_3CH_2O^-$. Attack occurs at the carbon attached to bromine in the substrate molecule, CH_3CH_2Br:

$$CH_3CH_2O^- \quad \overset{H_3C}{\underset{H}{\overset{}{C}}}{-}Br \longrightarrow CH_3CH_2O{-}CH_2CH_3 + Br^-$$

The products are bromide ion and ethoxyethane, $CH_3CH_2OCH_2CH_3$, an ether.

b. How would the preceding reaction be affected by each of the following changes?
   1. Replace bromoethane with fluoroethane.
   2. Replace bromoethane with bromomethane.
   3. Replace sodium ethoxide with sodium ethanethiolate, $NaSCH_2CH_3$.
   4. Replace ethanol with dimethylformamide (DMF).

SOLUTION

1. Table 6-4 tells us that fluoride is a stronger base than bromide and, therefore, a poorer leaving group. The reaction would still take place but would be very much slower. (The actual rate decrease is on the order of 10^{-4}.)
2. The carbon containing the leaving group in bromomethane is less sterically hindered than that in bromoethane, so the rate of reaction would increase (Section 6-10). The product of the reaction would be $CH_3OCH_2CH_3$, methoxyethane.
3. Both ethoxide and ethanethiolate are negatively charged. Oxygen in ethoxide is more basic than sulfur in ethanethiolate (Table 6-4), but the sulfur atom in the latter is larger, more polarizable, and less tightly solvated in the hydrogen-bonding ethanol solvent (compare Figure 6-6). We know that strong bases are good nucleophiles, but base strength is outweighed by the increased polarizability and reduced solvation of the larger atoms within the same column of the periodic table (Section 6-9). Ethanethiolate reacts hundreds of times as fast, giving as a product $CH_3CH_2SCH_2CH_3$, an example of a sulfide (Section 9-10).
4. Conversion from a protic, hydrogen-bonding solvent into a polar, aprotic one accelerates the reaction enormously by reducing solvation of the negatively charged oxygen atom (compare Table 6-6).

c. Which of the following compounds would be expected to react in an S_N2 manner at a reasonable rate with sodium azide, NaN_3, in ethanol? Which will not? Why not?

(i) $\diagdown\diagdown\diagdown$ NH$_2$ (ii) $\diagup\!\!\!\times\!\!\!_{I}$ (iii) $\diagdown\diagup\diagdown$ Br

(iv) $\diagup\!\!\!\times$ OH   (v) (cyclopentyl)CH$_2$CH$_2$Cl   (vi) $\diagdown\diagup$ CN

SOLUTION

What are the structural requirements for S_N2 reactions to occur? We need a strong nucleophile, a suitable substrate, and a good leaving group. The nucleophile is al-

ready specified, so we turn to the other two ingredients, ruling out systems that lack one or the other. Substrates (i), (iv), and (vi) lack good leaving groups: $^-NH_2$, ^-OH, and ^-CN are too strongly basic for this purpose (Table 6-4). These compounds will not undergo S_N2 displacement. Substrate (ii) contains a good leaving group, but the reaction site is a tertiary carbon and incapable of following the S_N2 mechanism. That leaves substrates (iii) and (v), both of which are primary haloalkanes with good leaving groups and no branching at the position adjacent to the site of displacement (Section 6-10). They will transform readily by the S_N2 mechanism.

IMPORTANT CONCEPTS

1. A **haloalkane,** commonly termed an alkyl halide, consists of an alkyl group and a halogen.
2. The physical properties of the haloalkanes are strongly affected by the polarization of the C–X bond and the polarizability of X.
3. Reagents bearing lone electron pairs are called **nucleophilic** when they attack positively polarized centers (other than protons). The latter are called **electrophilic.** When such a reaction leads to displacement of a substituent, it is a **nucleophilic substitution.** The group being displaced by the nucleophile is the **leaving group.**
4. The kinetics of the reaction of nucleophiles with primary (and most secondary) haloalkanes are second order, indicative of a **bimolecular** mechanism. This process is called **bimolecular nucleophilic substitution (S_N2 reaction).** It is a **concerted reaction,** one in which bonds are simultaneously broken and formed. Curved arrows are typically used to depict the flow of electrons as the reaction proceeds.
5. The S_N2 reaction is **stereospecific** and proceeds by **backside displacement,** thereby producing **inversion of configuration** at the reacting center.

6. An orbital description of the S_N2 transition state includes an sp^2-hybridized carbon center, partial bond-making between the nucleophile and the electrophilic carbon, and simultaneous partial bond-breaking between that carbon and the leaving group. Both the nucleophile and the leaving group bear partial charges.
7. **Leaving group ability,** a measure of the ease of displacement, is roughly proportional to the strength of the conjugate acid. Especially good leaving groups are weak bases such as chloride, bromide, iodide, and the sulfonates.
8. **Nucleophilicity** increases (a) with negative charge, (b) for elements farther to the left and down the periodic table, and (c) in polar aprotic solvents.
9. **Polar aprotic solvents** accelerate S_N2 reactions because the nucleophiles are well separated from their counterions but are not tightly solvated.
10. **Branching** at the reacting carbon or at the carbon next to it in the substrate leads to steric hindrance in the S_N2 transition state and decreases the rate of bimolecular substitution.

PROBLEMS

23. Name the following molecules according to the IUPAC system.

(a) CH_3CH_2Cl (b) $BrCH_2CH_2Br$ (c) $CH_3CH_2CHCH_2F$
$\qquad\qquad\qquad\qquad\qquad\qquad\qquad\qquad\qquad$ |
$\qquad\qquad\qquad\qquad\qquad\qquad\qquad\qquad\qquad CH_2CH_3$

(d) $(CH_3)_3CCH_2I$ (e) ⬡—CCl_3 (f) $CHBr_3$

24. Draw structures for each of the following molecules. (a) 3-ethyl-2-iodopentane; (b) 3-bromo-1,1-dichlorobutane; (c) *cis*-1-(bromomethyl)-2-(2-chloroethyl)cyclobutane; (d) (trichloromethyl)cyclopropane; (e) 1,2,3-trichloro-2-methylpropane.

25. Draw and name all possible structural isomers having the formula C_3H_6BrCl.

26. Draw and name all structurally isomeric compounds having the formula $C_5H_{11}Br$.

27. For each structural isomer in Problems 25 and 26, identify all stereocenters and give the total number of stereoisomers that can exist for the structure.

28. For each reaction in Table 6-3, identify the nucleophile, its nucleophilic atom (draw its Lewis structure first), the electrophilic atom in the organic substrate, and the leaving group.

29. A second Lewis structure can be drawn for one of the nucleophiles in Problem 28. **(a)** Identify it and draw its alternate structure (which is simply a second resonance form). **(b)** Does this second resonance form predict the presence of another nucleophilic atom in the nucleophile? If so, rewrite the reaction of Problem 28, using the new nucleophilic atom, and write a correct Lewis structure for the product.

30. For each reaction shown here, identify the nucleophile, its nucleophilic atom, the electrophilic atom in the substrate molecule, and the leaving group. Write the organic product of the reaction.

(a) $CH_3I + NaNH_2 \rightarrow$

(b) 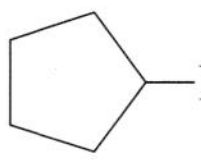 ⬠—$Br + H_2S \rightarrow$

(c) 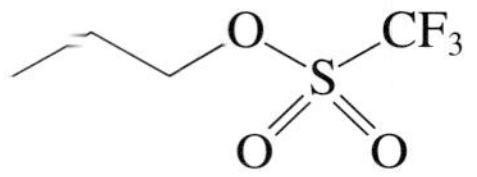 ‿‿O‿S‿CF_3 + NaI → (with O, O on the S)

(d) 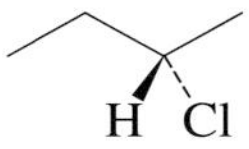 ‿‿⟨H Cl⟩ + $NaN_3 \rightarrow$

(e) CH_3Cl + ‿‿N(CH$_3$)‿‿ →

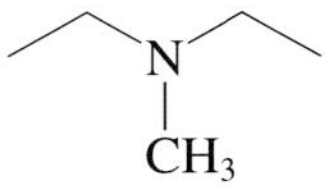

(f) 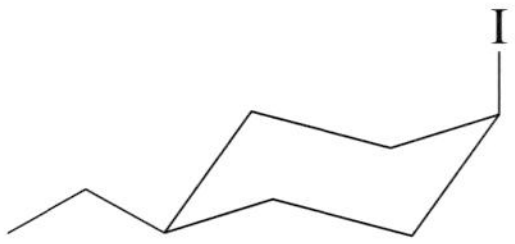 cyclohexane–I + $KSeCN \rightarrow$

31. A solution containing 0.1 M CH_3Cl and 0.1 M KSCN in DMF reacts to give CH_3SCN and KCl with an initial rate of 2×10^{-8} mol L^{-1} s^{-1}. **(a)** What is the rate constant for this reaction? **(b)** Calculate the initial reaction rate for each of the following sets of reactant concentrations: (i) $[CH_3Cl]$ = 0.2 M, $[KSCN]$ = 0.1 M; (ii) $[CH_3Cl]$ = 0.2 M, $[KSCN]$ = 0.3 M; (iii) $[CH_3Cl]$ = 0.4 M, $[KSCN]$ = 0.4 M.

32. Write the product of each of the following bimolecular substitutions. The solvent is indicated above the reaction arrow.

(a) $CH_3CH_2CH_2Br + Na^+I^-$ $\xrightarrow{\text{Propanone (acetone)}}$

(b) $(CH_3)_2CHCH_2I + Na^{+-}CN$ $\xrightarrow{\text{DMSO}}$

(c) $CH_3I + Na^{+-}OCH(CH_3)_2$ $\xrightarrow{(CH_3)_2CHOH}$

(d) $CH_3CH_2Br + Na^{+-}SCH_2CH_3$ $\xrightarrow{CH_3OH}$

(e) 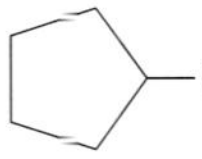 ⬠—$CH_2Cl + CH_3CH_2SeCH_2CH_3$ $\xrightarrow{\text{Propanone (acetone)}}$

(f) $(CH_3)_2CHOSO_2CH_3 + N(CH_3)_3$ $\xrightarrow{(CH_3CH_2)_2O}$

33. Determine the *R/S* designations for both starting materials and products in the following S$_N$2 reactions. Which of the products are optically active?

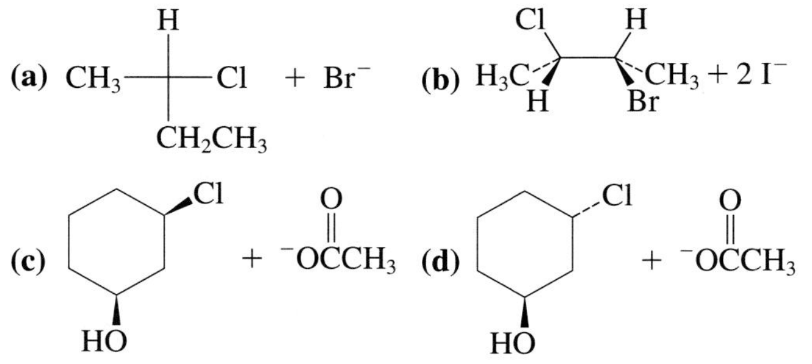

(a) CH$_3$—Cl + Br$^-$ (with H up, CH$_2$CH$_3$ down)

(b) H$_3$C—(Cl, H)—(H)—CH$_3$ + 2 I$^-$ (with Br)

(c) cyclohexane with Cl and HO substituents + $^-$OCCH$_3$ (with C=O)

(d) cyclohexane with Cl and HO substituents + $^-$OCCH$_3$ (with C=O)

34. List the product(s) of the reaction of 1-bromopropane with each of the following reagents. Write "no reaction" where appropriate. (**Hint:** Carefully evaluate the nucleophilic potential of each reagent.)
(a) H$_2$O   (b) H$_2$SO$_4$ (c) KOH   (d) CsI   (e) NaCN
(f) HCl   (g) (CH$_3$)$_2$S   (h) NH$_3$ (i) Cl$_2$ (j) KF

35. Formulate the potential product of each of the following reactions. As you did in Problem 34, write "no reaction" where appropriate. (**Hint:** Identify the expected leaving group in each of the substrates and evaluate its ability to undergo displacement.)

(a) CH$_3$CH$_2$CH$_2$CH$_2$Br + K^{+-}OH $\xrightarrow{CH_3CH_2OH}$

(b) CH$_3$CH$_2$I + K$^+$Cl$^-$ $\xrightarrow{DMF}$

(c) 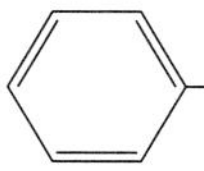 C$_6$H$_5$—CH$_2$Cl + Li^{+-}OCH$_2$CH$_3$ $\xrightarrow{CH_3CH_2OH}$

(d) (CH$_3$)$_2$CHCH$_2$Br + Cs$^+$I$^-$ $\xrightarrow{CH_3OH}$

(e) CH$_3$CH$_2$CH$_2$Cl + K^{+-}SCN $\xrightarrow{CH_3CH_2OH}$

(f) CH$_3$CH$_2$F + Li$^+$Cl$^-$ $\xrightarrow{CH_3OH}$

(g) CH$_3$CH$_2$CH$_2$OH + K$^+$I$^-$ $\xrightarrow{DMSO}$

(h) CH$_3$I + Na^{--}SCH$_3$ $\xrightarrow{CH_3OH}$

(i) CH$_3$CH$_2$OCH$_2$CH$_3$ + Na^{+-}OH $\xrightarrow{H_2O}$

(j) CH$_3$CH$_2$I + K^{+-}OCCH$_3$ (with C=O) $\xrightarrow{DMSO}$

36. Show how each of the following transformations might be achieved.

(a) (*R*)-CH$_3$CHCH$_2$CH$_3$ (with OSO$_2$CH$_3$) $\longrightarrow$ (*S*)-CH$_3$CHCH$_2$CH$_3$ (with N$_3$)

(b) 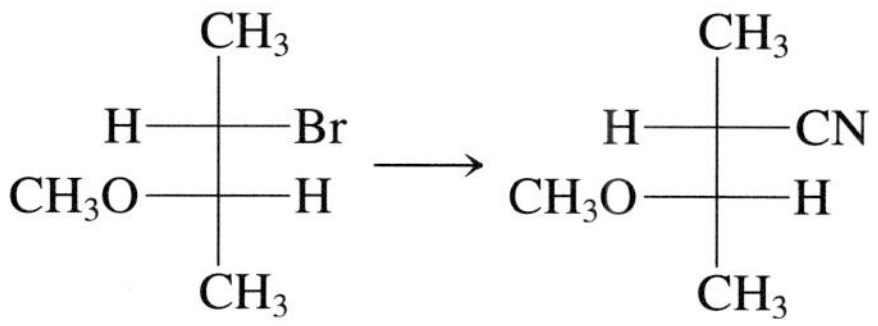 H—Br, CH$_3$O—H, CH$_3$ $\longrightarrow$ H—CN, CH$_3$O—H, CH$_3$ (Fischer projections with CH$_3$)

(c) 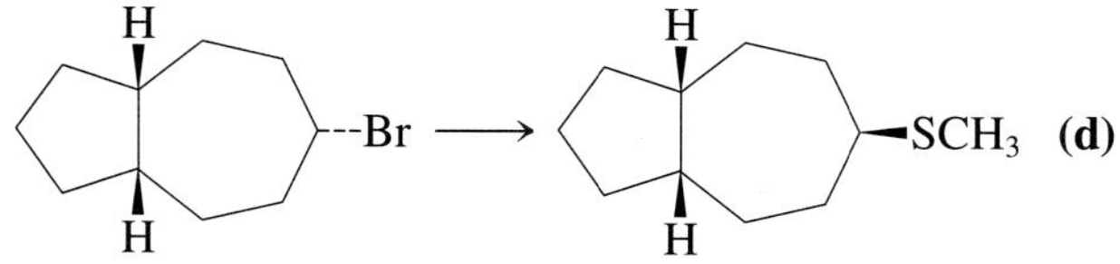 bicyclic structure with ---Br $\longrightarrow$ bicyclic structure with —SCH$_3$

(d) 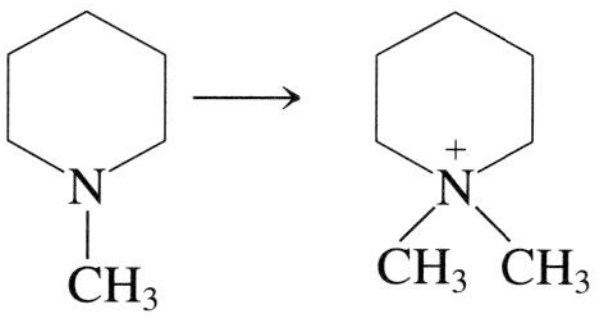 piperidine with N—CH$_3$ $\longrightarrow$ N$^+$(CH$_3$)(CH$_3$)

37. Rank the members of each of the following groups of species in the order of basicity, nucleophilicity, and leaving-group ability. Briefly explain your answers. (a) H$_2$O, HO$^-$, CH$_3$CO$_2^-$; (b) Br$^-$, Cl$^-$, F$^-$, I$^-$; (c) $^-$NH$_2$, NH$_3$, $^-$PH$_2$; (d) $^-$OCN, $^-$SCN; (e) F$^-$, HO$^-$, $^-$SCH$_3$; (f) H$_2$O, H$_2$S, NH$_3$.

38. Write the product(s) of each of the following reactions. Write "no reaction" as your answer, if appropriate.

(a) $CH_3CH_2CH_2CH_3 + Na^+Cl^- \xrightarrow{CH_3OH}$

(b) $CH_3CH_2Cl + Na^{+-}OCH_3 \xrightarrow{CH_3OH}$

(c) 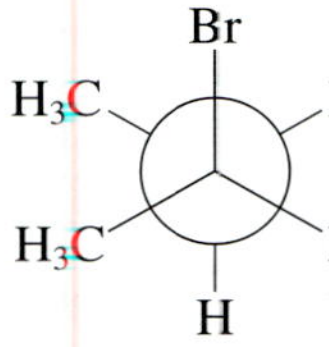 [Newman projection with Br, H_3C, H, H_3C, H, H] $+ Na^+I^- \xrightarrow{\text{Propanone (acetone)}}$

(d) 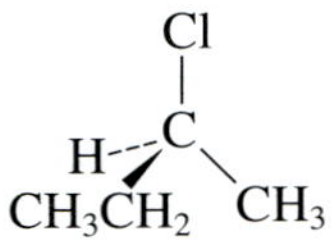 [structure with Cl, H—C, CH_3CH_2, CH_3] $+ Na^{+-}SCH_3 \xrightarrow{\text{Propanone (acetone)}}$

(e) $CH_3\overset{\overset{\displaystyle OH}{|}}{C}HCH_3 + Na^{+-}CN \longrightarrow$

(f) $CH_3\overset{\overset{\displaystyle OSO_2CH_3}{|}}{C}HCH_3 + HCN \xrightarrow{CH_3CH_2OH}$

(g) $CH_3\overset{\overset{\displaystyle OSO_2CH_3}{|}}{C}HCH_3 + Na^{+-}CN \xrightarrow{CH_3CH_2OH}$

(h) 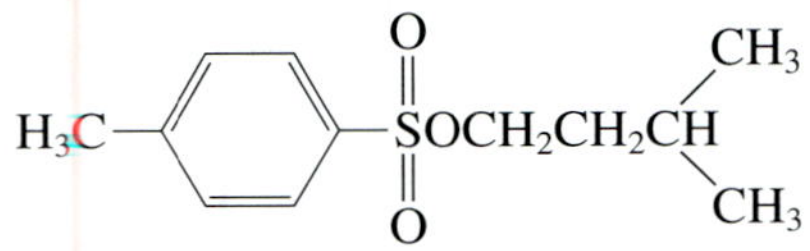 H_3C- [benzene ring] $-\overset{\overset{\displaystyle O}{||}}{\underset{\underset{\displaystyle O}{||}}{S}}OCH_2CH_2\overset{\overset{\displaystyle CH_3}{|}}{C}H\underset{CH_3}{} + K^{+-}SCN \xrightarrow{CH_3OH}$

(i) $CH_3CH_2NH_2 + Na^+Br^- \xrightarrow{DMSO}$

(j) $CH_3I + Na^{+-}NH_2 \xrightarrow{NH_3}$

(k) Product of (j) + more $CH_3I \longrightarrow$

(l) 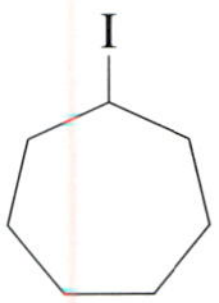 [cycloheptane ring with I] $+ Na^{+-}SH \xrightarrow{CH_3OH}$

(m) 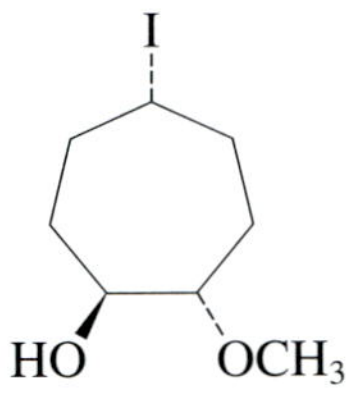 [cycloheptane ring with I, HO, OCH_3] $+ Na^{+-}SH \xrightarrow{CH_3OH}$

(n) $CH_3\overset{\overset{\displaystyle CH_3}{|}}{C}HCH_2Br +$ [triphenylphosphine structure with P] $\xrightarrow{CH_3CH_2OH}$

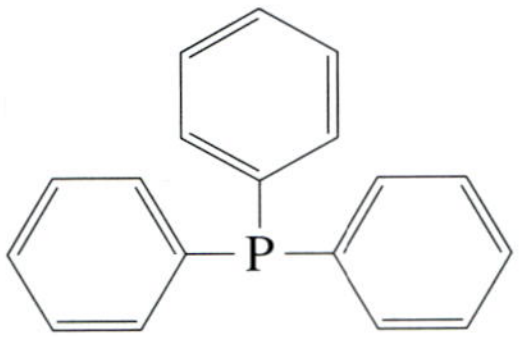

39. Using the information in Chapters 3 and 6, propose the best possible synthesis of each of the following compounds with propane as your organic starting material and any other reagents needed. [**Hint:** On the basis of the information in Section 3-7, you should not expect to find very good answers for (a), (c), and (e). One general approach is best, however.]
(a) 1-Chloropropane **(b)** 2-Chloropropane **(c)** 1-Bromopropane
(d) 2-Bromopropane **(e)** 1-Iodopropane **(f)** 2-Iodopropane

40. Propose two syntheses of *trans*-1-methyl-2-(methylthio)cyclohexane (shown in the margin), beginning with the starting compound **(a)** *cis*-1-chloro-2-methylcyclohexane; **(b)** *trans*-1-chloro-2-methylcyclohexane.

[cyclohexane structure with SCH_3 and CH_3]

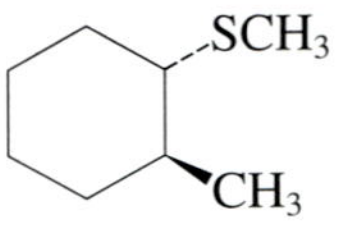

41. Rank each of the following sets of molecules in order of increasing S_N2 reactivity.

(a) CH_3CH_2Br, CH_3Br, $(CH_3)_2CHBr$

(b) $(CH_3)_2CHCH_2CH_2Cl$, $(CH_3)_2CHCH_2Cl$, $(CH_3)_2CHCl$

(c) CH_3CH_2Cl, CH_3CH_2I, ⬡—Cl

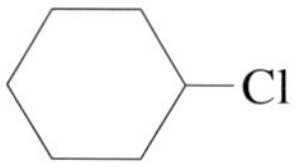

(d) $(CH_3CH_2)_2CHCH_2Br$, $CH_3CH_2CH_2CHBr$, $(CH_3)_2CHCH_2Br$
$\overset{|}{C}H_3$

42. Predict the effect of the changes given below on the rate of the reaction

$CH_3Cl + {}^-OCH_3 \xrightarrow{CH_3OH} CH_3OCH_3 + Cl^-$. (a) Change substrate from CH_3Cl to CH_3I; (b) change nucleophile from CH_3O^- to CH_3S^-; (c) change substrate from CH_3Cl to $(CH_3)_2CHCl$; (d) change solvent from CH_3OH to $(CH_3)_2SO$.

43. The following table presents rate data for the reactions of CH_3I with three different nucleophiles in two different solvents. What is the significance of these results regarding relative reactivity of nucleophiles under different conditions?

Nucleophile	k_{rel}, CH_3OH	k_{rel}, DMF
Cl^-	1	1.2×10^6
Br^-	20	6×10^5
$NCSe^-$	4000	6×10^5

44. Rings are readily prepared by means of intramolecular S_N2 reactions. An example and its mechanism are:

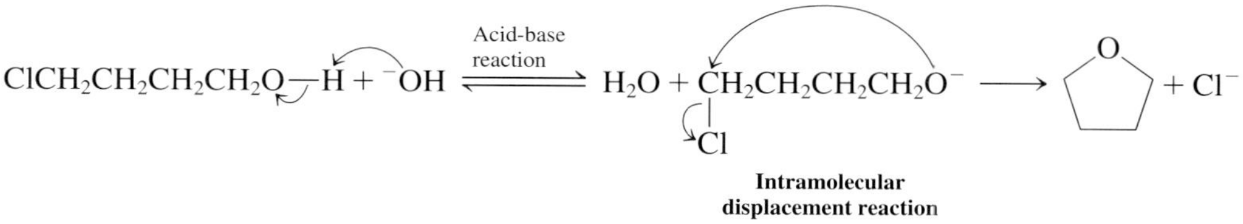

Intramolecular
displacement reaction

Explain the outcome of the following transformations mechanistically. (**Hint:** Notice in the example how an acid-base reaction leads to a stronger nucleophile at one end of the molecule. Use this in your answers.)

(a) $HSCH_2CH_2Br + NaOH \xrightarrow{CH_3CH_2OH}$ ▷S

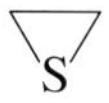

(b) $BrCH_2CH_2CH_2CH_2CH_2Br + NaOH \xrightarrow{CH_3OH}$ (hexagon with O)

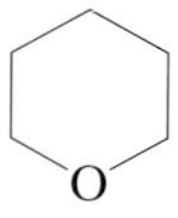

(c) $BrCH_2CH_2CH_2CH_2CH_2Br + NH_3 \xrightarrow{CH_3CH_2OH}$ (hexagon with N–H)

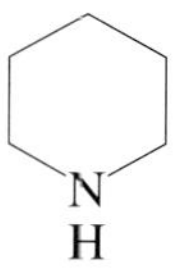

45. S_N2 reactions of halocyclopropane and halocyclobutane substrates are very much slower than those of analogous acyclic secondary haloalkanes. Suggest an explanation for this finding. (**Hint:** Consider the effect of bond-angle strain on the energy of the transition state; see Figure 6-4.)

46. Nucleophilic attack on halocyclohexanes is also somewhat retarded compared with that on acyclic secondary haloalkanes, even though in this case bond-angle strain is *not* an important factor. Explain. (**Hint:** Make a model, and refer to Chapter 4 and Section 6-10.)

Team Problem

47. Compounds A through H are isomeric bromoalkanes with the molecular formula $C_5H_{11}Br$. With your team, draw all eight constitutional isomers. Indicate any stereocenter(s), but do not label it (them) as R or S until you have completed your analysis. Using the data below, assign structures to A through H. Divide the problem into equal parts to share the effort of finding a solution. Reconvene and discuss your analysis. At this point, you should indicate the stereochemistry with wedged and dashed lines as appropriate.

- Treatment of compounds A through G with NaCN in DMF followed second-order kinetics and showed the following relative rates:

$$A \cong B > C > D \cong E > F >>> G$$

- Compound H does not undergo the S_N2 reaction under the preceding conditions.
- Compounds C, D, and F were found to be optically active, each having S absolute configuration at the stereocenter. Substitution reactions of D and F with NaCN in DMF proceeded with inversion of configuration, while treatment of C in the same way proceeded with retention of configuration.

Preprofessional Problems

48. The S_N2 reaction mechanism best applies to
(**a**) cyclopropane and H_2 (**b**) 1-chlorobutane and aqueous NaOH
(**c**) KCH and NaOH   (**d**) ethane and H_2O

49. The reaction $CH_3Cl + OH^- \longrightarrow CH_3OH + Cl^-$ is first order in both chloromethane and hydroxide. Given the rate constant $k = 3.5 \times 10^{-3}$ mol $L^{-1}s^-$, what is the observed rate at the following concentrations?

$$[CH_3Cl] = 0.50 \text{ mol } L^{-1}; [OH^-] = 0.015 \text{ mol } L^{-1}$$

(**a**) 2.6×10^{-5} mol $L^{-1} s^{-1}$ (**b**) 2.6×10^{-6} mol $L^{-1} s^{-1}$ (**c**) 2.6×10^{-3} mol $L^{-1} s^{-1}$
(**d**) 1.75×10^{-3} mol $L^{-1} s^{-1}$ (**e**) 1.75×10^{-5} mol $L^{-1} s^{-1}$

50. Which ion is the strongest nucleophile in aqueous solution?
(**a**) F^- (**b**) Cl^- (**c**) Br^- (**d**) I^- (**e**) all of these are equally strong

51. Only one of the following processes will occur measurably at room temperature. Which one?

(a) 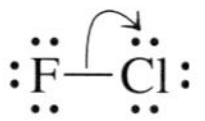 $:\ddot{F} - \ddot{C}l:$

(b) $:N \equiv C:^- \quad CH_3 - I$

(c) $:N \equiv N: \quad CH_3 - I$

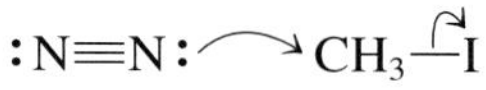

(d) 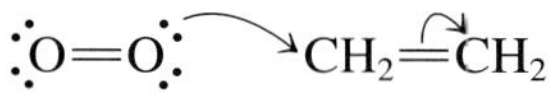 $:\ddot{O} = \ddot{O}: \quad CH_2 = CH_2$

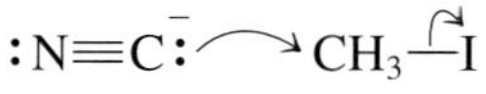

Further Reactions of Haloalkanes

7

Unimolecular Substitution and Pathways of Elimination

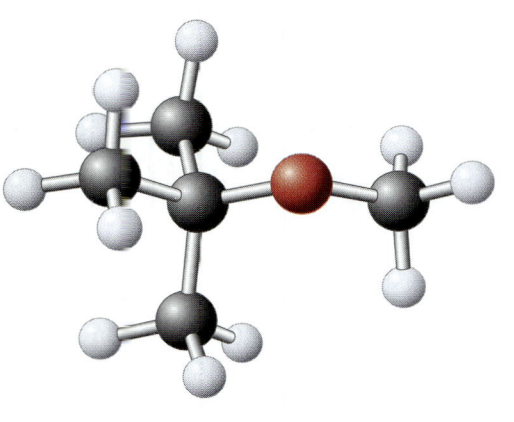

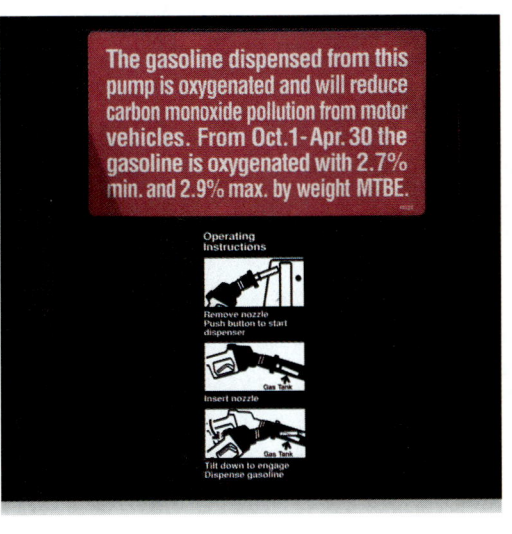

2-Methoxy-2-methylpropane, also known as methyl *tert*-butyl ether (MTBE), may be prepared by solvolysis; it replaces lead as an octane-rating enhancer in gasoline.

Is the S_N2 displacement process the only reaction available to haloalkanes? Are other mechanisms for substitution possible? Finally, are there other, fundamentally different types of transformations that haloalkanes undergo? In this chapter we shall see that haloalkanes can indeed follow reaction pathways other than S_N2 displacement, especially if the haloalkanes are tertiary or secondary. We shall see that bimolecular substitution is only one of *four* possible modes of reaction. The other three modes are unimolecular substitution and two types of elimination processes. The elimination processes give rise to double bonds through loss of HX and serve as our first method of entry into multiply bonded organic compounds.

7-1 Solvolysis of Tertiary and Secondary Haloalkanes

We have seen that the rate of the S_N2 reaction diminishes drastically when the reacting center changes from primary to secondary to tertiary. For example, although the S_N2 reactivity of bromomethane and bromoethane with iodide ion in propanone (acetone) is high, 2-bromopropane is much less reactive, and 2-bromo-2-methylpropane is essentially inert. However, these observations pertain only to *bimolecular* substitution. Secondary and tertiary halides do undergo substitution, but by another mechanism. This section will show that, in fact, these substrates transform readily, even in the presence of weak nucleophiles, to give substitution products.

247

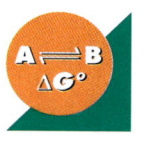

For example, when 2-bromo-2-methylpropane (*tert*-butyl bromide) is mixed with water, it is rapidly converted into 2-methyl-2-propanol (*tert*-butyl alcohol) and hydrogen bromide. Water is the nucleophile here, even though it is poor in this capacity. Such a transformation, in which a substrate undergoes substitution by *solvent* molecules, is called **solvolysis.** When the solvent is water, the term **hydrolysis** is applied.

An Example of Solvolysis: Hydrolysis

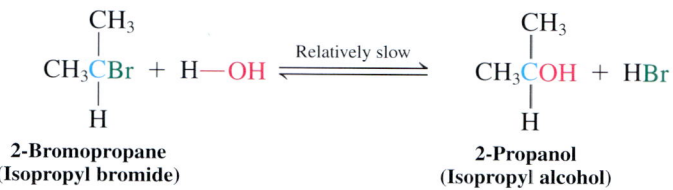

$$\underset{\substack{\text{2-Bromo-2-methylpropane}\\(\textit{tert}\text{-Butyl bromide})}}{\text{CH}_3\text{CBr}} + \text{H—OH} \underset{}{\overset{\text{Relatively fast}}{\rightleftharpoons}} \underset{\substack{\text{2-Methyl-2-propanol}\\(\textit{tert}\text{-Butyl alcohol})}}{\text{CH}_3\text{COH}} + \text{HBr}$$

> **Reminder**
> Nucleophile: red
> Electrophile: blue
> Leaving group: green

Methyl and Primary Haloalkanes: Unreactive in Solvolysis

CH_3Br
$\text{CH}_3\text{CH}_2\text{Br}$
$\text{CH}_3\text{CH}_2\text{CH}_2\text{Br}$

Essentially no reaction with H_2O at room temperature

2-Bromopropane is hydrolyzed similarly, albeit much more slowly, whereas 1-bromopropane, bromoethane, and bromomethane are relatively unaffected by these conditions.

Hydrolysis of a Secondary Haloalkane

$$\underset{\substack{\text{2-Bromopropane}\\(\text{Isopropyl bromide})}}{\text{CH}_3\text{CBr}} + \text{H—OH} \overset{\text{Relatively slow}}{\rightleftharpoons} \underset{\substack{\text{2-Propanol}\\(\text{Isopropyl alcohol})}}{\text{CH}_3\text{COH}} + \text{HBr}$$

Solvolysis also takes place in alcohol solvents.

Solvolysis of 2-Chloro-2-methylpropane in Methanol

$$\underset{\substack{\text{2-Chloro-}\\\text{2-methylpropane}}}{\text{CH}_3\text{CCl}} + \underset{\text{Solvent}}{\text{CH}_3\text{OH}} \rightleftharpoons \underset{\substack{\text{2-Methoxy-}\\\text{2-methylpropane}}}{\text{CH}_3\text{COCH}_3} + \text{HCl}$$

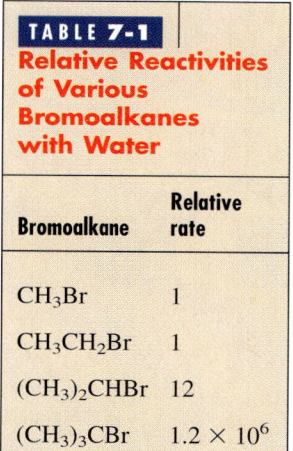

TABLE 7-1

Relative Reactivities of Various Bromoalkanes with Water

Bromoalkane	Relative rate
CH_3Br	1
$\text{CH}_3\text{CH}_2\text{Br}$	1
$(\text{CH}_3)_2\text{CHBr}$	12
$(\text{CH}_3)_3\text{CBr}$	1.2×10^6

The relative rates of reaction of 2-bromopropane and 2-bromo-2-methylpropane with water to give the corresponding alcohols are shown in Table 7-1 and are compared with the corresponding rates of hydrolysis of their unbranched counterparts. Although the process gives the products expected from an S_N2 reaction, the order of reactivity is *reversed* from that found under typical S_N2 conditions. Thus, primary halides are very slow in their reactions with water, secondary halides are more reactive, and tertiary systems are about *1 million times* as fast as primary ones.

These observations suggest that the mechanism of solvolysis of secondary and, especially, tertiary haloalkanes must be different from that of bimolecular substitution. To understand the details of this transformation, we will use the same methods that we used to study the S_N2 process: kinetics, stereochemistry, and the effect of substrate structure and solvent on reaction rates.

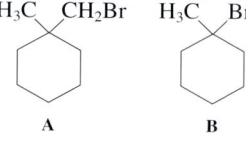

A   B

EXERCISE 7-1

Whereas compound A (shown in the margin) is completely stable in ethanol, B is rapidly converted into another compound. Explain.

7-2 Unimolecular Nucleophilic Substitution

In this section we shall learn about a new pathway for nucleophilic substitution. Recall that the S_N2 reaction has second-order kinetics, generates products stereospecifically, and is fastest with halomethanes, successively slower with primary and secondary halides, and does not take place with tertiary substrates at all. In contrast, solvolyses follow a *first-order* rate law, are *not* stereospecific, and are characterized by the *opposite* order of reactivity. Let us see how these findings can be accommodated mechanistically.

Solvolysis follows first-order kinetics

In Chapter 6 the reaction's kinetics revealed a bimolecular transition state: The rate of the S_N2 reaction is proportional to the concentration of both the haloalkane and the nucleophile. Similar studies have been carried out by varying the concentrations of 2-bromo-2-methylpropane and water in formic acid (a polar solvent of very low nucleophilicity) and measuring the rates of solvolysis. The results of these experiments show that *the rate of hydrolysis of the bromide is proportional to the concentration of only the starting halide, **not** the water.*

$$\text{Rate} = k[(CH_3)_3CBr] \text{ mol L}^{-1} \text{ s}^{-1}$$

What does this observation mean? First, it is clear that the haloalkane has to undergo some transformation on its own before anything else takes place. Second, because the final product contains a hydroxy group, water (or, in general, any nucleophile) must enter the reaction, but at a later stage and not in a way that will affect the rate law. The only way to explain this behavior is to postulate that any steps that follow the initial reaction of the halide are relatively fast. In other words, *the observed rate is that of the slowest step in the sequence:* the **rate-determining step.** It follows that only those species taking part in the transition state of this step enter into the rate expression: in this case, only the starting haloalkane.

In analogy, think of the rate-determining step as a bottleneck. Imagine a water hose with several attached clamps restricting the flow (Figure 7-1). We can see that the rate at which the water will spew out of the end is controlled by the narrowest constriction. If we were to reverse the direction of flow (to model the reversibility of a reaction), again the rate of flow would be controlled by this point. Such is the case in transformations consisting of more than one step—for example, solvolysis. What, then, are the steps in our example?

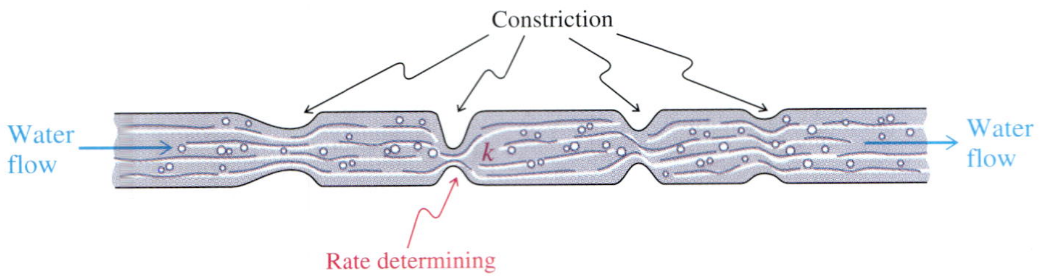

FIGURE 7-1
The rate, k, at which water flows through a hose is controlled by the narrowest constriction.

The mechanism of solvolysis includes carbocation formation

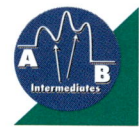

The hydrolysis of 2-bromo-2-methylpropane is said to proceed by **unimolecular nucleophilic substitution,** abbreviated $S_N1.$ The number 1 indicates that only one molecule, the haloalkane, participates in the rate-determining step: The rate of the reaction does *not* depend on the concentration of the nucleophile. The mechanism consists of three steps.

STEP 1. The rate-determining step is the dissociation of the haloalkane to an alkyl cation and bromide.

Dissociation of Halide to Form a Carbocation

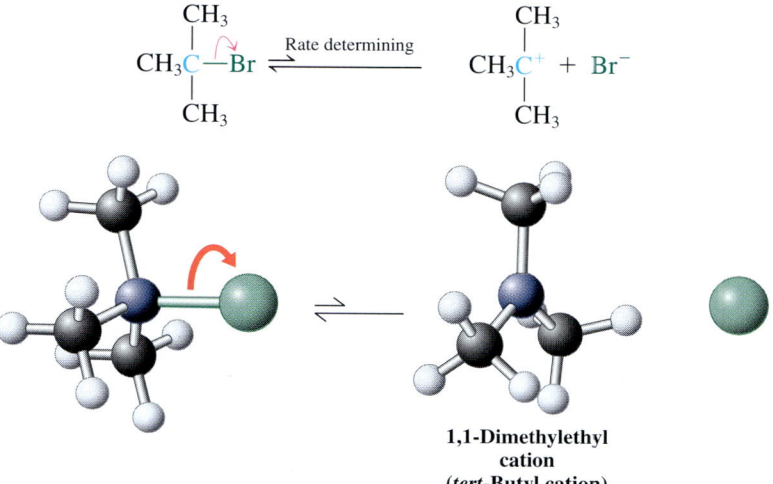

1,1-Dimethylethyl cation
(*tert*-Butyl cation)

This conversion is an example of heterolytic cleavage. The hydrocarbon product contains a positively charged central carbon atom attached to three other groups and bearing only an electron sextet. Such a structure is called a **carbocation.**

STEP 2. The 1,1-dimethylethyl (*tert*-butyl) cation formed in step 1 is a powerful electrophile that is immediately trapped by the surrounding water. This process can be viewed as a nucleophilic attack by the solvent on the electron-deficient carbon.

Nucleophilic Attack by Water

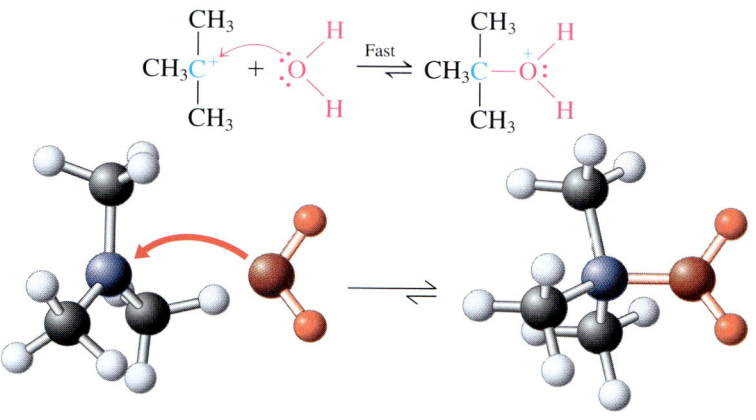

An alkyloxonium ion

The resulting species is an example of an **alkyloxonium ion,** the conjugate acid of an alcohol—in this case 2-methyl-2-propanol, the eventual product of the sequence.

STEP 3. Like the hydronium ion, H_3O^+, the first member of the series of oxonium ions, all alkyloxonium ions are strong acids. They are therefore readily deprotonated by the water in the reaction medium to furnish the final alcohol.

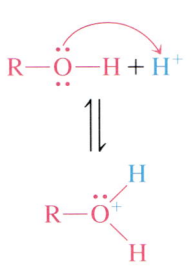

Alkyloxonium ion

Deprotonation

$$CH_3\overset{\overset{\displaystyle CH_3}{|}}{\underset{\underset{\displaystyle CH_3}{|}}{C}}\overset{+}{-}\ddot{O}: \quad + \quad \ddot{O}H_2 \underset{}{\overset{Fast}{\rightleftharpoons}} \quad CH_3\overset{\overset{\displaystyle CH_3}{|}}{\underset{\underset{\displaystyle CH_3}{|}}{C}}\ddot{O}H \quad + \quad H\overset{+}{\ddot{O}}H_2$$

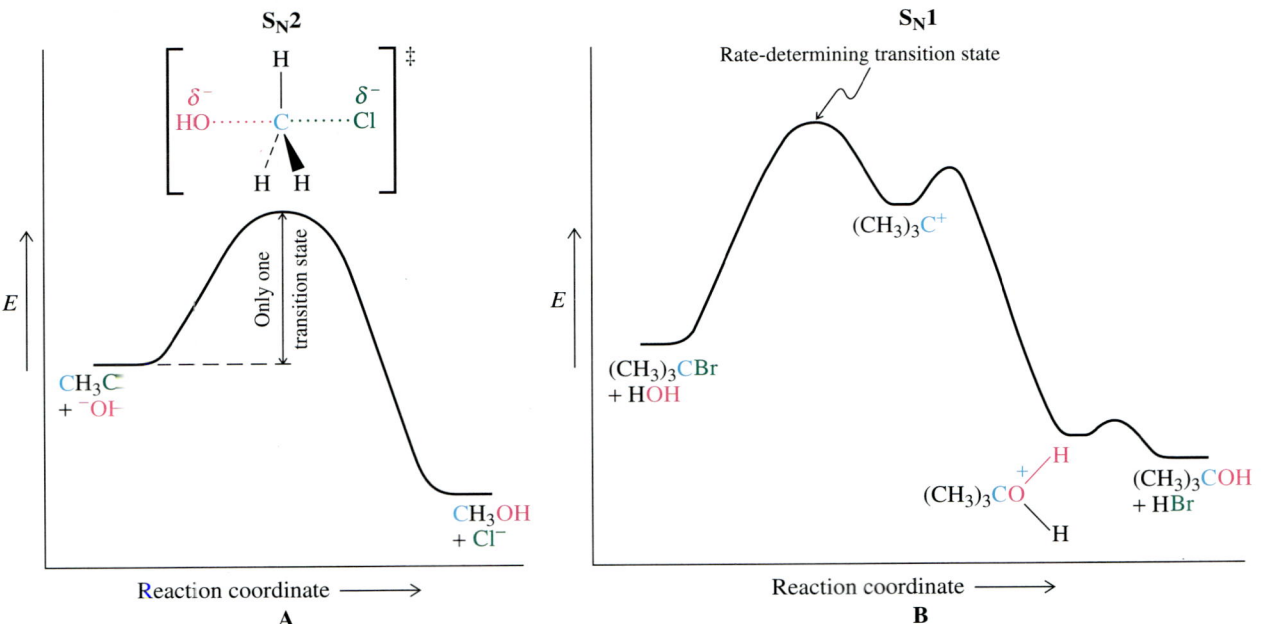

Alkyloxonium ion
(Strongly acidic)

2-Methyl-2-propanol

Figure 7-2 compares the potential-energy diagrams for the S_N2 reaction of chloromethane with hydroxide ion and the S_N1 reaction of 2-bromo-2-methylpropane

FIGURE 7-2

Potential-energy diagrams for (A) S_N2 reaction of chloromethane with hydroxide and (B) S_N1 hydrolysis of 2-bromo-2-methylpropane. Whereas the S_N2 process takes place in a single step, the S_N1 mechanism consists of three distinct events: rate-determining dissociation of the haloalkane into a halide ion and a carbocation, nucleophilic attack by water on the carbocation to give an alkyloxonium ion, and proton loss to furnish the final product. Note: For clarity, inorganic species have been omitted from the intermediate stages of (B).

with water. The latter exhibits three transition states, one for each step in the mechanism. The first has the highest energy—and thus is rate determining—because it requires the separation of opposite charges.

All three steps of the mechanism of solvolysis are reversible. The overall equilibrium can be driven in either direction by the suitable choice of reaction conditions. Thus, a large excess of nucleophilic solvent ensures complete solvolysis. In Chapter 9 we shall see how this reaction can be reversed to permit the synthesis of tertiary haloalkanes from alcohols.

In summary, the kinetics of haloalkane solvolysis leads us to a mechanism in which initial dissociation to form a carbocation is the crucial, rate-determining step. Can we back up our mechanistic hypothesis with other experimental observations?

7-3 Stereochemical Consequences of S_N1 Reactions

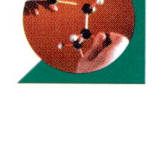

The proposed mechanism of unimolecular nucleophilic substitution has predictable stereochemical consequences because of the structure of the intermediate carbocation. To minimize electron repulsion, the positively charged carbon assumes trigonal planar geometry, the result of sp^2 hybridization (Sections 1-3 and 1-8). Such an intermediate is therefore achiral (make a model). Hence, starting with an optically active tertiary (or secondary) haloalkane in which the stereocenter bears the departing halogen, we should obtain racemic S_N1 products (Figure 7-3). This result is, in fact, observed in many solvolyses. In general, the formation of racemic products from optically active substrates is strong evidence for the intermediacy of a symmetrical, achiral species, such as a carbocation, in the course of a reaction.

FIGURE 7-3

The mechanism of hydrolysis of (R)-3-bromo-3-methylhexane predicts the stereochemistry of the reaction. Initial ionization furnishes a planar, achiral carbocation. This ion, when trapped with water, yields racemic alcohol.

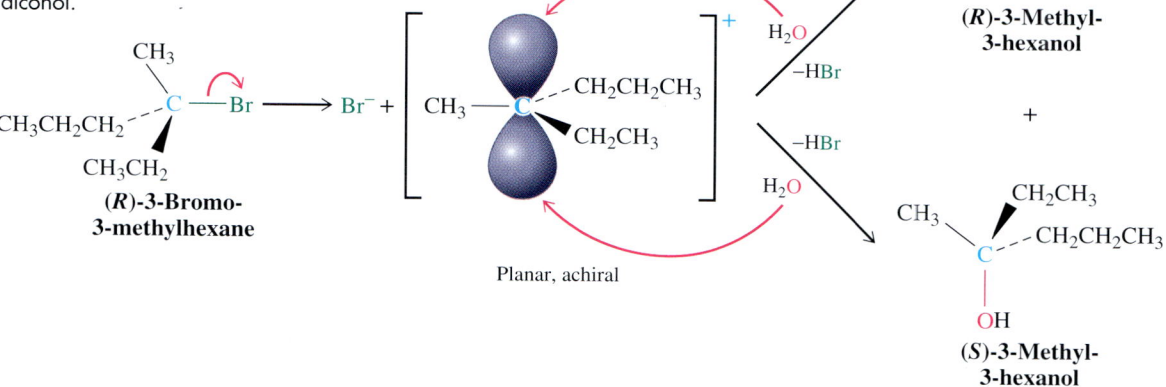

CHEMICAL HIGHLIGHT 7-1 | **Incomplete Racemization in S$_N$1 Reactions**

Many solvolyses of optically pure compounds in which the leaving group is attached to a stereocenter give racemic mixtures, but many others do not. It is often found that unequal amounts of *R* and *S* products are obtained from S$_N$1 reactions, with the major enantiomer being the one in which inversion of the reacting center has taken place. The hydrolyses of optically pure samples of (1-chloro-) and (1-bromoethyl)benzene illustrate this phenomenon. The bromo compound reacts with water to give racemic 1-phenylethanol. In contrast, the chloro analog gives a 15–20% excess of the alcohol enantiomer with inverted configuration at the site of displacement.

Why does this result occur? Chloride ion is a good, but not particularly outstanding, leaving group. Its departure to give the secondary carbocation is sluggish. As a consequence, it has a tendency to

hover in the vicinity of the cationic carbon, partially shielding one face of the latter from attack by solvent. Attack by water on the opposite face of the carbocation is not inhibited, leading to a net excess of inverted product. In contrast, the halogen in the bromo compound is an excellent leaving group; in its departure, it does not shield the nearer face of the carbocation. Formation of a completely racemized hydrolysis product is observed.

Interestingly, this result is noticeable, at least qualitatively, without sophisticated equipment: The two enantiomeric 1-phenylethanols have noticeably different odors. If we start with (*R*)-chloride, the odor of the hydrolysis product mixture, containing excess (*S*)-alcohol, differs from that of the product mixture from hydrolysis of the (*S*)-chloride, in which (*R*)-alcohol is present in excess.

When X is Br:	50%	50%	Racemic
When X is Cl:	41%	59%	Excess of inversion

EXERCISE 7-3

(*R*)-3-Bromo-3-methylhexane loses its optical activity when dissolved in nitromethane, a highly polar but nonnucleophilic solvent. Explain.

EXERCISE 7-4

Hydrolysis of molecule A (shown in the margin) gives two alcohols. Explain.

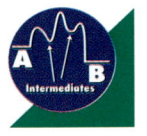

A

7-4 Effects of Solvent, Leaving Group, and Nucleophile on Unimolecular Substitution

As in S$_N$2 reactions, varying the solvent, the leaving group, and the nucleophile greatly affects unimolecular substitution.

Polar solvents accelerate the S$_N$1 reaction

Heterolytic cleavage of the C–X bond in the rate-determining step of the S$_N$1 reaction entails a transition-state structure that is highly polarized (Figure 7-4; see page 254), eventually leading to two fully charged ions. In contrast, in a typical S$_N$2 transition state, charges are not created; rather, they are dispersed (see Figure 6-4).

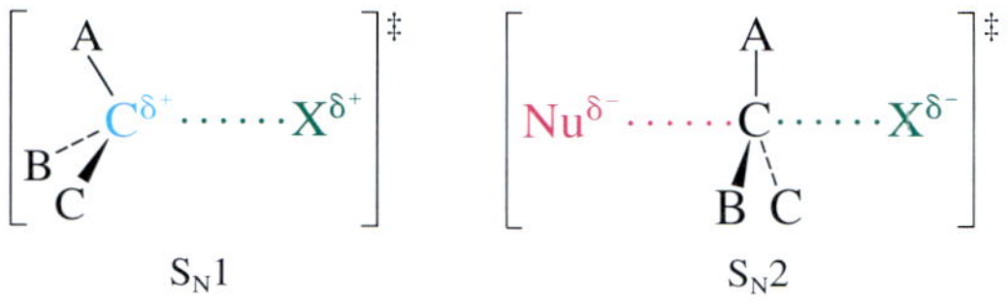

$$
\left[\begin{array}{c} A \\ \diagdown \\ B \diagup C^{\delta+} \cdots\cdots X^{\delta+} \\ C \end{array} \right]^{\ddagger} \qquad \left[\begin{array}{c} A \\ | \\ Nu^{\delta-} \cdots\cdots C \cdots\cdots X^{\delta-} \\ \diagup \diagdown \\ B \quad C \end{array} \right]^{\ddagger}
$$

S_N1 $\qquad\qquad\qquad$ S_N2

FIGURE 7-4

The respective transition states for the S_N1 and S_N2 reactions explain why the former is strongly accelerated by polar solvents. Heterolytic cleavage entails charge separation, a process aided by polar solvation.

Because of this polar transition state, the rate of an S_N1 reaction increases as solvent polarity is increased. The effect is particularly striking when the solvent is changed from aprotic to protic. For example, hydrolysis of 2-bromo-2-methylpropane is much faster in pure water than in a 9:1 mixture of propanone (acetone) and water. The protic solvent accelerates the S_N1 reaction, because it stabilizes the transition state shown in Figure 7-4 by hydrogen bonding with the leaving group. Remember that, in contrast, the S_N2 reaction is accelerated in polar *aprotic* solvents, mainly because of a solvent effect on the reactivity of the nucleophile and *not* of the substrate.

Effect of Solvent on the Rate of an S_N1 Reaction

Relative rate

$$(CH_3)_3CBr \xrightarrow{100\% \ H_2O} (CH_3)_3COH \ + \ HBr \qquad 400,000$$

$$(CH_3)_3CBr \xrightarrow{90\% \ propanone \ (acetone), \ 10\% \ H_2O} (CH_3)_3COH \ + \ HBr \qquad 1$$

The solvent nitromethane, CH_3NO_2 (see Table 6-5), is exceptional in being both highly polar and essentially nonnucleophilic. It therefore is useful in studies of S_N1 reactions with nucleophiles other than solvent molecules.

The S_N1 reaction speeds up with better leaving groups

Because the leaving group departs in the rate-determining step of the S_N1 reaction, it is not surprising that the rate of the reaction increases as the leaving-group ability of the departing group improves. Thus, tertiary iodoalkanes are more readily solvolyzed than are the corresponding bromides, and the latter are in turn more reactive than chlorides. Sulfonates are particularly prone to departure.

Relative Rate of Solvolysis of RX (R = Tertiary Alkyl)

$$X = -OSO_2R' > -I > -Br > -Cl$$

The strength of the nucleophile affects the product distribution but not the reaction rate

Does changing the nucleophile affect the rate of S_N1 reaction? The answer is no. Recall that, in the S_N2 process, the rate of reaction increases significantly as the nucleophilicity of the attacking species improves. However, because the rate-determining step of unimolecular substitution does *not* include the nucleophile, changing its structure (or concentration) should *not* alter the rate of disappearance of the haloalkane.

Nevertheless, when two or more nucleophiles compete for capture of the intermediate carbocation, their relative strengths and concentrations may greatly affect the *product distribution.*

For example, solvolysis of a 0.1 M solution of 2-chloro-2-methylpropane in methanol gives the expected 2-methoxy-2-methylpropane, with a rate constant k_1. Quite a different result is obtained when the same experiment is carried out in the presence of an equivalent amount of sodium azide: The product is 1,1-dimethylethyl (*tert*-butyl) azide, still formed at the *same* rate. In this case, the much more powerful nucleophile N_3^- (see Table 6-7) wins out in competition with methanol. The rate of disappearance of 2-chloro-2-methylpropane is determined by k_1 (regardless of the product eventually formed), but the relative yields of the *products* depend on the relative reactivities of the competing nucleophiles (k_{CH_3OH} is much smaller than $k_{N_3^-}$).

Competing Nucleophiles in the S_N1 Reaction

$$(CH_3)_3COCH_3 + HCl$$

**2-Methoxy-
2-methylpropane**

$(CH_3)_3CCl$

$\xrightarrow{k_{CH_3OH}}$

$+$

CH_3OH $\xrightarrow[\text{Rate determining}]{k_1}$ $(CH_3)_3C^+ + Cl^-$

$+$

$\xrightarrow{k_{N_3^-}}$

NaN_3

$$(CH_3)_3CN_3 + NaCl$$

**1,1-Dimethylethyl
azide
(*tert*-Butyl azide)**

EXERCISE 7-5

A solution of 2-methyl-2-propyl methanesulfonate in polar aprotic solvent containing equal amounts of sodium fluoride and sodium bromide produces 75% 2-fluoro-2-methylpropane and only 25% 2-bromo-2-methylpropane. Explain. (**Hint:** Refer to Section 6-9 and Problem 43 in Chapter 6 for information regarding relative nucleophilic strengths of the halide ions in aprotic solvents.)

To summarize, we have seen further evidence supporting the S_N1 mechanism for the reaction of tertiary (and secondary) haloalkanes with certain nucleophiles. The stereochemistry of the process, the effects of the solvent and the leaving-group ability on the rate, and the absence of such effects when the strength of the nucleophile is varied are consistent with the unimolecular route. The next question to be answered is, Why? What is so special about tertiary haloalkanes that they undergo conversion by the S_N1 pathway, whereas primary systems follow S_N2? How do secondary haloalkanes fit into this scheme?

7-5 Effect of the Alkyl Group on the S_N1 Reaction: Carbocation Stability

Somehow, the degree of substitution at the reacting carbon must control the pathway followed in the reaction of haloalkanes (and related derivatives) with nucleophiles. We shall see that only secondary and tertiary systems can form carbocations. For this

reason, tertiary halides, *whose steric bulk prevents them from undergoing S_N2 reactions,* transform solely by the S_N1 mechanism, primary haloalkanes only by S_N2, and secondary haloalkanes by either route, depending on conditions.

Carbocation stability increases from primary to secondary to tertiary

We have learned that primary haloalkanes undergo *only* direct nucleophilic substitution. In contrast, secondary systems often transform through carbocation intermediates, and tertiary systems always do. The reasons for this difference are twofold: First, steric hindrance increases along the series, thereby slowing down S_N2; and, second, increasing alkyl substitution stabilizes carbocationic centers. Only secondary and tertiary cations are energetically feasible under the conditions of the S_N1 reaction.

Relative Stability of Carbocations

$$CH_3CH_2CH_2\overset{+}{C}H_2 \quad < \quad CH_3CH_2\overset{+}{C}HCH_3 \quad < \quad (CH_3)_3\overset{+}{C}$$

Primary $\quad < \quad$ **Secondary** $\quad < \quad$ **Tertiary**

Now we can see why tertiary haloalkanes solvolyze so readily. Because tertiary carbocations are more stable than their less substituted relatives, they form more easily. But what is the reason for this order of stability?

Hyperconjugation stabilizes positive charge

Note that the order of carbocation stability parallels that of the corresponding radicals. Both trends have their roots in the same phenomenon: *hyperconjugation.* Recall from Section 3-2 that hyperconjugation is the result of overlap of a *p* orbital with a neighboring bonding molecular orbital, such as that of a C–H or a C–C bond. In a radical, the *p* orbital is singly filled; in a carbocation, it is empty. In both cases, the alkyl group donates electron density to the electron-deficient center and thus stabilizes it. Figure 7-5 compares the methyl cation, devoid of hyperconjugation, with the much more stable 1,1-dimethylethyl (*tert*-butyl) cation. Figure 7-6 shows the structure of the latter as derived from X-ray diffraction measurements.

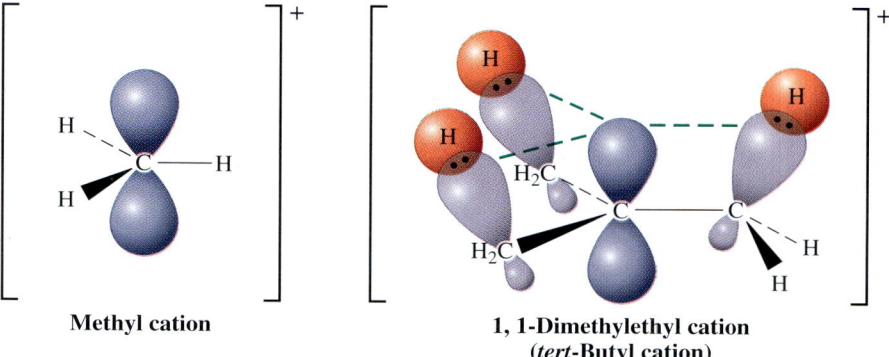

Methyl cation

1, 1-Dimethylethyl cation
(*tert*-Butyl cation)

FIGURE 7-5
The methyl cation is not stabilized by hyperconjugation (left), whereas the 1,1-dimethylethyl (*tert*-butyl) cation benefits from three hyperconjugative interactions (right).

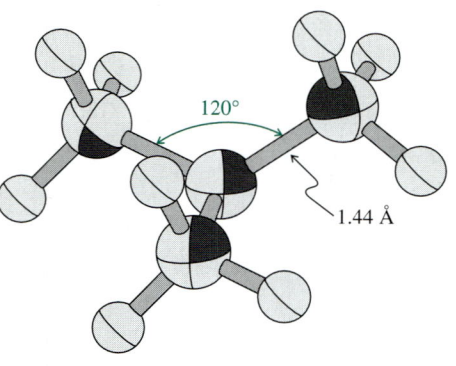

FIGURE 7-6

X-ray crystal structure determination for the 1,1-dimethylethyl (*tert*-butyl) cation. The four carbons lie in a plane with 120° C–C–C bond angles, consistent with sp^2 hybridization at the central carbon. The C–C bond length is 1.44 Å, shorter than normal single bonds, a consequence of hyperconjugative overlap.

Secondary systems undergo both S_N1 and S_N2 reactions

As you will have gathered from the preceding discussion, secondary haloalkanes exhibit the most varied substitution behavior. Do they prefer a bimolecular substitution pathway or are they more likely to enter into carbocation formation? *Both* are possible: Steric hindrance slows but does not preclude direct nucleophilic attack. At the same time, dissociation becomes competitive because of the relative stability of secondary carbocations. The pathway chosen depends on the reaction conditions: the solvent, the leaving group, and the nucleophile.

If we use a substrate carrying a very good leaving group, a nucleophile that is poor, and a polar protic solvent (S_N1 conditions), *unimolecular* substitution is favored. If we employ a high concentration of a good nucleophile, a polar aprotic solvent, and a haloalkane bearing a reasonable leaving group (S_N2 conditions), *bimolecular* substitution predominates. Table 7-2 summarizes our observations regarding the reactivity of haloalkanes toward nucleophiles.

TABLE 7-2	Reactivity of R–X in Nucleophilic Substitutions: R–X + Nu⁻ ⟶ R–Nu + X⁻	
R	**S_N1**	**S_N2**
CH_3	Not observed in solution (methyl cation too high in energy)	Frequent; fast with good nucleophiles and good leaving groups
Primary	Not observed in solution (primary carbocations too high in energy)[a]	Frequent; fast with good nucleophiles and good leaving groups, slow when branching at C2 is present in R
Secondary	Relatively slow; best with good leaving groups in polar protic solvents	Relatively slow; best with high concentrations of good nucleophiles in polar aprotic solvents
Tertiary	Frequent; particularly fast in polar, protic solvents and with good leaving groups	Extremely slow

[a]Exceptions are resonance-stabilized carbocations; see Chapter 14.

Substitution of a Secondary Haloalkane Under S_N2 Conditions

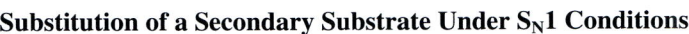

Substitution of a Secondary Substrate Under S_N1 Conditions

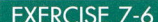

EXERCISE 7-6

Explain the following results.

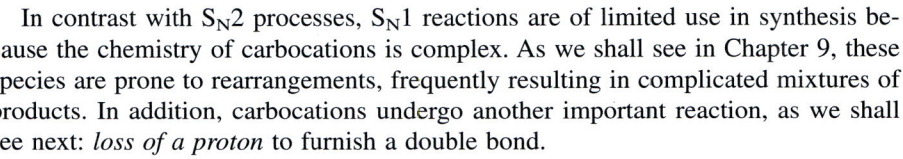

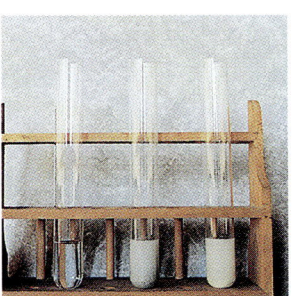

A visual demonstration of relative S_N1 reactivity. The three test tubes contain, from left to right, solutions of 1-bromobutane, 2-bromopropane, and 2-bromo-2-methylpropane in ethanol, respectively. Addition of a few drops of AgNO$_3$ solution to each causes immediate formation of a heavy AgBr precipitate from the *tert*-bromoalkane (right), less AgBr precipitation from the secondary substrate (center), and very little AgBr formation from the primary halide (left).

In contrast with S_N2 processes, S_N1 reactions are of limited use in synthesis because the chemistry of carbocations is complex. As we shall see in Chapter 9, these species are prone to rearrangements, frequently resulting in complicated mixtures of products. In addition, carbocations undergo another important reaction, as we shall see next: *loss of a proton* to furnish a double bond.

To summarize, tertiary haloalkanes are reactive in the presence of nucleophiles even though they are too sterically hindered to undergo S_N2 reactions: The tertiary carbocation is readily formed because it is stabilized by hyperconjugation. Subsequent trapping by a nucleophile, such as a solvent (solvolysis), results in the product of nucleophilic substitution. Primary haloalkanes do not react in this manner: The primary cation is too highly energetic (unstable) to be formed in solution. The primary substrate follows the S_N2 route. Secondary systems are converted into substitution products through either pathway, depending on the nature of the leaving group, the solvent, and the nucleophile.

7-6 Unimolecular Elimination: E1

We know that carbocations are readily trapped by nucleophiles through attack at the positively charged carbon. This is not their only mode of reaction, however. An alternative is deprotonation, furnishing a new class of compounds, the alkenes. Starting from a branched haloalkane, the overall transformation constitutes the removal of HX with the simultaneous generation of a double bond. The general term for such a process is **elimination,** abbreviated **E.**

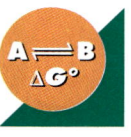

Elimination

$$\overset{H}{\underset{X}{\cdots}}\text{C}-\text{C}\cdots \xrightarrow{\text{Base:}\ :B^-} \text{C}=\text{C} + \text{H}-\text{B} + \text{X}^-$$

Eliminations can take place by several mechanisms. Let us establish the one that is followed in solvolysis.

When 2-bromo-2-methylpropane is dissolved in methanol, it rapidly disappears. As expected, the major product, 2-methoxy-2-methylpropane, arises by solvolysis. However, there is also a significant amount of another compound, 2-methylpropene, the product of *elimination* of HBr from the original substrate. Thus, in competition with the S_N1 process, which leads to displacement of the leaving group, another mechanism transforms the tertiary halide, giving rise to the alkene. What is it? Is it related to the S_N1 reaction? Once again we turn to a kinetic analysis and find that the rate of alkene formation depends on the concentration of *only* the starting halide; the reaction is first order. Because they are unimolecular, eliminations of this type are labeled **E1.** *The rate-determining step in the E1 process is the same as that in S_N1 reactions: dissociation to a carbocation. This intermediate then has a second pathway at its disposal along with nucleophilic trapping: loss of a proton from a carbon adjacent to the one bearing the positive charge.*

$$(CH_3)_3CBr \underset{}{\overset{CH_3OH}{\rightleftharpoons}} H_3C-\overset{+}{\underset{CH_3}{\overset{CH_3}{C}}} + Br^-$$

2-Bromo-2-methyl-propane

E1 ↓ S_N1 ↘ CH₃OH

$$H_2C=\overset{CH_3}{\underset{CH_3}{C}} + H^+ + Br^-$$

20%

2-Methylpropene

$$(CH_3)_3COCH_3 + H^+ + Br^-$$

80%

2-Methoxy-2-methylpropane

How exactly is the proton lost? Figure 7-7 (see page 260) depicts this process with orbitals. Although we often show protons that evolve in chemical processes by using the notation H^+, "free" protons do not participate under the conditions of ordinary organic reactions. A Lewis base (Section 2-9) typically removes the proton. In aqueous solution, water plays this role, giving H_3O^+; here, the proton is carried off by CH_3OH as $CH_3OH_2^+$, an alkyloxonium ion. The carbon left behind rehybridizes from sp^3 to sp^2. As the C–H bond breaks, its electrons shift to overlap in a π fashion with the vacant p orbital at the neighboring cationic center. The result is a hydrocarbon containing a double bond: an alkene. The complete mechanism is as follows.

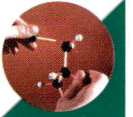

The E1 Reaction Mechanism

$$\overset{CH_3}{\underset{CH_3}{\overset{|}{CH_3C}}}\!\!-Br \underset{}{\overset{CH_3OH}{\rightleftharpoons}} Br^- + \overset{H_3C}{\underset{H_3C}{C}}\!\!\overset{+}{\overset{}{-}}\!\!\overset{H}{\underset{H}{C}} \xrightarrow{\ H\ddot{O}CH_3\ } \overset{H_3C}{\underset{H_3C}{C}}\!\!=\!\!\overset{H}{\underset{H}{C}} + \overset{H}{\underset{H}{\overset{+}{O}CH_3}}$$

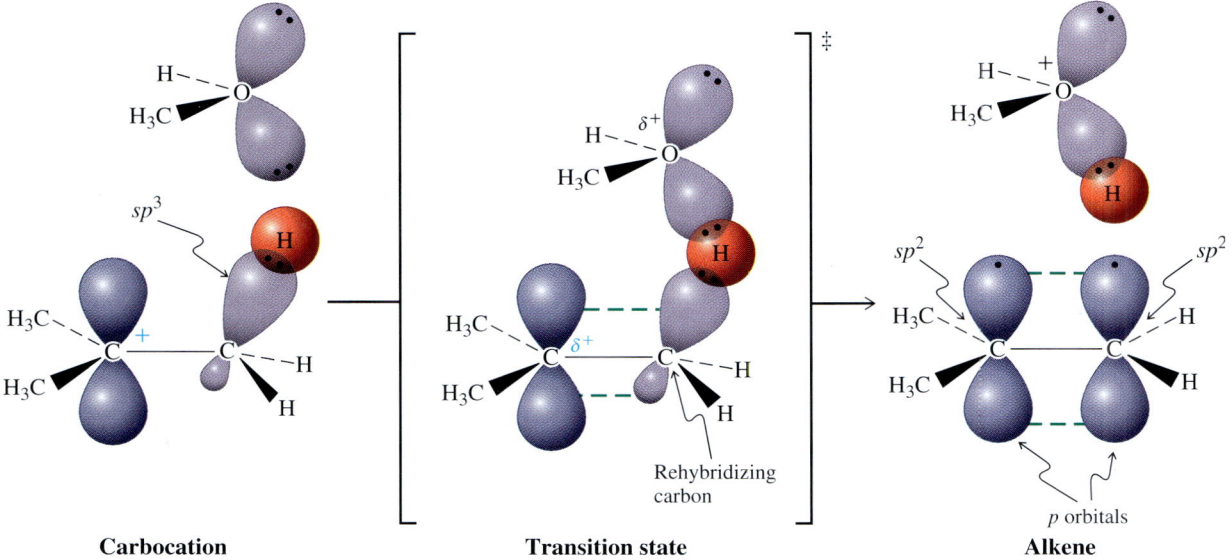

Carbocation   **Transition state**   **Alkene**

FIGURE 7-7

The alkene-forming step in unimolecular elimination (E1): deprotonation of a 1,1-dimethylethyl (*tert*-butyl) cation by the solvent methanol. In an orbital description of proton abstraction, an electron pair on the oxygen atom in the solvent attacks a hydrogen on a carbon adjacent to that bearing the positive charge. The proton is transferred, leaving an electron pair behind. As the carbon rehybridizes from sp^3 to sp^2, these electrons redistribute over the two *p* orbitals of the new double bond.

Any hydrogen positioned on *any carbon next to the center bearing the leaving group* can participate in the E1 reaction. The 1,1-dimethylethyl (*tert*-butyl) cation has nine such hydrogens, each of which is equally reactive. In this case, the product is the same regardless of the identity of the proton lost. In other cases, more than one product may be obtained. These pathways will be discussed in more detail in Chapter 11.

The E1 Reaction Can Give Product Mixtures

$$(CH_3CH_2)_2CH{-}\overset{\overset{\displaystyle CH_3}{|}}{\underset{\underset{\displaystyle Cl}{|}}{C}}{-}CH(CH_3)_2 \xrightarrow[-HCl^*]{CH_3OH,\ \Delta} (CH_3CH_2)_2CH{-}\overset{\overset{\displaystyle CH_3}{|}}{\underset{\underset{\displaystyle OCH_3}{|}}{C}}{-}CH(CH_3)_2$$

S$_N$1 product

$$+ \quad \underset{(CH_3CH_2)_2CH \qquad CH(CH_3)_2}{\overset{\overset{\displaystyle CH_2}{\|}}{C}} \quad + \quad \underset{(CH_3CH_2)_2CH \qquad\qquad CH_3}{\overset{CH_3 \qquad\qquad CH_3}{C{=}C}} \quad + \quad \underset{CH_3CH_2 \qquad\qquad CH(CH_3)_2}{\overset{CH_3CH_2 \qquad\qquad CH_3}{C{=}C}}$$

E1 products

The nature of the leaving group should have no effect on the ratio of substitution to elimination, because the carbocation formed is the same in either case. This is in-

---

*This notation indicates that the elements of the acid have been removed from the starting material. In reality, the proton ends up protonating the base. This system will be used occasionally in other elimination reactions in this book.

deed observed qualitatively (Table 7-3). The product ratio may be affected by the addition of base, but at low base concentration this effect is usually small. Recall that strong bases are usually strong nucleophiles as well (Section 6-9), so addition of a base will generally not greatly favor deprotonation of the carbocation at the expense of nucleophilic attack, and the ratio of E1 to S_N1 products remains approximately constant. However, at high concentrations of strong base, the proportion of elimination rises dramatically. This effect is not the consequence of a change in the E1 : S_N1 ratio, however. Instead, a new pathway for elimination becomes important. This reaction is the subject of the next section.

TABLE 7-3 Ratio of S_N1 to E1 Products in the Hydrolyses of 2-Halo-2-methyl-propanes at 25°C	
X in $(CH_3)_3CX$	Ratio $S_N1 : E1$
Cl	95:5
Br	95:5
I	96:4

EXERCISE 7-7

When 2-bromo-2-methylpropane is dissolved in aqueous ethanol at 25°C, a mixture of $(CH_3)_3COCH_2CH_3$ (30%), $(CH_3)_3COH$ (60%), and $(CH_3)_2C=CH_2$ (10%) is obtained. Explain.

To summarize, carbocations formed in solvolysis reactions are not only trapped by nucleophiles to give S_N1 products but also deprotonated in an elimination (E1) reaction. In this process, the nucleophile (usually the solvent) acts as a base.

7-7 Bimolecular Elimination: E2

In addition to S_N2, S_N1, and E1 reactions, there is a fourth pathway by which haloalkanes may react with nucleophiles *that are also strong bases:* elimination by a *bimolecular* mechanism.

Strong bases effect bimolecular elimination

The preceding section taught us that unimolecular elimination may compete with substitution. A dramatic change of the kinetics is observed at higher concentrations of strong base, however. The rate of alkene formation becomes proportional to the concentrations of both the starting halide *and* the base: The kinetics of elimination are now second order, and the process is called **bimolecular elimination,** abbreviated **E2.**

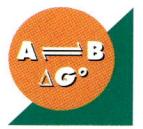

Kinetics of the E2 Reaction of 2-Chloro-2-methylpropane

$$(CH_3)_3CCl + Na^{+-}OH \xrightarrow{k} CH_2=C(CH_3)_2 + NaCl + H_2O$$
$$Rate = k[(CH_3)_3CCl][^-OH] \text{ mol L}^{-1} \text{ s}^{-1}$$

What causes this change in mechanism? Strong bases (such as hydroxide, HO^-, and alkoxides, RO^-) can attack haloalkanes before carbocation formation. The target is a hydrogen on a carbon atom *next to* the one carrying the leaving group. This reaction pathway is not restricted to tertiary halides, although, in secondary and primary systems, it must compete with the S_N2 process.

Competition Between E2 and S_N2 Reactions

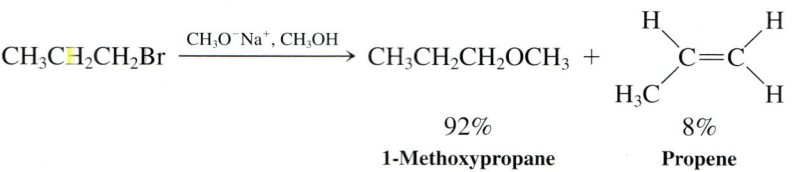

92%   8%

1-Methoxypropane   Propene

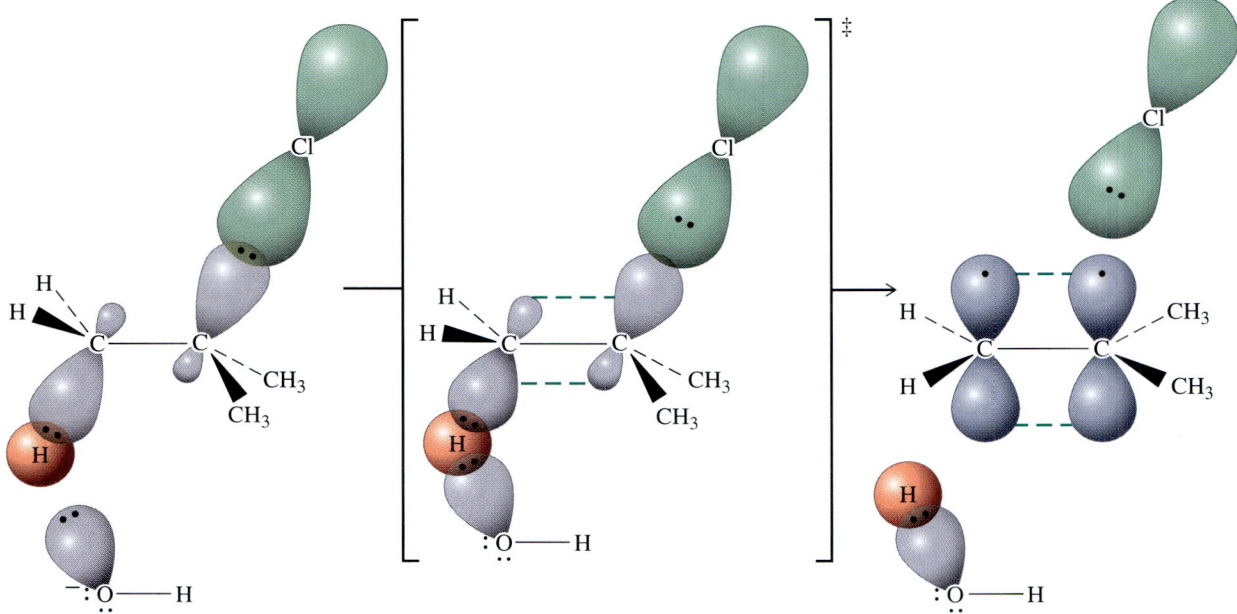

FIGURE 7-8
Orbital description of the E2 reaction of 2-chloro-2-methylpropane with hydroxide ion.

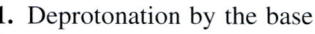

EXERCISE 7-8

What products do you expect from the reaction of bromocyclohexane with hydroxide ion?

EXERCISE 7-9

Give the products (if any) of the E2 reaction of the following substrates: CH_3CH_2I; CH_3I; $(CH_3)_3CCl$; $(CH_3)_3CCH_2I$.

E2 reactions proceed in one step

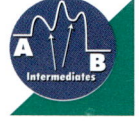

The bimolecular elimination mechanism consists of a *single step*. The bonding changes that occur in its transition state are shown here with electron pushing arrows; in Figure 7-8, they are shown with orbitals. Three changes take place:

1. Deprotonation by the base
2. Departure of the leaving group

The E2 Reaction Mechanism

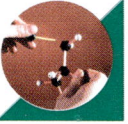

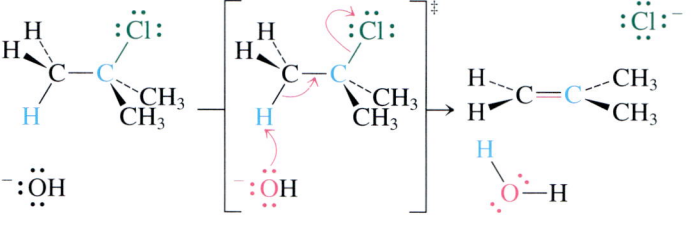

3. Rehybridization of the reacting carbon centers from sp^3 to sp^2 to furnish the two p orbitals of the emerging double bond

All three take place *simultaneously*: The E2 reaction is a one-step, *concerted* process.

Notice that the E1 (Figure 7-7) and E2 mechanisms are very similar, differing only in the sequence of events. In the bimolecular reaction, proton abstraction and leaving-group departure are simultaneous. In the E1 process, the halide leaves first, to be followed by an attack by the base. A good way of thinking about the difference is to imagine that the strong base participating in the E2 reaction is more aggressive. It does not wait for the tertiary or secondary halide to dissociate but attacks the substrate directly.

Experiments elucidate the detailed structure of the E2 transition state

What is the experimental evidence in support of a one-step process with a transition state like that depicted in Figure 7-8? There are three pieces of relevant information. First, the second-order rate law requires that both the haloalkane and the base take part in the rate-determining step. Second, better leaving groups result in faster eliminations. This observation implies that the bond to the leaving group is partially broken in the transition state.

Relative Reactivity in the E2 Reaction

RCl < RBr < RI

Explain the result in the reaction shown below.

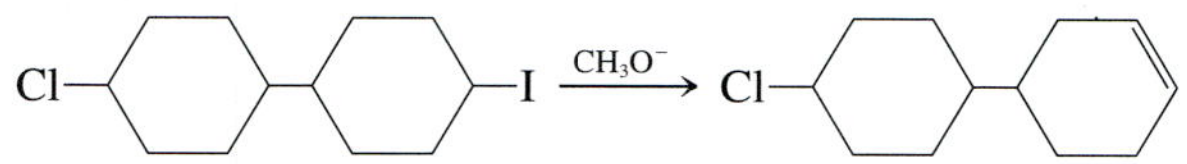

$$Cl{-}\langle\;\rangle{-}\langle\;\rangle{-}I \xrightarrow{CH_3O^-} Cl{-}\langle\;\rangle{-}\langle\;\rangle$$

The third observation is one that not only strongly suggests that both the C–H and the C–X bonds are broken in the transition state, but also describes their relative orientation in space when this event takes place. Figure 7-8 illustrates a characteristic feature of the E2 reaction: its stereochemistry. The substrate is pictured as reacting in a conformation that places the breaking C–H and C–X bonds in an *anti* relation. How can we establish the structure of the transition state with such precision? For this purpose, we can use the principles of conformation and stereochemistry. Treatment of *cis*-1-bromo-4-(1,1-dimethylethyl)cyclohexane with strong base leads to rapid bimolecular elimination to the corresponding alkene. In contrast, under the same conditions the *trans* isomer reacts only very slowly. Why? When we examine the most stable chair conformation of the *cis* compound, we find that two hydrogens are located *anti* to the axial bromine substituent. This geometry is very similar to that required by the E2 transition state, and consequently elimination is easy. Conversely, the *trans* system has no C–H bonds aligned *anti* to the equatorial leaving group (make a model). E2 elimination in this case would require either ring-flip to a diaxial conformer (see Section 4-4) or removal of a hydrogen *gauche* to the bromine, both energetically costly. The latter would be an example of an elimination proceeding through an unfavorable *syn* transition state (*syn*, Greek, together). We will return to E2 elimination in further detail in Chapter 11.

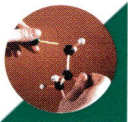

Anti **Elimination Occurs Readily for *cis*- but Not for**
trans-1-Bromo-4-(1,1-dimethylethyl)cyclohexane

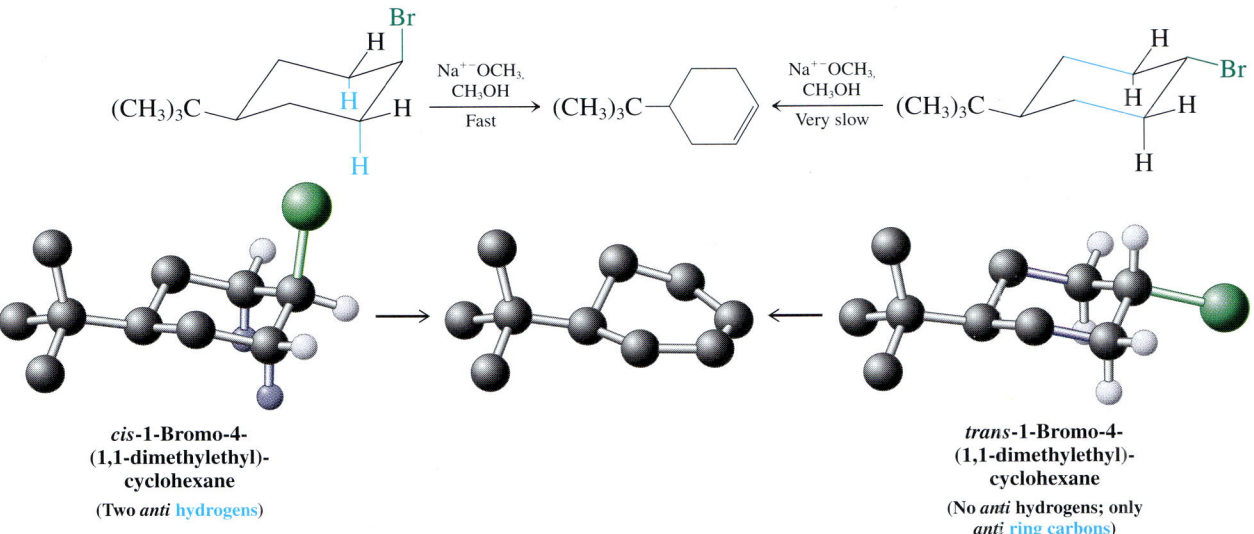

cis-1-Bromo-4-
(1,1-dimethylethyl)-
cyclohexane
(Two *anti* hydrogens)

trans-1-Bromo-4-
(1,1-dimethylethyl)-
cyclohexane
(No *anti* hydrogens; only
***anti* ring carbons)**

EXERCISE 7-11

The rate of elimination of *cis*-1-bromo-4-(1,1-dimethylethyl)cyclohexane is proportional to the concentration of both substrate and base, but that of the *trans* isomer is proportional *only* to the concentration of the substrate. Explain.

EXERCISE 7-12

The isomer of 1,2,3,4,5,6-hexachlorocyclohexane shown in the margin undergoes E2 elimination 7000 times *more slowly* than any of its stereoisomers. Explain.

In summary, strong bases react with haloalkanes not only by substitution, but also by elimination. The kinetics of these reactions are second order, an observation pointing to a bimolecular mechanism. An *anti* transition state is preferred, in which the base abstracts a proton at the same time as the leaving group departs.

7-8 Competition Between Substitution and Elimination

The multiple reaction pathways, S_N2, S_N1, E2, and E1, that haloalkanes may follow in the presence of nucleophiles may seem confusing. Given the many parameters that affect the relative importance of these transformations, are there some simple guidelines that might allow us to predict, at least roughly, what the outcome of any particular reaction will be? The answer is a cautious yes. This section will explain how consideration of *base strength* and *steric bulk* of the reacting species can help us decide whether substitution or elimination will predominate.

Weakly basic nucleophiles give substitution

Good nucleophiles that are weaker bases than hydroxide give good yields of S_N2 products with primary and secondary halides and of S_N1 products with tertiary substrates. Examples include I^-, Br^-, RS^-, N_3^-, $RCOO^-$, and PR_3. Thus, 2-bromopropane reacts with both iodide and acetate ions cleanly through the S_N2 pathway, with virtually no competing elimination.

$$CH_3\underset{\underset{H}{|}}{\overset{\overset{CH_3}{|}}{C}}Br + Na^+ I^- \xrightarrow{\text{Propanone (acetone)}} CH_3\underset{\underset{H}{|}}{\overset{\overset{CH_3}{|}}{C}}Cl + Na^+ Br^-$$

$$CH_3\underset{\underset{H}{|}}{\overset{\overset{CH_3}{|}}{C}}Br + CH_3\overset{\overset{O}{||}}{C}O^- Na^+ \xrightarrow{\text{Propanone (acetone)}} CH_3\underset{\underset{H}{|}}{\overset{\overset{CH_3}{|}}{C}}O\overset{\overset{O}{||}}{C}CH_3 + Na^+ Br^-$$
$$100\%$$

2-Chloro-2-methylpropane transforms with sodium azide in methanol to 1,1-dimethylethyl (*tert*-butyl) azide (Section 7-4) through the S_N1 mechanism.

$$CH_3\underset{\underset{CH_3}{|}}{\overset{\overset{CH_3}{|}}{C}}Cl \xrightarrow{\text{NaN}_3, \text{CH}_3\text{OH}} CH_3\underset{\underset{CH_3}{|}}{\overset{\overset{CH_3}{|}}{C}}N_3 + CH_3\underset{\underset{CH_3}{|}}{\overset{\overset{CH_3}{|}}{C}}OCH_3 + \underset{CH_3 \quad CH_3}{\overset{\overset{CH_2}{||}}{C}}$$

$$\textbf{Major} \qquad\qquad \textbf{Minor}$$

Weak nucleophiles such as water and alcohols react at appreciable rates only with secondary and tertiary halides, substrates capable of following the S_N1 pathway. Unimolecular elimination is usually only a minor side reaction.

$$CH_3CH_2\overset{\overset{Br}{|}}{C}HCH_2CH_3 \xrightarrow{\text{H}_2\text{O, CH}_3\text{OH, 80°C}} CH_3CH_2\overset{\overset{OH}{|}}{C}HCH_2CH_3 + CH_3CH=CHCH_2CH_3$$
$$85\% \qquad\qquad 15\%$$

Strongly basic nucleophiles give more elimination as steric bulk increases

We have seen (Section 7-7) that strong bases may give rise to elimination through the E2 pathway. Is there some straightforward way to predict how much elimination will occur in competition with substitution in any particular situation? Yes, but other factors need to be considered. Let us examine the reactions of sodium ethoxide, a strong base, with several halides, measuring the relative amounts of ether and alkene produced in each case.

Ether	Alkene
(Substitution product)	(Elimination product)

$$CH_3CH_2CH_2Br \xrightarrow[-\text{HBr}]{\text{CH}_3\text{CH}_2\text{O}^-\text{Na}^+, \text{CH}_3\text{CH}_2\text{OH}} CH_3CH_2CH_2OCH_2CH_3 + \underset{H \quad\quad H}{\overset{H_3C \quad\quad H}{C=C}}$$
$$91\% \qquad\qquad 9\%$$

$$CH_3\underset{\underset{H}{|}}{\overset{\overset{CH_3}{|}}{C}}CH_2Br \xrightarrow[-\text{HBr}]{\text{CH}_3\text{CH}_2\text{O}^-\text{Na}^+, \text{CH}_3\text{CH}_2\text{OH}} CH_3\underset{\underset{H}{|}}{\overset{\overset{CH_3}{|}}{C}}CH_2OCH_2CH_3 + \underset{H_3C \quad\quad H}{\overset{H_3C \quad\quad H}{C=C}}$$
$$40\% \qquad\qquad 60\%$$

$$\underset{\substack{\text{CH}_3\text{CBr} \\ | \\ \text{H}}}{\overset{\substack{\text{CH}_3 \\ |}}{}} \xrightarrow[\text{– HBr}]{\text{CH}_3\text{CH}_2\text{O}^-\text{Na}^+,\ \text{CH}_3\text{CH}_2\text{OH}} \underset{\substack{\text{CH}_3\text{COCH}_2\text{CH}_3 \\ | \\ \text{H} \\ 13\%}}{\overset{\substack{\text{CH}_3 \\ |}}{}} + \underset{87\%}{\overset{\text{H}_3\text{C} \qquad \text{H}}{\text{C}=\text{C}}}$$

Reactions of simple primary halides with strongly basic nucleophiles give mostly S_N2 products. As steric bulk is increased around the carbon bearing the leaving group, substitution is retarded relative to elimination because an attack at carbon is subject to more steric hindrance than is an attack on hydrogen. Thus, branched primary substrates give about equal amounts of S_N2 and E2 reaction, whereas E2 is the major outcome with secondary substrates.

The S_N2 mechanism is not an option for tertiary halides. S_N1 and E1 pathways compete under neutral or weakly basic conditions. However, high concentrations of strong base give exclusive E2 reaction.

Sterically hindered basic nucleophiles favor elimination

We have seen that primary haloalkanes react by substitution with good nucleophiles, including strong bases. The situation changes when the steric bulk of the nucleophile hinders attack at the electrophilic carbon. In this case, elimination may predominate, even with primary systems, through deprotonation at the less hindered periphery of the molecule.

$$\text{CH}_3\text{CH}_2\text{CH}_2\text{CH}_2\text{Br} \xrightarrow[\text{– HBr}]{(\text{CH}_3)_3\text{CO}^-\text{K}^+,\ (\text{CH}_3)_3\text{COH}} \underset{85\%}{\text{CH}_3\text{CH}_2\text{CH}=\text{CH}_2} + \underset{15\%}{\text{CH}_3\text{CH}_2\text{CH}_2\text{CH}_2\text{OC}(\text{CH}_3)_3}$$

Two examples of sterically hindered bases are potassium *tert*-butoxide and lithium diisopropylamide (LDA). When used in elimination reactions, they are frequently dissolved in their conjugate acids, 2-methyl-2-propanol and *N*-(1-methylethyl)-1-methylethanamine (diisopropylamine), respectively.

Sterically Hindered Bases

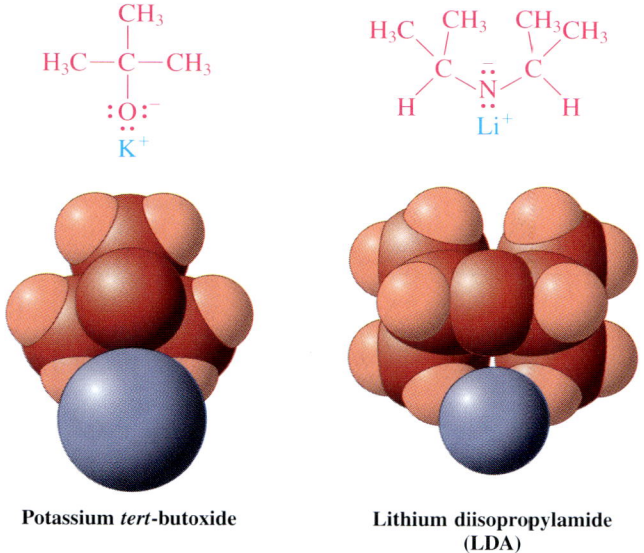

Potassium *tert*-butoxide

Lithium diisopropylamide
(LDA)

In summary, we have identified three principal factors that affect the competition between substitution and elimination: basicity of the nucleophile, steric hindrance in the haloalkane, and steric bulk around the nucleophilic (basic) atom.

FACTOR 1. Base strength of the nucleophile

Weak Bases	**Strong Bases**
H_2O*, ROH*, PR_3, halides, RS^-, N_3^-, NC^-, $RCOO^-$	HO^-, RO^-, H_2N^-, R_2N^-
Substitution more likely	Likelihood of elimination increased

FACTOR 2. Steric hindrance around the reacting carbon

Sterically Unhindered	**Sterically Hindered**
Primary haloalkanes	Branched primary, secondary, tertiary haloalkanes
Substitution more likely	Likelihood of elimination increased

FACTOR 3. Steric hindrance in the nucleophile (strong base)

Sterically Unhindered	**Sterically Hindered**
HO^-, CH_3O^-, $CH_3CH_2O^-$, H_2N^-	$(CH_3)_3CO^-$, $[(CH_3)_2CH]_2N^-$
Substitution may occur	Elimination strongly favored

For simple predictive purposes, we assume that their relative importance is equal in determining the ratio of elimination to substitution. Thus, the "majority rules." This method of analysis is quite reliable. Verify that it applies to the examples of this section and the summary section that follows.

EXERCISE 7-13

Which nucleophile in each of the following pairs will give a higher elimination:substitution product ratio in reaction with 1-bromo-2-methylpropane?

$$CH_3$$
$$|$$

(a) $N(CH_3)_3$, $P(CH_3)_3$ (b) H_2N^-, $(CH_3CH)_2N^-$ (c) I^-, Cl^-

---

EXERCISE 7-14

In all cases where substitution and elimination compete, higher reaction temperatures lead to greater proportions of elimination products. Thus, the amount of elimination accompanying hydrolysis of 2-bromo-2-methylpropane doubles as the temperature is raised from 25° to 65°C, and that from reaction of 2-bromopropane with ethoxide rises from 80% at 25°C to nearly 100% at 55°C. Explain.

7-9 Summary of Reactivity of Haloalkanes

Primary, secondary, and tertiary haloalkanes may react with nucleophiles through different pathways.

PRIMARY HALOALKANES. Unhindered primary alkyl substrates always react in a bimolecular way and almost always give predominantly substitution products, except when sterically hindered strong bases, such as potassium *tert*-butoxide, are employed. In these cases, the S_N2 pathway is slowed down sufficiently for steric reasons to allow the E2 mechanism to take over. Another way of reducing substitution is to in-

*Reacts only with S_N1 substrates; no reaction with simple primary halides.

troduce branching. However, even in these cases, good nucleophiles still furnish predominantly substitution products. Only strong bases, such as alkoxides, RO⁻, or amides, R₂N⁻, tend to react by elimination.

Reactivity of Primary Haloalkanes R–X with Nucleophiles (Bases)

For unhindered primary R–X:

S_N2 with good nucleophiles that are not strongly basic

$$CH_3CH_2CH_2Br + {}^-CN \xrightarrow{\text{Propanone (acetone)}} CH_3CH_2CH_2CN + Br^-$$

S_N2 with good nucleophiles that are also strong bases

$$CH_3CH_2CH_2Br + CH_3O^- \xrightarrow{CH_3OH} CH_3CH_2CH_2OCH_3 + Br^-$$

But E2 with strong, hindered base

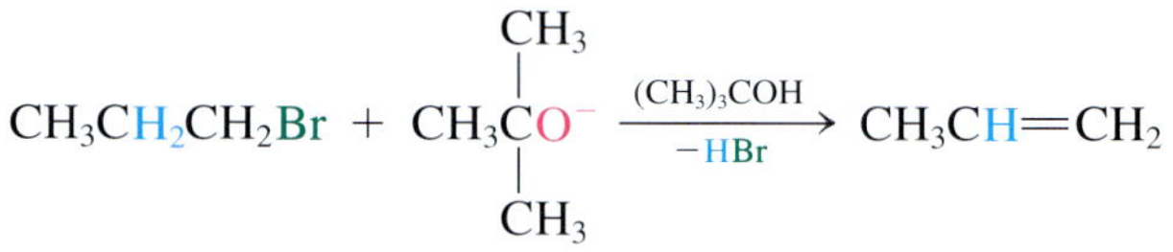

$$CH_3CH_2CH_2Br + CH_3\overset{\overset{\displaystyle CH_3}{|}}{\underset{\underset{\displaystyle CH_3}{|}}{C}}O^- \xrightarrow[-HBr]{(CH_3)_3COH} CH_3CH=CH_2$$

No (or exceedingly slow) reaction with poor nucleophiles (CH_3OH)

For branched primary R–X:

S_N2 with good nucleophiles (although slow compared with unhindered R–X)

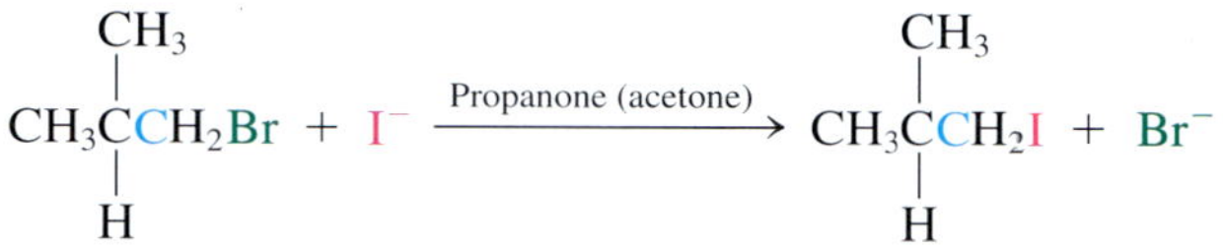

$$CH_3\overset{\overset{\displaystyle CH_3}{|}}{\underset{\underset{\displaystyle H}{|}}{C}}CH_2Br + I^- \xrightarrow{\text{Propanone (acetone)}} CH_3\overset{\overset{\displaystyle CH_3}{|}}{\underset{\underset{\displaystyle H}{|}}{C}}CH_2I + Br^-$$

E2 with strong base (not necessarily hindered)

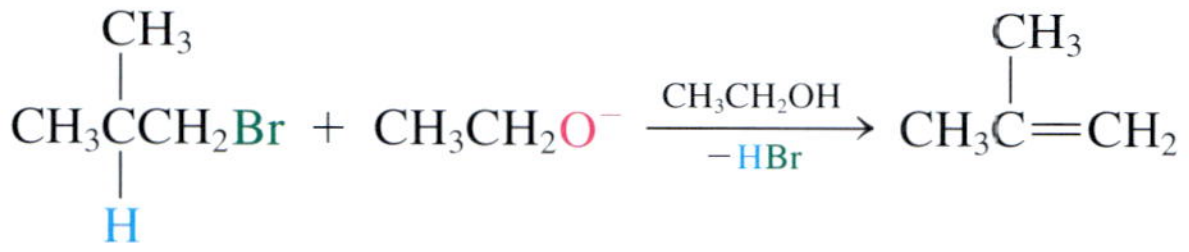

$$CH_3\overset{\overset{\displaystyle CH_3}{|}}{\underset{\underset{\displaystyle H}{|}}{C}}CH_2Br + CH_3CH_2O^- \xrightarrow[-HBr]{CH_3CH_2OH} CH_3\overset{\overset{\displaystyle CH_3}{|}}{C}=CH_2$$

No (or exceedingly slow) reaction with poor nucleophiles

SECONDARY HALOALKANES. Secondary alkyl systems undergo, depending on conditions, both eliminations and substitutions by either possible pathway: uni- or bimolecular. Good nucleophiles favor S_N2, strong bases result in E2, and weakly nucleophilic polar media give mainly S_N1 and E1.

Reactivity of Secondary Haloalkanes R–X with Nucleophiles (Bases)

S_N1 and E1 when X is a good leaving group in a highly polar medium with weak nucleophiles

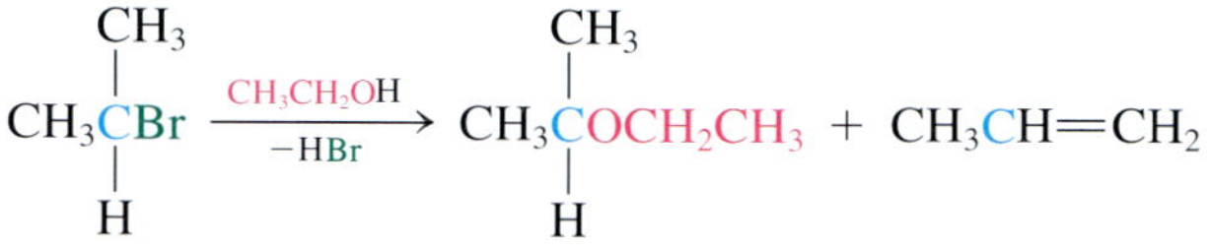

$$CH_3\overset{\overset{\displaystyle CH_3}{|}}{\underset{\underset{\displaystyle H}{|}}{C}}Br \xrightarrow[-HBr]{CH_3CH_2OH} CH_3\overset{\overset{\displaystyle CH_3}{|}}{\underset{\underset{\displaystyle H}{|}}{C}}OCH_2CH_3 + CH_3CH=CH_2$$

$$\qquad\qquad\qquad\qquad\qquad \textbf{Major} \qquad\qquad\qquad \textbf{Minor}$$

S_N2 with high concentrations of good, weakly basic nucleophiles

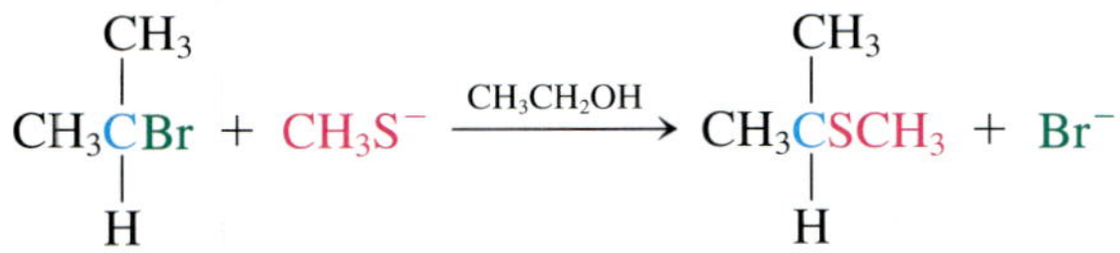

$$\underset{\underset{H}{|}}{\overset{\overset{CH_3}{|}}{CH_3CBr}} + CH_3S^- \xrightarrow{CH_3CH_2OH} \underset{\underset{H}{|}}{\overset{\overset{CH_3}{|}}{CH_3CSCH_3}} + Br^-$$

E2 with high concentrations of strong base (for example, HO^- or RO^- in alcohol solvent)

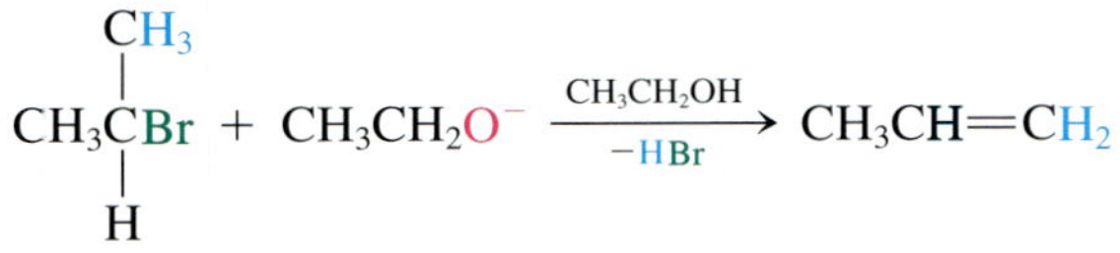

$$\underset{\underset{H}{|}}{\overset{\overset{CH_3}{|}}{CH_3CBr}} + CH_3CH_2O^- \xrightarrow[-HBr]{CH_3CH_2OH} CH_3CH{=}CH_2$$

TERTIARY HALOALKANES. Tertiary systems eliminate (E2) with concentrated strong base and are substituted in nonbasic media (S_N1). Bimolecular substitution is not observed, but elimination by E1 accompanies S_N1.

Reactivity of Tertiary Haloalkanes R–X
with Nucleophiles (Bases)

S_N1 and E1 in polar solvents when X is a good leaving group and dilute or no base is present

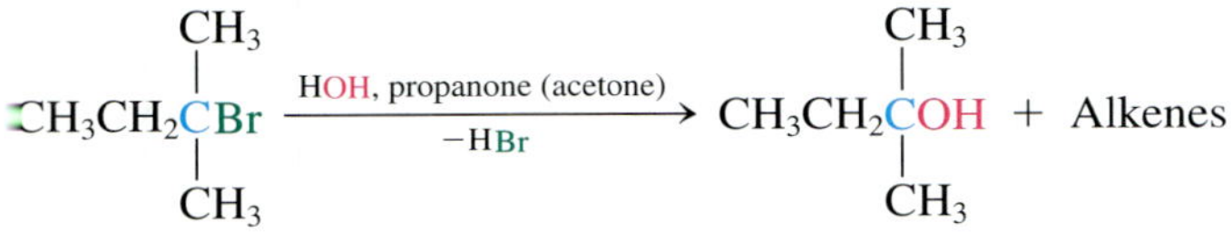

$$\underset{\underset{CH_3}{|}}{\overset{\overset{CH_3}{|}}{CH_3CH_2CBr}} \xrightarrow[-HBr]{HOH,\ propanone\ (acetone)} \underset{\underset{CH_3}{|}}{\overset{\overset{CH_3}{|}}{CH_3CH_2COH}} + \text{Alkenes}$$

E2 with high concentrations of strong base

$$\underset{\underset{\underset{\underset{CH_3}{|}}{CH_2}}{|}}{\overset{\overset{\overset{\overset{CH_3}{|}}{CH_2}}{|}}{CH_3CH_2CCl}} \xrightarrow[-HCl]{CH_3O^-,\ CH_3OH} \underset{\underset{CH_3}{|}}{\overset{\overset{\overset{CH_3}{|}}{CH_2}}{CH_3CH_2C}}{=}CHCH_3$$

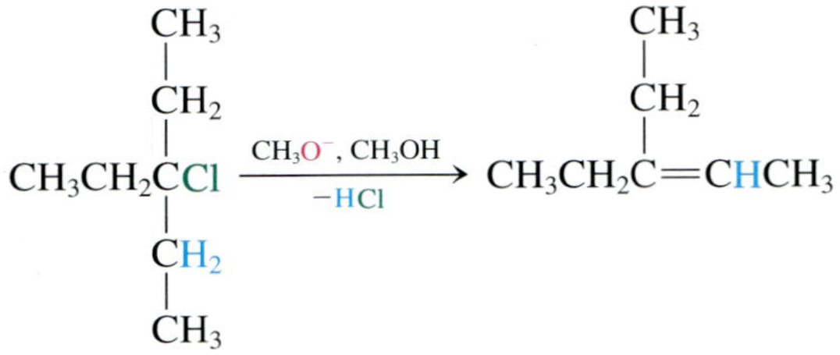

Table 7-4 (see page 270) summarizes the most likely mechanisms by which the haloalkanes undergo substitution and elimination.

EXERCISE 7-15

Predict which reaction in each of the following pairs will have a higher E2:E1 product ratio and explain why.

(a) 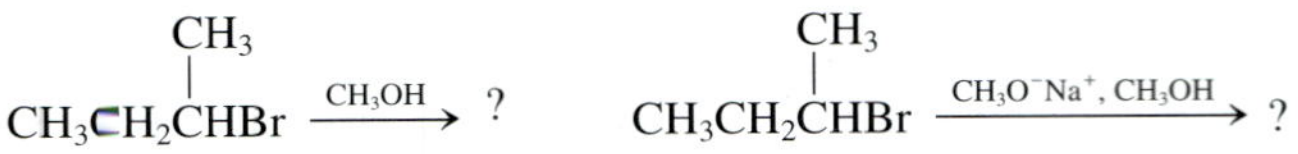 $\underset{\underset{CH_3}{|}}{CH_3CH_2CHBr} \xrightarrow{CH_3OH} ?$ $\underset{\underset{CH_3}{|}}{CH_3CH_2CHBr} \xrightarrow{CH_3O^-Na^+,\ CH_3OH} ?$

(b) 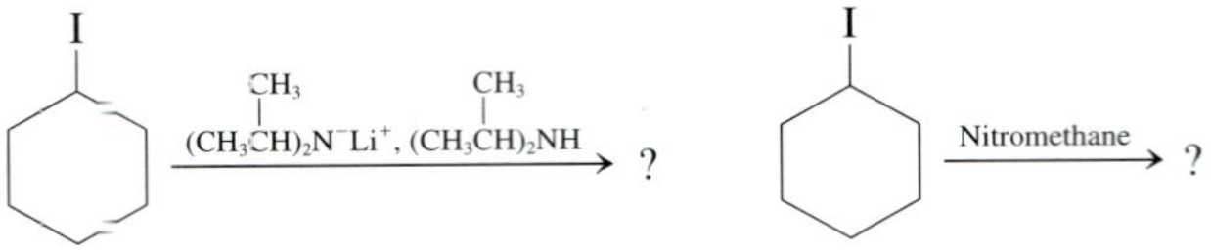 [cyclohexane ring with I substituent] $\xrightarrow{(CH_3CH)_2N^-Li^+,\ (CH_3CH)_2NH} ?$ [cyclohexane ring with I substituent] $\xrightarrow{Nitromethane} ?$

TABLE 7-4	Likely Mechanisms by Which Haloalkanes React with Nucleophiles (Bases)			
	Type of nucleophile (base)			
Type of haloalkane	**Poor nucleophile (e.g., H_2O)**	**Weakly basic, good nucleophile (e.g., I^-)**	**Strongly basic, unhindered nucleophile (e.g., CH_3O^-)**	**Strongly basic, hindered nucleophile (e.g., $(CH_3)_3CO^-$)**
Methyl	No reaction	S_N2	S_N2	S_N2
Primary				
Unhindered	No reaction	S_N2	S_N2	E2
Branched	No reaction	S_N2	E2	E2
Secondary	Slow S_N1, E1	S_N2	E2	E2
Tertiary	S_N1, E1	S_N1, E1	E2	E2

CHAPTER INTEGRATION PROBLEM

a. 2-Bromo-2-methylpropane (*tert*-butyl bromide) reacts readily in nitromethane with chloride and iodide ions.
1. Write the structures of the substitution products, and write the complete mechanism by which one of them is formed.
2. Assume equal concentrations of all reactants, and predict the relative rates of these two reactions.
3. Which reaction will give more elimination? Write its mechanism.

SOLUTION

1. We begin by analyzing the participating species and then recognizing just what kind of reaction is likely to take place. The substrate is a haloalkane with a good leaving group attached to a *tertiary* carbon. According to Table 7-4, displacement by the S_N2 mechanism is not an option, but S_N1, E1, and E2 processes are possibilities. Chloride and iodide are good nucleophiles and weak bases, suggesting that substitution should predominate to give $(CH_3)_3CCl$ and $(CH_3)_3CI$ as the products, respectively (Section 7-8). The mechanism (S_N1) is as shown in Section 7-2 except that, subsequent to initial ionization of the C–Br bond to give the carbocation, halide ion attacks at carbon to give the final product directly. The very polar nitromethane is a good solvent for S_N1 reactions (Section 7-4).
2. We learned in Section 7-4 that different nucleophilic power has no effect on the rates of unimolecular processes. The rates should be (and, experimentally, are) identical.
3. This part requires a bit more thought. According to Table 7-4 and Section 7-8, elimination by the E1 pathway always accompanies S_N1 displacement. However, increasing the base strength of the nucleophile may "turn on" the E2 mechanism, increasing the proportion of elimination product. Referring to Tables 6-4 and 6-7, we see that chloride is more basic (and less nucleophilic) than iodide. More elimination is indeed observed with chloride than with iodide. The mechanism is as shown in Figure 7-7, with chloride acting as the base to remove a proton from the carbocation.

b. The table in the margin presents data for the reactions that take place when the chloro compound shown here is dissolved in propanone (acetone) containing varying quantities of water and sodium azide, NaN_3:

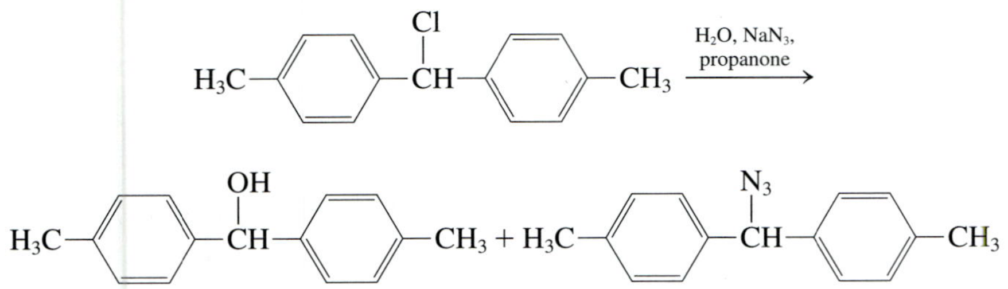

% H_2O	$[N_3^-]$	% RN_3	k_{rel}
10	0 M	0	1
10	0.05 M	60	1.5
15	0.05 M	60	7
20	0.05 M	60	22
50	0.05 M	60	*
50	0.10 M	75	*
50	0.20 M	85	*
50	0.50 M	95	*

*Too fast to measure.

In the table, % H_2O is the percentage of water by volume in the solvent, $[N_3^-]$ is the initial concentration of sodium azide, % RN_3 is the percentage of organic azide in the product mixture (the remainder is the alcohol), and k_{rel} is the relative rate constant for the reaction, derived from the rate at which the starting material is consumed. The initial concentration of substrate is 0.04 M in all experiments. Answer the following questions.

1. Describe and explain the effects of changing the percentage of H_2O on the rate of the reaction and on the product distribution.

2. Do the same for the effects of changing $[N_3^-]$. Additional information: The reaction rates shown are the same when other ions—for example, Br^- or I^-—are used instead of azide.

SOLUTION

1. We begin by examining the data in the table, specifically lines 2-5, which compare reactions in the presence of different amounts of water at constant azide concentration. The rate of substitution increases rapidly as the proportion of water goes up, but the ratio of the two products stays the same: 60% azide and 40% alcohol. These two results suggest that the only effect of increasing the amount of water is to make the solvent environment more polar, thereby speeding up the initial ionization of the substrate. Even when the proportion of water is only 10%, it is present in great excess and is trapping carbocations as fast as it can, relative to the rate that azide ion reacts with the same intermediates (Section 7-2).

2. We note from lines 1 and 2 in the table that the reaction rate rises by about 50% when NaN_3 is added. Without further information, we might assume that this effect is a consequence of the occurrence of the S_N2 mechanism. Were that to be the case, however, other anions should affect the rate differently. But we were told that bromide and iodide, far more powerful nucleophiles, affect the measured rate in exactly the same way as does azide. We can explain this observation only by assuming that displacement is entirely by the S_N1 mechanism, and added ions affect the rate only by increasing the polarity of the solution and speeding up ionization (Section 7-4).

In lines 5–8 of the table we note that increasing the amount of azide ion increases the amount of azide-containing product that is formed. At the higher concentrations, azide, a better nucleophile than water, is better able to complete for reaction with the carbocation intermediate.

NEW REACTIONS

1. Bimolecular Substitution—S$_N$2 (Sections 6-3 through 6-10, 7-5)

Primary and secondary substrates only

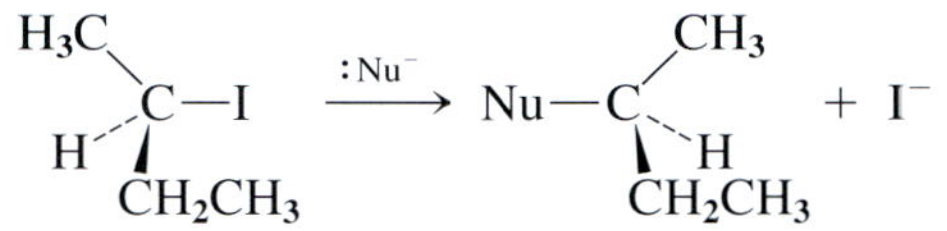

Direct backside displacement with 100% inversion of configuration

2. Unimolecular Substitution—S$_N$1 (Sections 7-1 through 7-5)

Secondary and tertiary substrates only

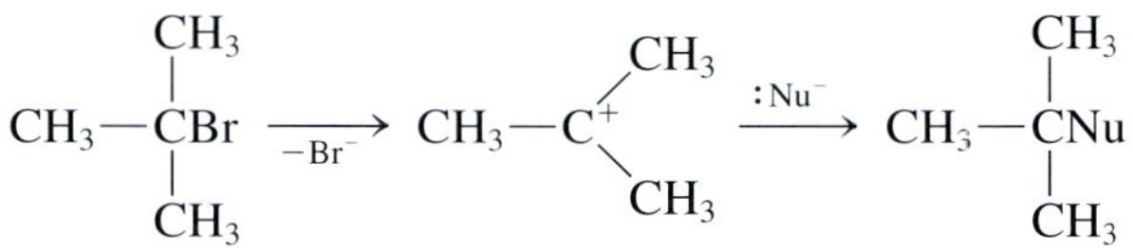

Through carbocation: Chiral systems are racemized

3. Unimolecular Elimination—E1 (Section 7-6)

Secondary and tertiary substrates only

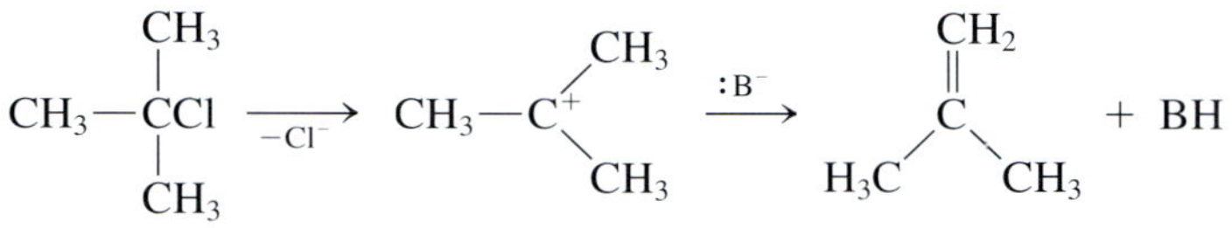

Through carbocation

4. Bimolecular Elimination—E2 (Section 7-7)

$$CH_3CH_2CH_2I \xrightarrow{\text{:B}^-} CH_3CH=CH_2 + BH + I^-$$

Simultaneous elimination of leaving group and neighboring proton

IMPORTANT CONCEPTS

1. Secondary haloalkanes undergo slow and tertiary haloalkanes fast **unimolecular substitution** in polar media. When the solvent serves as the nucleophile, the process is called **solvolysis.**

2. The slowest, or rate-determining, step in unimolecular substitution is dissociation of the C–X bond to form a **carbocation** intermediate. Added strong nucleophile changes the product but not the reaction rate.

3. Carbocations are stabilized by **hyperconjugation:** Tertiary are the most stable, followed by secondary. Primary and methyl cations are too unstable to form in solution.

4. **Racemization** often results when unimolecular substitution takes place at a chiral carbon.

5. **Unimolecular elimination** to form an alkene accompanies substitution in secondary and tertiary systems.

6. High concentrations of strong base may bring about **bimolecular elimination.** Expulsion of the leaving group accompanies removal of a hydrogen from the neighboring carbon by the base. The stereochemistry indicates an *anti* conformational arrangement of the hydrogen and the leaving group.

7. Substitution is favored by unhindered substrates and small, less basic nucleophiles.

8. Elimination is favored by hindered substrates and bulky, more basic nucleophiles.

PROBLEMS

16. What is the major substitution product of each of the following solvolysis reactions?

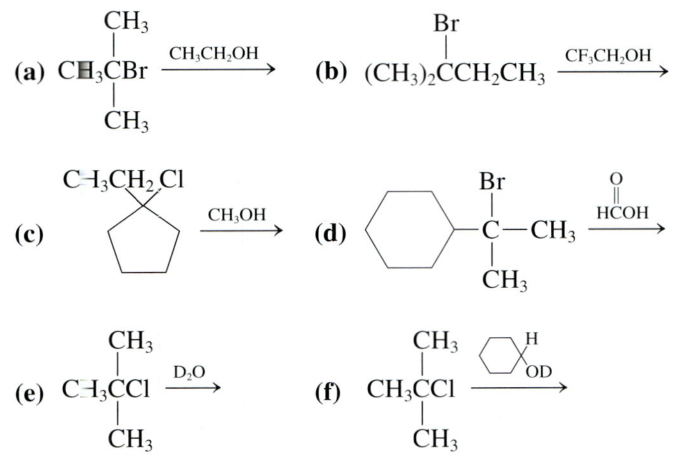

(a) CH_3CBr (with two CH_3 groups) $\xrightarrow{CH_3CH_2OH}$

(b) $(CH_3)_2CCH_2CH_3$ (with Br) $\xrightarrow{CF_3CH_2OH}$

(c) cyclopentane with CH_3CH_2Cl substituent $\xrightarrow{CH_3OH}$

(d) cyclohexyl-$\overset{Br}{\underset{CH_3}{C}}$-$CH_3$ $\xrightarrow{HCOH}$ (O double bond on C)

(e) CH_3CCl (with two CH_3 groups) $\xrightarrow{D_2O}$

(f) CH_3CCl (with two CH_3 groups) $\xrightarrow{\text{cyclohexyl OD, H}}$

17. Write the two major substitution products of the reaction shown in the margin. **(a)** Write a mechanism to explain the formation of each of them. **(b)** Monitoring the reaction mixture reveals that an *isomer* of the starting material is generated as an intermediate. Draw its structure and explain how it is formed.

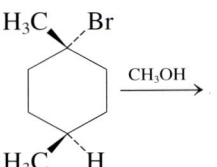

18. Give the two major substitution products of the following reaction.

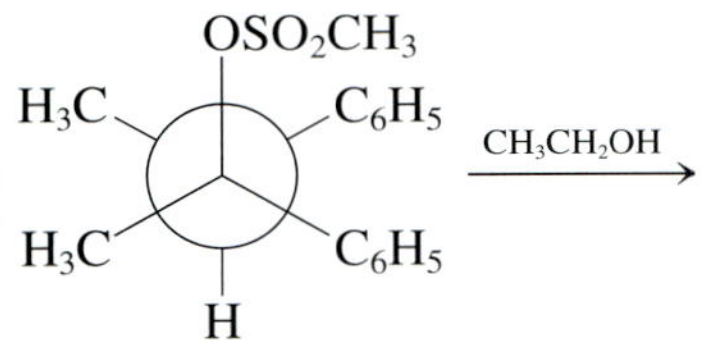

$$\overset{OSO_2CH_3}{\underset{H}{\underset{H_3C}{\overset{H_3C}{\diagdown}}}}\quad \begin{array}{c} C_6H_5 \\ C_6H_5 \end{array} \xrightarrow{CH_3CH_2OH}$$

19. How would each reaction in Problem 16 be affected by the addition of each of the following substances to the solvolysis mixture?
(a) H_2O **(b)** KI
(c) NaN_3 **(d)** $CH_3CH_2OCH_2CH_3$ (**Hint:** Low polarity.)

20. Rank the following carbocations in decreasing order of stability.

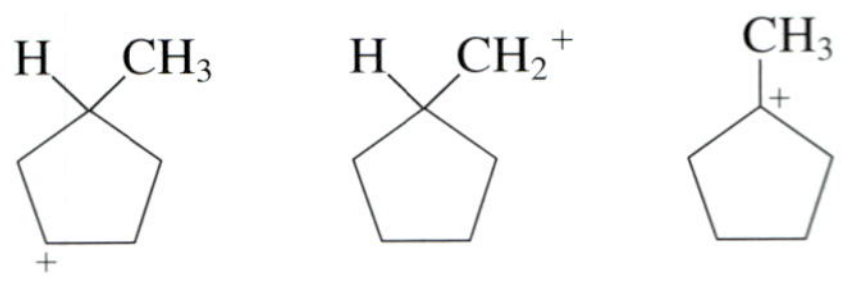

$$\overset{H\quad CH_3}{\text{cyclopentane}^+}\qquad \overset{H\quad CH_2{}^+}{\text{cyclopentane}}\qquad \overset{CH_3}{\text{cyclopentane}^+}$$

21. Rank the compounds in each of the following groups in order of decreasing rate of solvolysis in aqueous propanone (acetone).

(a) $CH_3CHCH_2CH_2Cl$ (with CH_3) $CH_3CHCHCH_3$ (with CH_3 and Cl) $CH_3CCH_2CH_3$ (with CH_3 and Cl)

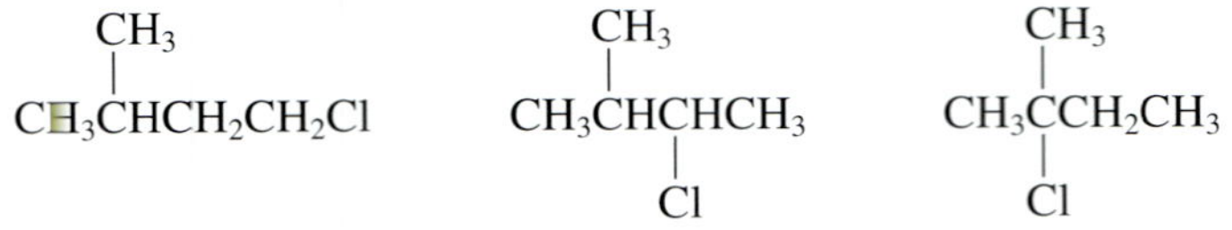

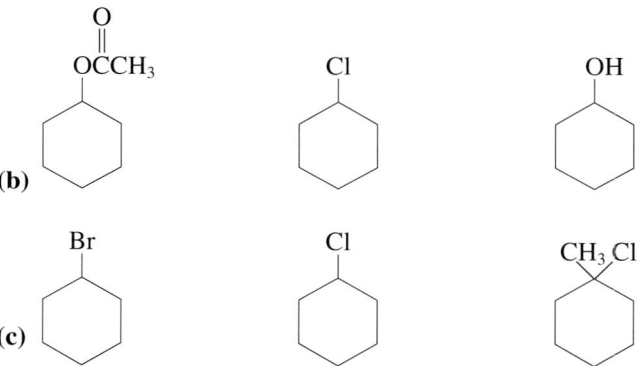

22. Give the products of the following substitution reactions. Indicate whether they arise through the S_N1 or the S_N2 process. Formulate the detailed mechanisms of their generation.

(a) $(CH_3)_2CHOSO_2CF_3 \xrightarrow{CH_3CH_2OH}$

(b) 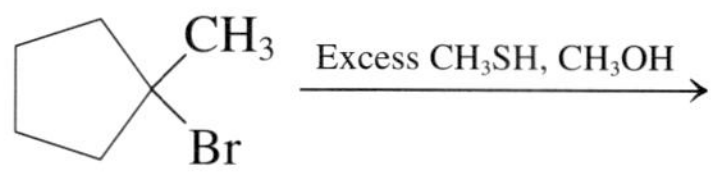 (cyclopentane ring with CH_3 and Br) $\xrightarrow{\text{Excess } CH_3SH, \, CH_3OH}$

(c) $CH_3CH_2CH_2CH_2Br \xrightarrow{(C_6H_5)_3P, \, DMSO}$

(d) $CH_3CH_2CHClCH_2CH_3 \xrightarrow{NaI, \, \text{propanone (acetone)}}$

23. Give the product of each of the following substitution reactions. Which of these transformations should proceed faster in a polar, aprotic solvent (such as propanone or DMSO) than in a polar, protic solvent (such as water or CH_3OH)? Explain your answer on the basis of the mechanism that you expect to be operating in each case.
(a) $CH_3CH_2CH_2Br + Na^+ \, ^-CN \longrightarrow$
(b) $(CH_3)_2CHCH_2I + Na^+N_3^- \longrightarrow$
(c) $(CH_3)_3CBr + HSCH_2CH_3 \longrightarrow$
(d) $(CH_3)_2CHOSO_2CH_3 + HOCH(CH_3)_2 \longrightarrow$

24. Propose a synthesis of (R)-$CH_3CHN_3CH_2CH_3$, starting from (R)-2-chloro-butane.

25. Two substitution reactions of (S)-2-bromobutane are shown here. Show their stereochemical outcomes.

$$(S)\text{-}CH_3CH_2CHBrCH_3 \xrightarrow{\overset{\overset{O}{\|}}{HCOH}}$$

$$(S)\text{-}CH_3CH_2CHBrCH_3 \xrightarrow{\overset{\overset{O}{\|}}{HCO^-Na^+, \, DMSO}}$$

26. Write all possible E1 products of each reaction in Problem 16.

27. Write the products of the following elimination reactions. Specify the predominant mechanism (E1 or E2) and formulate it in detail.

(a) $(CH_3CH_2)_3CBr$ $\xrightarrow{NaNH_2, \ NH_3}$

(b) $CH_3CH_2CH_2CH_2Cl$ $\xrightarrow{KOC(CH_3)_3, \ (CH_3)_3COH}$

(c) 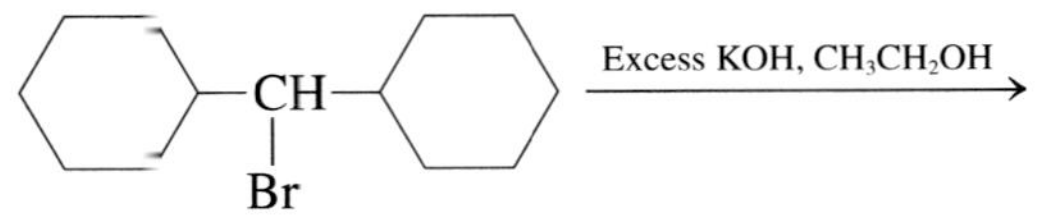 $\xrightarrow{Excess \ KOH, \ CH_3CH_2OH}$

(d) 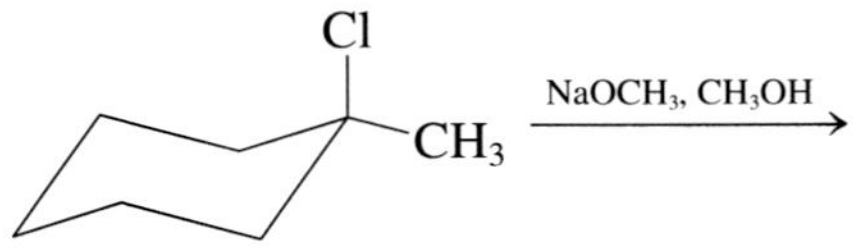 $\xrightarrow{NaOCH_3, \ CH_3OH}$

28. Predict the major product(s) that should form from reaction between 1-bromobutane and each of the following substances. By which reaction mechanism is each formed—S_N1, S_N2, E1, or E2? If it appears that a reaction will either not take place or be exceedingly slow, write "no reaction." Assume that each reagent is present in large excess. The solvent for each reaction is given.
(a) KCl in DMF   **(b)** KI in DMF   **(c)** KCl in CH_3NO_2
(d) NH_3 in CH_3CH_2OH **(e)** $NaOCH_2CH_3$ in CH_3CH_2OH **(f)** CH_3CH_2OH
(g) $KOC(CH_3)_3$ in $(CH_3)_3COH$ **(h)** $(CH_3)_3P$ in CH_3OH **(i)** CH_3CO_2H

29. Predict the major product(s) and mechanism(s) for reaction between 2-bromobutane (*sec*-butyl bromide) and each of the reagents in Problem 28.

30. Predict the major product(s) and mechanism(s) for reaction between 2-bromo-2-methylpropane (*tert*-butyl bromide) and each of the reagents in Problem 28.

31. Three reactions of 2-chloro-2-methylpropane are shown here. **(a)** Write the major product of each transformation. **(b)** Compare the rates of the three reactions. Assume identical solution polarities and reactant concentrations. Explain mechanistically.

$(CH_3)_3CCl$ $\xrightarrow{H_2S, \ CH_3OH}$

$(CH_3)_3CCl$ $\xrightarrow{CH_3\overset{O}{\overset{\|}{C}}O^- K^+, \ CH_3OH}$

$(CH_3)_3CCl$ $\xrightarrow{CH_3O^- K^+, \ CH_3OH}$

32. Give the major product(s) of the following reactions. Indicate which of the following mechanism(s) is in operation: S_N1, S_N2, E1, or E2. If no reaction takes place, write "no reaction."

(a) 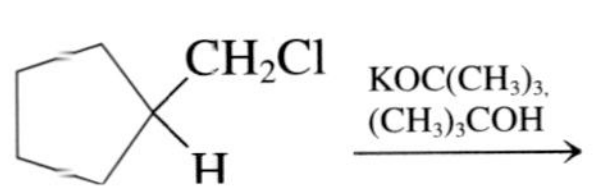 $\xrightarrow{KOC(CH_3)_3, \ (CH_3)_3COH}$

(b) $CH_3\overset{F}{\underset{|}{C}}HCH_2CH_3$ $\xrightarrow{KBr, \ propanone \ (acetone)}$

(c) 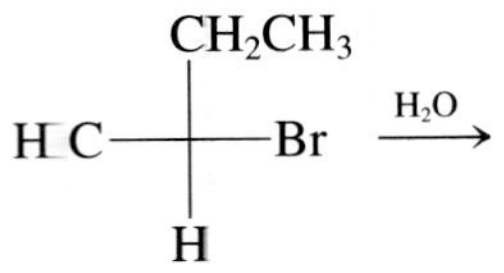 $H_3C\overset{CH_2CH_3}{\underset{H}{\overset{|}{\underset{|}{C}}}}Br$ $\xrightarrow{H_2O}$

(d) 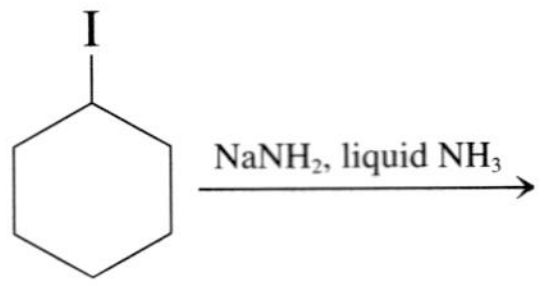 $\xrightarrow{NaNH_2, \ liquid \ NH_3}$

(e) $(CH_3)_2CHCH_2CH_2CH_2Br$ $\xrightarrow{NaOCH_2CH_3, CH_3CH_2OH}$

(f) 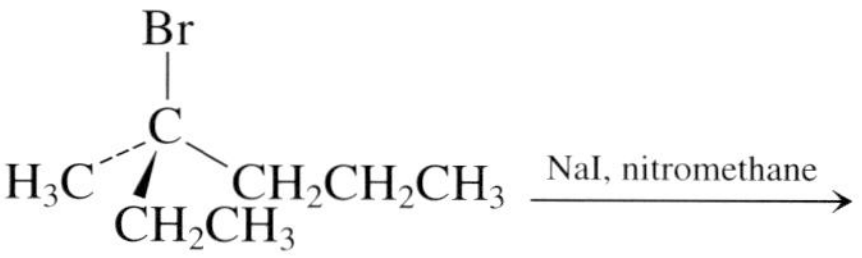 H$_3$C—C(Br)(CH$_2$CH$_3$)—CH$_2$CH$_2$CH$_3$ $\xrightarrow{NaI, nitromethane}$

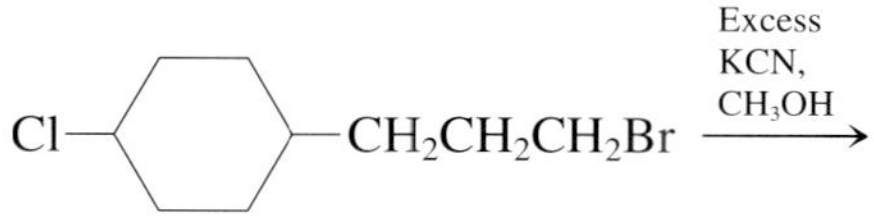

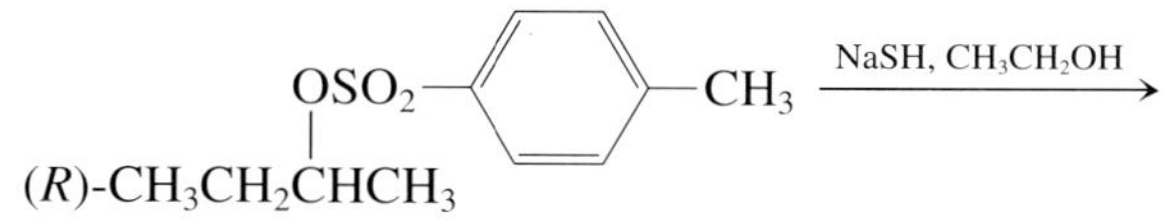

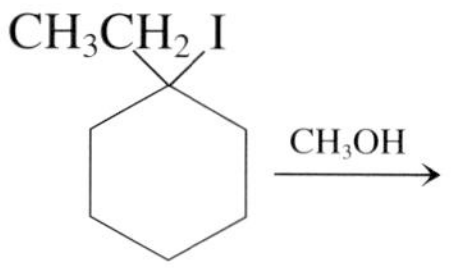

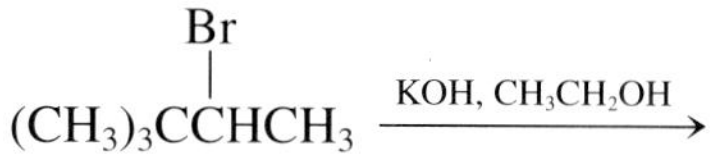

(g) 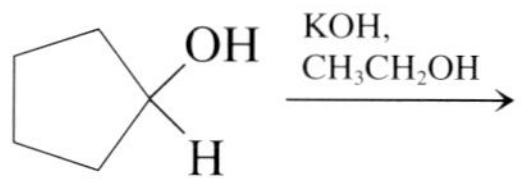 cyclopentane-OH, H $\xrightarrow[CH_3CH_2OH]{KOH,}$ **(h)** Cl—C$_6$H$_{10}$—CH$_2$CH$_2$CH$_2$Br $\xrightarrow[CH_3OH]{Excess KCN,}$

(i) (R)-CH$_3$CH$_2$CHCH$_3$—OSO$_2$—C$_6$H$_4$—CH$_3$ $\xrightarrow{NaSH, CH_3CH_2OH}$

(j) cyclohexane with CH$_3$CH$_2$ and I $\xrightarrow{CH_3OH}$ **(k)** $(CH_3)_3CCHCH_3$ with Br $\xrightarrow{KOH, CH_3CH_2OH}$

(l) CH$_3$CH$_2$Cl $\xrightarrow{CH_3COH, O}$

33. Consider the reaction shown here. Will it proceed by substitution or by elimination? What factors determine the most likely mechanism? Write the expected
product. (**Hint:** Draw the chair conformation of the substrate or, better yet,
make a model.)

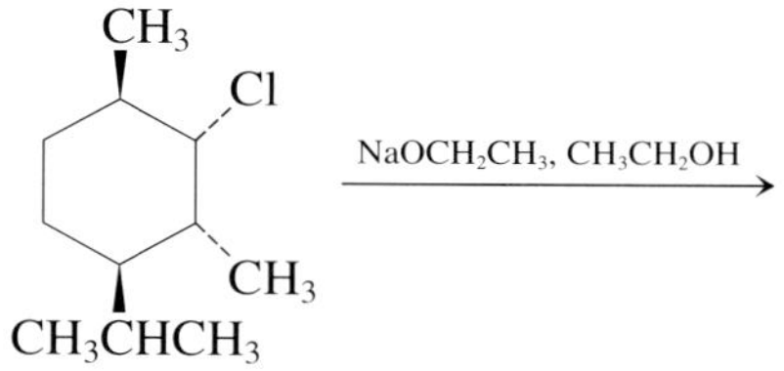

cyclohexane with CH$_3$, Cl, CH$_3$, CH$_3$CHCH$_3$ $\xrightarrow{NaOCH_2CH_3, CH_3CH_2OH}$

34. Fill the blanks in the following table with the major product(s) of the reaction
of each haloalkane with the reagents shown.

Haloalkane	Reagent			
	H$_2$O	NaSeCH$_3$	NaOCH$_3$	KOC(CH$_3$)$_3$
CH$_3$Cl	_____	_____	_____	_____
CH$_3$CH$_2$CH$_2$Cl	_____	_____	_____	_____
(CH$_3$)$_2$CHCl	_____	_____	_____	_____
(CH$_3$)$_3$CCl	_____	_____	_____	_____

35. Indicate the major mechanism(s) (simply specify S$_N$2, S$_N$1, E2, or E1) required
for the formation of each product that you wrote in Problem 34.

36. For each of the following reactions, indicate whether the reaction would work well, poorly, or not at all. Formulate alternative products, if appropriate.

(a) $CH_3CH_2CHCH_3$ $\xrightarrow{\text{NaOH, propanone (acetone)}}$ $CH_3CH_2CHCH_3$
 ‌   ‌   ‌   ‌   ‌ | |
 ‌   ‌   ‌   ‌   ‌ Br   OH

(b) $CH_3\overset{\displaystyle H_3C}{\overset{|}{C}}HCH_2Cl$ $\xrightarrow{CH_3OH}$ $CH_3\overset{\displaystyle H_3C}{\overset{|}{C}}HCH_2OCH_3$

(c) 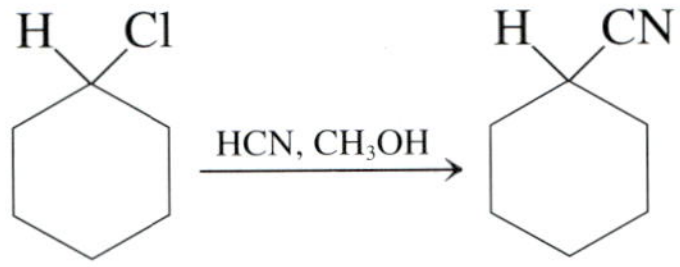 $\xrightarrow{\text{HCN, } CH_3OH}$

(d) 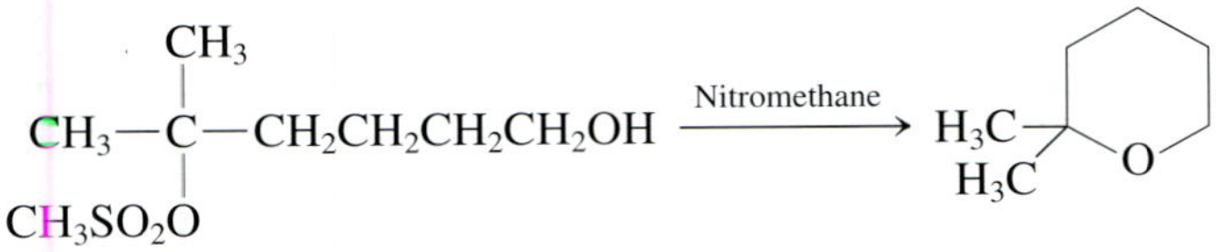 $CH_3-\overset{\displaystyle CH_3}{\overset{|}{\underset{|}{C}}}-CH_2CH_2CH_2CH_2OH$ $\xrightarrow{\text{Nitromethane}}$
 ‌   ‌   ‌   ‌   ‌ CH_3SO_2O

(e) 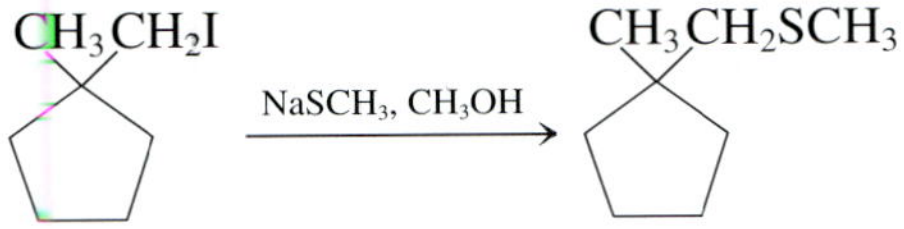 $\xrightarrow{\text{NaSCH}_3,\ CH_3OH}$

(f) $CH_3CH_2CH_2Br$ $\xrightarrow{\text{NaN}_3,\ CH_3OH}$ $CH_3CH_2CH_2N_3$

(g) $(CH_3)_3CCl$ $\xrightarrow{\text{NaI, nitromethane}}$ $(CH_3)CI$

(h) $(CH_3CH_2)_2O$ $\xrightarrow{CH_3I}$ $(CH_3CH_2)_2\overset{+}{O}CH_3 + I^-$

(i) CH_3I $\xrightarrow{CH_3OH}$ CH_3OCH_3

(j) $(CH_3CH_2)_3COCH_3$ $\xrightarrow{\text{NaBr, } CH_3OH}$ $(CH_3CH_2)_3CBr$

(k) $CH_3\overset{\displaystyle CH_3}{\overset{|}{C}}HCH_2CH_2Cl$ $\xrightarrow{\text{NaOCH}_2CH_3,\ CH_3CH_2OH}$ $CH_3\overset{\displaystyle CH_3}{\overset{|}{C}}HCH=CH_2$

(l) $CH_3CH_2CH_2CH_2Cl$ $\xrightarrow{\text{NaOCH}_2CH_3,\ CH_3CH_2OH}$ $CH_3CH_2CH=CH_2$

37. Propose syntheses of the following molecules from the indicated starting materials. Make use of any other reagents or solvents that you need. In some cases, there may be no alternative but to employ a reaction that results in a mixture of products. If so, use reagents and conditions that will maximize the yield of the desired material (compare Problem 39 in Chapter 6).
(a) $CH_3CH_2CHICH_3$, from butane
(b) $CH_3CH_2CH_2CH_2I$, from butane
(c) $(CH_3)_3COCH_3$, from methane and 2-methylpropane
(d) Cyclohexene, from cyclohexane
(e) Cyclohexanol, from cyclohexane
(f) 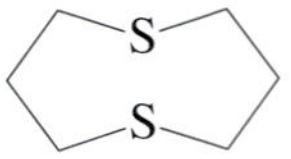 , from 1,3-dibromopropane

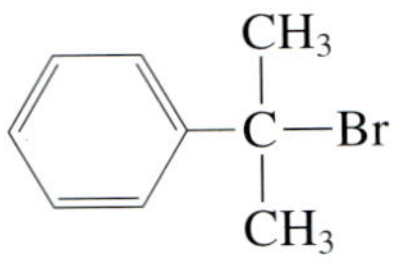

38. [(1-Bromo-1-methyl)ethyl]benzene, shown in the margin, undergoes solvolysis in a unimolecular, strictly first order process. The reaction rate for [RBr] = 0.1 M RBr in 9:1 propanone (acetone):water is measured to be 2×10^{-4} mol L^{-1} s^{-1}. **(a)** Calculate the rate constant k from these data. What is the product of this reaction? **(b)** In the presence of 0.1 M LiCl, the rate is found to increase to 4×10^{-4} mol L^{-1} s^{-1}, although the reaction still remains strictly first order. Calculate the new rate constant k_{LiCl} and suggest an explanation. **(c)** When 0.1 M LiBr is present instead of LiCl, the measured rate

drops to 1.6×10^{-4} mol L^{-1} s^{-1}. Explain this observation, and write the appropriate chemical equations to describe the reactions.

39. The stabilities of three cyclic cations illustrated here differ greatly. Predict their order of stability and provide a rationalization for your assignments.

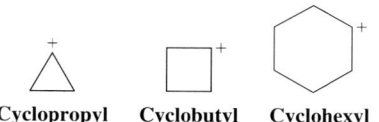

Cyclopropyl   **Cyclobutyl**   **Cyclohexyl**

40. Match each of the following transformations to the correct reaction profile shown here, and draw the structures of the species present at all points on the energy curves marked by capital letters.

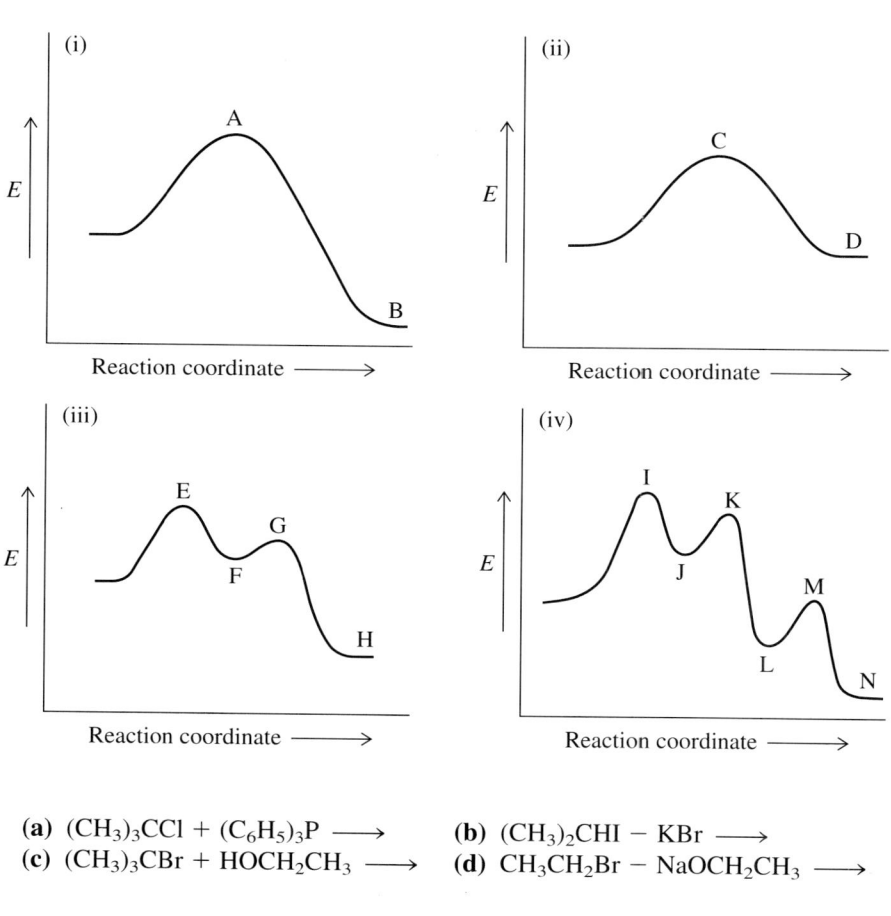

(a) $(CH_3)_3CCl + (C_6H_5)_3P \longrightarrow$ **(b)** $(CH_3)_2CHI - KBr \longrightarrow$
(c) $(CH_3)_3CBr + HOCH_2CH_3 \longrightarrow$ **(d)** $CH_3CH_2Br - NaOCH_2CH_3 \longrightarrow$

41. Formulate the structure of the most likely product of the following reaction of 4-chloro-4-methyl-1-pentanol in neutral polar solution.

$$(CH_3)_2\overset{\overset{\displaystyle Cl}{|}}{C}CH_2CH_2CH_2OH \longrightarrow HCl + C_6H_{12}O$$

In strongly *basic* solution, the starting material again converts into a molecule with the molecular formula $C_6H_{12}O$, but with a completely different structure. What is it? Explain the difference between the two results.

42. The following reaction can proceed through both E1 and E2 mechanisms.

$$C_6H_5CH_2\underset{\underset{\displaystyle CH_3}{|}}{\overset{\overset{\displaystyle CH_3}{|}}{C}}Cl \xrightarrow{\text{NaOCH}_3,\ \text{CH}_3\text{OH}} C_6H_5CH{=}C(CH_3)_2 + C_6H_5CH_2\underset{\underset{}{}}{\overset{\overset{\displaystyle CH_3}{|}}{C}}{=}CH_2$$

The E1 rate constant $k_{E1} = 1.4 \times 10^{-4}\ s^{-1}$ and the E2 rate constant $k_{E2} = 1.9 \times 10^{-4}\ L\ mol^{-1}\ s^{-1}$; 0.02 M haloalkane. **(a)** What is the predominant elimination mechanism with 0.5 M $NaOCH_3$? **(b)** What is the predominant elimination mechanism with 2.0 M $NaOCH_3$? **(c)** At what concentration of base does exactly 50% of the starting material react by an E1 route and 50% by an E2 pathway?

43. When 2-methyl-2-propanol is shaken with concentrated aqueous HBr, 2-bromo-2-methylpropane (margin) rapidly forms, which is the reverse of S_N1 hydrolysis (Section 7-1). Propose a detailed mechanism for this process.

$$(CH_3)_3COH$$
$$\downarrow \text{Conc. HBr}$$
$$(CH_3)_3CBr$$

44. Give the mechanism and major product for the reaction of a secondary haloalkane in a polar aprotic solvent with the following nucleophiles. The pK_a value of the conjugate acid of the nucleophile is given in parentheses.

(a) N_3^- (4.6) **(b)** H_2N^- (35) **(c)** NH_3 (9.5)
(d) HSe^- (3.7) **(e)** F^- (3.2) **(f)** $C_6H_5O^-$ (9.9)
(g) PH_3 (-12) **(h)** NH_2OH (6.0) **(i)** NCS^- (-0.7)

45. Cortisone is an important steroidal anti-inflammatory agent. Cortisone can be synthesized efficiently from the alkene shown here.

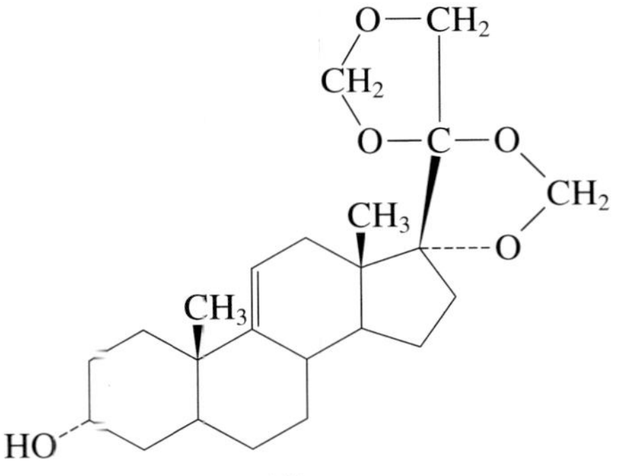

Alkene 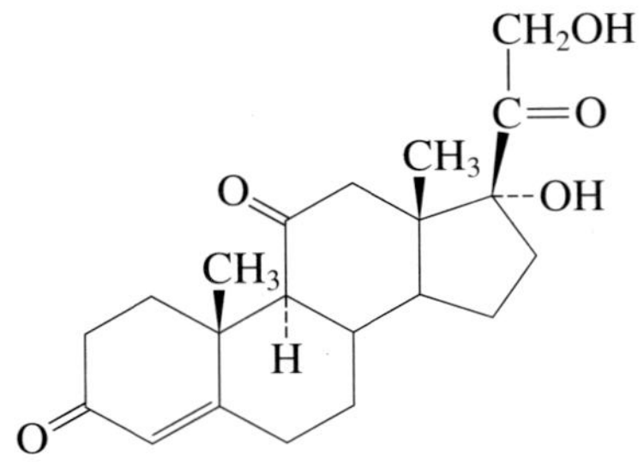   **Cortisone**

Of the following three chlorinated compounds, two give reasonable yields of the alkene shown above by E2 elimination with base, but one does not. Which one does not work well, and why? What does it give during attempted E2 elimination? (**Hint:** Consider the geometry of each system.)

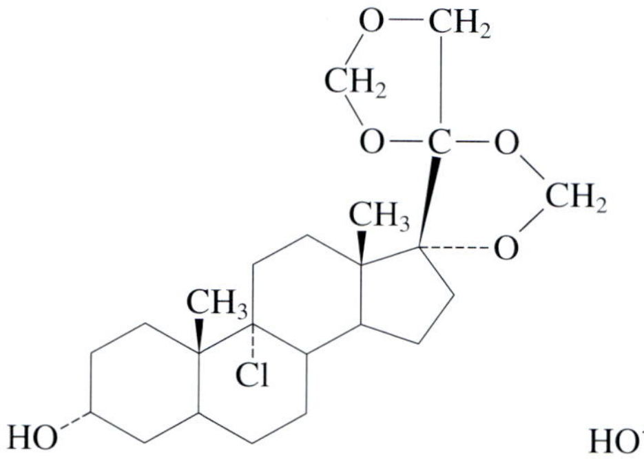

A

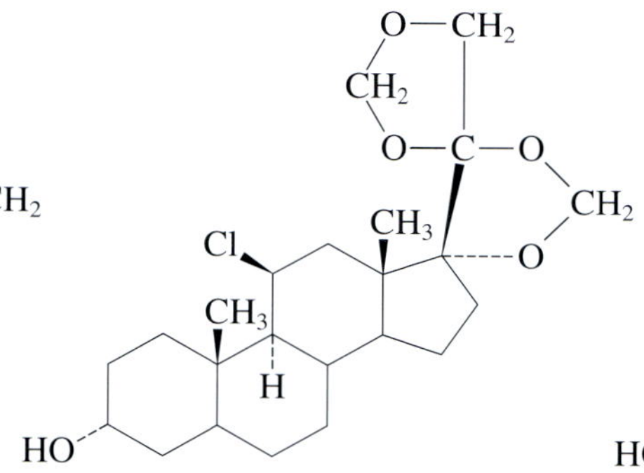

B

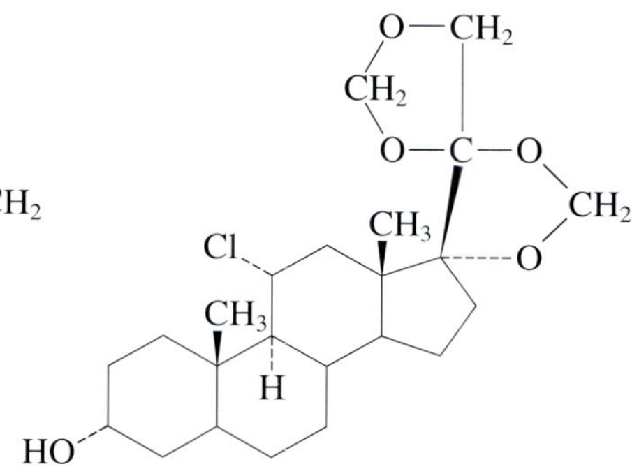

C

46. The chemistry of derivatives of *trans*-decalin is of interest because this ring system is part of the structure of steroids. Make models of the brominated systems (i and ii) to help you answer the following questions.

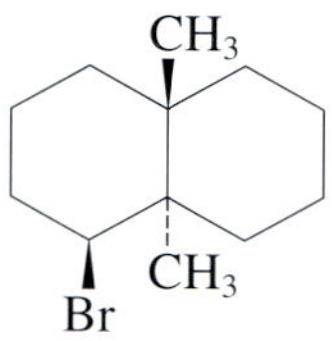

i

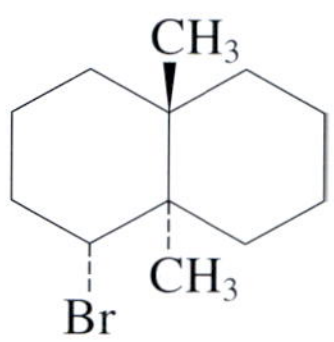

ii

(a) One of the molecules undergoes E2 reaction with $NaOCH_2CH_3$ in CH_3CH_2OH considerably faster than does the other. Which molecule is which? Explain. **(b)** The following deuterated analogs of systems i and ii react with base to give the products shown.

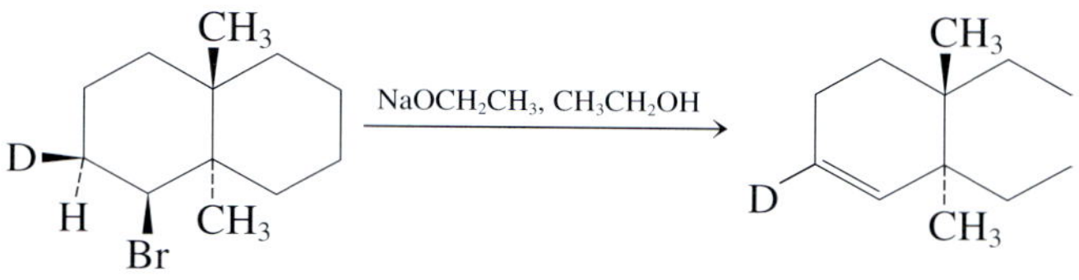

i-deuterated

(All D retained)

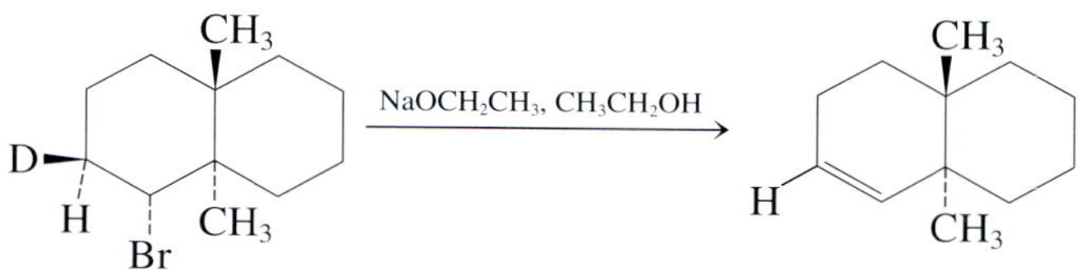

ii-deuterated

(All D lost)

Specify whether *anti* or *syn* eliminations have taken place. Draw the conformations that the molecules must adopt for elimination to occur. Does your answer to (b) help you in solving (a)?

Team Problem

47. Consider the general substitution-elimination reactions of the bromoalkanes.

$$R-Br \xrightarrow{\text{Nu/Base}} R-Nu + \text{alkene}$$

How do the reaction mechanisms and product formation differ when the structure of the substrate and reaction conditions change? To begin to unravel the nuances of bimolecular and unimolecular substitution and elimination reactions, focus on the treatment of bromoalkanes A through D under conditions (a) through (e). Divide the problem evenly among yourselves so that each of you tackles the questions of reaction mechanism(s) and qualitative distribution of product(s), if any. Reconvene to discuss your conclusions and come to a consensus. When you are explaining a reaction mechanism to the rest of the team, use curved arrows to show the flow of electrons. Label the stereochemistry of starting materials and products as *R* or *S*, as appropriate.

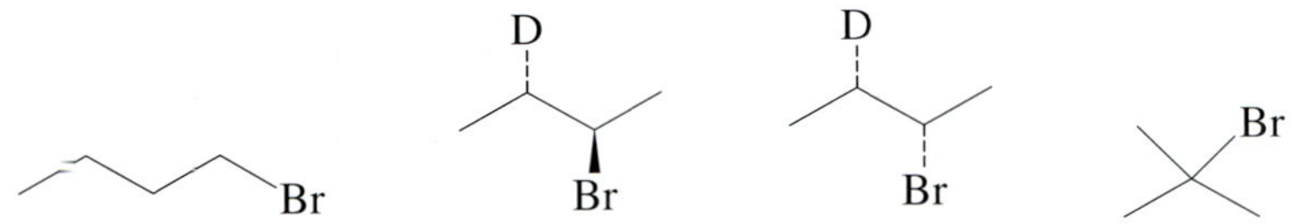

(a) $N \equiv N_3$, DMF   **(b)** LDA, DMF   **(c)** NaOH, DMF   **(d)** $CH_3CO^- Na^+$, CH_3COH **(e)** CH_3OH

Preprofessional Problems

48. Which of the following haloalkanes will undergo hydrolysis most rapidly?
   (a) $(CH_3)_3CF$ **(b)** $(CH_3)_3Cl$ **(c)** $(CH_3)_3CBr$ **(d)** $(CH_3)_3CI$

49. The reaction

$$(CH_3)_3CCl \xrightarrow{CH_3O^-} \underset{CH_3}{\overset{CH_3}{\diagdown}} C{=}CH_2$$

is an example of which of the following processes?
   (a) E1   **(b)** E2   **(c)** S_N1 **(d)** S_N2

50. In this transformation

$$A \xrightarrow{H_2O, \text{ propanone (acetone)}} \underset{OH}{\overset{}{CH_3CH_2C(CH_3)_2}}$$

what is the best structure for A?

   (a) $BrCH_2CH_2CH(CH_3)_2$ **(b)** $CH_3CH_2\underset{CH_3}{\overset{CH_3}{\underset{|}{\overset{|}{C}}}}Br$

   (c) $CH_3CH_2\underset{CH_2Br}{\overset{CH_3}{\underset{|}{\overset{|}{CH}}}}$ **(d)** $CH_3CHCH(CH_3)_2$, Br

51. Which of the following isomeric carbocations is the most stable?

(a)

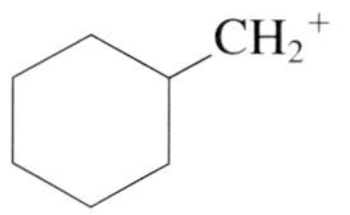

(b)

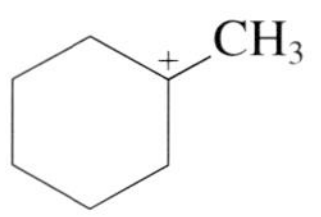

(c)

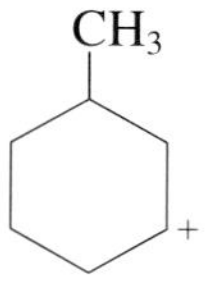

(d)

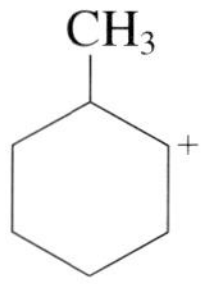

52. Which reaction intermediate is involved in the following reaction?

$$\text{2-methylbutane} \xrightarrow{\text{Br}_2,\ h\nu} \text{2-bromo-3-methylbutane}$$
(not the major product)

(a) A secondary radical   **(b)** A tertiary radical

(c) A secondary carbocation   **(d)** A tertiary carbocation

Hydroxy Functional Group

Properties of the Alcohols and Strategy in Synthesis

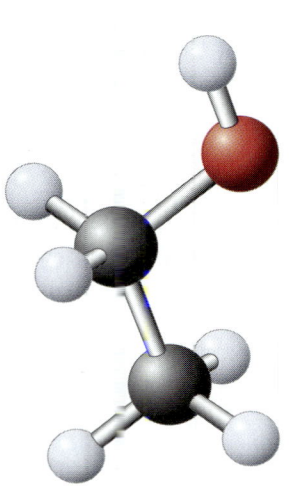

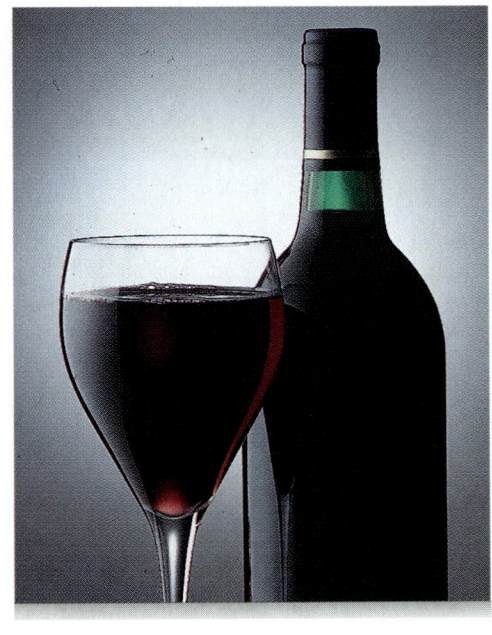

What is your first thought when you hear the word "alcohol"? Undoubtedly, whether pleasant or not, it is connected in some way to ethanol, as contained in alcoholic beverages. The euphoric effects of (limited) ethanol consumption have been known and purposely used for thousands of years. This is perhaps not surprising, because ethanol is naturally generated by the fermentation of carbohydrates. For example, the addition of yeast to an aqueous sugar solution leads to the evolution of CO_2 and the formation of ethanol.

$$C_6H_{12}O_6 \xrightarrow{\text{Yeast enzymes}} 2\ CH_3CH_2OH\ +\ 2\ CO_2$$
$$\textbf{Sugar} \qquad\qquad\qquad \textbf{Ethanol}$$

Ethanol is a member of a large family of compounds called **alcohols.** This chapter introduces you to some of their chemistry. From Chapter 2, we know that alcohols have carbon backbones bearing the substituent OH, the **hydroxy** group. They may be viewed as derivatives of water in which one hydrogen has been replaced by an alkyl group. Replacement of the second hydrogen gives an **ether** (Chapter 9).

Alcohols are abundant in nature and varied in structure (see, e.g., Section 4-7). Simple alcohols are used as solvents; others aid in the synthesis of more complex molecules. They are a good example of how functional groups shape the properties and applications of organic compounds. Our discussion will begin with the naming

$$H\!-\!O\!-\!H$$
Water

$$CH_3\!-\!O\!-\!H$$
Methanol
(An alcohol)

$$CH_3\!-\!O\!-\!CH_3$$
Methoxymethane
(Dimethyl ether)
(An ether)

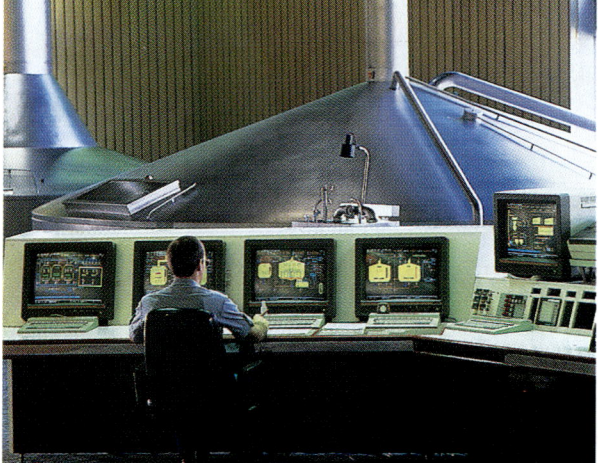

The ancient art of brewing beer by the yeast fermentation of cereal grains is now practiced on a large scale and controlled by computers.

of alcohols, followed by a brief description of their structures and other physical properties, particularly in comparison with those of the alkanes and haloalkanes. Finally, their preparation will introduce us to strategies for efficiently synthesizing new organic compounds.

8-1 Naming the Alcohols

Like other compounds, alcohols may have both systematic and common names. Systematic nomenclature treats alcohols as derivatives of alkanes. The ending *-e* of the alkane is replaced by **-ol.** Thus, an alkane is converted into an **alkanol.** For example, the simplest alcohol is derived from methane. It is methanol. Ethanol stems from ethane, propanol from propane, and so on. In more complicated, branched systems, the name of the alcohol is based on the longest chain *containing the OH substituent—* not necessarily the longest chain in the molecule.

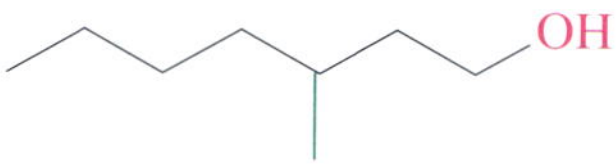

A methyl heptanol   **A methyl propyl octanol**

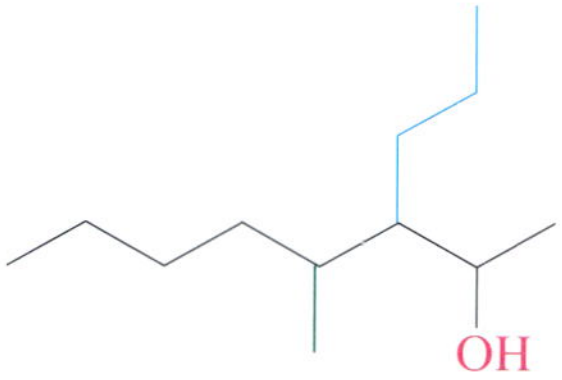

To locate positions along the chain, number each carbon atom beginning from the end closest to the OH group. The names of other substituents along the chain can then be added to the alkanol stem as prefixes. Complex alkyl appendages are named according to the IUPAC rules for hydrocarbons (Section 2-3).

$$\overset{3}{C}H_3\overset{2}{C}H_2\overset{1}{C}H_2OH$$

1-Propanol

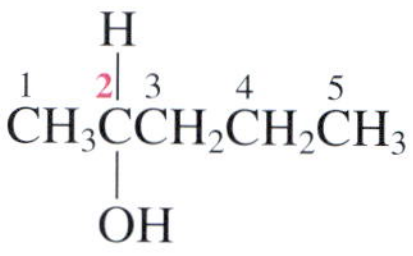

$$\overset{1}{C}H_3\overset{2}{\underset{OH}{C}}\overset{3}{C}H_2\overset{4}{C}H_2\overset{5}{C}H_3$$

2-Pentanol

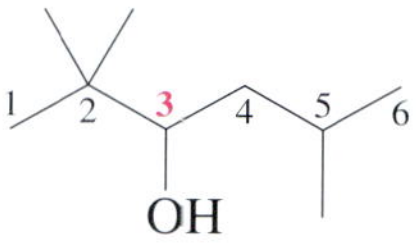

2,2,5-Trimethyl-3-hexanol

Cyclic alcohols are called **cycloalkanols.** Here the carbon carrying the functional group automatically receives the number 1.

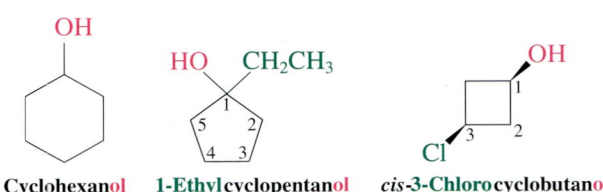

| **Cyclohexanol** | **1-Ethylcyclopentanol** | *cis*-**3-Chlorocyclobutanol** |

When named as a substituent, the OH group is called *hydroxy.* Like haloalkanes, alcohols can be classified as primary, secondary, or tertiary.

$$
\begin{array}{ccc}
 & \overset{\displaystyle OH}{\underset{\displaystyle H}{RCR'}} & \overset{\displaystyle OH}{\underset{\displaystyle R''}{RCR'}} \\
RCH_2OH & & \\
\text{A primary alcohol} & \text{A secondary alcohol} & \text{A tertiary alcohol}
\end{array}
$$

EXERCISE 8-1

Draw the structures of the following alcohols. **(a)** (*S*)-3-methyl-3-hexanol; **(b)** *trans*-2-bromocyclopentanol; **(c)** 2,2-dimethyl-1-propanol (neopentyl alcohol).

EXERCISE 8-2

Name the following compounds.

(a) $CH_3CHCH_2CHCH_3$ with CH_3 and OH groups

(b) 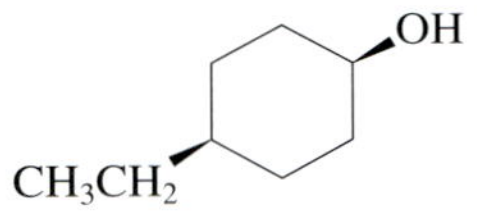 CH_3CH_2—cyclohexane with OH

(c) $CH_3CHCHCH_2OH$ with Br and Cl

In common nomenclature, the name of the alkyl group is followed by the word *alcohol,* written separately. Common names are found in the older literature; although it is best not to use them, we should be able to recognize them.

CH_3OH
Methyl alcohol

$$CH_3\overset{\displaystyle CH_3}{\underset{\displaystyle OH}{CH}}$$
Isopropyl alcohol

$$CH_3\overset{\displaystyle CH_3}{\underset{\displaystyle CH_3}{COH}}$$
tert-**Butyl alcohol**

In summary, alcohols can be named as alkanols (IUPAC) or alkyl alcohols. In IUPAC nomenclature, the name is derived from the chain bearing the hydroxy group, whose position is given the lowest possible number.

8-2 Structural and Physical Properties of Alcohols

The hydroxy functional group strongly shapes the physical characteristics of the alcohols. It affects their molecular structure and allows them to enter into hydrogen bonding. As a result, it raises their boiling points and increases their solubilities in water.

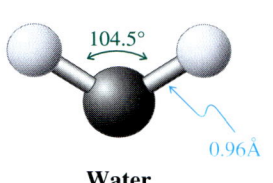

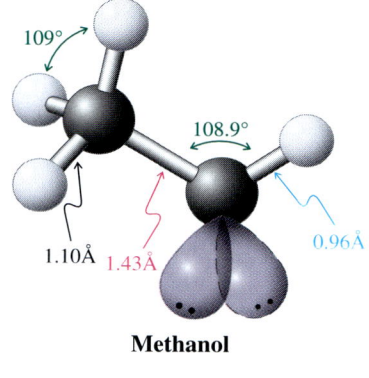

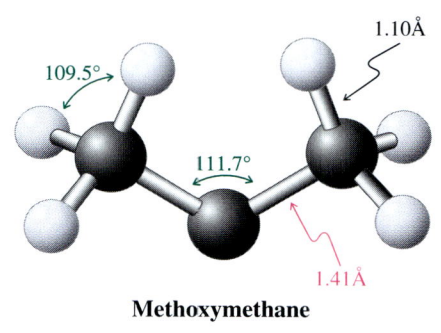

Water   Methanol   Methoxymethane

FIGURE 8-1 _____

The similarity in structure of water, methanol, and methoxymethane. The oxygens are approximately sp^3 hybridized, so all three structures exhibit nearly tetrahedral bond angles around the heteroatom. Remember that the oxygen bears two lone electron pairs in two nonbonding sp^3 hybrid orbitals (shown in the center for methanol).

The structure of alcohols resembles that of water

Figure 8-1 shows how closely the structure of methanol resembles those of water and of methoxymethane (dimethyl ether). In all three, the oxygen atoms are roughly sp^3 hybridized and their bond angles nearly tetrahedral. The minor differences are due to the steric effect produced by the replacement of hydrogen atoms by alkyl groups. The O–H bond is considerably shorter than the C–H bond, in part because of the high electronegativity of oxygen relative to that of carbon. Consistent with this bond shortening is the order of bond strengths: $DH^\circ_{O-H} = 104$ kcal mol^{-1}; $DH^\circ_{C-H} = 98$ kcal mol^{-1}.

The electronegativity of oxygen causes an unsymmetrical distribution of charge in alcohols. This effect polarizes the O–H bond so that the hydrogen has a partial positive charge and gives rise to a molecular dipole (Section 1-3) similar to that observed for water.

TABLE 8-1	Physical Properties of Alcohols and Selected Analogous Haloalkanes and Alkanes				
Compound	IUPAC name	Common name	Melting point (°C)	Boiling point (°C)	Solubility in H$_2$O at 23°C
CH$_3$OH	Methanol	Methyl alcohol	−97.8	65.0	Infinite
CH$_3$Cl	Chloromethane	Methyl chloride	−97.7	−24.2	0.74 g/100 mL
CH$_4$	Methane		−182.5	−161.7	3.5 mL (gas)/100 mL
CH$_3$CH$_2$OH	Ethanol	Ethyl alcohol	−114.7	78.5	Infinite
CH$_3$CH$_2$Cl	Chloroethane	Ethyl chloride	−136.4	12.3	0.447 g/100 mL
CH$_3$CH$_3$	Ethane		−183.3	−88.6	4.7 mL (gas)/100 mL
CH$_3$CH$_2$CH$_2$OH	1-Propanol	Propyl alcohol	−126.5	97.4	Infinite
CH$_3$CH$_2$CH$_3$	Propane		−187.7	−42.1	6.5 mL (gas)/100 mL
CH$_3$CH$_2$CH$_2$CH$_2$OH	1-Butanol	Butyl alcohol	−89.5	117.3	8.0 g/100 mL
CH$_3$(CH$_2$)$_4$OH	1-Pentanol	Pentyl alcohol	−79	138	2.2 g/100 mL

Bond and Molecular Dipoles of Water and Methanol

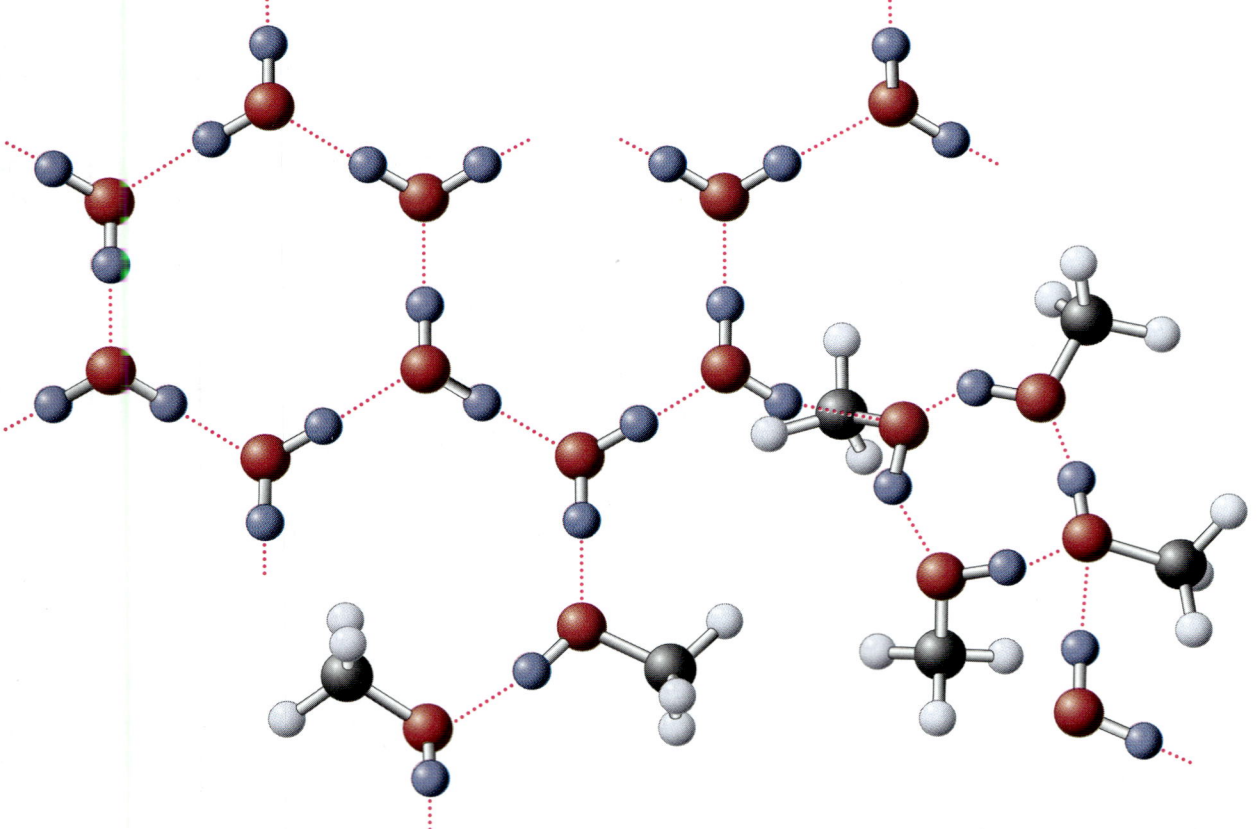

Hydrogen bonding raises the boiling points and water solubilities of alcohols

In Section 6-2 we invoked the polarity of the haloalkanes to explain why their boiling points are higher than those of the corresponding nonpolar alkanes. The polarity of alcohols is similar to that of the haloalkanes. Does this mean that the boiling points of haloalkanes and alcohols should correspond? Inspection of Table 8-1 shows that they do not: Alcohols have unusually high boiling points, much higher than those of comparable alkanes and haloalkanes.

The explanation lies in hydrogen bonding. Hydrogen bonds may form between the oxygen atoms of one alcohol molecule and the hydroxy hydrogen atoms of another. Alcohols build up an extensive network of these interactions (Figure 8-2). Although hydrogen bonds are longer and much weaker ($DH° \sim$ 5–6 kcal mol^{-1}) than the covalent O–H linkage ($DH° \sim$ 104 kcal mol^{-1}), so many of them form that their combined strength makes it difficult for molecules to escape the liquid. The result is a higher boiling point.

FIGURE 8-2
Hydrogen bonding in an aqueous solution of methanol. The molecules form a complex three-dimensional array, and only one layer is depicted here. Pure water (in ice) tends to arrange itself in cyclic hexamer units (top left); neat small alcohols prefer a cyclic tetramer structure (bottom right).

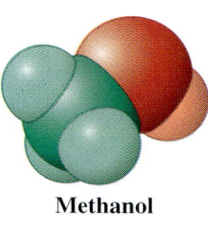
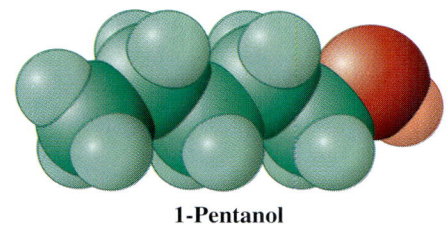

Methanol   1-Pentanol

FIGURE 8-3
The hydrophobic (green) and hydrophilic (red) parts of methanol and 1-pentanol (space-filling models). The polar functional group dominates the physical properties of methanol: The molecule is completely soluble in water but only partially so in hexane. Conversely, the increased size of the hydrophobic part in the higher alcohol leads to infinite solubility in hexane but reduced solubility in water (Table 8-1).

The effect is even more pronounced in water, which has two hydrogens available for hydrogen bonding (see Figure 8-2). This phenomenon explains why water, with a molecular weight of only 18, has a boiling point of 100°C. Without this property, water would be a gas at ordinary temperatures. Considering the importance of water in all living organisms, imagine how the absence of liquid water would have affected the development of life on our planet.

Hydrogen bonding in water and alcohols is responsible for another property: Many alcohols are appreciably water soluble (Table 8-1). This behavior contrasts with that of the nonpolar alkanes, which are poorly solvated by this medium. Because of their characteristic insolubility in water, alkanes are said to be **hydrophobic** (*hydro,* Greek, water; *phobos,* Greek, fear). So are most alkyl chains. Conversely, the OH group and other polar substituents, such as COOH and NH$_2$, are **hydrophilic:** They enhance water solubility.

As the values in Table 8-1 show, the larger the alkyl (hydrophobic) part of an alcohol, the lower its solubility in water. At the same time, the alkyl part increases the solubility of the alcohol in nonpolar solvents (Figure 8-3). (For example, alcohols are effective at removing greasy deposits from the tape heads of cassette decks and VCRs.) The "waterlike" structure of the lower alcohols, particularly methanol and ethanol, also makes them excellent solvents for polar compounds and even salts. It is not surprising, then, that alcohols are popular solvents in the S$_N$2 reaction (Chapter 6).

In summary, the oxygen in alcohols (and ethers) is tetrahedral and *sp^3* hybridized. The covalent O–H bond is shorter and stronger than the C–H bond. Because of the electronegativity of the oxygen, alcohols, like water and ethers, exhibit appreciable molecular polarity. The hydroxy hydrogen enters into hydrogen bonding with other alcohol molecules. These properties lead to a substantial increase in the boiling points and in the solubilities of alcohols in polar solvents relative to those of the alkanes and haloalkanes.

8-3 Alcohols as Acids and Bases

Many applications of the alcohols depend on their ability to act both as acids and as bases. (See the review of these concepts in Section 6-8.) Thus, deprotonation gives alkoxide ions. We shall see how structural features affect their pK_a values. The lone electron pairs on oxygen render alcohols basic as well, and protonation results in alkyloxonium ions.

The acidity of alcohols resembles that of water

The acidity of alcohols in water is expressed by the equilibrium constant K.

$$ROH + H_2O \xrightleftharpoons{K} H_3O^+ + RO^-$$

**Alkoxide
ion**

Making use of the constant concentration of water (55 mol L^{-1}; Section 6-8), we derive a new equilibrium constant K_a.

$$K_a = K[H_2O] = \frac{[H_3O^+][RO^-]}{[ROH]} \text{ mol } L^{-1}, \text{ and } pK_a = -\log K_a$$

Table 8-2 lists the pK_a values of several alcohols and related compounds. A comparison of these values with those given in Table 2-6 for mineral and other strong acids shows that alcohols, like water, are fairly weak acids. Their acidity is far greater, however, than that of alkanes and haloalkanes.

Why are alcohols acidic, whereas alkanes and haloalkanes are not? The answer lies in the relatively strong electronegativity of the oxygen to which the proton is attached, which stabilizes the negative charge of the alkoxide ion.

To drive the equilibrium between alcohol and alkoxide to the side of the conjugate base, it is necessary to use a base *stronger* than the alkoxide formed (i.e., a base derived from a conjugate acid *weaker* than the alcohol; see also Section 9-1). An example is the reaction of sodium amide, $NaNH_2$, with methanol to furnish sodium methoxide and ammonia.

$$CH_3OH + Na^+ \, ^-NH_2 \xrightleftharpoons{K} Na^+ \, ^-OCH_3 + NH_3$$

$pK_a = 15.5$ $pK_a = 35$

**Sodium
amide**   **Sodium
methoxide**

This equilibrium lies well to the right ($K \sim 10^{35-15.5} = 10^{19.5}$), because methanol is a much stronger acid than is ammonia, or, conversely, because amide is a much stronger base than is methoxide.

TABLE 8-2	pK$_a$ Values of Alcohols and Related Compounds in Water		
Compound	**pK_a**	**Compound**	**pK_a**
H_2O	15.7	HOCl	7.53
CH_3OH	15.5	$ClCH_2CH_2OH$	14.3
CH_3CH_2OH	15.9	CF_3CH_2OH	12.4
$(CH_3)_2CHOH$	17.1	$CF_3CH_2CH_2OH$	14.6
$(CH_3)_3COH$	18	$CF_3CH_2CH_2CH_2OH$	15.4
H_2O_2	11.64		

It is sometimes sufficient to generate alkoxides in less than stoichiometric equilibrium concentrations. For this purpose, we may add an alkali metal hydroxide to the alcohol.

$$CH_3CH_2OH + Na^+ {}^-OH \; \underset{}{\overset{K}{\rightleftharpoons}} \; CH_3CH_2O^- Na^+ + H_2O$$
$$pK_a = 15.9 \qquad\qquad\qquad\qquad\qquad\qquad pK_a = 15.7$$

With this base present, approximately one-half of the alcohol will exist as the alkoxide, if we assume equimolar concentrations of starting materials. If the alcohol is the solvent (i.e., present in large excess), however, all of the base will exist in the form of the alkoxide.

EXERCISE 8-3

Which of the following bases are strong enough to cause essentially complete deprotonation of methanol? The pK_a of the conjugate acid is given in parentheses.
(a) KCN (9.2); **(b)** $CH_3CH_2CH_2CH_2Li$ (50); **(c)** CH_3CO_2Na (4.7); **(d)** $LiN[CH(CH_3)_2]_2$ (LDA, 40); **(e)** KH (38); **(f)** CH_3SNa (10).

Steric disruption and inductive effects control the acidity of alcohols

Table 8-2 shows a large variation in the pK_a values of the alcohols. A closer look at the first column reveals that the acidity decreases (pK_a increases) from methanol to primary, secondary, and finally tertiary systems.

> **Relative pK_a Values of Alcohols (in Solution)**
>
> CH_3OH < primary < secondary < tertiary
> **Strongest acid** **Weakest acid**

This ordering has been ascribed to steric disruption of solvation and to hydrogen bonding in the alkoxide (Figure 8-4). Because solvation and hydrogen bonding stabilize the negative charge on oxygen, interference with these processes leads to an increase in pK_a.

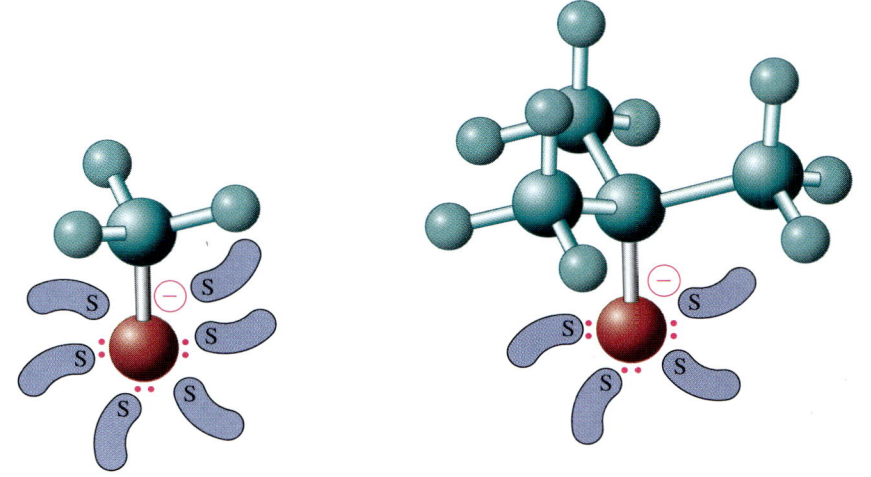

FIGURE 8-4
The smaller methoxide is better solvated than is the larger tertiary butoxide ion. (S, solvent molecules).

The second column in Table 8-2 reveals another contribution to the pK_a of alcohols: The presence of halogens increases acidity. Recall that the carbon of the C–X bond is positively polarized as a result of the high electronegativity of X (Sections 1-3 and 6-2). Electron withdrawal by the halogen also causes atoms farther away to be slightly positively charged. This phenomenon of transmission of charge, both negative and positive, through the σ bonds in a chain of atoms is called an **inductive effect.** Here it stabilizes the negative charge on the alkoxide oxygen by electrostatic attraction. The inductive effect in alcohols increases with the number of electronegative groups but decreases with distance from the oxygen.

Inductive Effect of the Chlorine in 2-Chloroethoxide

$$\overset{\leftarrow+}{Cl}-\overset{\leftarrow+}{CH_2}-\overset{\leftarrow\,..}{CH_2}-\overset{..}{\underset{..}{O}}:^-$$

EXERCISE 8-4

Rank the following alcohols in order of increasing acidity.

EXERCISE 8-5

Which side of the following equilibrium reaction is favored (assume equimolar concentrations of starting materials)?

$$(CH_3)_3CO^- + CH_3OH \rightleftharpoons (CH_3)_3COH + CH_3O^-$$

The lone electron pairs on oxygen make alcohols basic

Alcohols may also be basic, although weakly so. Very strong acids are required to protonate the OH group, as indicated by the low pK_a values (strong acidity) of their conjugate acids, the alkyloxonium ions (Table 8-3). Molecules that may be both acids and bases are called **amphoteric** (*ampho,* Greek, both).

The amphoteric nature of the hydroxy functional group characterizes the chemical reactivity of alcohols. In strong acids, they exist as alkyloxonium ions, in neutral media as alcohols, and in strong bases as alkoxides.

Alcohols Are Amphoteric

| | Alkyloxonium ion | Alcohol | Alkoxide ion |

In summary, alcohols are amphoteric. They are acidic by virtue of the electronegativity of the oxygen and are converted into alkoxides by strong bases. In solution, the steric bulk of branching inhibits solvation of the alkoxide, thereby raising the pK_a of the corresponding alcohol. Electron-withdrawing substituents close to the functional group lower the pK_a. Alcohols are also weakly basic and can be protonated by strong acids to furnish alkyloxonium ions.

TABLE 8-3 pK_a Values of Four Protonated Alcohols

Compound	pK_a
$CH_3\overset{+}{O}H_2$	-2.2
$CH_3CH_2\overset{+}{O}H_2$	-2.4
$(CH_3)_2CH\overset{+}{O}H_2$	-3.2
$(CH_3)_3C\overset{+}{O}H_2$	-3.8

8-4 Industrial Sources of Alcohols: Carbon Monoxide and Ethene

Let us now turn to the preparation of alcohols. We start in this section with methods of special importance in industry. Subsequent sections will take up procedures that are used more generally in the synthetic laboratory to introduce the hydroxy functional group into a wide range of organic molecules.

Methanol is made on a large scale (more than 11 billion pounds [5 billion kg] in the United States in 1995) from a pressurized mixture of CO and H_2 called **synthesis gas.** The reaction makes use of a catalyst consisting of copper, zinc oxide, and chromium(III) oxide.

$$CO + 2\ H_2 \xrightarrow{\text{Cu-ZnO-Cr}_2\text{O}_3,\ 250°\text{C, 50–100 atm}} CH_3OH$$

Changing the catalyst to rhodium or ruthenium leads to 1,2-ethanediol (ethylene glycol), an important industrial chemical that is the principal component of automobile antifreeze.

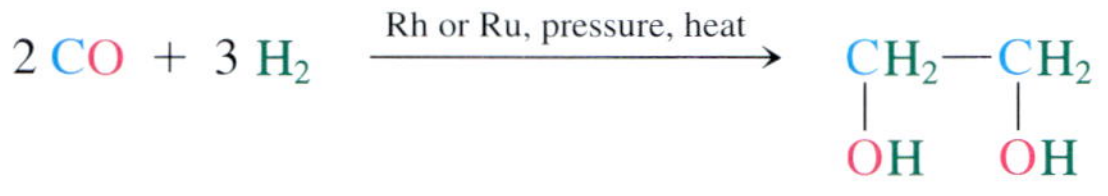

$$2\ CO + 3\ H_2 \xrightarrow{\text{Rh or Ru, pressure, heat}} \underset{\substack{| \quad\ | \\ \text{OH} \quad \text{OH}}}{CH_2-CH_2}$$

**1,2-Ethanediol
(Ethylene glycol)**

Other reactions that would permit the selective formation of a given alcohol from synthesis gas are the focus of much current research, because synthesis gas is readily available by the gasification of coal in the presence of water.

$$\text{Coal} \xrightarrow{\text{Air, H}_2\text{O, }\Delta} x\ CO + y\ H_2$$

Ethanol is prepared in large quantities by fermentation of sugars or by the phosphoric acid–catalyzed hydration of ethene (ethylene). The United States produced about 620 million pounds (280 million kg) of ethanol in 1995. The hydration (and other addition reactions) of alkenes are considered in detail in Chapter 12.

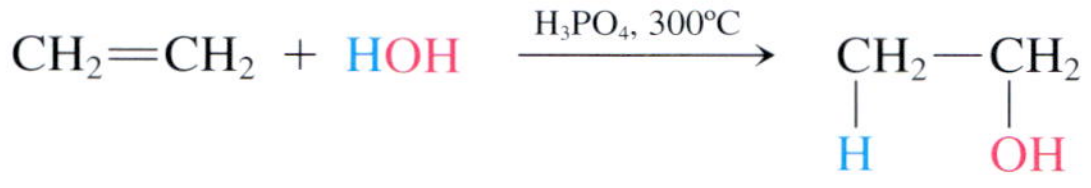

$$CH_2{=}CH_2 + HOH \xrightarrow{\text{H}_3\text{PO}_4,\ 300°\text{C}} \underset{\substack{| \quad\ | \\ \text{H} \quad \text{OH}}}{CH_2-CH_2}$$

In summary, the industrial preparation of methanol and 1,2-ethanediol proceeds by reduction of carbon monoxide with hydrogen, and that of ethanol by the acid-catalyzed hydration of ethene (ethylene).

8-5 Synthesis of Alcohols by Nucleophilic Substitution

On a smaller than industrial scale, we can prepare alcohols from a wide variety of starting materials. For example, conversions of haloalkanes into alcohols by S_N2 and S_N1 processes featuring hydroxide and water, respectively, as nucleophiles were de-

scribed in Chapters 6 and 7. These methods are not as widely used as one might think, however, because the required halides are often accessible only from the corresponding alcohols (Chapter 9). They also suffer from the usual drawbacks of nucleophilic substitution: Bimolecular elimination can be a major side reaction of hindered systems, and tertiary halides form carbocations that may undergo E1 reactions. Nevertheless, a number of alcohol syntheses have made use of such simple nucleophilic substitution reactions, two of which are shown here.

Alcohols by Nucleophilic Substitution

95%

Steroid found in human
umbilical cord blood

86%

Precursor for syntheses
of antitumor antibiotics

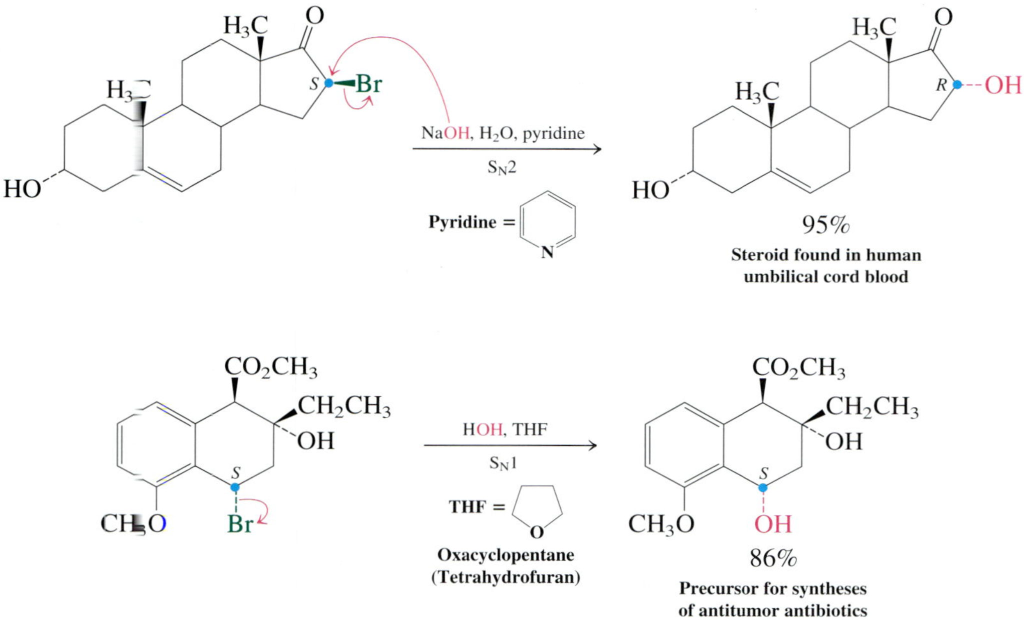

In the first example, stereochemical inversion was achieved by an S_N2 reaction with hydroxide. The second reaction is a solvolysis that follows the S_N1 mechanism. Stereoselectivity in this case is observed because the bottom face of the molecule is considerably less sterically hindered than the top. Both nucleophilic substitutions use organic **cosolvents** (pyridine and THF, respectively) to improve the solubility of the substrates.

EXERCISE 8-6

Show how you might convert the following haloalkanes into alcohols.
(a) bromoethane; **(b)** chlorocyclohexane; **(c)** 3-chloro-3-methylpentane.

A way around the problem of elimination in S_N2 reactions of oxygen nucleophiles with secondary or sterically encumbered, branched primary substrates is the use of less basic functional equivalents of water, such as acetate (Section 6-9). The resulting alkyl acetate can then be converted into the desired alcohol by aqueous hydroxide. In a second step, the carbonyl oxygen bond is broken, thereby leaving the alkoxy residue unchanged. We shall consider this reaction, known as *ester hydrolysis,* in Chapter 20.

Alcohols from Haloalkanes by Acetate Substitution—Hydrolysis

STEP 1. Acetate formation (S_N2 reaction)

$$CH_3CH_2CHCH_2CH_2\!-\!Br + CH_3CO^-Na^+ \xrightarrow{\text{DMF, 80°C}} CH_3CH_2CHCH_2CH_2OCCH_3 + Na^+Br^-$$

with CH_3 substituent on both structures, O on carbonyls

95%

1-Bromo-3-methylpentane   **3-Methylpentyl acetate**

STEP 2. Conversion into the alcohol (ester hydrolysis)

$$CH_3CH_2CHCH_2CH_2O\!-\!CCH_3 + Na^+\ ^-OH \xrightarrow[{-CH_3CO^-Na^+}]{H_2O} CH_3CH_2CHCH_2CH_2OH$$

with CH_3 substituent, O on carbonyl

85%

3-Methyl-1-pentanol

In summary, alcohols may be prepared from haloalkanes by nucleophilic substitution, provided the haloalkane is readily available and side reactions such as elimination do not interfere.

8-6 Synthesis of Alcohols: Oxidation–Reduction Relation Between Alcohols and Carbonyl Compounds

This section describes an important synthesis of alcohols: reduction of aldehydes and ketones. Later, we shall see that these compounds may be converted into alcohols by addition of organometallic reagents, with concomitant formation of a new carbon–carbon bond. Because of the versatility of aldehydes and ketones in synthesis, we shall also illustrate their preparation by alcohol oxidation.

Oxidation and reduction have special meanings in organic chemistry

We can readily recognize inorganic oxidation and reduction processes as the loss and gain of electrons, respectively. With organic compounds, it is often less clear whether electrons are being gained or lost in a reaction. Hence, organic chemists find it more useful to define oxidation and reduction in other terms. A process that adds electronegative atoms such as halogen or oxygen to, or removes hydrogen from, a molecule constitutes an **oxidation;** conversely, the removal of halogen or oxygen or the addition of hydrogen is defined as **reduction.** You can readily visualize this definition in the step-by-step oxidation of methane, CH_4, to carbon dioxide, CO_2.

Step-by-Step Oxidation of CH_4 to CO_2

$$CH_4 \xrightarrow{+O} CH_3OH \xrightarrow{-2H} H_2C\!=\!O \xrightarrow{+O} HCOH \xrightarrow{-2H} CO_2$$

This definition of an oxidation–reduction relation allows us to connect alcohols to aldehydes and ketones. Addition of two hydrogen atoms to the double bond of a carbonyl group constitutes reduction to the corresponding alcohol. Aldehydes give primary alcohols; ketones give secondary alcohols. The reverse process, removal of hy-

drogen to furnish carbonyl compounds, is an example of oxidation. Together, these processes are referred to as **redox reactions.**

The Redox Relation Between Alcohols and Carbonyl Compounds

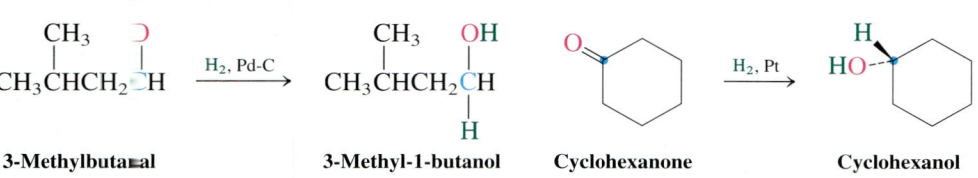

| Aldehyde | Primary alcohol | Ketone | Secondary alcohol |

Reduction of carbonyl compounds can be carried out either by addition of molecular hydrogen or by exposure to hydride reagents.

Alcohols can be made by catalytic hydrogenation of aldehydes and ketones

The addition of gaseous hydrogen to a double bond, called **hydrogenation,** requires a catalyst (see Chemical Highlight 3-1) and, in many cases, must be carried out under high pressure in order to proceed at a useful rate. Most such hydrogenations utilize **heterogeneous catalysts** (*héteros,* Greek, other; *génos,* Greek, kind) that are insoluble in the reaction solvent. (In contrast, soluble catalysts are called **homogeneous.**) The reaction takes place on the surface of the suspended particles, which typically consist of finely divided metals such as platinum, palladium, or nickel, often deposited on a supporting material, such as carbon, to maximize the surface area. Catalytic hydrogenation is also an important reaction of alkenes; its mechanism will be discussed in Chapter 12.

Hydrogenation of an Aldehyde

CH₃   O
 | ||
CH₃CHCH₂CH $\xrightarrow{H_2,\ Pd-C}$

CH₃   OH
 | |
CH₃CHCH₂CH
           |
           H

3-Methylbutanal   **3-Methyl-1-butanol**

Hydrogenation of a Ketone

Cyclohexanone $\xrightarrow{H_2,\ Pt}$ Cyclohexanol

Alcohols can form by hydride reduction of the carbonyl group

The electrons in the **carbonyl group,** C=O, are not distributed evenly between the two component atoms. Because oxygen is more electronegative than carbon, the carbon of a carbonyl group is electrophilic, the oxygen nucleophilic. This polarization can be represented by a charge-separated resonance form.

Polar Character of the Carbonyl Function

Because the carbonyl carbon is electrophilic, nucleophilic hydride, H⁻, may be delivered to it by **hydride reagents.** Two such commercial reagents are sodium boro-

CHEMICAL HIGHLIGHT **8-1** | **Biological Oxidation and Reduction**

Alcohols are metabolized by oxidation to carbonyl compounds. In biological systems, ethanol is converted into acetaldehyde by the cationic oxidizing agent *nicotinamide adenine dinucleotide* (abbreviated as NAD^+; for its structure, see Chapter 25). The process is catalyzed by the enzyme *alcohol dehydrogenase*. (The latter also catalyzes the reverse process, reduction of aldehydes and ketones to alcohols; see Problems 47 and 48 at the end of this chapter.) When the two enantiomers of 1-deuterioethanol were subjected to the enzyme, the

biochemical oxidation was found to be stereospecific, NAD^+ removing only the hydrogen marked by the solid arrowhead in the first reaction below from C1 of the alcohol (see Box 25-4).

Other alcohols are similarly oxidized biochemically. The relatively high toxicity of methanol ("wood alcohol") is due largely to its oxidation to formaldehyde. The latter specifically interferes with a system responsible for the transfer of one-carbon fragments between nucleophilic sites in biomolecules.

$$CH_3 \underset{\underset{D}{\overset{}{\rvert}}}{\overset{}{C}}{\overset{H}{\diagup}} - OH + NAD^+ \xrightarrow[-NAD-H]{\text{Alcohol dehydrogenase}} \underset{CH_3}{\overset{O}{\overset{\|}{C}}}{\diagdown}_D$$

(S)-1-Deuterioethanol

$$CH_3 \underset{\underset{H}{\overset{}{\rvert}}}{\overset{}{C}}{\overset{D}{\diagup}} - OH + NAD^+ \xrightarrow[-NAD-D]{\text{Alcohol dehydrogenase}} \underset{CH_3}{\overset{O}{\overset{\|}{C}}}{\diagdown}_H$$

(R)-1-Deuterioethanol

The capability of alcohols to undergo enzymatic oxidation makes them important relay stations in metabolism. One of the most important functions of the metabolic degradation of the food that we eat is its controlled "burning" (i.e., combustion; see Section 3-10) to release the heat and chemical energy required to run our bodies. Another is the selective introduction of functional groups, especially hydroxy groups, into unfunctionalized molecules or parts of molecules—in other words, alkanes or alkyl substituents. The *cytochrome* proteins are crucial biomolecules that enable nature to accomplish this task. These molecules are present in almost all living cells and emerged almost 1.5 billion years ago, before the development of plants and animals as separate species. Cytochrome P-450 (see Section 22-9) uses O_2 to accomplish the direct hydroxylation of organic molecules. In the liver,

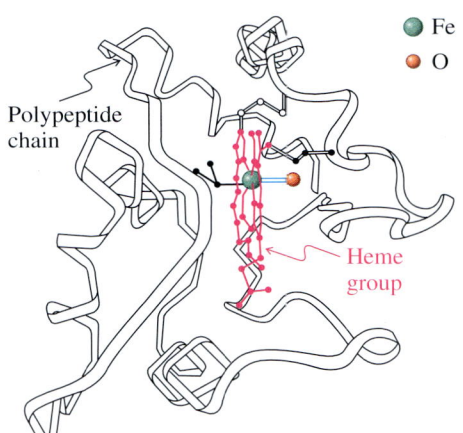

● Fe
● O

Polypeptide chain

Heme group

Cytochrome model.

this process serves to detoxify substances foreign to the body (xenobiotic), many of which are the medicines that we take. Often, the primary effect of hydroxylation is simply to impart greater water solubility, thereby accelerating the excretion of a drug and thus preventing its accumulation to toxic levels.

Selective hydroxylation is important in steroid synthesis (Section 4-7). For example, progesterone is converted by triple hydroxylation at C17, C21, and C11 into cortisol. Not only does the protein pick specific positions as targets for functionalization with complete stereoselectivity, but it also controls the sequence in which these reactions take place. You can get an inkling of the origin of this selectivity when you inspect the cytochrome model shown on the opposite page.

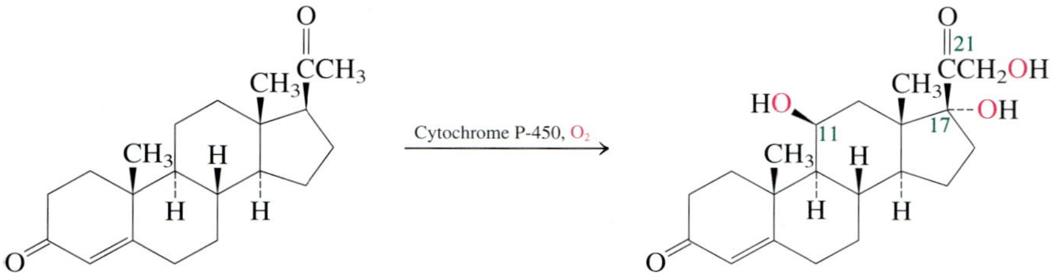

Progesterone → Cytochrome P-450, O_2 → **Cortisol**

The active site is an Fe atom tightly held by a strongly bound heme group (see Section 26-8) embedded in the cloak of a polypeptide (protein) chain. The Fe center binds O_2 to generate an Fe–O_2 species, which is then reduced to H_2O and the Fe=O unit shown. This oxide reacts as a radical (Section 3-4) with R–H, producing an Fe–OH intermediate in the presence of R· . The carbon-based radical then abstracts OH to furnish the alcohol.

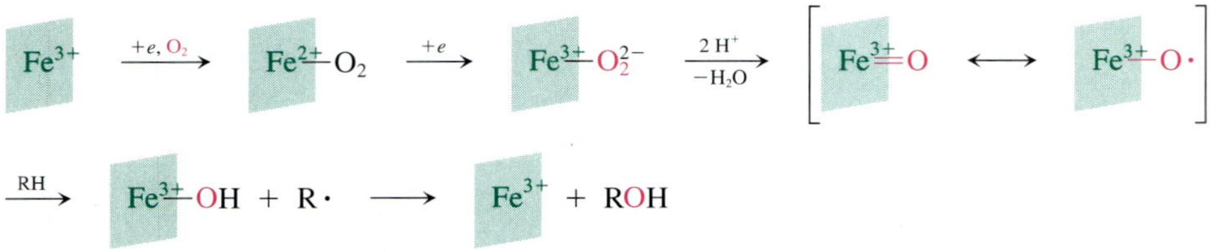

The steric and electronic environment provided by the polypeptide mantle allows substrates, such as progesterone, to approach the active iron site only in very specific orientations, leading to preferential oxidation at only certain positions, such as C17, C21, and C11.

hydride, $NaBH_4$, and lithium aluminum hydride, $LiAlH_4$. These two compounds possess the advantage of higher solubility in common organic solvents than do simpler analogs such as LiH and NaH.

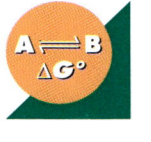

General Hydride Reductions of Aldehydes and Ketones

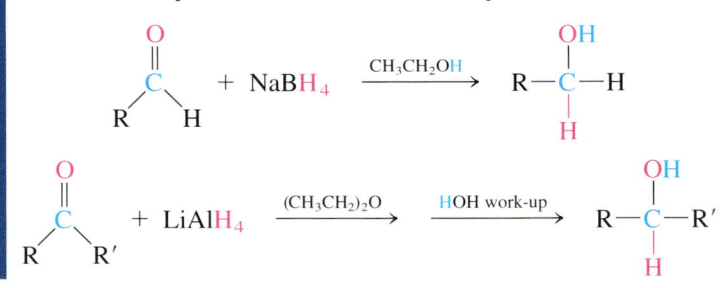

The chemistry of these reagents is dominated by the hydridic (H^-) character of the hydrogen atoms. Reduction of a carbonyl group is achieved by addition of a hydride to carbon and a proton to oxygen.

Examples of Hydride Reductions of Aldehydes and Ketones

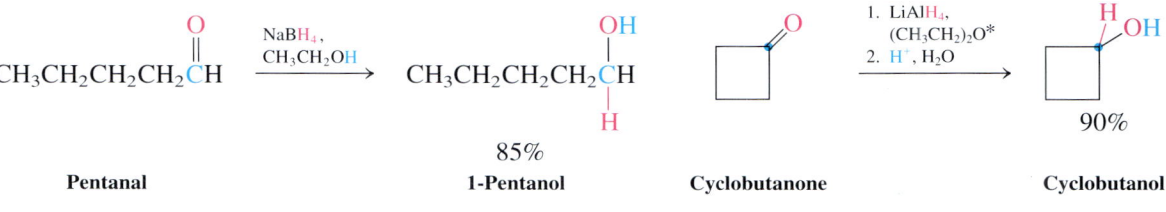

	85%		90%
Pentanal	**1-Pentanol**	**Cyclobutanone**	**Cyclobutanol**

EXERCISE 8-7

Formulate all of the expected products of $NaBH_4$ reduction of the following compounds. (**Hint:** Remember the possibility of stereoisomerism.)

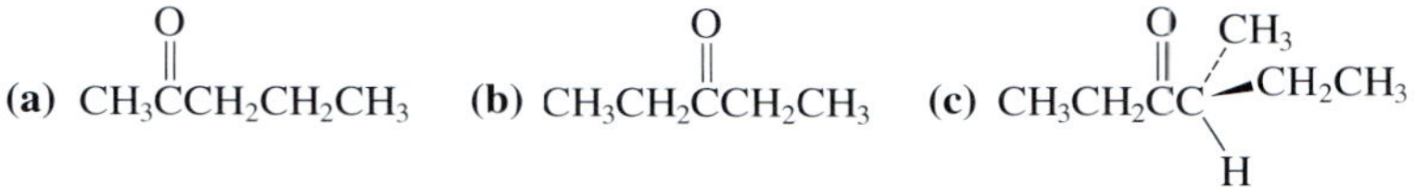

(a) $CH_3\overset{O}{\overset{||}{C}}CH_2CH_2CH_3$ (b) $CH_3CH_2\overset{O}{\overset{||}{C}}CH_2CH_3$ (c) $CH_3CH_2\overset{O}{\overset{||}{C}}\overset{CH_3}{\underset{H}{\overset{|}{C}}}CH_2CH_3$

EXERCISE 8-8

Hydride reductions are often highly stereoselective, with the delivery of hydrogen from the less hindered side of the substrate molecule. Predict the likely stereochemical outcome of the treatment of compound A with $NaBH_4$.

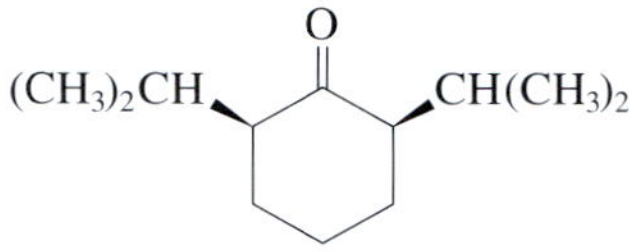

$(CH_3)_2CH$ — [cyclohexanone ring with O] — $CH(CH_3)_2$

A

Although free hydride ion is a powerful base that is immediately protonated by protic solvents [see Exercise 8-3(e)], attachment to boron in BH_4^- moderates its re-

---

*The numbers refer to reagents that are used *sequentially*. Thus, in the first reaction, the substrate at the left of the arrow is treated with the reagents listed after the number 1. The product of this transformation then undergoes a reaction with the reagents listed after the number 2, and so on. The last reaction gives the final product shown on the right.

activity considerably, thus allowing $NaBH_4$ to be used in solvents such as ethanol. When an aldehyde or a ketone is exposed to borohydride, the reagent donates an H^- to the carbonyl carbon, with simultaneous protonation of the carbonyl oxygen by the solvent. The ethoxide by-product combines with the remaining boron fragment, giving ethoxyborohydride.

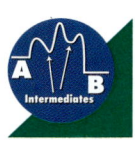

Mechanism of $NaBH_4$ Reduction

$$Na^+ \ H_3\bar{B}\!-\!H \quad \overset{}{\underset{}{C}}\!=\!O \quad H\!-\!\bar{O}CH_2CH_3 \quad \longrightarrow \quad H\!-\!\overset{|}{\underset{|}{C}}\!-\!OH \quad + \quad Na^+ \ H_3\bar{B}OCH_2CH_3$$

Ethanol solvent   Product alcohol   Sodium ethoxyborohydride

The resulting ethoxyborohydride may attack three more carbonyl substrates before all the hydride atoms of the original reagent have been used up. As a result, one equivalent of borohydride is capable of reducing *four* equivalents of aldehyde or ketone to alcohol. The boron reagent is finally converted into tetraethoxyborate, $^-B(OCH_2CH_3)_4$.

Lithium aluminum hydride is more reactive than sodium borohydride, because its hydrogens are less strongly bound to the metal and more negatively polarized (Table 1-2). They are thus much more basic (as well as nucleophilic) and are attacked vigorously by water and alcohols to give hydrogen gas. Reductions utilizing lithium aluminum hydride are therefore carried out in aprotic solvents, such as ethoxyethane (diethyl ether).

Reaction of Lithium Aluminum Hydride with Protic Solvents

$$LiAlH_4 \ + \ 4\,CH_3OH \quad \xrightarrow{\text{Fast}} \quad LiAl(OCH_3)_4 \ + \ 4\,H\!-\!H\uparrow$$

Addition of lithium aluminum hydride to an aldehyde or ketone initially furnishes alkoxyaluminum hydride, which continues to deliver H^- to three more carbonyl groups, thus reducing a total of four equivalents of aldehyde or ketone. Addition of aqueous acid (aqueous, or HOH, work-up) consumes excess reagent, hydrolyzes the tetraalkoxyaluminate, and releases the product alcohol.

Mechanism of $LiAlH_4$ Reduction

Repeat three times:
React with three more $C\!=\!O$

$$Li^+ \ H_3\bar{Al}\!-\!H \quad \overset{}{\underset{}{C}}\!=\!O \quad \longrightarrow \quad H\!-\!\overset{|}{\underset{|}{C}}\!-\!\bar{O}AlH_3 \ Li^+ \quad \xrightarrow{\hspace{2cm}}$$

Lithium
alkoxyaluminum
hydride

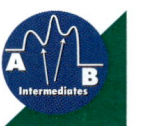

$$(H\!-\!\overset{|}{\underset{|}{C}}\!-\!O)_4Al^- \ Li^+ \quad \xrightarrow{\text{HOH work-up}} \quad 4\,H\!-\!\overset{|}{\underset{|}{C}}\!-\!OH \ + \ Al(OH)_3 \ + \ LiOH$$

Lithium tetraalkoxy-
aluminate   Product alcohol

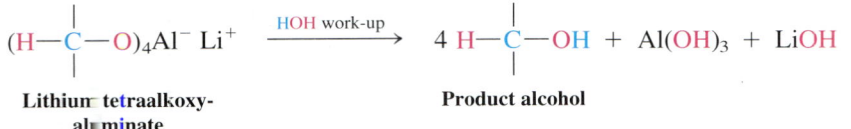

EXERCISE 8-9

Formulate reductions that would give rise to the following alcohols. **(a)** 1-decanol; **(b)** 4-methyl-2-pentanol; **(c)** cyclopentylmethanol; **(d)** 1,4-cyclohexanediol.

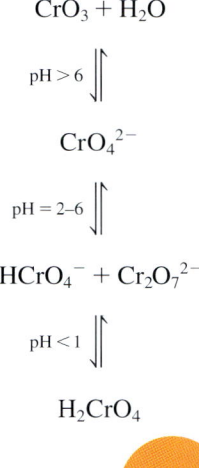

$CrO_3 + H_2O$

$pH > 6 \updownarrow$

CrO_4^{2-}

$pH = 2-6 \updownarrow$

$HCrO_4^- + Cr_2O_7^{2-}$

$pH < 1 \updownarrow$

H_2CrO_4

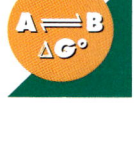

$A \rightleftharpoons B$
$\Delta G°$

Alcohol synthesis by reduction can be reversed: chromium reagents

We have seen several methods for synthesizing alcohols from aldehydes and ketones by reduction with hydrogen or hydride reagents. The reverse is also possible: Alcohols may be oxidized to produce aldehydes and ketones. A useful reagent for this purpose is a transition metal in a high oxidation state: chromium(VI). In this form, chromium has a yellow-orange color. On exposure to an alcohol, the Cr(VI) species is reduced to the deep green Cr(III). The reagent is usually supplied as a dichromate salt ($K_2Cr_2O_7$ or $Na_2Cr_2O_7$) or as CrO_3. Oxidation of secondary alcohols to ketones is often carried out in aqueous acid solution, in which all of the chromium reagents are generating varying amounts of chromic acid, H_2CrO_4, depending on pH.

Oxidation of a Secondary Alcohol to a Ketone with Aqueous Cr(VI)

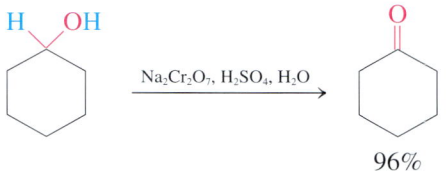

96%

EXERCISE 8-10

Write a balanced equation for the preceding redox process. The inorganic products are $Cr_2(SO_4)_3$ and Na_2SO_4. (**Hint:** Review the appropriate section of your general chemistry textbook on balancing half reactions.)

Under these conditions, primary alcohols tend to *overoxidize* to carboxylic acids, as shown for 1-propanol.

$$CH_3CH_2CH_2OH \xrightarrow[H_2SO_4, H_2O]{K_2Cr_2O_7,} CH_3CH_2\overset{\overset{\textstyle O}{\|}}{C}H \xrightarrow{\text{Overoxidation}} CH_3CH_2\overset{\overset{\textstyle O}{\|}}{C}OH$$

Propanal **Propanoic acid**

In the absence of water, however, aldehydes are not susceptible to overoxidation. Therefore, a water-free form of Cr(VI) has been developed by reaction of CrO_3 with HCl, followed by the addition of the organic base pyridine. The result is the oxidizing agent **pyridinium chlorochromate,** abbreviated as $pyH^+ CrO_3Cl^-$ or just **PCC** (margin), which gives excellent yields of aldehydes on exposure to primary alcohols in dichloromethane solvent.

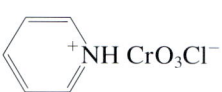

$^+NH \ CrO_3Cl^-$

Pyridinium chlorochromate
(PCC or pyH$^+$CrO$_3$Cl$^-$)

PCC Oxidation of a Primary Alcohol to an Aldehyde

$$CH_3(CH_2)_8CH_2OH \xrightarrow{pyH^+ CrO_3Cl^-, CH_2Cl_2} CH_3(CH_2)_8\overset{\overset{\textstyle O}{\|}}{C}H$$

92%

PCC oxidation conditions are often also used with secondary alcohols, because the relatively nonacidic reaction conditions minimize side reactions (e.g., carbocation formation; Sections 7-2, 7-3, and 9-3) and often give better yields than does the aqueous chromate method. Tertiary alcohols are unreactive toward oxidation by Cr(VI), because they do not carry hydrogens next to the OH function and therefore cannot readily form a carbon–oxygen double bond.

Chromic esters are intermediates in alcohol oxidation

What is the mechanism of the chromium(VI) oxidation of alcohols? The first step is formation of an intermediate called a **chromic ester;** the oxidation state of chromium stays unchanged in this process.

Chromic Ester Formation from an Alcohol

$$RCH_2\ddot{O}H + H\ddot{O}-\overset{\overset{:O:}{\|}}{\underset{\underset{:O:}{\|}}{Cr}}^{VI}-\ddot{O}H \;\rightleftharpoons\; RCH_2\ddot{O}-\overset{\overset{:O:}{\|}}{\underset{\underset{:O:}{\|}}{Cr}}^{VI}-\ddot{O}H + H_2\ddot{O}$$

Chromic acid   Chromic ester

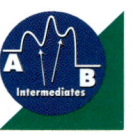

The next step in alcohol oxidation is equivalent to an E2 reaction. Here water (or pyridine, in the case of PCC) acts as a mild base, removing the proton next to the alcohol oxygen; $HCrO_3^-$ functions as a leaving group. The donation of an electron pair to chromium changes its oxidation state by two units, yielding Cr(IV).

Aldehyde Formation from a Chromic Ester

$$R-\overset{\overset{\displaystyle H}{|}}{\underset{\underset{\displaystyle H}{}}{C}}-\ddot{O}\!:\quad \ddot{O}\!: \xrightarrow{} \overset{\displaystyle H}{\underset{\displaystyle R}{C}}\!\!=\!\ddot{O}\!: + H_3\overset{+}{O}\!: + \;^-O_3\overset{IV}{Cr}H$$

In contrast with the kinds of E2 reactions considered so far, this elimination furnishes a carbon–oxygen instead of a carbon–carbon double bond. The Cr(IV) species formed undergoes a redox reaction with itself to Cr(III) and Cr(V), and the latter may function as an oxidizing agent independently. Eventually all Cr(VI) is reduced to Cr(III).

EXERCISE 8-11

Formulate a synthesis of each of the following carbonyl compounds from the corresponding alcohol.

(a) $CH_3CH_2\overset{\overset{\displaystyle O}{\|}}{C}CH(CH_3)_2$ (b) [structure with –CHO] (c) CH_3CH_2 [cyclohexanone with CH_3, O]

To summarize, reductions of aldehydes and ketones constitute general syntheses of primary and secondary alcohols, respectively. Either catalytic hydrogenation or hydride reagents may be employed in these processes. The reverse reactions, oxidations of primary alcohols to aldehydes and secondary alcohols to ketones, are achieved with chromium(VI) reagents. Use of pyridinium chlorochromate (PCC) prevents overoxidation of primary alcohols to carboxylic acids.

8-7 Organometallic Reagents: Sources of Nucleophilic Carbon for Alcohol Synthesis

The reduction of aldehydes and ketones with hydride reagents is a useful way of synthesizing alcohols. This approach would be even more powerful if, instead of hydride,

CHEMICAL HIGHLIGHT 8-2 | **The Breath Analyzer Test**

The color change from Cr(VI) (orange) in the presence of alcohols to Cr(III) (green) is used in a preliminary determination of the ethanol level in the breath (and therefore blood) of suspected alcohol-intoxicated persons, especially drivers. (For the physiological effects of ethanol, see Section 9-11.) If this test is positive, it is taken as a justification by law enforcement officers to administer a more accurate blood or urine screening. It works because of the diffusion of blood alcohol through the lung into the breath, with a measured distribution ratio of roughly 2100:1 (i.e., 2100 ml of breath contains as much ethanol as 1 ml of blood). In the simplest version of this test, the participant is asked to blow into a tube containing $K_2Cr_2O_7$ and H_2SO_4 supported on powdered silica gel (SiO_2) for a duration of 10–20 seconds (as indicated by the inflation of a plastic bag attached at the end of the tube). Any alcohol present in the breath is oxidized to acetic (ethanoic) acid, a reaction signaled by the progressive color change from orange to green along the tube, according to the following balanced equation.

$$2 \underset{\text{Orange}}{K_2Cr_2O_7} + 8 H_2SO_4 + 3 CH_3CH_2OH \longrightarrow$$

$$2 \underset{\text{Green}}{Cr_2(SO_4)_3} + 2 K_2SO_4 + 3 CH_3\overset{\displaystyle O}{\overset{\|}{C}}OH + 11 H_2O$$

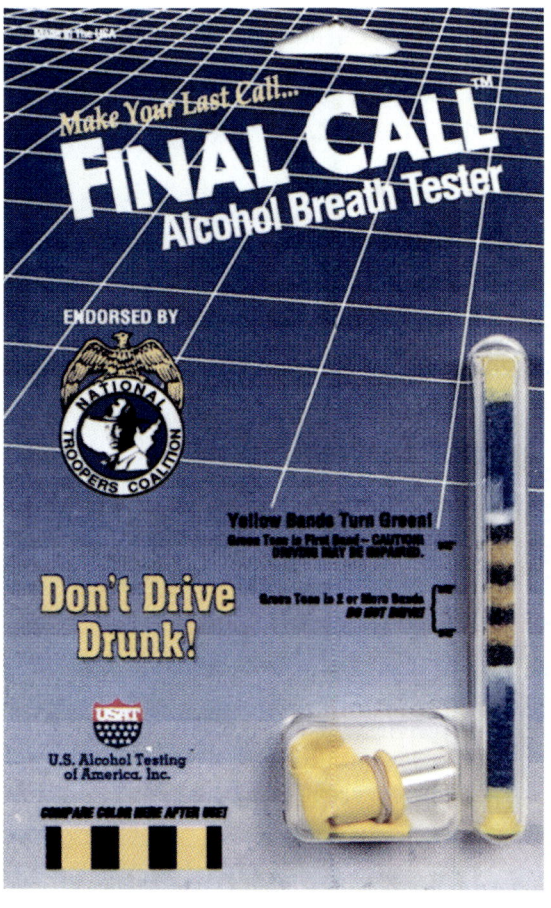

Kit for testing alcohol in breath.

If green develops beyond the halfway mark, a blood alcohol concentration greater than 0.08% is indicated, which is considered a criminal offense in many countries. A more sophisticated version of this simple procedure uses spectrophotometric techniques to quantify the extent of oxidation, the so-called *breath analyzer test,* and more recent developments include the use of mini-gas chromatographs, electrochemical analyzers, and infrared spectrometers (Section 11-5). Some people claim that a breath analyzer can be tricked into producing a "false negative" by smoking, chewing coffee beans, eating garlic, or ingesting chlorophyll preparations beforehand: *These claims are false.*

we could use a source of *nucleophilic carbon.* Attack by a carbon nucleophile on a carbonyl group would give an alcohol and simultaneously form a carbon–carbon bond. *We would thus have constructed a product with more carbon atoms and, therefore, more complexity than that present in the starting materials.*

To achieve such transformations, we need to find a way of making carbon-based nucleophiles, R^-. This section will describe how this goal can be reached. Metals, particularly lithium and magnesium, act on haloalkanes to generate new compounds,

called **organometallic reagents,** in which a carbon atom of an organic group is bound to a metal. These species are strong bases and good nucleophiles, and they are extremely useful in organic syntheses.

Alkyllithium and alkylmagnesium reagents are prepared from haloalkanes

Organometallic compounds of lithium and magnesium are most conveniently prepared by direct reaction of a haloalkane with the metal suspended in ethoxyethane (diethyl ether) or oxacyclopentane (tetrahydrofuran, THF). The reactivity of the haloalkanes increases in the order $Cl < Br < I$; fluorides are not normally used as starting materials in these reactions.

Alkyllithium Synthesis

$$CH_3Br + 2 Li \xrightarrow{(CH_3CH_2)_2O, 0°-10°C} CH_3Li + LiBr$$

Methyl-
lithium

Alkylmagnesium (Grignard) Synthesis

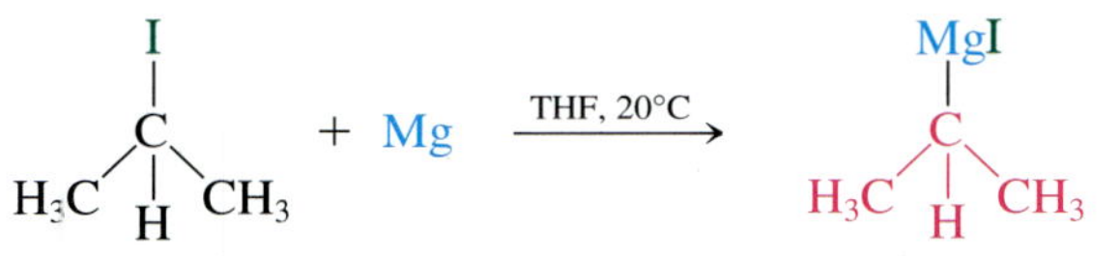

1-Methylethyl-
magnesium **iodide**

Organomagnesium compounds, RMgX, are also called **Grignard reagents,** named after their discoverer, F. A. Victor Grignard.* They form in reactions starting with primary, secondary, and tertiary haloalkanes (and, as we shall see in later chapters, with haloalkenes and halobenzenes).

Alkyllithium compounds and Grignard reagents are rarely isolated; they are formed in solution and used immediately in the desired reaction. Sensitive to air and moisture, they must be prepared and handled under rigorously air-free and dry conditions.

The formulas RLi and RMgX oversimplify the true structures of these reagents. Thus, as written, the metal ions are highly electron deficient. To make up the desired electron octet, the metals function as Lewis acids (Section 2-9) and attach themselves to the Lewis basic solvent molecules. For example, alkylmagnesium halides are stabilized by bonding to two ether molecules. The solvent is said to be **coordinated** to the metal. When the structures of Grignard reagents are written, this coordination is rarely shown. It is crucial, however, because their formation is very difficult in its absence.

Grignard Reagents Are Coordinated to Solvent

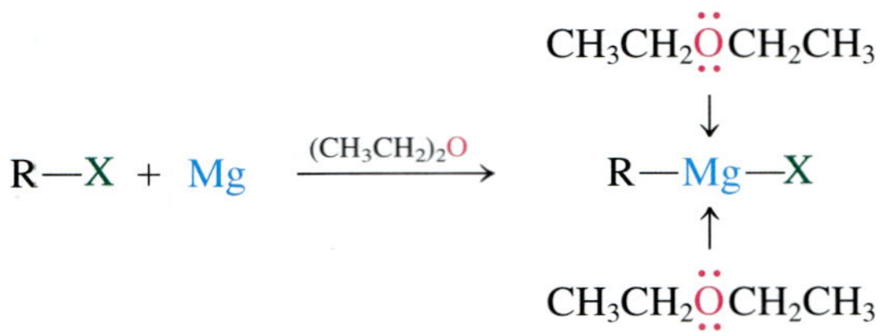

$$CH_3CH_2\overset{..}{\underset{..}{O}}CH_2CH_3$$
$$\downarrow$$
$$R{-}X + Mg \xrightarrow{(CH_3CH_2)_2O} R{-}Mg{-}X$$
$$\uparrow$$
$$CH_3CH_2\overset{..}{\underset{..}{O}}CH_2CH_3$$

*Professor François Auguste Victor Grignard (1871–1935), University of Lyon, France, Nobel Prize 1912 (chemistry).

The alkylmetal bond is strongly polar

Alkyllithium and alkylmagnesium reagents have strongly polarized carbon–metal bonds; the strongly electropositive metal (Table 1-2) is the positive end of the dipole. The degree of polarization is sometimes referred to as "percentage of ionic bond character." The carbon–lithium bond, for example, has about 40% ionic character and the carbon–magnesium bond 35%. Such systems react chemically as if they contained a negatively charged carbon. To symbolize this behavior, we can show the carbon–metal bond with a resonance form that places the full negative charge on the carbon atom: a **carbanion.**

Carbon–Metal Bond
in Alkyllithium and Alkylmagnesium Compounds

$$\left[\quad \overset{|}{\underset{|}{C}}\overset{\delta^-}{-}\overset{\delta^+}{M} \quad \longleftrightarrow \quad \overset{|}{\underset{|}{C}}{:}^- \; M^+ \quad \right]$$

Polarized   Charge separated

M = metal

    The preparation of alkylmetals from haloalkanes illustrates an important principle in synthetic organic chemistry: **reverse polarization.** In a haloalkane, the presence of the electronegative halogen turns the carbon into an electrophilic center. On treatment with a metal, the $C^{\delta+}$–$X^{\delta-}$ unit is converted into $C^{\delta-}$–$M^{\delta+}$. In other words, the direction of polarization is reversed. Metallation has turned an electrophilic carbon into a nucleophilic center.

The alkyl group in alkylmetals is strongly basic

Carbanions are very strong bases. In fact, alkylmetals are much more basic than are amides or alkoxides, because carbon is considerably less electronegative than either nitrogen or oxygen (Table 1-2) and much less capable of supporting a negative charge. Recall (Table 2-6, Section 2-9) that alkanes are *extremely* weak acids: The pK_a of methane is estimated to be 50. It is not surprising, therefore, that carbanions are such strong bases: They are, after all, the *conjugate bases of alkanes.* Their basicity makes organometallic reagents moisture sensitive and incompatible with OH or similarly acidic functional groups. In the presence of water, hydrolysis—often violent—furnishes the metal hydroxide and an alkane. The outcome of this transformation is predictable on purely electrostatic grounds.

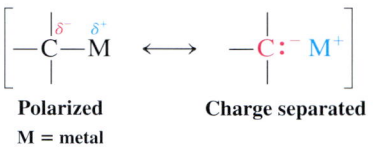

$$\overset{\delta-}{R}\overset{\delta+}{-M} + \overset{\delta+}{H}\overset{\delta-}{-OH} \longrightarrow R-H + M-OH$$

Organo-   Alkane   Metal
metal   hydroxide

Hydrolysis of an Organometallic Reagent

$$\underset{\substack{\text{3-Methylpentylmagnesium} \\ \text{bromide}}}{CH_3CH_2\overset{\overset{\displaystyle CH_3}{|}}{C}HCH_2CH_2MgBr} + HOH \longrightarrow \underset{\substack{\text{3-Methylpentane}}}{\underset{100\%}{CH_3CH_2\overset{\overset{\displaystyle CH_3}{|}}{C}HCH_2CH_2H}} + BrMgOH$$

The metallation–hydrolysis sequence affords a means by which a haloalkane can be converted into an alkane. A more direct way of achieving the same goal is the reaction of a haloalkane with the powerful hydride donor lithium aluminum hydride, an S_N2 displacement of halide by H^-. The less reactive $NaBH_4$ is incapable of performing this substitution.

$$CH_3(CH_2)_7CH_2-Br \xrightarrow[-LiBr]{LiAlH_4, (CH_3CH_2)_2O} CH_3(CH_2)_7CH_2-H$$

1-Bromononane   **Nonane**

A useful application of metallation–hydrolysis is the introduction of hydrogen isotopes, such as deuterium, into a molecule by exposure of the organometallic compound to labeled water.

**Introduction of Deuterium by Reaction
of an Organometallic Reagent
with D_2O**

$$(CH_3)_3CCl \xrightarrow[2. D_2O]{1. Mg} (CH_3)_3CD$$

EXERCISE 8-12

Show how you would prepare monodeuteriocyclohexane from cyclohexane.

In summary, haloalkanes can be converted into organometallic compounds of lithium or magnesium (Grignard reagents) by reaction with the respective metals in ether solvents. In these compounds, the alkyl group is negatively polarized, a charge distribution opposite that found in the haloalkane. Although the alkyl–metal bond is to a large extent covalent, the carbon attached to the metal behaves as a strongly basic carbanion, exemplified by its ready protonation.

8-8 Organometallic Reagents in the Synthesis of Alcohols

Among the most useful applications of organometallic reagents of magnesium and lithium are those in which the alkyl group reacts as a nucleophile. Like the hydrides, these reagents can attack the carbonyl group of an aldehyde or ketone to produce an alcohol. The difference is that a new carbon–carbon bond is formed in the process.

**Alcohol Syntheses from
Aldehydes, Ketones, and Organometallics**

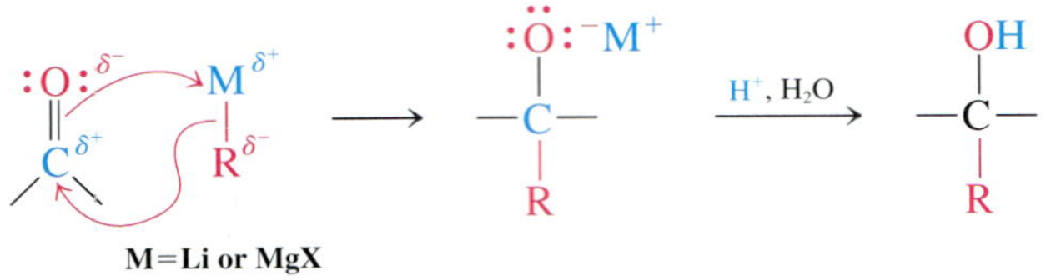

M = Li or MgX

Following the flow of electrons can help us understand the reaction. In the first step, the nucleophilic alkyl group in the organometallic compound attacks the car-

bonyl carbon. As an electron pair from the alkyl group shifts to generate the new carbon–carbon linkage, it "pushes" two electrons from the double bond onto the oxygen, thus producing a metal alkoxide. The addition of a dilute aqueous acid furnishes the alcohol by hydrolyzing the metal–oxygen bond, another example of aqueous work-up.

The reaction of organometallic compounds with *formaldehyde* results in *primary alcohols.*

Formation of a Primary Alcohol from a Grignard Reagent and Formaldehyde

NOTE: Sequential arrows denote the execution of two (or more) steps in sequence. In the present examples, they indicate that first the Grignard addition reaction is carried out in the ether solvent, followed by acidic aqueous work-up.

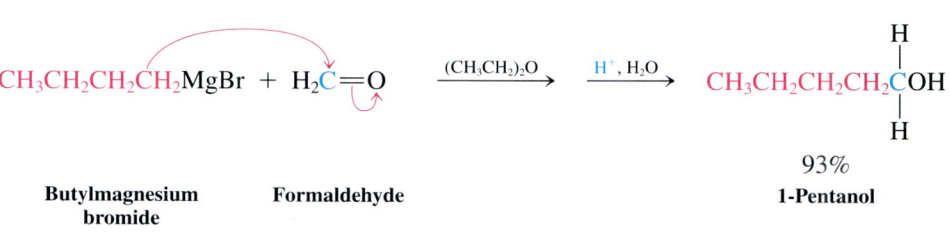

However, *aldehydes* other than formaldehyde convert into *secondary alcohols.*

Formation of a Secondary Alcohol from a Grignard Reagent and an Aldehyde

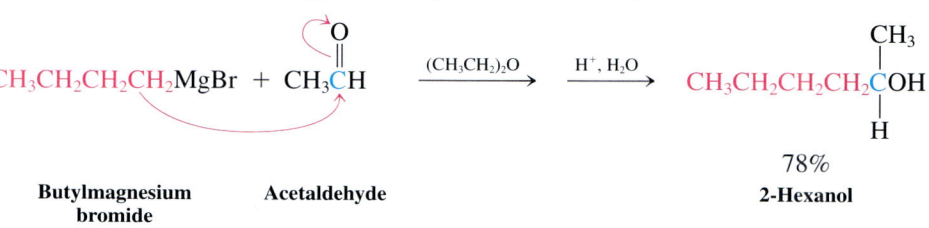

Ketones furnish *tertiary alcohols.*

Formation of a Tertiary Alcohol from a Grignard Reagent and a Ketone

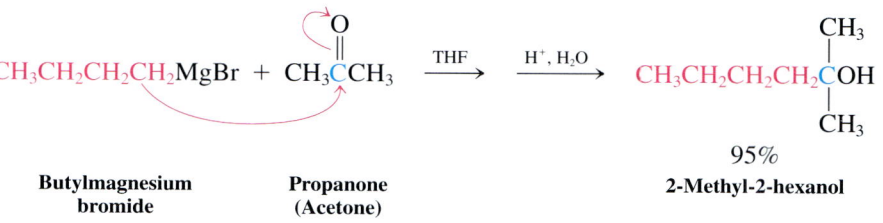

EXERCISE 8-13

Write a synthetic scheme for the conversion of 2-bromopropane, $(CH_3)_2CHBr$, into 2-methyl-1-propanol, $(CH_3)_2CHCH_2OH$.

EXERCISE 8-14

Propose efficient syntheses of the following products from starting materials containing no more than four carbons.

(a) $CH \equiv CH_2)_4OH$

(b)
$$CH_3CH_2CH_2\overset{\overset{\displaystyle OH}{|}}{C}HCH_2CH_2CH_3$$

(c)
$$\overset{\displaystyle C(CH_3)_3}{\underset{}{\boxed{}}} \!\!-\!OH$$

(d)
$$CH_3CH_2CH_2\overset{\overset{\displaystyle OH}{|}}{\underset{\underset{\displaystyle CH_3}{|}}{C}}CH_2CH_3$$

Although the nucleophilic addition of alkyllithium and Grignard reagents to the carbonyl group provides us with a powerful C–C bond–forming transformation, such nucleophilic attack is too slow on haloalkanes and related electrophiles, such as those encountered in Chapter 6. This kinetic problem is what enables us to make the organometallic reagents described in Section 8-7: The product alkylmetal does not attack the haloalkane from which it is made. To achieve such coupling reactions, we must resort to copper-based reagents (Section 13-10).

In summary, alkyllithium and alkylmagnesium reagents add to aldehydes and ketones to give alcohols in which the alkyl group of the organometallic reagent has formed a bond to the original carbonyl carbon.

8-9 Complex Alcohols: An Introduction to Synthetic Strategy

The reactions introduced so far are part of the "vocabulary" of organic chemistry; unless we know the vocabulary, we cannot speak the language of organic chemistry. These reactions allow us to manipulate molecules and interconvert functional groups, so it is important to become familiar with these transformations—their types, the reagents used, the conditions under which they occur (especially when the conditions are crucial to the success of the process), and the limitations of each type.

This task may seem monumental, one that will require much memorization. But *it is made easier by an understanding of the reaction mechanisms.* We already know that reactivity can be predicted from a small number of factors, such as electronegativity, Coulombic forces, and bond strengths. Let us see how organic chemists apply this understanding to devise useful synthetic strategies.

Let us begin with a few examples in which we predict reactivity on mechanistic grounds. Then we shall turn to synthesis—the making of molecules. How do chemists develop new synthetic methods, and how can we make a "target" molecule as efficiently as possible? The two topics are closely related. The second, known as **total synthesis,** usually requires a series of reactions. In studying these tasks, therefore, we will also be reviewing much of the reaction chemistry that we have considered so far.

Mechanisms help in predicting the outcome of a reaction

First, recall how we predict the outcome of a reaction. What are the factors that let a particular mechanism go forward? Here are three examples.

How to Predict the Outcome of a Reaction on Mechanistic Grounds

$$ICH_2CH_2CH_2Br \overset{I^-}{\longleftrightarrow\!\!\!\!\!\times} FCH_2CH_2CH_2Br \overset{I^-}{\longrightarrow} FCH_2CH_2CH_2I$$

Not formed

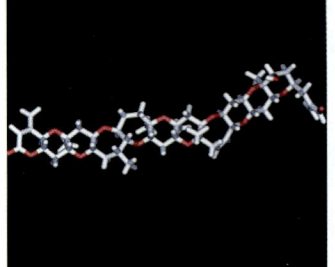

Brevetoxin, the molecule on the front cover, was synthesized in 1994 after 12 years of effort in 83 steps and an overall yield of 0.043% (average yield for each step: 91%) from the simple sugar 2-deoxyribose (see Chapter 24).

Explanation. Bromide is a better leaving group than fluoride.

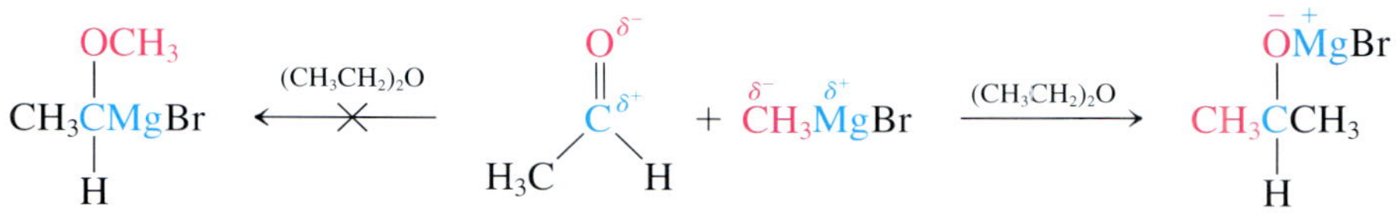

Explanation. The positively polarized carbonyl carbon forms a bond to the negatively polarized alkyl group of the organometallic reagent.

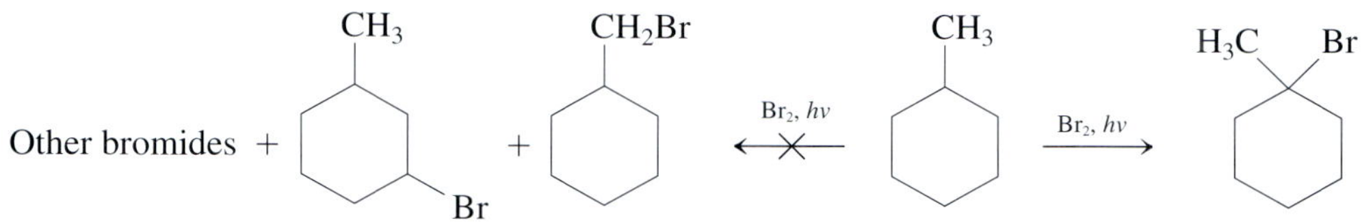

Explanation. The tertiary C–H bond is weaker than a primary or secondary C–H bond, and Br_2 is quite selective in radical halogenations.

EXERCISE 8-15

Predict and explain the outcome of each of the following reactions on mechanistic grounds.

(a) $\overset{\displaystyle Br}{ClCH_2CH_2CH_2\overset{|}{C}(CH_3)_2}$ + CH_3CH_2OH $\longrightarrow$

(b) $\overset{\displaystyle CH_2Cl}{ClCH_2CH_2CH_2\overset{|}{C}(CH_3)_2}$ + $(CH_3)_3CO^-$ ^+K $\xrightarrow{(CH_3)_3COH}$

(c) $\overset{\displaystyle OH}{HOCH_2CH_2CH_2\overset{|}{C}(CH_3)_2}$ $\xrightarrow{PCC, CH_2Cl_2}$

New reactions lead to new synthetic methods

New reactions are found by design or by accident. For example, consider how two different students might discover the reactivity of a Grignard reagent with a ketone to give an alcohol. The first student, knowing about electronegativity and the electronic makeup of ketones, would predict that the nucleophilic alkyl group of the Grignard species should attach itself to the electrophilic carbonyl carbon. This student would be pleased by the successful outcome of the experiment, verifying chemical principles in practice. The second student, with less knowledge, might attempt to dilute a particularly concentrated solution of a Grignard reagent with what one might conceive to be a perfectly good polar solvent: propanone (acetone). A violent reaction would immediately reveal that this notion is incorrect, and further investigation would uncover the powerful potential of the reagent in alcohol synthesis.

When a reaction has been discovered, it is important to show its scope and its limitations. For this purpose, many different substrates are tested, side products (if any)

noted, new functional groups subjected to the reaction conditions, and mechanistic studies carried out. Should these investigations prove the new reaction to be generally applicable, it will be added as a new synthetic method to the organic chemist's arsenal.

Because a reaction leads to a very specific change in a molecule, it is frequently useful to emphasize this "molecular alteration." A simple example is the addition of a Grignard reagent to formaldehyde. What is the structural change in this transformation? A one-carbon unit is added to an alkyl group. The method is valuable because it allows a straightforward one-carbon extension, also called a *homologation*.

Even though our synthetic vocabulary at this stage is relatively limited, we already have quite a number of molecular alterations at our disposal. For example, bromoalkanes are excellent starting points for numerous transformations.

R—M
Alkyl group
+
$H_2C=O$
One-carbon unit
↓
R—CH_2—OH

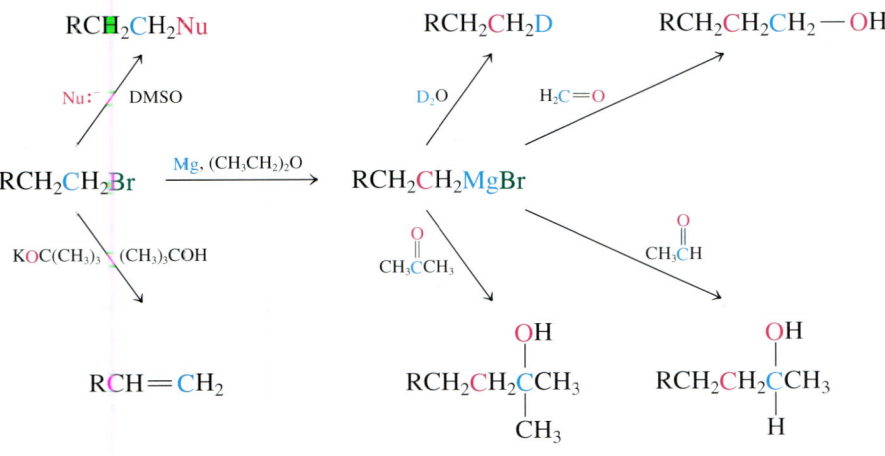

Each one of the products in the scheme can enter into further transformations of its own, thereby leading to more complicated products.

When we ask, "What good is a reaction? What sort of structures can we make by applying it?" we address a problem of *synthetic methodology*. Let us ask a different question. Suppose that we want to prepare a specific target molecule. How would we go about devising an efficient route to it? How do we find suitable starting materials? The problem with which we are dealing now is *total synthesis*.

Organic chemists want to make complex molecules for specific purposes. For example, certain compounds might have valuable medicinal properties but are not readily available from natural sources. Biochemists need a particular isotopically labeled molecule to trace metabolic pathways. Physical organic chemists frequently design novel structures to study. There are many reasons for the total synthesis of organic molecules.

Whatever the final target, a successful synthesis is characterized by brevity and high overall yield. The starting materials should be readily available, preferably commercially, and inexpensive. Moreover, safety and environmental concerns demand that, ideally, the reagents used be relatively nontoxic and easy to handle.

Retrosynthetic analysis simplifies synthesis problems

Many compounds that are commercially available and inexpensive are also small, containing six or fewer carbon atoms. Therefore, the most frequent task facing the

synthetic planner is that of building up a larger, complicated molecule from smaller, simple fragments. The best approach to the preparation of the target is to work its synthesis *backward* on paper, an approach called **retrosynthetic analysis*** (*retro,* Latin, backward). In this analysis, strategic carbon–carbon bonds are "broken" at points at which their formation seems possible. This way of thinking backward may seem strange to you at first, because you are accustomed to learning reactions in a forward way—for example, "A plus B *gives* C." Retrosynthesis requires that you think of this process in the reverse manner—for example, "C is *derived* from A plus B."

Why retrosynthesis? The answer is that, in any "building" of a complex framework from simple building blocks, the number of possibilities of adding pieces increases drastically when going forward and includes myriad "dead end" options. In contrast, in working backward, complexity decreases and unworkable solutions are minimized. A simple analogy is a jigsaw puzzle: It is clearly easier to dismantle step by step than it is to assemble. For example, a retrosynthetic analysis of the synthesis of 3-hexanol from two three-carbon units would suggest its formation from a propyl organometallic compound and propanal.

Retrosynthetic Analysis of 3-Hexanol Synthesis from Two Three-Carbon Fragments

$$\underset{\text{CH}_3\text{CH}_2\text{CH}_2\overset{\overset{\displaystyle \text{OH}}{|}}{\text{CH}}\text{CH}_2\text{CH}_3}{} \Longrightarrow \underset{\textbf{Propylmagnesium bromide}}{\text{CH}_3\text{CH}_2\text{CH}_2\text{MgBr}} + \underset{\textbf{Propanal}}{\overset{\overset{\displaystyle \text{O}}{\|}}{\text{H}}\text{C}\text{CH}_2\text{CH}}$$

The double-shafted arrow indicates the so-called **strategic disconnection.** We recognize that the bond "broken" in this analysis, that between C3 and C4 in the product, is one that we can construct by using a transformation that we know, $\text{CH}_3\text{CH}_2\text{CH}_2\text{MgBr} + \text{CH}_3\text{CH}_2\text{CHO}$. In this case, only one reaction is necessary to achieve the connection; in others, it might require several steps. Two alternate, but inferior, retrosyntheses of 3-hexanol are

$$\underset{\text{CH}_3\text{CH}_2\text{CH}_2\overset{\overset{\displaystyle \text{OH}}{|}}{\text{CH}}\text{CH}_2\text{CH}_3}{} \Longrightarrow \text{NaBH}_4 + \underset{}{\text{CH}_3\text{CH}_2\text{CH}_2\overset{\overset{\displaystyle \text{O}}{\|}}{\text{C}}\text{CH}_2\text{CH}_3}$$

$$\underset{\text{CH}_3\text{CH}_2\text{CH}_2\overset{\overset{\displaystyle \text{OH}}{|}}{\text{CH}}\text{CH}_2\text{CH}_3}{} \Longrightarrow \text{Na}\overset{\overset{\displaystyle \text{O}}{\|}}{\text{O}}\text{C}\text{CH}_3 + \underset{}{\text{CH}_3\text{CH}_2\text{CH}_2\overset{\overset{\displaystyle \text{Br}}{|}}{\text{CH}}\text{CH}_2\text{CH}_3}$$

They are not as good as the first because they do not significantly *simplify* the target structure: No carbon–carbon bonds are "broken."

Retrosynthetic analysis aids in alcohol construction

Let us apply retrosynthetic analysis to the preparation of a tertiary alcohol, 4-ethyl-4-nonanol. Because of their steric encumbrance and hydrophobic nature, this alcohol and its homologs have important industrial applications as cosolvents and additives

---

*Pioneered by Professor Elias J. Corey (b. 1928), Harvard University, Nobel Prize 1990 (chemistry).

in certain polymerization processes (Section 12-13). There are two steps to follow at each stage of the process. First, we identify all possible strategic disconnections, "breaking" all bonds that can be formed by reactions that we know. Second, we evaluate the relative merits of these disconnections, seeking the one that best simplifies the target structure. The strategic bonds in 4-ethyl-4-nonanol are those around the functional group. There are three disconnections leading to simpler precursors. Path *a* cleaves the ethyl group from C4, suggesting as the starting materials for its construction ethylmagnesium bromide and 4-nonanone. Cleavage *b* is an alternative possibility leading to a propyl Grignard reagent and 3-octanone as precursors. Finally, disconnection *c* reveals a third synthesis route derived from the addition of pentylmagnesium bromide to 3-hexanone.

Partial Retrosynthetic Analysis of the Synthesis of 4-Ethyl-4-nonanol

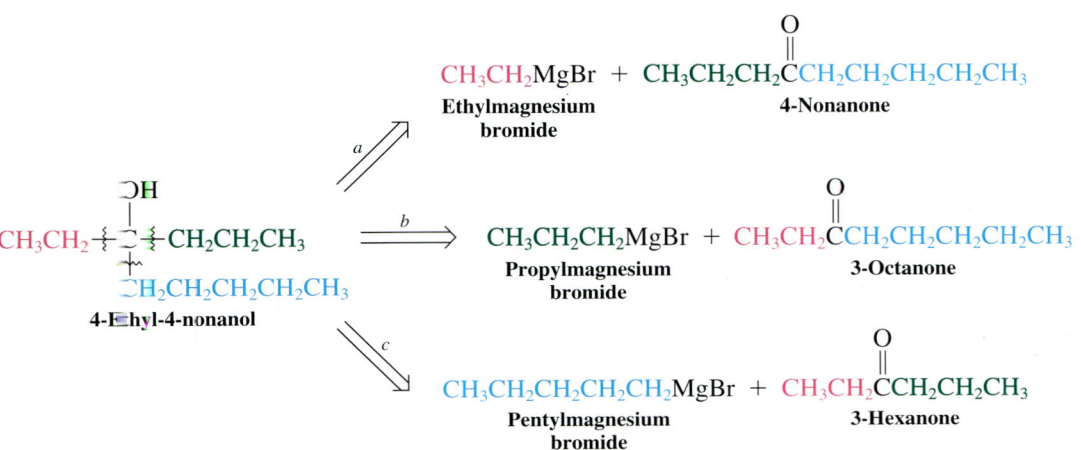

Evaluation reveals that pathway *c* is best: The necessary building blocks are almost equal in size, containing five and six carbons; thus, this disconnection provides the greatest simplification in structure.

EXERCISE 3-16

Apply retrosynthetic analysis to 4-ethyl-4-nonanol, disconnecting the carbon–*oxygen* bond. Does this lead to an efficient synthesis? Explain.

Can we pursue either of the fragments arising from disconnection by pathway *c* to even simpler starting materials? Yes; recall (Section 8-6) that ketones are obtained from the oxidation of secondary alcohols by Cr(VI) reagents. We may therefore envision preparation of 3-hexanone from the corresponding alcohol, 3-hexanol.

$$CH_3CH_2CH_2\overset{\displaystyle O}{\overset{\displaystyle \|}{C}}CH_2CH_3 \Longrightarrow Na_2Cr_2O_7 + CH_3CH_2CH_2\overset{\displaystyle OH}{\underset{}{\overset{|}{C}H}}CH_2CH_3$$

3-Hexanone   3-Hexanol

Because we earlier identified an efficient disconnection of the latter into three-carbon fragments, we are now in a position to present our complete synthetic scheme (see page 312):

Synthesis of 4-Ethyl-4-nonanol

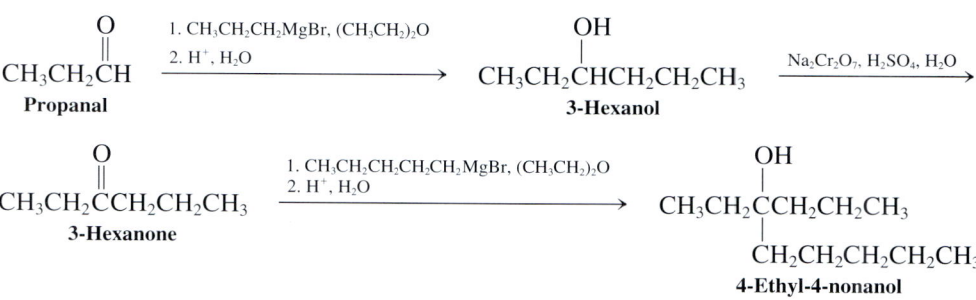

This example illustrates a very powerful general sequence for the construction of complex alcohols: first, Grignard or organolithium addition to an aldehyde to give a secondary alcohol; then oxidation to a ketone; and finally, addition of another organo-metallic reagent to give a tertiary alcohol.

Utility of Alcohol Oxidations in Synthesis

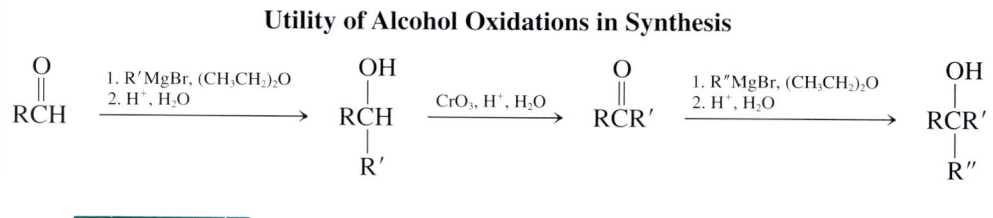

EXERCISE 8-17

Write an economical retrosynthetic analysis of 3-cyclobutyl-3-octanol. (**Hint:** Consult the Chapter Integration Problem.)

EXERCISE 8-18

Show how you would prepare 2-methyl-2-propanol from methane as the only organic starting material.

Watch out for pitfalls in planning syntheses

There are several considerations to keep in mind when practicing synthetic chemistry that will help to avoid designing unsuccessful or low-yielding approaches to a target molecule. First, *try to minimize the total number of transformations required to convert the initial starting material into the desired product.* This point is so important that, in some cases, it is worthwhile to accept a low-yield step if it allows a significant shortening of the synthetic sequence. For example (with the assumption that all starting materials are of comparable cost), a seven-step synthesis in which each step has an 85% yield is less productive than a four-step synthesis with three yields at 95% and one at 45%. The overall efficiency in the first sequence comes to $(0.85 \times 0.85 \times 0.85 \times 0.85 \times 0.85 \times 0.85 \times 0.85) \times 100 = 32\%$, whereas the second synthesis, in addition to being three steps shorter, gives $(0.95 \times 0.95 \times 0.95 \times 0.45) \times 100 = 39\%$. In these examples, all steps take place consecutively, a procedure called **linear synthesis.** In general, it is better to approach complex targets through two or more concurrent routes, as long as the overall number of steps is the same, a strategy described as **convergent synthesis.** Although a simple overall-yield calculation

is not possible for a convergent strategy, you can readily convince yourself of its increased efficiency by comparing the actual *amounts* of starting materials required by the two approaches to make the same amount of product. In the following example, 10 g of a product H is prepared in three steps (50% each) by a linear sequence A → B → C → H and a convergent one starting from D and F, respectively, through E and G. If we assume (for the sake of simplicity) that the molecular weights of these compounds are all the same, the first preparation requires 80 g and, the second only (a combined) 40 g of starting materials.

$$\underset{80 \text{ g}}{A} \xrightarrow{50\%} \underset{40 \text{ g}}{B} \xrightarrow{50\%} \underset{20 \text{ g}}{C} \xrightarrow{50\%} \underset{10}{H}$$

Linear synthesis of H

$$\underset{20 \text{ g}}{D} \xrightarrow{50\%} \underset{10 \text{ g}}{E}$$
$$\underset{20 \text{ g}}{F} \xrightarrow{50\%} \underset{10 \text{ g}}{G}$$
$$\xrightarrow{50\%} \underset{10 \text{ g}}{H}$$

Convergent synthesis of H

Second, *do not use reagents whose molecules have functional groups that would interfere with the desired reaction.* For example, treating a hydroxyaldehyde with a Grignard reagent leads to an acid–base reaction, destroying the Grignard, and not to carbon–carbon bond formation.

$$\underset{\underset{CH_3}{|}}{\overset{\overset{OH}{|}}{HOCH_2CH_2CH}} \xleftarrow{\times} \overset{O}{\overset{||}{HOCH_2CH_2CH}} + CH_3MgBr \longrightarrow \overset{O}{\overset{||}{BrMgOCH_2CH_2CH}} + \overset{H}{\underset{}{CH_3}}$$

A possible solution to this problem would be to add two equivalents of Grignard reagent: *one* to react with the acidic hydrogen as shown, the *other* to achieve the desired addition to the carbonyl group.

Do not try to make a Grignard reagent from a bromoketone. Such a reagent is not stable and will, as soon as it is formed, decompose by reacting with its own carbonyl group (in the same or another molecule).

Neopentyl-like Hindered Haloalkanes

$$CH_3CH_2\underset{\underset{CH_3}{|}}{\overset{\overset{CH_3}{|}}{C}}CH_2Br$$

Third, *take into account any mechanistic and structural constraints affecting the reactions under consideration.* For example, radical brominations are more selective than chlorinations. Keep in mind the structural limitations on nucleophilic reactions, and do not forget the lack of reactivity of the 2,2-dimethyl-1-halopropanes (neopentyl halides). Although sometimes difficult to recognize, many haloalkanes have "neopentyl-like" structures and are similarly unreactive. Nevertheless, such systems do form organometallic reagents and may be further functionalized in this manner.

For example, treatment of the Grignard reagent made from 1-bromo-2,2-dimethyl-propane with formaldehyde leads to the corresponding alcohol.

$$(CH_3)_3CCH_2Br \xrightarrow[\substack{3.\ H^+,\ H_2O}]{\substack{1.\ Mg \\ 2.\ CH_2=O}} (CH_3)_3CCH_2CH_2OH$$

1-Bromo-2,2-dimethylpropane **3,3-Dimethyl-1-butanol**

Tertiary halides, if incorporated into a more complex framework, also are sometimes difficult to recognize. Remember that tertiary halides do not undergo S_N2 reactions but eliminate in the presence of bases.

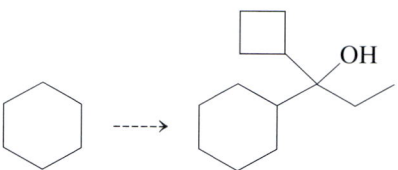

Expertise in synthesis, as in many other aspects of organic chemistry, develops largely from practice. Planning the synthesis of complex molecules requires a review of the reactions and mechanisms covered in earlier sections. The knowledge thus acquired can then be applied to the solution of synthetic problems.

CHAPTER INTEGRATION PROBLEM

Tertiary alcohols are important additives in some industrial processes utilizing Lewis acidic metal compounds (Sections 2-9 and 6-4) as catalysts. The alcohol provides the metal with a sterically protecting and hydrophobic environment (see Figure 8-3; see also Chemical Highlight 8-1), which ensures solubility in organic solvents, longer lifetimes, and selectivity in substrate activation. Preparation of these tertiary alcohols typically follows the synthetic principles outlined in Section 8-9.

Starting from cyclohexane and using any other building blocks containing four or fewer carbons, in addition to any necessary reagents, formulate a synthesis of tertiary alcohol A.

A

SOLUTION

Before we start a random trial and error approach to solving this problem, it is better to take an inventory of what is given. First, we are given cyclohexane, and we note that this unit shows up as a substituent in tertiary alcohol A. Second, a total of seven additional carbons appears in the product, so our synthesis will require some additional stitching together of smaller fragments because we cannot use com-

pounds containing more than four carbons. Third, target A is a tertiary alcohol, which should be amenable to the retrosynthetic analysis introduced in Section 8-9 (M = metal):

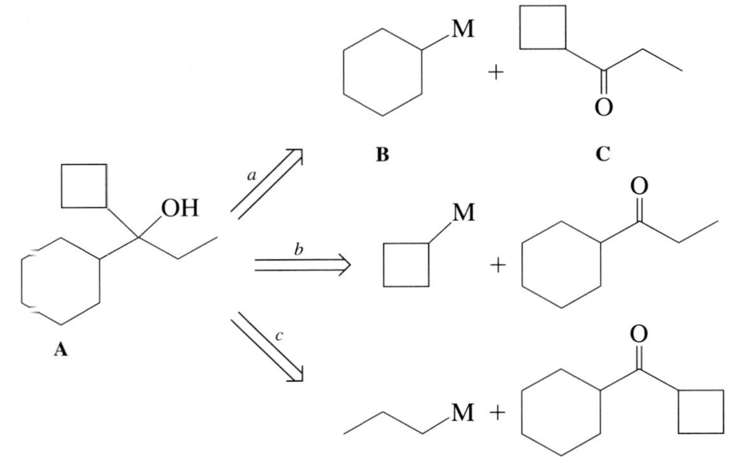

Approach *a* is clearly the one of choice because it breaks up tertiary alcohol A into evenly sized fragments B and C.

Having chosen route *a* as the most suitable way to find direct precursors to A, our analysis proceeds to working backward further: What are the appropriate precursors to B and C, respectively? The former is readily traced by retrosynthesis to our starting hydrocarbon, cyclohexane: The precursor to the organometallic compound B must be a halocyclohexane, which in turn can be made from cyclohexane by free-radical halogenation:

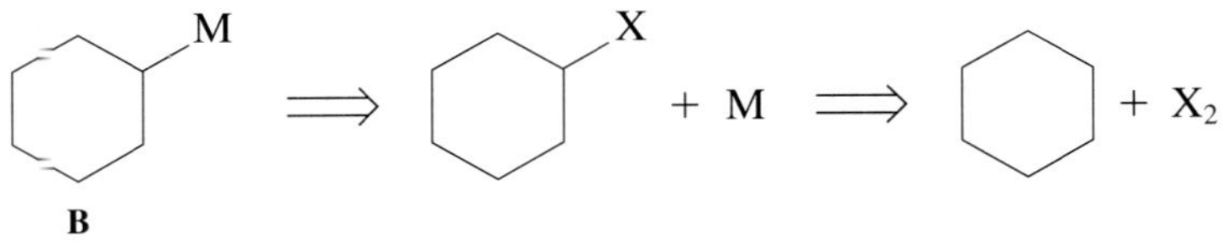

Ketone C must be broken into two smaller components; the best breakdown would be a "four + three" carbon combination: It is the most evenly sized solution, and it suggests the use of cyclobutyl intermediates. Because the only C–C disconnection that we know at this stage is that of alcohols, the first retrosynthetic step from C is its precursor alcohol (plus a chromium oxidant). Further retrosynthesis then provides the required starting pieces D and E.

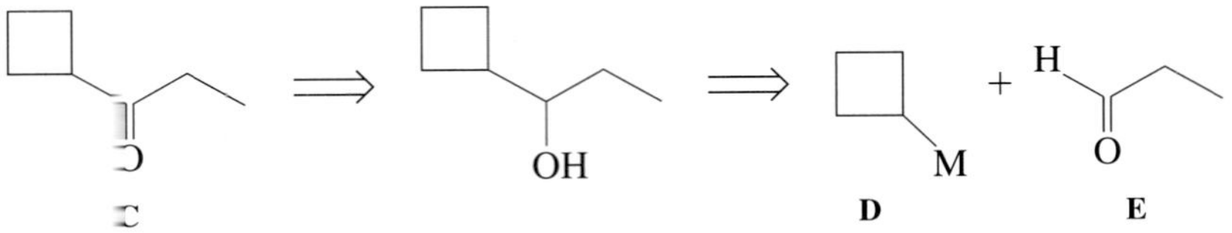

Now we can write the detailed synthetic scheme in a forward mode, with cyclohexane and pieces D and E as our starting materials:

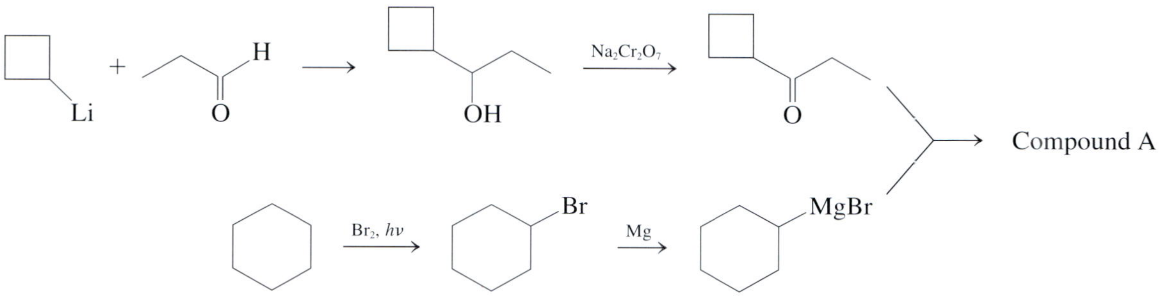

A final note: In this and subsequent synthetic exercises in this book, retrosynthetic analysis requires that you have command of all reactions not only in a forward fashion (i.e., starting material + reagent → product) but also in reverse (i.e., product ← starting material + reagent). The two sets address two different questions. The first asks: What are all the possible products that I can make from my starting material in the presence of all the reagents that I know? The second asks: What are all the conceivable starting materials that, with the appropriate reagents, will lead to my product? The two types of schematic summaries of reactions that you will see at the end of this and subsequent chapters emphasize this point.

NEW REACTIONS

1. Acid-Base Properties of Alcohols (Section 8-3)

Alkyloxonium ion **Alcohol** **Alkoxide**

Acidity: $RO-H \approx HO-H > H_2N-H > H_3C-H$
Basicity: $RO^- \approx HO^- < H_2N^- < H_3C^-$

Industrial Preparation of Alcohols

2. Synthesis Gas (Section 8-4)

$$Coal \xrightarrow{Air, H_2O, \Delta} x\,CO + y\,H_2$$

Synthesis gas

3. Methanol Synthesis from Synthesis Gas (Section 8-4)

$$CO + 2\,H_2 \xrightarrow{Cu\text{-}ZnO\text{-}Cr_2O_3,\ 250°C,\ 50–100\ atm} CH_3OH$$

4. Ethanol by Hydration of Ethene (Section 8-4)

$$CH_2{=}CH_2 + HOH \xrightarrow{H_3PO_4,\ 300°C} CH_3CH_2OH$$

Laboratory Preparation of Alcohols

5. Nucleophilic Displacement of Halides and Other Leaving Groups by Hydroxide (Section 8-5)

$$RCH_2X + HO^- \xrightarrow[S_N2]{H_2O} RCH_2OH + X^-$$

X = halide, sulfonate

Primary, secondary (tertiary undergoes elimination)

$$\underset{\overset{|}{R}}{RCHX} + CH_3\overset{\overset{\displaystyle O}{\|}}{C}O^- \xrightarrow{S_N2} \underset{\overset{|}{R'}}{RCH}O\overset{\overset{\displaystyle O}{\|}}{C}CH_3 \xrightarrow[\text{Ester hydrolysis}]{HO^-} \underset{\overset{|}{R'}}{RCHOH}$$

$$\underset{\overset{|}{R''}}{\overset{\overset{|}{R}}{R'CX}} \xrightarrow[S_N1]{H_2O, \text{ propanone (acetone)}} \underset{\overset{|}{R''}}{\overset{\overset{|}{R}}{R'COH}}$$

Best method for tertiary

6. Catalytic Hydrogenation of Aldehydes and Ketones (Section 8-6)

$$\underset{\textbf{Aldehyde}}{\overset{\overset{\displaystyle O}{\|}}{RCH}} \xrightarrow{H_2, Pt} \underset{\substack{\textbf{Primary}\\\textbf{alcohol}}}{RCH_2OH} \qquad \underset{\textbf{Ketone}}{\overset{\overset{\displaystyle O}{\|}}{RCR'}} \xrightarrow{H_2, \text{ catalyst}} \underset{\substack{\textbf{Secondary}\\\textbf{alcohol}}}{\overset{\overset{\displaystyle OH}{|}}{\underset{\overset{|}{H}}{RCR'}}}$$

7. Reduction of Aldehydes and Ketones by Hydrides (Section 8-6)

$$\overset{\overset{\displaystyle O}{\|}}{RCH} \xrightarrow{NaBH_4, CH_3CH_2OH} RCH_2OH \qquad \overset{\overset{\displaystyle O}{\|}}{RCR'} \xrightarrow{NaBH_4, CH_3CH_2OH} \overset{\overset{\displaystyle OH}{|}}{\underset{\overset{|}{H}}{RCR'}}$$

$$\underset{\textbf{Aldehyde}}{\overset{\overset{\displaystyle O}{\|}}{RCH}} \xrightarrow[\substack{2. H^+, H_2O}]{1. LiAlH_4, (CH_3CH_2)_2O} \underset{\substack{\textbf{Primary}\\\textbf{alcohol}}}{RCH_2OH} \qquad \underset{\textbf{Ketone}}{\overset{\overset{\displaystyle O}{\|}}{RCR'}} \xrightarrow[\substack{2. H^+, H_2O}]{1. LiAlH_4, (CH_3CH_2)_2O} \underset{\substack{\textbf{Secondary}\\\textbf{alcohol}}}{\overset{\overset{\displaystyle OH}{|}}{\underset{\overset{|}{H}}{RCR'}}}$$

Oxidation of Alcohols

8. Chromium Reagents (Section 8-6)

$$\underset{\textbf{Primary alcohol}}{RCH_2OH} \xrightarrow{PCC, CH_2Cl_2} \underset{\textbf{Aldehyde}}{\overset{\overset{\displaystyle O}{\|}}{RCH}} \qquad \underset{\textbf{Secondary alcohol}}{\overset{\overset{\displaystyle OH}{|}}{RCHR'}} \xrightarrow{Na_2Cr_2O_7, H_2SO_4} \underset{\textbf{Ketone}}{\overset{\overset{\displaystyle O}{\|}}{RCR'}}$$

Organometallic Reagents

9. Reaction of Metals with Haloalkanes (Section 8-7)

$$RX + Li \xrightarrow{(CH_3CH_2)_2O} RLi$$

Alkyllithium reagent

$$RX + Mg \xrightarrow{(CH_3CH_2)_2O} RMgX$$

Grignard reagent

R cannot contain acidic groups such as O–H or electrophilic groups such as C=O

10. Hydrolysis (Section 8-7)

$$RLi \quad or \quad RMgX + H_2O \longrightarrow RH$$
$$RLi \quad or \quad RMgX + D_2O \longrightarrow RD$$

11. Addition of Organometallic Compounds to Aldehydes and Ketones (Section 8-8)

$$RLi \quad or \quad RMgX + CH_2{=}O \longrightarrow RCH_2OH$$

Formaldehyde   **Primary alcohol**

$$RLi \quad or \quad RMgX + R'\overset{\displaystyle O}{\overset{\|}{C}}H \longrightarrow R\overset{\displaystyle OH}{\underset{\displaystyle H}{\overset{|}{\underset{|}{C}}}}R'$$

Aldehyde   **Secondary alcohol**

$$RLi \quad or \quad RMgX + R'\overset{\displaystyle O}{\overset{\|}{C}}R'' \longrightarrow R\overset{\displaystyle OH}{\underset{\displaystyle R''}{\overset{|}{\underset{|}{C}}}}R'$$

Ketone   **Tertiary alcohol**

Aldehyde or ketone cannot contain other groups that react with organometallic reagents such as O–H or other C=O groups

12. Alkanes from Haloalkanes and Lithium Aluminum Hydride (Section 8-7)

$$RX + LiAlH_4 \xrightarrow{(CH_3CH_2)_2O} RH$$

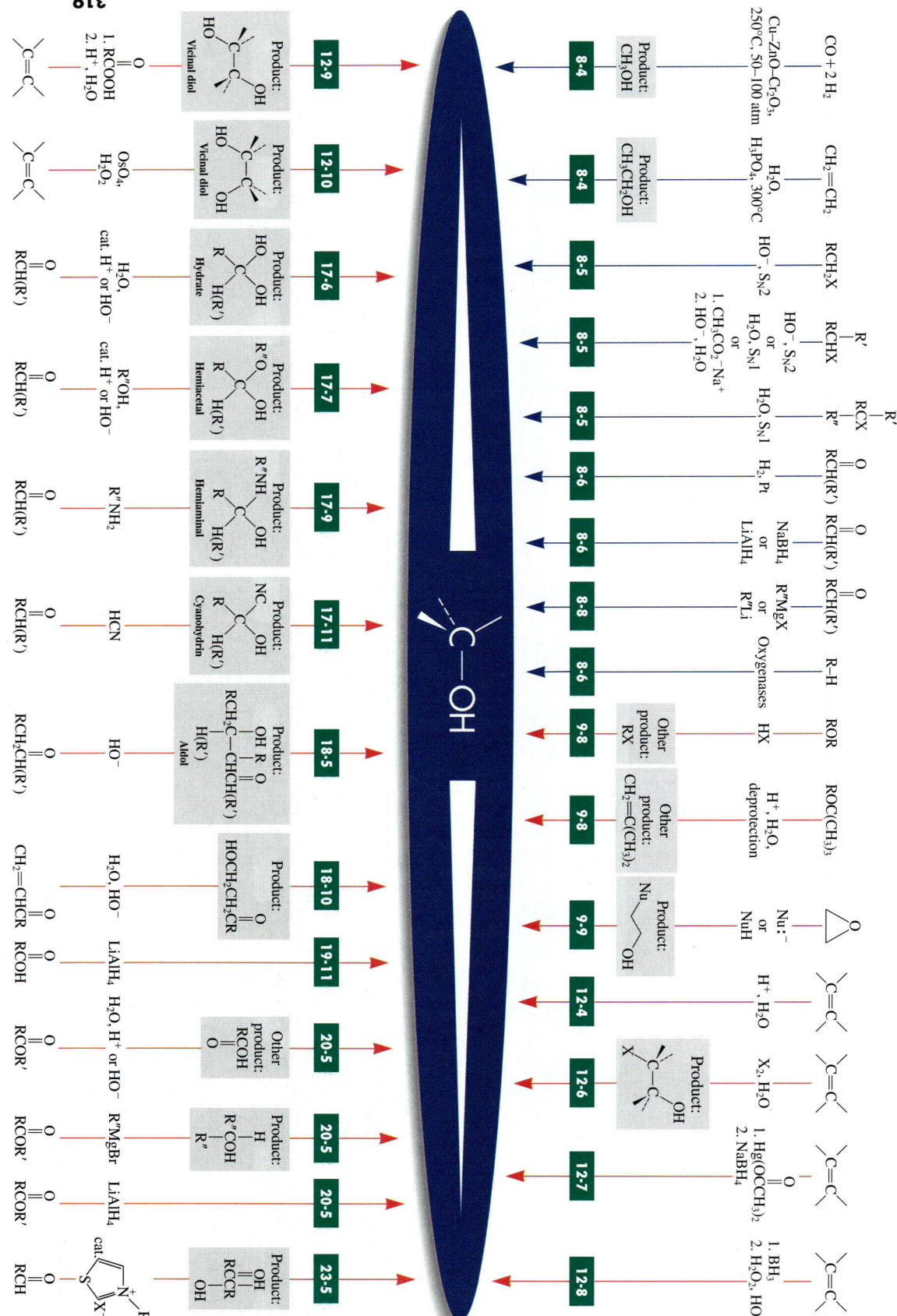

Preparation of Alcohols   section number

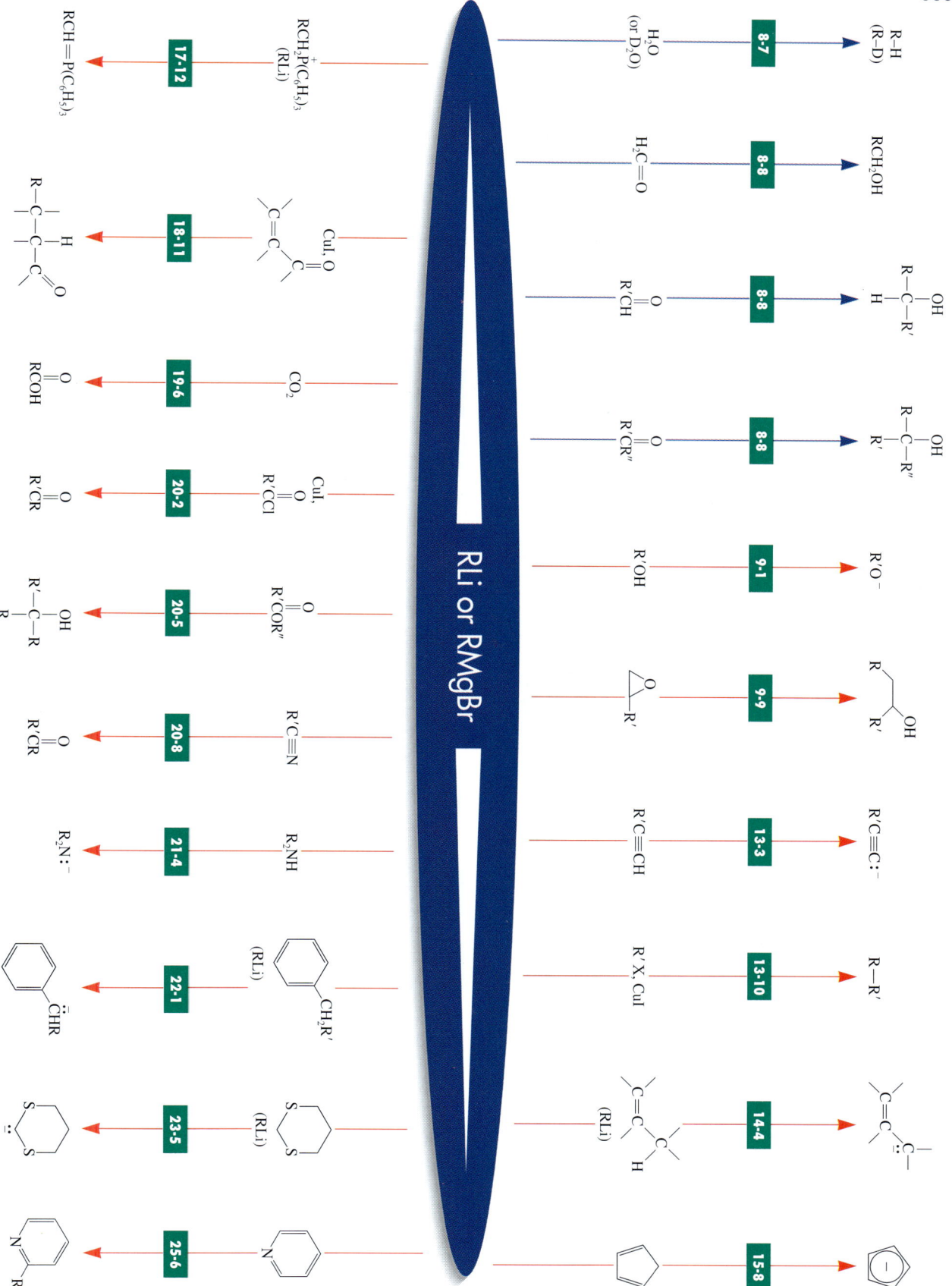

Reactions of Alkyllithium and Grignard Reagents

section number

IMPORTANT CONCEPTS

1. Alcohols are **alkanols** in IUPAC nomenclature. The stem containing the functional group gives the alcohol its name. Alkyl and halo substituents are added as prefixes.

2. Like water, alcohols have a **polarized** and short O–H bond. The proton is **hydrophilic** and enters into **hydrogen bonding.** Consequently, alcohols have unusually high boiling points and, in many cases, appreciable water solubility. The alkyl part of the molecule is **hydrophobic.**

3. Again like water, alcohols are **amphoteric:** They are both acidic and basic. Complete deprotonation to an **alkoxide** takes place with bases whose conjugate acids are considerably weaker than the alcohol. Protonation gives an **alkyloxonium ion.** In solution, the order of acidity is primary > secondary > tertiary alcohol. Electron-withdrawing substituents increase the acidity (and reduce the basicity).

4. The **hydrogenation** of aldehydes and ketones furnishes alcohols and requires a catalyst.

5. The conversion of the electrophilic alkyl group in a haloalkane, $C^{\delta+}-X^{\delta-}$, into its nucleophilic analog in an **organometallic compound,** $C^{\delta-}-M^{\delta+}$, is an example of **reverse polarization.**

6. The carbon atom in the **carbonyl group,** C=O, of an aldehyde or a ketone is electrophilic and therefore subject to attack by nucleophiles, such as hydride in **hydride reagents** or alkyl in organometallic compounds. Subsequent to aqueous work-up, the products of such transformations are alcohols.

7. The **oxidation** of alcohols to aldehydes and ketones by chromium(VI) reagents opens up important synthetic possibilities based on further reactions with organometallic reagents.

8. **Retrosynthetic analysis** aids in planning the synthesis of complex organic molecules by identifying strategic bonds that may be constructed in an efficient sequence of reactions.

PROBLEMS

19. Name the following alcohols according to the IUPAC nomenclature system. Indicate stereochemistry (if any) and label the hydroxy groups as primary, secondary, or tertiary.

(a) $CH_3CH_2CHCH_3$ with OH

(b) $CH_3CHCH_2CHCH_2CH_3$ with Br and OH

(c) $HOCH_2CH(CH_2CH_2CH_3)_2$

(d) structure with CH_2Cl, C, H, OH

(e) cyclobutane with CH_2CH_3 and OH

(f) cyclic structure with OH and Br

(g) $C(CH_2OH)_4$

(h) Fischer projection: CH_2OH / H—OH / H—OH / CH_2OH

(i) cyclopentane with OH and CH_2CH_2OH

(j) CH_3CH_2—C(Cl)(CH_3)—CH_2OH

20. Draw the structures of the following alcohols. (a) 2-(trimethylsilyl)ethanol; (b) 1-methylcyclopropanol; (c) 3-(1-methylethyl)-2-hexanol; (d) (R)-2-pentanol; (e) 3,3-dibromocyclohexanol.

21. Rank each group of compounds in order of increasing boiling point. (a) cyclohexane, cyclohexanol, chlorocyclohexane; (b) 2,3-dimethyl-2-pentanol, 2-methyl-2-hexanol, 2-heptanol.

22. Explain the order of water solubilities for the compounds in each of the following groups. (a) ethanol > chloroethane > ethane; (b) methanol > ethanol > 1-propanol.

23. 1,2-Ethanediol exists to a much greater extent in the *gauche* conformation than does 1,2-dichloroethane. Explain. Would you expect the *gauche:anti* conformational ratio of 2-chloroethanol to be similar to that of 1,2-dichloroethane or more like that of 1,2-ethanediol?

24. Rank the compounds in each group in order of decreasing acidity.
    (a) $CH_3CHClCH_2OH$, $CH_3CHBrCH_2OH$, $BrCH_2CH_2CH_2OH$
    (b) $CH_3CCl_2CH_2OH$, CCl_3CH_2OH, $(CH_3)_2CClCH_2OH$
    (c) $(CH_3)_2CHOH$, $(CF_3)_2CHOH$, $(CCl_3)_2CHOH$

25. Write an appropriate equation to show how each of the following alcohols acts as, first, a base, and, second, an acid in solution. How do the base and acid strengths of each compare with those of methanol? (a) $(CH_3)_2CHOH$; (b) CH_3CHFCH_2OH; (c) CCl_3CH_2OH.

26. Given the pK_a values of -2.2 for $CH_3\overset{+}{O}H_2$ and 15.5 for CH_3OH, calculate the pH at which (a) methanol will contain exactly equal amounts of $CH_3\overset{+}{O}H_2$ and CH_3O^-; (b) 50% CH_3OH and 50% $CH_3\overset{+}{O}H_2$ will be present; (c) 50% CH_3OH and 50% CH_3O^- will be present.

27. Do you expect hyperconjugation to be important in the stabilization of alkyloxonium ions (e.g., $R\overset{+}{O}H_2$, $R_2\overset{+}{O}H$)? Explain your answer.

28. Evaluate each of the following possible alcohol syntheses as being good (the desired alcohol is the major or only product), not so good (the desired alcohol is a minor product), or worthless. (**Hint:** Refer to Section 7-9 if necessary.)

(a) $CH_3CH_2CH_2CH_2Cl$ $\xrightarrow{H_2O, \ CH_3\overset{\overset{\displaystyle O}{\|}}{C}CH_3}$ $CH_3CH_2CH_2CH_2OH$

(b) 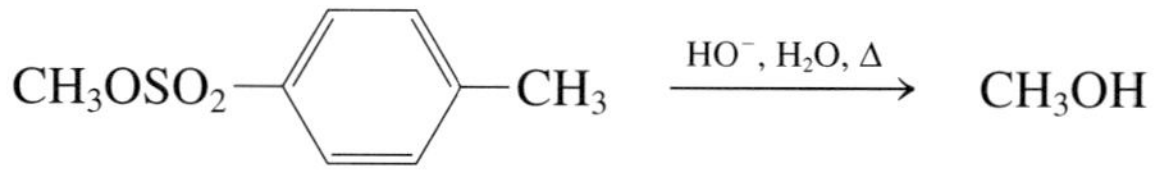 CH_3OSO_2—⟨benzene ring⟩—CH_3 $\xrightarrow{HO^-, \ H_2O, \ \Delta}$ CH_3OH

(c) 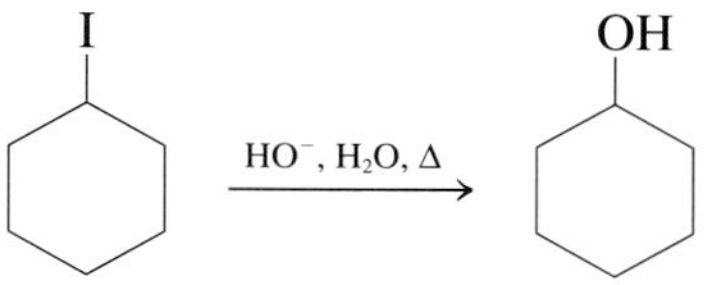 $\xrightarrow{HO^-, \ H_2O, \ \Delta}$

(d) CH₃CHCH₂CH₂CH₃ $\xrightarrow{H_2O, \Delta}$ CH₃CHCH₂CH₂CH₃
(with I and OH substituents)

(e) CH₃CHCH₃ $\xrightarrow{HO^-, H_2O, \Delta}$ CH₃CHCH₃
(with CN and OH substituents)

(f) CH₃OCH₃ $\xrightarrow{HO^-, H_2O, \Delta}$ CH₃OH

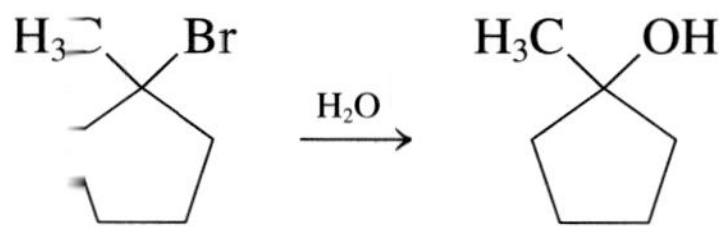

(g) $\xrightarrow{H_2O}$

(h) CH₃CHCH₂Cl $\xrightarrow{HO^-, H_2O, \Delta}$ CH₃CHCH₂OH
(with CH₃ substituents)

29. Give the major product(s) of each of the following reactions. Aqueous work-up steps (when necessary) have been omitted.

(a) CH₃CH=CHCH₃ $\xrightarrow[\text{(Hint: See Section 8-4)}]{H_3PO_4, H_2O, \Delta}$

(b) CH₃CCH₂CH₂CCH₃ $\xrightarrow{H_2, Pt}$
(with two C=O groups)

(c) $\xrightarrow{NaBH_4, CH_3CH_2OH}$

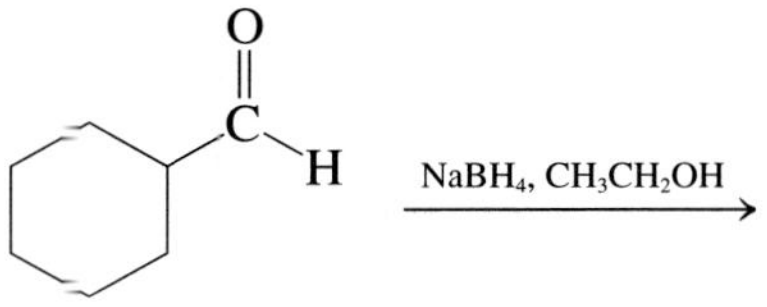

(d) $\xrightarrow{LiAlH_4, (CH_3CH_2)_2O}$

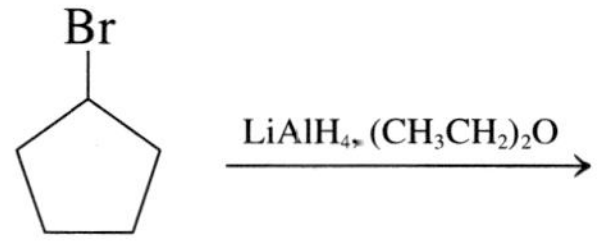

(e) $\xrightarrow{NaBH_4, CH_3CH_2OH}$

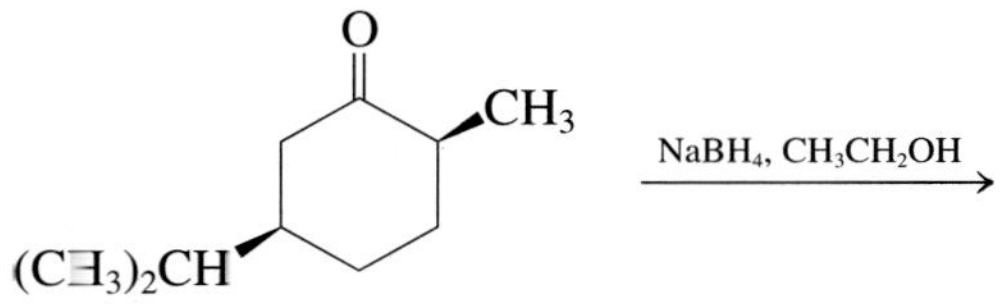

(f) $\xrightarrow{NaBH_4, CH_3CH_2OH}$

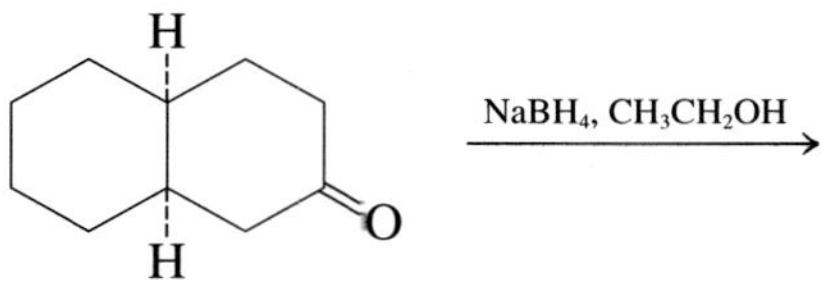

30. What is the direction of the following equilibrium? (**Hint:** The pK_a for H_2 is about 38.)

$$H^- + H_2O \rightleftharpoons H_2 + HO^-$$

31. Formulate the product of each of the following reactions. The solvent in each case is (CH₃CH₂)₂O.

(a) CH₃CH $\xrightarrow[\text{2. H}^+, \text{H}_2\text{O}]{\text{1. LiAlD}_4}$
(with C=O)

(b) CH₃CH $\xrightarrow[\text{2. D}^+, \text{D}_2\text{O}]{\text{1. LiAlH}_4}$
(with C=O)

(c) CH₃CH₂I $\xrightarrow{LiAlD_4}$

32. Give the major product(s) of each of the following reactions [after work-up with aqueous acid in (d), (f), and (h)].

(a) $CH_3(CH_2)_5CHClCH_3$ $\xrightarrow{Mg, (CH_3CH_2)_2O}$

(b) Product of (a) $\xrightarrow{D_2O}$

(c) 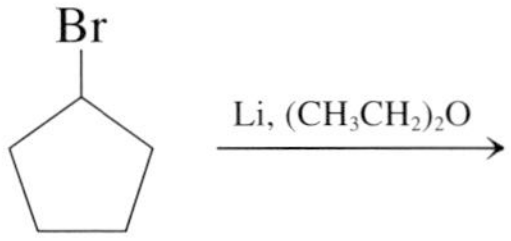 (bromocyclopentane) $\xrightarrow{Li, (CH_3CH_2)_2O}$

(d) Product of (c) 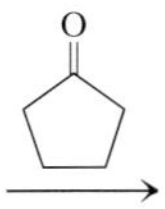 + (cyclopentanone) $\longrightarrow$

(e) $CH_3CH_2CH_2Cl + Mg$ $\xrightarrow{(CH_3CH_2)_2O}$

(f) Product of (e) + (phenyl methyl ketone, $C_6H_5\overset{O}{\overset{\|}{C}}CH_3$) $\longrightarrow$

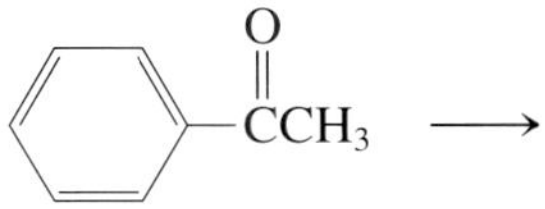

(g) 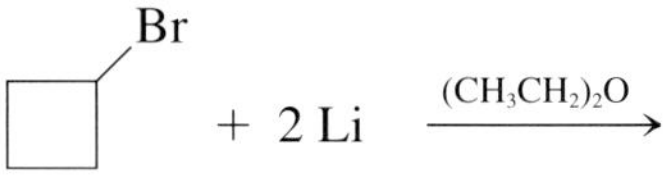 (bromocyclobutane) $+ 2 Li$ $\xrightarrow{(CH_3CH_2)_2O}$

(h) 2 mol product of (g) + 1 mol $CH_3\overset{O}{\overset{\|}{C}}CH_2CH_2\overset{O}{\overset{\|}{C}}CH_3$ $\longrightarrow$

33. The common practice of washing laboratory glassware with propanone (acetone) can lead to unintended consequences. For example, a student plans to carry out the preparation of methylmagnesium iodide, CH_3MgI, which he will add to benzaldehyde, C_6H_5CHO. What compound is he intending to synthesize after aqueous work-up? Using his freshly washed glassware, he carries out the procedure and finds that he has produced an unexpected tertiary alcohol as a product. What substance did he make? How did it form?

34. Give the major product(s) of each of the following reactions (after aqueous work-up). The solvent in each case is ethoxyethane (diethyl ether).

(a) 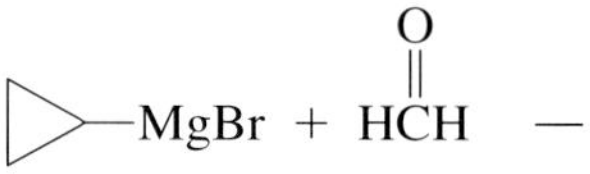 (cyclopropyl)$-MgBr + H\overset{O}{\overset{\|}{C}}H$ $-$

(b) $CH_3\overset{CH_3}{\overset{|}{C}}HCH_2MgCl + CH_3\overset{O}{\overset{\|}{C}}H$ $\longrightarrow$

(c) $C_6H_5CH_2Li + C_6H_5\overset{O}{\overset{\|}{C}}H$ $\longrightarrow$

(d) $CH_3\overset{MgBr}{\overset{|}{C}}HCH_3 +$ (cyclohexanone) $\longrightarrow$

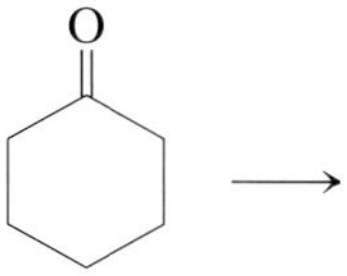

(e) 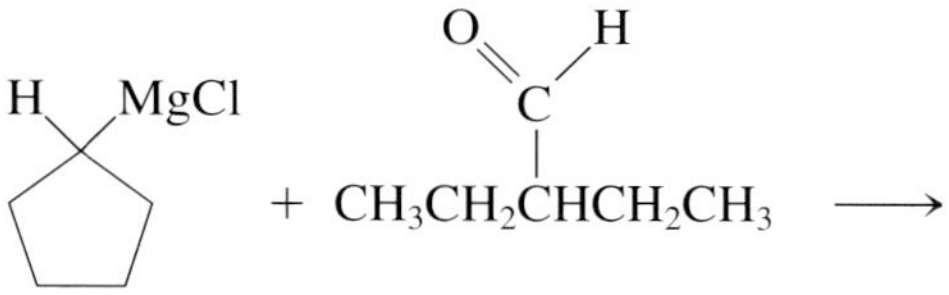 (cyclopentyl, $H\,\overset{}{\underset{}{C}}\,MgCl$) $+ CH_3CH_2\overset{O\,\diagup H}{\overset{\diagdown\,\diagup}{\underset{|}{C}}}CHCH_2CH_3$ $\longrightarrow$

35. Write the structures of the products of reaction of ethylmagnesium bromide, CH_3CH_2MgBr, with each of the following carbonyl compounds. Identify any reaction that gives more than one stereoisomeric product, and indicate whether you would expect the products to form in identical or in differing amounts.

(a)

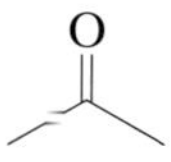

(b)

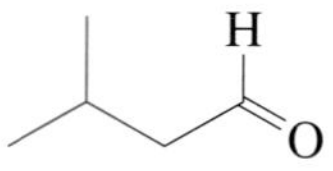

(c)

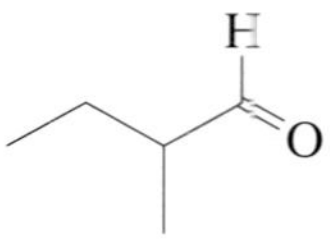

(d)

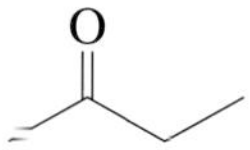

(e)

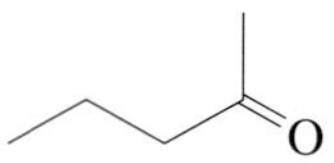

(f)

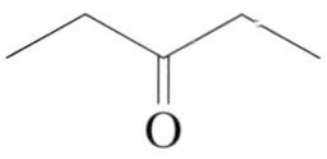

(g)

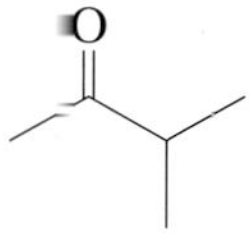

(h)

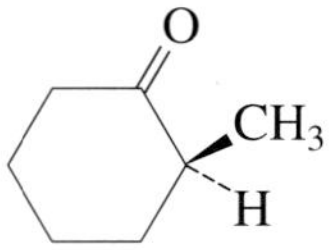

(i)

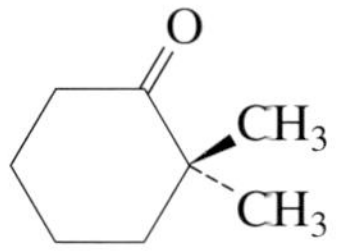

(j)

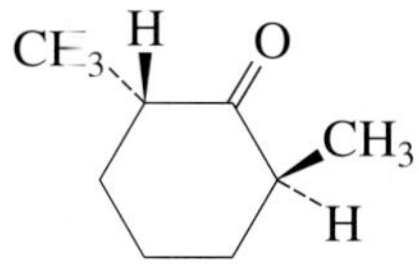

36. Give the expected major product of each of the following reactions. PCC is the abbreviation for pyridinium chlorochromate (Section 8-6).

(a) $CH_3CH_2CH_2OH$ $\xrightarrow{Na_2Cr_2O_7, H_2SO_4, H_2O}$

(b) $(CH_3)_2CHCH_2OH$ $\xrightarrow{PCC, CH_2Cl_2}$

(c) 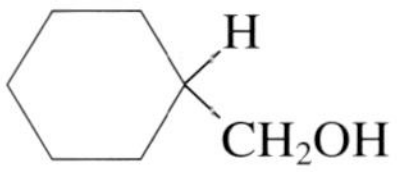 $\xrightarrow{Na_2Cr_2O_7, H_2SO_4, H_2O}$

(d) 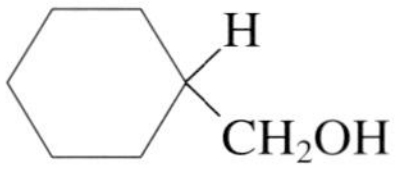 $\xrightarrow{PCC, CH_2Cl_2}$

(e) 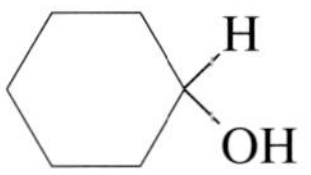 $\xrightarrow{PCC, CH_2Cl_2}$

37. Give the expected major product of each of the following reaction *sequences*. PCC refers to pyridinium chlorochromate.

(a) $(CH_3)_2CHOH$ $\xrightarrow{\begin{array}{l}\text{1. } CrO_3, H_2SO_4, H_2O \\ \text{2. } CH_3CH_2MgBr, (CH_3CH_2)_2O \\ \text{3. } H^+, H_2O\end{array}}$

(b) $CH_3CH_2CH_2CH_2Cl$ $\xrightarrow{\begin{array}{l}\text{1. } ^-OH, H_2O \\ \text{2. } PCC, CH_2Cl_2 \\ \text{3. } \text{⬠}-Li, (CH_3CH_2)_2O \\ \text{4. } H^+, H_2O\end{array}}$

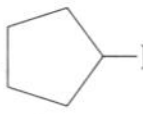

(c) Product of (b) $\xrightarrow{\begin{array}{l}\text{1. } CrO_3, H_2SO_4, H_2O \\ \text{2. } LiAlD_4, (CH_3CH_2)_2O \\ \text{3. } H^+, H_2O\end{array}}$

38. Unlike Grignard and organolithium reagents, organometallic compounds of the most electropositive metals (Na, K, etc.) react rapidly with haloalkanes. As a result, attempts to convert RX into RNa or RK by reaction with the corresponding metal lead to alkanes by a reaction called *Wurtz* coupling.

$$2 \text{ RX} + 2 \text{ Na} \longrightarrow \text{R—R} + 2 \text{ NaX}$$

which is the result of

$$R—X + 2\,Na \longrightarrow R—Na + NaX$$

followed rapidly by

$$R—Na + R—X \longrightarrow R—R + NaX$$

When it was still in use, the Wurtz coupling reaction was employed mainly for the preparation of alkanes by the coupling of two identical alkyl groups (e.g., equation 1 below). Suggest a reason why Wurtz coupling might not be a useful method for coupling two *different* alkyl groups (equation 2).

$$2\,CH_3CH_2CH_2Cl + 2\,Na \longrightarrow CH_3CH_2CH_2CH_2CH_2CH_3 + 2\,NaCl \qquad (1)$$

$$CH_3CH_2Cl + CH_3CH_2CH_2Cl + 2\,Na \longrightarrow CH_3CH_2CH_2CH_2CH_3 + 2\,NaCl \qquad (2)$$

39. The reaction of two equivalents of Mg with 1,4-dibromobutane produces compound A. The reaction of A with two equivalents of CH_3CHO (acetaldehyde), followed by work-up with dilute aqueous acid, produces compound B, having the formula $C_8H_{18}O_2$. What are the structures of A and B?

40. Suggest the best synthetic route to each of the following simple alcohols, using in each case a simple alkane as your initial starting molecule. What are some disadvantages of beginning syntheses with alkanes?
(a) Methanol   (b) Ethanol   (c) 1-Propanol
(d) 2-Propanol   (e) 1-Butanol   (f) 2-Butanol
(g) 2-Methyl-2-propanol

41. For each alcohol in Problem 40, suggest (if possible) a synthetic route that starts with, first, an aldehyde and, second, a ketone.

42. Outline the best method for preparing each of the following compounds from an appropriate alcohol.

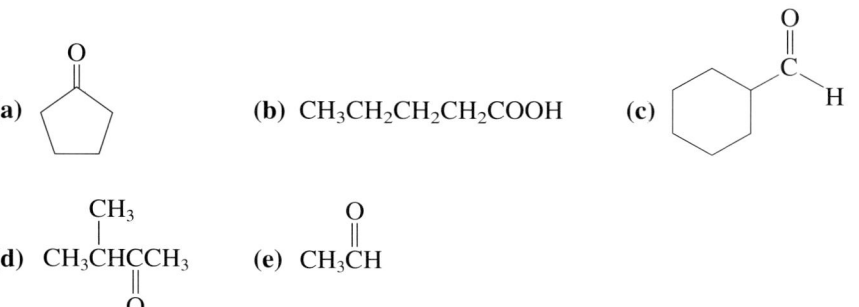

(a)   (b) $CH_3CH_2CH_2CH_2COOH$ (c)

(d) $CH_3\overset{\displaystyle CH_3}{\underset{}{CH}}\overset{}{\underset{\displaystyle O}{C}}CH_3$ (e) $CH_3\overset{\displaystyle O}{C}H$

43. Suggest three different syntheses of 2-methyl-2-hexanol. Each route should utilize one of the following starting materials. Then use any number of steps and any other reagents needed.

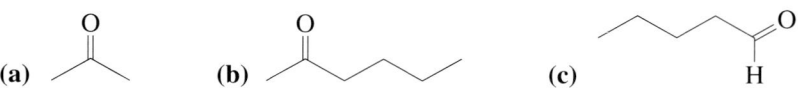

(a)   (b)   (c)

44. Devise three different syntheses of 3-octanol starting with **(a)** a ketone; **(b)** an aldehyde; **(c)** an aldehyde different from that employed in (b).

45. Propose sensible synthetic schemes for the preparation of each of the following compounds, using only the organic starting material(s) indicated. Use any organic solvents or inorganic reagents necessary.

(a) [structure] from ethane and propane

(b) [structure] from butane, ethane, and formaldehyde $\left(\begin{matrix} & O \\ & \parallel \\ & HCH \end{matrix}\right)$

46. Waxes are naturally occurring esters (alkyl alkanoates) containing long, straight alkyl chains. Whale oil contains the wax 1-hexadecyl hexadecanoate, as shown in the margin. How would you synthesize this wax, using an S_N2 reaction?

$$CH_3(CH_2)_{14}\overset{\displaystyle O}{\overset{\displaystyle \parallel}{C}}O(CH_2)_{15}CH_3$$
1-Hexadecyl hexadecanoate

47. The B vitamin commonly known as niacin is used by the body to synthesize the coenzyme nicotinamide adenine dinucleotide (NAD^+; Chapter 25). In the presence of a variety of enzyme catalysts, the reduced form of this substance (NADH) acts as a biological hydride donor, capable of reducing aldehydes and ketones to alcohols, according to the general formula

$$\overset{\displaystyle O}{\overset{\displaystyle \parallel}{RCR}} + NADH + H^+ \xrightarrow{\text{Enzyme}} \overset{\displaystyle OH}{\overset{\displaystyle |}{RCHR}} + NAD^+$$

The COOH functional group of carboxylic acids is not reduced. Write the products of the NADH reduction of each of the molecules below.

(a) $CH_3\overset{\displaystyle O}{\overset{\displaystyle \parallel}{C}}H + NADH \xrightarrow{\text{Alcohol dehydrogenase}}$

(b) $CH_3\overset{\displaystyle OO}{\overset{\displaystyle \parallel\parallel}{CC}}OH + NADH \xrightarrow{\text{Lactate dehydrogenase}}$ **Lactic acid**
2-Oxopropanoic acid
(Pyruvic acid)

(c) $HO\overset{\displaystyle O}{\overset{\displaystyle \parallel}{C}}CH_2\overset{\displaystyle OO}{\overset{\displaystyle \parallel\parallel}{CC}}OH + NADH \xrightarrow{\text{Malate dehydrogenase}}$ **Malic acid**
2-Oxobutanedioic acid
(Oxaloacetic acid)

48. Reductions by NADH (Problem 47) are stereospecific, with the stereochemistry of the product controlled by an enzyme (see Chemical Highlight 8-1). The common forms of lactate and malate dehydrogenases produce exclusively the *S* stereoisomers of lactic and malic acids, respectively. Draw these stereoisomers.

49. Chemically modified steroids have become increasingly important in medicine. Give the possible product(s) of the following reactions. In each case, identify the major stereoisomer formed on the basis of delivery of the attacking reagent from the less hindered side of the substrate molecule. (**Hint:** Make models and refer to Section 4-7.)

(a)

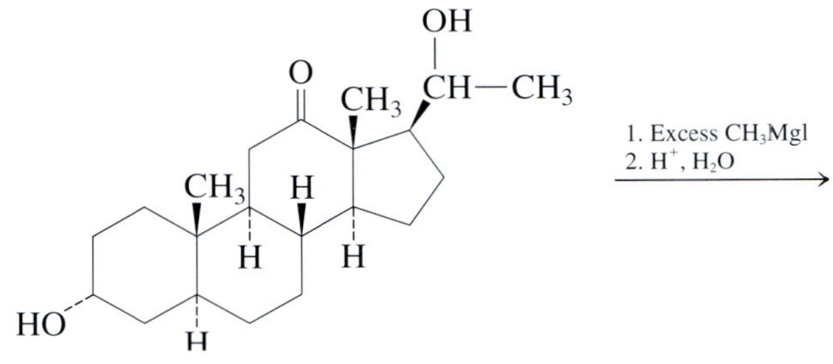
OH
CH—CH3
CH3
O
CH3
H
H
H
HO
H
1. Excess CH3MgI
2. H+, H2O

(b)

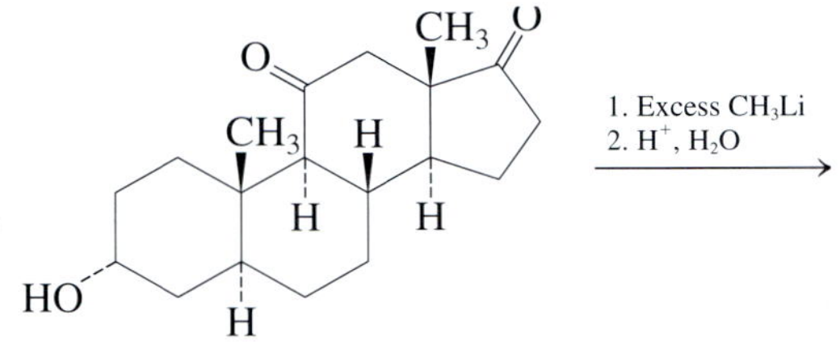
CH3
O
O
CH3
H
H
H
HO
H
1. Excess CH3Li
2. H+, H2O

Team Problem

50. Your team has been asked to devise a synthesis of the tertiary alcohol 2-cyclohexyl-2-butanol, A. Your laboratory is well stocked with the usual organic and inorganic reagents and solvents. An inventory check reveals that there are many appropriate bromoalkanes and alcohols on hand. As a group, analyze alcohol A retrosynthetically and propose all possible strategic disconnections. Check the inventory to see if a particular route is feasible in regard to available starting materials. Then divide the proposed routes evenly among yourselves to evaluate the merits or pitfalls of these strategies. Write a detailed synthetic plan based on your chosen retrosynthesis for the synthesis of 2-cyclohexyl-2-butanol. Reconvene to defend or reject these plans. Finally, take into consideration the prices of your starting materials. Which one of your routes to A is the cheapest?

Target Molecule	Inventory (Price)	
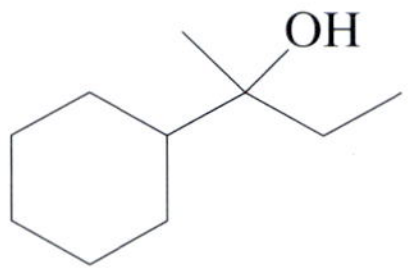	2-Bromobutane ($31/kg)	Cyclohexanol ($14/kg)
	Bromocyclohexane ($50/kg)	1-Cyclohexylethanol ($120/25 g)
	Bromoethane ($20/kg)	Cyclohexylmethanol ($42/100 g)
2-Cyclohexyl-2-butanol **A**	Bromomethane ($400/kg)	(Bromomethyl)cyclohexane ($86/100 g)
	2-Butanol ($23/kg)	

Preprofessional Problems

51. A compound known to contain only C, H, and O gives the following upon microanalysis (atomic weights: C = 12.0, H = 1.00, O = 16.0): 52.1% C, 13.1% H. It is found to have a boiling point of 78°C. Its structure is
(a) CH_3OCH_3 **(b)** CH_3CH_2OH **(c)** $HOCH_2CH_2CH_2CH_2OH$
(d) $HOCH_2CH_2CH_2OH$ **(e)** none of these

52. The compound whose structure is $(CH_3)_2CHCH_2\underset{\mid \atop OH}{C}HCH_2CH_3$ is best named (IUPAC):

(a) 2-methyl-4-hexanol **(b)** 5-methyl-3-hexanol **(c)** 1,4,4-trimethyl-2-hexanol **(d)** 1-isopropyl-2-hexanol

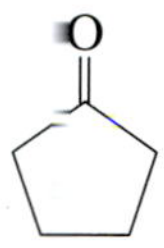

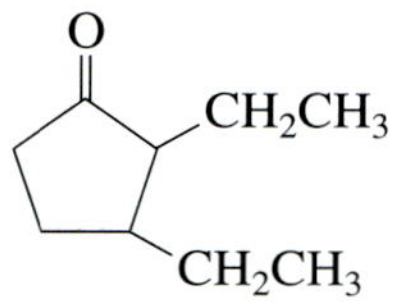

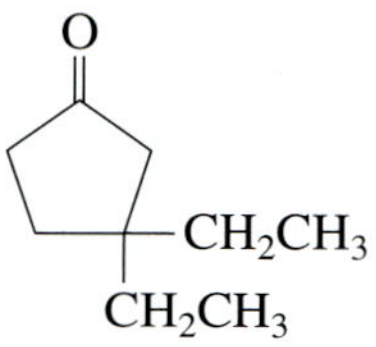

53. In this transformation, what is the best structure for "A"?

(a)

(b)

(c)

(d)

54. Ester hydrolysis is best illustrated by

(a) $CH_3O\overset{O}{\overset{\|}{C}}CH_3 \xrightarrow{H^+,\ H_2O} CH_3OCH_3 + CO$ **(b)** $CH_3O\overset{O}{\overset{\|}{C}}CH_3 \xrightarrow{H^+,\ H_2O} CH_3OH + HO\overset{O}{\overset{\|}{C}}CH_3$

(c) $CH_3OCH_2OH \xrightarrow{H_2O} CH_3OH + H{-}\underset{\overset{\|}{O}}{C}{-}H$ **(d)** $CH_3OH + CH_3CO_2H \xrightarrow{H_2O} CH_3O\underset{\overset{\|}{O}}{C}CH_3$

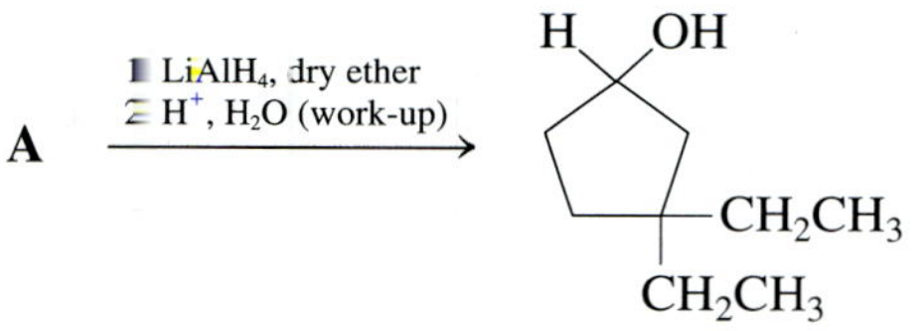

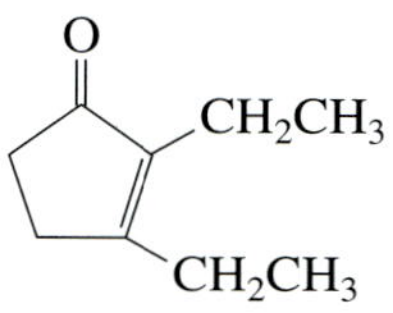